Uni

Maria Welleda Baldoni · Ciro Ciliberto
Giulia Maria Piacentini Cattaneo

# Elementary Number Theory, Cryptography and Codes

 Springer

Maria Welleda Baldoni
Ciro Ciliberto
Giulia Maria Piacentini Cattaneo
Università di Roma - Tor Vergata
Dipartimento di Matematica
Via della Ricerca Scientifica, 1
00133 Roma
Italy
baldoni@mat.uniroma2.it
cilibert@mat.uniroma2.it
piacentini@mat.uniroma2.it

Translation from the Italian language edition:
*Aritmetica, crittografia e codici* by W.M. Baldoni, C. Ciliberto, and G.M. Piacentini Cattaneo
Copyright © 2006 Springer-Verlag Italia
Springer is a part of Springer Science+Business Media
All Rights Reserved

ISBN 978-3-540-69199-0          e-ISBN 978-3-540-69200-3

Library of Congress Control Number: 2008938959

Mathematics Subject Classification (2000): 11G05, 14G50, 94B05

© 2009 Springer-Verlag Berlin Heidelberg

Cover figure from *Balla, Ciacomo* © VG Bild-Kunst, Bonn 2008

Cover design: WMX Design GmbH, Heidelberg

Printed on acid-free paper

9 8 7 6 5 4 3 2 1

springer.com

# Introduction

Mathematics, possibly due to its intrinsic abstraction, is considered to be a merely intellectual subject, and therefore extremely remote from everyday human activities. Surprisingly, this idea is sometimes found not only among laymen, but among working mathematicians as well. So much so that mathematicians often talk about *pure mathematics* as opposed to *applied mathematics* and sometimes attribute to the former a questionable birthright.

On the other hand, it has been remarked that those two categories do not exist but, just as we have good and bad literature, or painting, or music, so we have *good* or *bad* mathematics: the former is applicable, even if at first sight this is not apparent, in any number of fields, while the latter is worthless, even within mathematics itself. However, one must recognise the truth in the interesting sentence with which two of our colleagues, experts about applications, begin the preface to the book [47]: *In theory there is no difference between theory and practice. In practice there is.*

We believe that this difference cannot be ascribed to the intrinsic nature of mathematical theories, but to the stance of each single mathematician who creates or uses these theories. For instance, until recently the branch of mathematics regarded as the closest to applications was undoubtedly mathematical analysis and especially the theory of differential equations. The branches of mathematics supposed to be farthest from applications were algebra and number theory. So much so that a mathematician of the calibre of G. H. Hardy claimed in his book [25] the supremacy of number theory, which was to be considered the true *queen* of mathematics, precisely due to its distance from the petty concerns of everyday life. This made mathematics, in his words, "gentle and clean". A strange opinion indeed, since the first developments of algebra and number theory among the Arabs and the European merchants in the Middle Ages find their motivation exactly in very concrete problems arising in business and accountancy.

Hardy's opinion, dating back to the 1940s, was based upon a prejudice, then largely shared among scientists. It is quite peculiar that Hardy did not know, or pretended not to know, that A. Turing, whom he knew very well, had used that very mathematics he considered so detached to break the Enigma code, working for English secret services, dealing a deadly blow to German espionage (cf. [28]). However, the role played by algebra and number theory in military and industrial cryptography is well known from time immemorial. Perhaps Hardy incorrectly believed that the mathematical tools then used in cryptography, though sometimes quite complex, were nevertheless essentially elementary, not more than combinatorial tricks requiring a measure of extemporaneous talent to be devised or cracked, but leading to no solid, important, and enduring theories.

The advances in computer science in the last sixty years have made cryptography a fundamental part of all aspects of contemporary life. More precisely, cryptography studies transmission of data, coded in such a way that authorised receivers only may decode them, and be sure about their provenience, integrity and authenticity. The development of new, non-classical cryptographic techniques, like public-key cryptography, have promoted and enhanced the applications of this branch of the so-called *discrete mathematics*, which studies, for instance, the enumeration of symbols and objects, the construction of complex structures starting with simpler ones, and so on. Algebra and number theory are essential tools for this branch of mathematics, which is in a natural way suitable for the workings of computers, whose language is intrinsically *discrete* rather than *continuous*, and is essential in the construction of all security systems for data transmission. So, even if we are not completely aware of it, each time we use credit cards, on-line bank accounts or e-mail, we are actually fully using algebra and numbers. But there is more: the same techniques have been applied since the 1940s to the transmission of data on channels where interference is present. This is the subject of the theory of error-correcting codes which, though unwittingly, we use daily in countless ways: for instance when we listen to music recorded on a CD or when surfing the Web.

This textbook originated from the teaching experience of the authors at the University of Rome "Tor Vergata" where, in the past years, they taught this subject to Mathematics, Computer Science, Electronic Engineering and Information Technology students, as well as for the "Scuola di Insegnamento a Distanza", and at several different levels. They gave courses with a strong algebraic or geometric content, but keeping in mind the algorithmic and constructive aspects of the theories and the applications we have been mentioning.

The point of view of this textbook is to be *friendly* and *elementary*. Let us try to explain what we mean by these terms.

By *friendly* we mean our attempt to always give motivations of the theoretical results we show to the reader, by means of examples we consider to be simple, meaningful, sometimes entertaining, and useful for the applications. Indeed, starting from the examples, we have expounded the general methods of resolution of problems that only apparently look different in form, setting and language. With this in mind, we have aimed to a simple and colloquial style, while never losing sight of the formal rigour required in a mathematical treatise.

By *elementary* we mean that we assume our readers to have a quite limited background in basic mathematical knowledge. As a rule of the thumb, a student having followed a good first semester in Mathematics, Physics, Computer Science or Engineering may confidently venture through this book. However, we have tried to make the treatment as self-contained as possible regarding the elements of algebra and number theory needed in cryptography and coding theory applications. *Elementary*, however, does not mean *easy*: we introduced quite advanced concepts, but did so gradually and always trying to accompany the reader, without assuming previous advanced knowledge.

The starting point of this book is the well-known set of integer numbers and their *arithmetic*, that is the study of the operations of addition e multiplication. Chapter 1 aims to make the reader familiar with integer numbers. Here mathematical induction and recursion are covered, giving applications to several concrete problems, such as the analysis of dynamics of populations with assigned reproduction rules, the computation of numbers of moves in several games, and so on. The next topics are divisions, the greatest common divisor and how to compute it using the well-known Euclidean algorithm, the resolution of Diophantine equations, and numeral systems in different bases. These basic notions are first presented in an elementary way and then a more general theoretical approach is given, by introducing the concept of Euclidean ring. The last part of the chapter is devoted to continued fractions.

One of the goals of Chapter 1 is to show how, in order to solve concrete problems using mathematical methods, the first step is to build a *mathematical model* that allows a translation into one or more mathematical problems. The next step is the determination of suitable *algorithms*, that is procedures consisting of a finite sequence of *elementary operations* yielding the solution to the mathematical problems describing the initial question. In Chapter 2 we discuss the fundamental concept of *computational complexity* of an algorithm, which basically counts the elementary operations an algorithm consists of, thus evaluating the time needed to execute it. The importance of this concept is manifest: among the algorithms we have to distinguish the feasible ones, that is those executable in a sufficiently short time, and the unfeasible ones, due to the time needed for their execution being too long independently of the computing device used. The algorithms of the first kind are the *polynomial* ones, while among those of the second kind there are, for instance, the *exponential* ones. We proceed then to calculate the complexity of some fundamental algorithms used to perform elementary operations with integer numbers.

In Chapter 3 we introduce the concept of congruence, which allows the passage from the infinite set of integer numbers to the finite set of residue classes. This passage from infinite to finite enables us to implement the elementary operations on integers in computer programming: a computer, in fact, can work on a finite number of data only.

Chapter 4 is devoted to the fundamental problem of factoring integer numbers. So we discuss prime numbers, which are the building blocks of the structure of integer numbers, in the sense that each integer number may be represented as a product of prime numbers: this is the so-called *factorisation* of an integer number. Factoring an integer number is an apparently harmless problem from a theoretical viewpoint: the factorisation exists, it is essentially unique, and it can be found by the famous *sieve of Eratosthenes*. We show, however, the unfeasibility of this exponential algorithm. For instance, in 1979 it has been proved that the number $2^{44497} - 1$, having 13395 decimal digits, is prime: by using the sieve of Eratosthenes, it would take a computer executing one million multiplications per second about $10^{6684}$ years to get this result! The modern public-key cryptography, covered in Chapter 7, basically relies on the difficulty of factoring an integer number. In Chapter 4 elements of the general theory of factorial rings can also be found, in particular as regards its application to polynomials.

In Chapter 5 finite fields are introduced; they are a generalisation of the rings of residue classes of integers modulo a prime number. Finite fields are fundamental for the applications to cryptography and codes. Here we present their main properties, expounded with several examples. We give an application of finite fields to the resolution of polynomial Diophantine equations. In particular, we prove the law of quadratic reciprocity, the key to solving second degree congruences.

In Chapter 6 most of the theory presented so far is applied to the search for *primality tests*, that is algorithms to determine whether a number is prime or not, and for factorisation methods more sophisticated than the sieve of Eratosthenes; even if they are in general exponential algorithms, just like Eratosthenes', in special situations they may become much more efficient. In particular, we present some primality tests of probabilistic type: they are able to discover in a very short time whether a number has a high probability of being a prime number. Moreover, we give the proof of a recent polynomial primality test due to M. Agrawal, N. Kayal and N. Saxena; its publication has aroused a wide interest among the experts.

Chapter 7 describes the applications to cryptography. Firstly, we describe several classical cryptographic methods, and discuss the general laying out of a cryptographic system and the problem of cryptanalysis, which studies the techniques to break such a system. We introduce next the revolutionary concept of public-key cryptography, on which the transmission of the bulk of confidential information, distinctive of our modern society, relies. We discuss several public-key ciphers, main among them the well-known *RSA* system, whose security relies on the computational difficulty of factoring large numbers, and some of its variants making it possible, for instance, the electronic authentication of signatures. Recently new frontiers for cryptography, especially regarding security, have been opened by the interaction of classical algebra and arithmetic with ideas and concepts originating from algebraic geometry, and especially the study of a class of plane curves known as *elliptic curves.* At the end of the chapter an introduction to these important developments is given.

Chapter 8 presents an introduction to coding theory, already mentioned above. This is a recent branch of mathematics in which sophisticated combinatorial, algebraic and geometric techniques converge, in order to study the mathematical aspects of the problem of transmitting data through noisy channels. In other words, coding theory studies techniques to send data through a channel when we give for granted that some errors will happen during transmission. These techniques enable us to correct the errors that might arise, as well as to quickly encode and decode the data we intend to send.

In Chapter 9 we give a quick glance at the new frontiers offered by *quantum cryptography*, which relies on ideas originating in *quantum mechanics*. This branch of physics makes the creation of a *quantum computer* at least conceivable; if such a computer were actually built, it could execute in polynomial time computations a usual computer would need an exponential time to perform. This would make all present cryptographic systems vulnerable, seriously endangering civil, military, financial security systems. This might result in the collapse of our civilisation, largely based on such systems. On the other hand, by its very nature, the concept of a quantum computer allows the design of absolutely unassailable *quantum cryptographic systems*, even by a quantum computer; furthermore, such systems have the astonishing property of being able to detect if eavesdroppers attempt, even unsuccessfully, to hear in on a restricted communication.

Each chapter is followed by an appendix containing:

- a list of exercises on the theory presented there, with several levels of difficulty; in some of them proofs of supplementary theorems or alternative proofs of theorems already proved in the text are given;
- a list of exercises from a computational viewpoint;
- suggestions for programming exercises.

The most difficult exercises are marked by an asterisk. At the end of the book many of the exercises are solved, especially the hardest theoretical ones.

Some sections of the text may be omitted in a first reading. They are set in a smaller type, and so are the appendices.

We wrote this book having in mind students of Mathematics, Physics, Computer Science, Engineering, as well as researchers who are looking for an introduction, without entering in too many details, to the themes we have quickly described above.

In particular, the book can be useful as a complementary text for first and second year students in Mathematics, Physics or Computer Science taking a course in Algebra or Discrete Mathematics. In Chapters 1, 3, and 4 they will find a concrete approach, with many examples and exercises, to some basic algebraic theories. Chapters 5 and 6, though more advanced, are in our opinion within the reach of a reader of this category.

The text is particularly suitable for a second or third year course giving an introduction to cryptography or to codes. Students of such a course will probably already have been exposed to the contents of Chapters 1, 3, and 4; so teachers can limit themselves to quick references to them, suggesting to the students only to solve some exercises. They can then devote more time to the material from Chapter 5 on, and particularly to Chapter 7, giving more or less space to Chapters 8 and 9.

The bibliography lists texts suggested for further studies in cryptography and codes, useful for more advanced courses.

A first version of this book, titled "Note di matematica discreta", was published in 2002 by Aracne; we are very grateful to the publishers for their permission for the publication of this book. This edition is widely expanded and modified: the material is presented differently, several new sections and in-depth analysis have been added, a wider selection of solved exercises is offered.

Lastly, we thank Dr Alberto Calabri for supervising the layout of the book and the editing of the text, especially as regards the exercise sections.

Rome,
August 2008

*M. Welleda Baldoni*
*Ciro Ciliberto*
*Giulia Maria Piacentini Cattaneo*

# Contents

# 1

## A round-up on numbers

This chapter rounds up some basic notions about numbers; we shall need them later on, and it is useful to fix the ideas on some concepts and techniques which will be investigated in this book. Some of what follows will be studied again in more detail, but we shall assume a basic knowledge about:

- some elements of set theory and logic (see for instance [43]);
- the construction of the fundamental number sets:

$$\mathbb{N} = \text{the set of natural numbers,}$$
$$\mathbb{Z} = \text{the set of integer numbers,}$$
$$\mathbb{Q} = \text{the set of rational numbers,}$$
$$\mathbb{R} = \text{the set of real numbers,}$$
$$\mathbb{C} = \text{the set of complex numbers,}$$

and of the operations on them (see [15] or [22]);
- the idea of limit and of numerical series (as given in any calculus text, for instance [12]);
- some elements of algebra (see [4], [15], [32] or [45]): in particular, the reader will need the definitions of the main algebraic structures, like *semigroups*, *groups*, *rings*, *integral domains*, *fields*;
- basic notions of linear algebra (see [13]): *vector spaces*, *matrices*, *eigenvalues*, and *eigenvectors*;
- elementary concepts of probability theory (see [5] or [29]).

## 1.1 Mathematical induction

In this section we shall fix our attention on the set $\mathbb{N} = \{0, 1, 2, 3, \ldots\}$ of *natural numbers* on which, as is well known, the operations $+$, the *addition*, and $\cdot$, the *multiplication*, as well as a natural order relation $\leq$ are given. Recall that both $(\mathbb{N}, +)$ and $(\mathbb{N}, \cdot)$ are semigroups, that is to say, the operations are associative, and admit an identity element.

On the set $\mathbb{N}$ the map

$$succ : n \in \mathbb{N} \to n + 1 \in \mathbb{N}$$

is defined, associating with each natural number its *successor*. This mapping is injective but not surjective, as 0 is not the successor of any natural number. The existence of such an injective but not surjective mapping of $\mathbb{N}$ in itself implies that it is an infinite set.

Furthermore, the following fundamental property holds in $\mathbb{N}$:

**Mathematical induction.** *Let $A$ be a subset of $\mathbb{N}$ satisfying the following two properties:*

(1) $n_0 \in A$;
(2) *if $n \in A$ then, for each $n$, $succ(n) = n + 1 \in A$.*

*Then $A$ includes all natural numbers greater or equal than $n_0$. In particular, if $n_0 = 0$, then $A$ coincides with $\mathbb{N}$.*

It is well known that the existence of the mapping *succ* and mathematical induction uniquely determine the set of natural numbers. Mathematical induction is important not only for the formal construction of the set $\mathbb{N}$, but is also a fundamental proof tool to which we want to draw the reader's attention.

Let us look at a simple example. Suppose we want to solve the following problem: *compute the sum of the first $n$ natural numbers*, that is to say compute the number

$$1 + 2 + \cdots + (n - 1) + n.$$

Some of the readers might already know that this problem, in the case $n = 100$, appears in an episode of Carl Friedrich Gauss's life. When he was six years old, his teacher gave it to his unruly pupils, in the hope that it would take them some time to solve it, to keep them quiet in the meantime. Unfortunately (for the teacher), Gauss noticed that

$$n + 1 = (n - 1) + 2 = (n - 2) + 3 = \cdots ,$$

that is, the sum of the last term and of the first one equals the sum of the last but one plus the second one, and so forth; so he guessed in a few seconds the general formula

$$1 + 2 + \cdots + (n - 1) + n = \frac{n(n + 1)}{2} \tag{1.1}$$

and immediately obtained

$$1 + 2 + \cdots + 99 + 100 = 5050.$$

But how may we *prove* that, as young Gauss guessed, formula (1.1) always holds? Of course, it is not possible to check it for each $n$ by actually summing up the terms, because we should verify an infinite number of cases. What *mathematical induction* allows us to do is precisely solving problems of this kind, even in more general cases.

Consider a set $X$ and a *sequence* $\{\mathcal{P}_n\}$ of propositions defined in $X$, that is, for each number $n \in \mathbb{N}$, $\mathcal{P}_n$ is a proposition about the elements of $X$. For instance, in the case $X = \mathbb{N}$, we may take

$$\mathcal{P}_n = \text{formula (1.1) holds,}$$

that is, $\mathcal{P}_n$ is the claim that for the number $n \in \mathbb{N}$ the sum $1+2+\cdots+(n-1)+n$ equals $n(n+1)/2$. Suppose we want to prove that the proposition $\mathcal{P}_n$ is true for each $n$. Thus, we have to prove infinitely many propositions. Consider the set

$$A := \{n \in \mathbb{N} \mid \mathcal{P}_n \text{ is true}\}.$$

We have to prove that $A$ coincides with $\mathbb{N}$. Applying mathematical induction it suffices to proceed as follows:

(1) *basis of the induction*: prove that $\mathcal{P}_0$ is true;
(2) *inductive step*: prove that, for each $k \geq 0$, from the truth of $\mathcal{P}_k$ *(induction hypothesis)*, it follows that $\mathcal{P}_{k+1}$ is true.

*Then we may conclude that $\mathcal{P}_n$ is true for each $n \in \mathbb{N}$.*

With a proof by induction we may obtain *infinitely many* results in just two steps. In this sense, it is a method of *reduction from infinite to finite*, and so it has a crucial importance, infinity being by its very nature intractable. Further on we shall show several methods, techniques and ideas in the same spirit of reducing from infinite to finite.

An apparently more restrictive, but actually equivalent (see Exercises A1.1–A1.3) formulation of the same principle is as follows:

**Complete induction (or Strong induction)** (CI). *Let $A$ be a subset of $\mathbb{N}$ satisfying the following properties:*

(1) $n_0 \in A$;
(2) *if $k \in A$ for each $k$ such that $n_0 \leq k < n$, then $n \in A$ as well.*

*Then $A$ includes all natural numbers greater than $n_0$. In particular, if $n_0 = 0$, then $A$ coincides with $\mathbb{N}$.*

This yields, as above, the following formulation:

(1) *basis of the induction*: prove that $\mathcal{P}_0$ is true;
(2) *inductive step*: prove that, for each $k \geq 0$, from the truth of $\mathcal{P}_h$ for each $h \leq k$, it follows that $\mathcal{P}_{k+1}$ is true.

*Then we may conclude that $\mathcal{P}_n$ is true for each $n \in \mathbb{N}$.*

Let the reader be warned that, as implicitely stated above, mathematical induction, in itself, *does not yield formulas*, but allows us to prove them *if we already know them*. In other words, if we *already are in possession* of the sequence of propositions $\mathcal{P}_n$ we may hope to prove their truth by mathematical induction, but this method in itself *will not give us* the sequence $\mathcal{P}_n$. In practice, if we have a problem like the one given to Gauss as a young boy, in order to guess the right sequence of propositions $\mathcal{P}_n$ it is necessary to study what happens for the first values of $n$ and, following Gauss's example, venture a conjecture about the general situation.

As an example, we prove by induction formula (1.1).

The basis of the induction lies just in observing that the formula is obviously true for $n = 1$. Suppose now that the formula is true for a particular value of $n$, and let us prove its truth for its successor $n + 1$. We have:

$$1 + 2 + \cdots + (n - 1) + n + (n + 1) =$$
$$= [1 + 2 + \cdots + (n - 1) + n] + (n + 1) = \quad \text{(by induction hypothesis)}$$
$$= \frac{n(n + 1)}{2} + (n + 1) = \frac{(n + 1)(n + 2)}{2}.$$

This proves the inductive step for each $n$, and so proves formula (1.1).

Other examples in which mathematical induction is used to prove formulas similar to (1.1) are given in the appendix at the end of this chapter (see Exercises B1.5–B1.11).

**Remark 1.1.1.** Before carrying on, it might be useful to warn readers of the snares deriving by erroneous applications of mathematical induction. In a proof by induction, *both steps*, the basis of the induction and the inductive step, are indispensable to a correct application of the procedure, and both are to be correctly carried out. Otherwise, we are in danger of making gross mistakes. For instance, an erroneous application of mathematical induction might yield a proof of the following ludicrous claim: *All cats are the same colour.*

Let us proceed by induction, by proving that for each $n \in \mathbb{N}$, any set of $n$ cats is made up of cats of the same colour:

- basis of the induction: It is obvious; indeed any set including a single cat is made up of cats of the same colour, that is, the colour of the unique cat in the set.
- inductive step: Suppose that every time we have $n - 1$ cats they are the same colour and let us prove that the same claim holds for $n$ cats. Examine the following picture, where the dots represent cats:

$$\overbrace{\bullet \bullet \bullet \bullet \cdots \bullet \bullet \bullet \bullet}^{n-1}_{\underbrace{\qquad\qquad\qquad}_{n-1}}. \tag{1.2}$$

By induction hypothesis, the first $n - 1$ cats are all the same colour. By the same reason, the last $n - 1$ cats are the same colour as well, this colour being a priori different from the colour of the first cats. But the common cats, that is the cats appearing both among the first $n - 1$ and the last $n - 1$, must be the same colour. So *all* the cats are the same colour.

Since, fortunately, there are cats of different colours, we are confident that we have made a mistake. Where is it? In the inductive step we used the fact that there are cats in common to the two sets we were considering, the first $n - 1$ cats and the last $n - 1$ cats. But this is true only if $n \geq 3$. So the inductive step does not hold *for each* $n$ because the implication from the case $n = 1$ to $n = 2$ does not hold.

Notice that if we want to prove a proposition $\mathcal{P}_n$ not for all values of $n$, but for all $n \geq n_0$, it is enough to prove as the basis for the induction the proposition $\mathcal{P}_{n_0}$ and then verifying the inductive step for each $n \geq n_0$. Studying again the example about cats, the inductive steps holds for $n \geq 2$, but the basis of the induction does not hold for $n = 2$, that is, it is not true that each pair of cats consists of cats of the same colour!

## 1.2 The concept of recursion

Recursion is a fundamental concept, strictly connected to mathematical induction. Suppose we have a function defined on the set $\mathbb{N}$ of natural numbers taking values in a set $X$. Such a function is commonly said to be a *sequence* in $X$ and denoted by $\{a_n\}_{n \in \mathbb{N}}$, or simply $\{a_n\}$, where $a_n$ is the value taken by the function on the integer $n$. The values $a_n$ are said to be the *terms* of the sequence.

Suppose now we have a method allowing us to determine the term $a_n$ for each integer $n$ greater or equal than a fixed integer $n_0$ when we know the term $a_{n-1}$. Suppose moreover we know the *initial terms* of the sequence, that is $a_0, a_1, a_2, \ldots, a_{n_0-1}, a_{n_0}$. We claim that, with these premises, we are able to compute the value of the sequence for *each* natural number $n$. This is a consequence of mathematical induction and its easy proof is left to the reader (see Exercise A1.10).

A particular but very interesting example of this procedure is the case of numeric sequences satisfying *linear* recurrence relations. Let us give a general definition:

**Definition 1.2.1.** *Let* $\{a_n\}_{n \in \mathbb{N}}$ *be a sequence of elements in a vector space* $V$ *on a field* $\mathbb{K}$. *A* linear recurrence relation, *or* formula, *for the sequence is a formula of the kind*

$$a_{n+k} = f_{k-1}(a_{n+k-1}) + f_{k-2}(a_{n+k-2}) + \cdots + f_0(a_n) + d_n, \qquad (1.3)$$

*holding for each integer* $n \geq 0$; *here* $k$ *is a positive integer,* $a_0, a_1, \ldots, a_{k-1}$ *are the* initial values *or* conditions, $f_0, f_1, \ldots, f_{k-1}$ *are linear maps of* $V$ *in itself, called* coefficients *of the recurrence relation, and* $\{d_n\}$ *is a (possibly constant) sequence of elements in* $V$ *said* constant term. *If* $d_n = 0$, *the relation is said to be* homogeneous.

So, formula (1.3) gives an expression for the $(n+k)$-th term of the sequence $\{a_n\}$ as a function of the $k$ preceding terms. We shall mostly consider the case where $\{d_n\}$ is a constant sequence with each term equal to $d$. The word *linear* refers to the fact that we are working in a vector space $V$. In particular, it is possible to consider sequences $\{a_n\}_{n\in\mathbb{N}}$ of elements of $\mathbb{K}$ verifying a recurrence relation. In this case $f_0$, $f_1$, ..., $f_{k-1}$ are the product by elements $b_0$, $b_1$, ..., $b_{k-1}$ of $\mathbb{K}$ and relation (1.3) is of the form

$$a_{n+k} = b_{k-1}a_{n+k-1} + b_{k-2}a_{n+k-2} + \cdots + b_0 a_n + d_n. \qquad (1.4)$$

A sequence $\{a_n\}_{n\in\mathbb{N}}$ is said to be a *solution* of a linear recurrence relation of the form (1.3) if the terms $a_n$ of the sequence satisfy the relation. It is obvious that the sequence is uniquely determined by relation (1.3) and by the initial terms $a_0$, $a_1$, ..., $a_{k-1}$.

On the other hand, if we know that a sequence $\{a_n\}_{n\in\mathbb{N}}$ of elements of the field $\mathbb{K}$ verifies a linear recurrence relation of the form (1.4), but we do not know the coefficients $b_0$, $b_1$, ..., $b_{k-1}$ and the constant term $d$, we may expect to be able to determine these coefficients, and then the whole sequence, if we know sufficiently many terms of the sequence (see, as a particular instance, Exercise A1.27).

Recurrence relations appear in a natural way when studying several different kinds of problems, like computing increments or decrements of populations with given reproduction rules, colouring pictures with just two colours, computing the number of moves in different games, computing compounded interests, solving geometrical problems and so forth. Some of these problems will be shown as examples or suggested as exercises in the appendix.

### 1.2.1 Fibonacci numbers

**Example 1.2.2.** Two newborn rabbits, a male and a female, are left on a desert island on the 1st of January. This couple becomes fertile after two months and, starting on the 1st of March, they give birth to two more rabbits, a male and a female, the first day of each month. Each couple of newborn rabbits, analogously, becomes fertile after two months and, starting on the first day of their third month, gives birth to a new couple of rabbits. How many couples are there on the island after $n$ months?

In order to answer this question, we must construct a mathematical model for the population increase of rabbits, as described in the example. Denote by $f_n$ the number of couples of rabbits, a male and a female, that are present in the island during the $n$th month. It is clear that $f_n$ is the sum of two numbers completely determined by the situation in the preceding months, that is $f_n$ is the sum

(1) of the number $f_{n-1}$ of the couples of rabbits in the island in the $(n-1)$-th month, as no rabbit dies;
(2) of the number of the couples of rabbits born on the first day of $n$-th month, which are as many as the couples of rabbits which are fertile on that day, and these in turn are as many as the $f_{n-2}$ couples of rabbits that were in the island two months before.

As a consequence, we may write for the sequence $\{f_n\}_{n \in \mathbb{N}}$ the following recurrence relation:
$$f_n = f_{n-1} + f_{n-2}$$
for each $n \geq 2$ with the obvious initial conditions $f_0 = 0$ e $f_1 = 1$.

The sequence $\{f_n\}$ of natural numbers satisfying the following recurrence relation with given initial conditions

$$\boxed{f_0 = 0, \quad f_1 = 1, \quad f_n = f_{n-1} + f_{n-2} \qquad \text{for } n > 1,} \tag{1.5}$$

is called *Fibonacci sequence*, and the terms of the sequence are called *Fibonacci numbers*. Each term of the sequence is the sum of the two preceding terms and knowing this sequence it is possible to give an answer to the problem described in Example 1.2.2. The first terms of the sequence are easy to compute:

$$0, \ 1, \ 1, \ 2, \ 3, \ 5, \ 8, \ 13, \ 21, \ 34, \ 55, \ 89, \ 144, \ 233, \ \ldots$$

Fibonacci numbers are not only related to population increase, but are often found in the description of several natural phenomenona. For instance, sunflowers' heads display florets in spirals which are generally arranged with 34 spirals in one direction and 55 in the other. If the sunflower is smaller, it has 21 spirals in one direction and 34 in the other, or 13 and 21. If it is very large, it has 89 and 144 spirals! In each case these numbers are, not by chance, Fibonacci numbers.

Fibonacci numbers were introduced by Leonardo Fibonacci, or Leonardo Pisano, in 1202, with the goal of describing the increase of a rabbit population. These numbers have many interesting mathematical properties, so much that along the centuries they have been, and still are, studied by many mathematicians. For instance, at the end of the 19th century Edouard Lucas used some properties of Fibonacci numbers to show that the 39-digit number

$$170141183460469231731687303715884105727 = 2^{127} - 1$$

is a prime number (see Chapter 4).

Let us remark that writing relation (1.5) is not an altogether satisfying way of answering the question posed in Example 1.2.2. We would like, in fact, to have a *solution* of the recurrence relation (1.5), that is a *closed formula* giving the $n$-th term of Fibonacci sequence, without having to compute all the preceding terms. In order to do so, we shall use matrix operations and some principles of linear algebra.

Consider the matrix on $\mathbb{R}$

$$A = \begin{pmatrix} 0 & 1 \\ 1 & 1 \end{pmatrix}. \tag{1.6}$$

We may rewrite conditions (1.5) in the following way:

$$A \begin{pmatrix} f_{n-2} \\ f_{n-1} \end{pmatrix} = \begin{pmatrix} f_{n-1} \\ f_n \end{pmatrix} \qquad \text{for all } n \geq 2,$$

that is, setting $X_n = \begin{pmatrix} f_{n-1} \\ f_n \end{pmatrix}$, consider the linear system

$$AX_{n-1} = X_n, \qquad \text{for all } n \geq 2,$$

and so

$$A^n X_0 = X_n.$$

Thus, if we know $A^n$, to find the closed formula expressing $f_n$ as a function of the initial conditions it suffices to multiply the second row of $A^n$ by $X_0$. In this case it is easy to prove by induction, using formula (1.5), that (see Exercise A1.28):

**Proposition 1.2.3.** *For each integer number $n \geq 1$ we have*

$$A^n = \begin{pmatrix} f_{n-1} & f_n \\ f_n & f_{n+1} \end{pmatrix},$$

*where $\{f_n\}$ is Fibonacci sequence.*

Unfortunately, in the general case it is not easy to compute the powers of a matrix: in Chapter 2 we shall fully appreciate this problem, when we study the computational complexity of some operations. In some cases, however, as in the present one, the computation is not difficult, as we are going to show.

If we have a *diagonal* matrix $D$, that is one of the form

$$D = \begin{pmatrix} a & 0 \\ 0 & b \end{pmatrix},$$

then computing $D^n$ is trivial, because we have

$$D^n = \begin{pmatrix} a^n & 0 \\ 0 & b^n \end{pmatrix}.$$

Let us recall that a matrix $B$ on a field $\mathbb{K}$ is said to be *diagonalisable* if there exists a matrix $C$ whose determinant is not equal to zero such that $B = C \cdot D \cdot C^{-1}$, where $D$ is a diagonal matrix. For diagonalisable matrices computing powers is also simple. In fact, if $B$ is as above, we trivially have $B^n = C \cdot D^n \cdot C^{-1}$. As $D^n$ is easy to compute, it suffices to know $D$ and $C$ in order to know the powers of $B$. Now, there is an easy criterion to ascertain whether a matrix is diagonalisable: *an $m \times m$ matrix $B$ is diagonalisable if its characteristic polynomial $P_B(t)$ has $m$ distinct roots in $\mathbb{K}$* (see the definitions recalled in § 1.3.6). Let us recall that $P_B(t)$ is the polynomial of degree $m$ on $\mathbb{K}$ defined as the determinant $|B - tI_m|$, where $I_m$ is *identity matrix*, that is the square $m \times m$ matrix with entries equal to 1 on the main diagonal and zero elsewhere. The roots of the characteristic polynomial $P_B(t)$ that are elements of $\mathbb{K}$ are called the *eigenvalues* of $B$. If $B = C \cdot D \cdot C^{-1}$ with diagonal $D$, the elements on the main diagonal of $D$ are the eigenvalues of $B$.

For the real matrix $A$ in (1.6) we have that

$$P_A(t) = \det \begin{pmatrix} -t & 1 \\ 1 & 1-t \end{pmatrix} = t^2 - t - 1$$

is a polynomial having two distinct real roots given by

$$\lambda_1 = \frac{1+\sqrt{5}}{2}, \qquad \lambda_2 = \frac{1-\sqrt{5}}{2}. \tag{1.7}$$

Thus $A$ is diagonalisable and as a consequence we have an expression of the form $A = C \cdot D \cdot C^{-1}$, with

$$D = \begin{pmatrix} (1+\sqrt{5})/2 & 0 \\ 0 & (1-\sqrt{5})/2 \end{pmatrix}. \tag{1.8}$$

The matrix $C$ is easy to write down. The reader may verify (see Exercises B1.12 and B1.13) that

$$C = \begin{pmatrix} 1 & 1 \\ (1+\sqrt{5})/2 & (1-\sqrt{5})/2 \end{pmatrix}, \quad C^{-1} = \frac{1}{\sqrt{5}} \begin{pmatrix} -(1-\sqrt{5})/2 & 1 \\ (1+\sqrt{5})/2 & -1 \end{pmatrix}. \tag{1.9}$$

In conclusion, by Proposition 1.2.3, we have the relation

$$\begin{pmatrix} f_{n-1} & f_n \\ f_n & f_{n+1} \end{pmatrix} = C \cdot \begin{pmatrix} ((1+\sqrt{5})/2)^n & 0 \\ 0 & ((1-\sqrt{5})/2)^n \end{pmatrix} \cdot C^{-1}.$$

Hence, by multiplying the matrices in the right-hand side, we get the following closed formula for the $n$-th Fibonacci number:

$$\boxed{f_n = \frac{1}{\sqrt{5}} \left[ \left(\frac{1+\sqrt{5}}{2}\right)^n - \left(\frac{1-\sqrt{5}}{2}\right)^n \right].} \tag{1.10}$$

We give the following proposition, which generalises what we have proved in the case of the recurrence relation (1.5).

**Proposition 1.2.4.** *Given a positive integer $k$, consider the homogeneous linear recurrence relation defined on a field $\mathbb{K}$*

$$a_{n+k} = b_{k-1}a_{n+k-1} + b_{k-2}a_{n+k-2} + \cdots + b_0 a_n, \quad \text{for } n \geq 0, \tag{1.11}$$

*where $b_0, b_1, \ldots, b_{k-1}$ are the coefficients and $a_0, a_1, \ldots, a_{k-1}$ the initial values. Consider the square $k \times k$ matrix defined by*

$$A = \begin{pmatrix} 0 & 1 & 0 & 0 & \cdots & 0 \\ 0 & 0 & 1 & 0 & \cdots & 0 \\ 0 & 0 & 0 & 1 & \cdots & 0 \\ \vdots & \vdots & \vdots & \vdots & \ddots & \vdots \\ 0 & 0 & 0 & 0 & \cdots & 1 \\ b_0 & b_1 & b_2 & b_3 & \cdots & b_{k-1} \end{pmatrix}$$

*whose characteristic polynomial is*

$$P_A(t) = t^k - b_{k-1}t^{k-1} - b_{k-2}t^{k-2} - \cdots - b_1 t - b_0.$$

*Suppose that $P_A(t)$ has k distinct roots $\lambda_i$, $1 \leq i \leq k$, in $\mathbb{K}$. Then the solutions of the recurrence relation are of the form*

$$a_n = \sum_{i=1}^{k} c_i \lambda_i^n, \qquad \text{for each } n \geq 0,$$

*with $c_i$ constants uniquely determined by the linear system*

$$\sum_{i=1}^{k} c_i \lambda_i^h = a_h, \qquad \text{with } 0 \leq h \leq k-1,$$

*determined by the initial values $a_0, a_1, \ldots, a_{k-1}$.*

So we can now solve homogeneous linear recurrence relations when the characteristic polynomial of the matrix associated with the recurrence relation has distinct roots.

**Remark 1.2.5.** The proof of Proposition 1.2.4 is not substantially different from the one that led us to formula (1.10). So we omit its proof, leaving to the interested readers the task of rediscovering it, by following the indications given above.

**Remark 1.2.6.** When the eigenvalues are not distinct, it is still possible, with analogous but less simple techniques, to find a formula giving the solution of the recurrence relation, but it has a more involved form.

**Remark 1.2.7.** The number

$$\frac{1 + \sqrt{5}}{2} \tag{1.12}$$

we have met when computing Fibonacci numbers is called *golden ratio* or *divine proportion*. The origin of this name lies in the fact that two quantities $a$ and $b$ were considered by the Greeks to be connected by the most harmonious ratio if the relation

$$\frac{a}{b} = \frac{a+b}{a} \tag{1.13}$$

held. For instance, the side lengths of the Parthenon's façade satisfy this property. The ratio (1.13) can be reworded by saying that the whole is to the larger part as the larger part is to the smaller one. By solving the ratio (1.13), we easily get

$$\frac{a}{b} = \frac{1 + \sqrt{5}}{2} = 1,618033989\ldots$$

In Exercise A1.29 we describe the geometric construction that, given a line segment of length $a$, determines a segment of length $b$ such that $a/b$ is the golden ratio.

The number (1.12) is sometimes denoted by the letter $\Phi$, from the name of the Greek artist Phidias who often used this ratio in his sculptures. The other root

$$\frac{1 - \sqrt{5}}{2} = -\frac{1}{\Phi} = 1 - \Phi = -0,618033989\ldots$$

shares many of the properties of $\Phi$, and is often denoted by $\hat{\Phi}$. For further information about golden ratio in ancient mathematics, the reader may refer to [10], [61].

Let us lastly remark that formula (1.10) says that $f_n$ is the nearest integer number to the irrational number $\Phi^n / \sqrt{5}$, as $|\hat{\Phi}^n / \sqrt{5}| < 1/2$ for each $n$.

In the rest of this section we shall give further examples of problems that may be solved by using recurrence formulas. These examples are interesting, but not strictly necessary for what follows.

### 1.2.2 Further examples of population dynamics

**Example 1.2.8.** An entomologist observes a population of beetles whose evolution is subject to the following rules:

- one half of the beetles die one year after their birth;
- 2/3 of the survivors die two years after their birth;
- in its third year each beetle spawns 6 beetles and dies.

Study the population's evolution.

Denote by:

$a_n$: the number of beetles between 0 and 1 year old, observed by the entomologist in the $n$-th year of his study of the population;

$b_n$: the number of beetles between 1 and 2 year old, observed by the entomologist in the $n$-th year;

$c_n$: the number of beetles between 2 and 3 year old, observed by the entomologist in the $n$-th year.

With these numbers we form a column vector

$$X_n = \begin{pmatrix} a_n \\ b_n \\ c_n \end{pmatrix}$$

describing the distribution of the ages in the population of beetles in the $n$-th year, which is what we intend to determine. The initial value we are assuming as known is the vector $X_0$.

We want to describe the evolution of the population by a recurrence formula of the form

$$X_{n+1} = A \cdot X_n \qquad \text{for each } n \geq 0,$$

where $A$ is a $3 \times 3$ matrix; so we have

$$X_n = A^n \cdot X_0 \qquad \text{for each } n \geq 1.$$

How can we determine $A$? It is sufficient to observe that the evolution rules are described by the following relations:

$$a_{n+1} = 6c_n, \quad b_{n+1} = \frac{a_n}{2}, \quad c_{n+1} = \left(1 - \frac{2}{3}\right) b_n = \frac{b_n}{3},$$

which can be written in matrix form as follows:

$$X_{n+1} = \begin{pmatrix} a_{n+1} \\ b_{n+1} \\ c_{n+1} \end{pmatrix} = \begin{pmatrix} 0 & 0 & 6 \\ 1/2 & 0 & 0 \\ 0 & 1/3 & 0 \end{pmatrix} \cdot \begin{pmatrix} a_n \\ b_n \\ c_n \end{pmatrix} = A \cdot X_n,$$

where

$$A = \begin{pmatrix} 0 & 0 & 6 \\ 1/2 & 0 & 0 \\ 0 & 1/3 & 0 \end{pmatrix}.$$

The characteristic polynomial $P_A(t)$ of $A$ is $1 - t^3$, having the three distinct complex roots $1$, $(-1 + i\sqrt{3})/2$, and $(-1 - i\sqrt{3})/2$, where $i$ is the imaginary unit. So, $A$ is diagonalisable on $\mathbb{C}$; this allows us to compute without difficulty the powers of $A$. This, in turn, yields a way of computing a closed formula for vector $X_n$; we leave this to the interested reader (see Exercise B1.17).

**Example 1.2.9.** Each year one tenth of Italian people living in an Italian region other than Liguria arrive in Liguria and start living there, and simultaneously one fifth of those living in Liguria depart from it. How does Liguria's population evolve?

Denote by:

$y_n$: the number of persons living outside of Liguria in the $n$-th year of our study of this region's population;

$z_n$: the number of persons living in Liguria in the $n$-th year.

By constructing the usual vector

$$X_n = \begin{pmatrix} y_n \\ z_n \end{pmatrix},$$

we see that the phenomenon may be recursively described by the formula

$$X_n = A \cdot X_{n-1}$$

for each $n \geq 1$, where $A$ is the matrix

$$A = \frac{1}{10} \begin{pmatrix} 9 & 2 \\ 1 & 8 \end{pmatrix}.$$

So, for each $n \geq 1$ the following holds

$$X_n = A^n \cdot X_0.$$

The eigenvalues of matrix $A$ are $7/10$ and $1$. So $A$ is diagonalisable and it is easy to compute its powers. This allows to readily compute a closed formula for the vector $X_n$. This is left to the reader (see Exercise B1.19).

### 1.2.3 The tower of Hanoi: a non-homogeneous linear case

**Example 1.2.10.** The *game of the tower of Hanoi* was invented by the mathematician E. Lucas in 1883. The tower of Hanoi consists of $n$ circular holed discs, with a vertical peg $A$ running through all of them; the discs are stacked with their diameters decreasing from bottom up.

The goal of the game is to transfer all discs, in the same order, that is to say, with their diameters decreasing from bottom up, on another peg $C$, by using a support peg $B$ (see figure 1.1) and observing the following rules:

(i) the discs must be transferred one at a time from one peg to another one;
(ii) never during the game, on any peg, a disc with a greater diameter may be located above a disc with a smaller diameter.

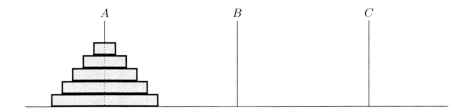

**Fig. 1.1.** The tower of Hanoi with $n = 5$ discs

We want to determine the number $M_n$ of moves necessary to conclude the game starting with $n$ discs.

This game apparently has the following origin. The priests of Brahma's temple were required to continuously transfer 64 gold discs placed on three gold pegs standing on diamond bases. According to a legend, were the transfer accomplished, the world would come to an end!

We shall proceed by induction on $n$. For $n = 1$, of course, one move is sufficient: $M_1 = 1$. Assume now $n$ discs are on peg $A$. By the inductive hypothesis, we may move the upper $n - 1$ discs from peg $A$ to peg $B$ with $M_{n-1}$ moves. In doing so, the largest disc on peg $A$ is never moved. With a single move we now transfer this largest disc from peg $A$ to peg $C$. Then we transfer with $M_{n-1}$ moves the $n - 1$ discs on peg $B$ to peg $C$, putting them on the larger disc. So we accomplished our task with $2M_{n-1} + 1$ moves, and it is plainly clear that it is not possible to solve the game with fewer moves.

So we have the following recurrence relation:

$$M_n = 2M_{n-1} + 1, \qquad M_1 = 1,$$

which we may solve to get a closed formula, as follows:

$$M_n = 2M_{n-1} + 1 = 2(2M_{n-2} + 1) + 1 = 2^2 M_{n-2} + 2 + 1 =$$
$$= 2^2(2M_{n-3} + 1) + 2 + 1 = 2^3 M_{n-3} + 2^2 + 2 + 1 =$$
$$= \cdots = 2^{n-1} M_1 + 2^{n-2} + 2^{n-3} + \cdots + 2^2 + 2 + 1 =$$
$$= 2^{n-1} + 2^{n-2} + 2^{n-3} + \cdots + 2^2 + 2 + 1 = 2^n - 1,$$

so the total number of moves is

$$M_n = 2^n - 1.$$

If it takes to the priests of Brahma's temple thirty seconds to move a disc and if they never make a mistake, it would take them $30 \cdot (2^{64} - 1)$ seconds to complete their task. The reader may give an estimate of the number of years before the end of the world: a very long time! (see Exercise B1.33).

## 1.3 The Euclidean algorithm

In this section we work in the set $\mathbb{Z} = \{\ldots, -3, -2, -1, 0, 1, 2, 3, \ldots\}$ of *integer numbers*. As is well known, on $\mathbb{Z}$ the two operations $+$ (*addition*) and $\cdot$ (*multiplication*) are defined; with these operations $\mathbb{Z}$ is a commutative ring with unity, with no zero-divisors, that is to say, a ring in which the *zero-product property* holds (saying that $ab = 0$ implies that either $a = 0$ or $b = 0$); so $\mathbb{Z}$ is an integral domain. Moreover, in $\mathbb{Z}$ there is a natural order relation $\leq$, allowing us to define the function *absolute value*

$$n \in \mathbb{Z} \to |n| \in \mathbb{Z},$$

where $|n| = n$ if $n \geq 0$, while $|n| = -n$ if $n \leq 0$.

### 1.3.1 Division

We begin by recalling a very simple fact, already learnt in primary school: we can perform division between integer numbers. This operation is made possible by an algorithm presented in the following proposition:

**Proposition 1.3.1.** *Let $a$ and $b$ be integer numbers, with $b \neq 0$. Then two integers $q$ and $r$ exist, and are uniquely determined, such that*

$$a = bq + r, \qquad with \quad 0 \leq r < |b|.$$

PROOF. Suppose initially that $b$ is a positive integer. Consider the set of all integer multiples of $b$:

$$\ldots, -kb, \ldots, -2b, -b, 0, b, 2b, \ldots, kb, \ldots,$$

where $k$ is a positive integer. There exists a unique $q \in \mathbb{Z}$ such that (see Exercise A1.8)

$$qb \le a < (q+1)b.$$

Define

$$r = a - qb;$$

this determines the two numbers $q$ and $r$, as required. Notice that $0 \le r < b$ by costruction and $q$ is unique because it is the greatest integer whose product by $b$ is less than or equal than $a$. Consequently, $r$ is unique too.

If $b$ is negative, by virtue of what we have just proved, we have, in a unique way, $a = q'(-b) + r$, with $0 \le r < -b = |b|$. So it is sufficient to define $q = -q'$ to find the numbers $q$ and $r$ as required; their uniqueness follows from what has been proved in the case $b > 0$.                                             □

Thus, the algorithm described in Proposition (1.3.1) allows us to determine the integers $q$ and $r$ starting from $a$ and $b$, and is called *division* of $a$ by $b$. The term $a$ will be called the *dividend*, $b$ the *divisor*, $q$ the *quotient* and $r$ the *remainder* of the division. For instance, dividing 34 by 8 or by $-8$, we get respectively

$$34 = 8 \cdot 4 + 2, \qquad 34 = (-8) \cdot (-4) + 2,$$

so the quotient and the remainder are 4 and 2 in the first case, $-4$ and 2 in the second one. On the other hand, dividing $-34$ by 8 or by $-8$, we get

$$-34 = 8 \cdot (-5) + 6, \qquad -34 = (-8) \cdot 5 + 6,$$

so the quotient and the remainder are $-5$ and 6 in the first case, 5 and 6 in the second one.

**Definition 1.3.2.** *A number $a$ is said to be* divisible *by a number $b \ne 0$ (or we say that $b$ is a* divisor *of $a$, or that $b$* divides *$a$, and we denote this by $b \mid a$), if the remainder of the division of $a$ by $b$ is zero. In other words, $a$ is divisible by $b$ if there exists an integer $m$ such that $a = mb$, that is if $a$ is an integer multiple of $b$.*

Each integer $a$ has, among its divisors, 1, $-1$, $a$ and $-a$. These are said to be the *trivial divisors* of $a$. The numbers $a$ and $-a$, which only differ by the sign, are said to be *associated* with $a$. Of course 1 and $-1$ have no divisors different from 1 and $-1$, so they are the only invertible numbers in $\mathbb{Z}$ (a number $a$ is said to be *invertible* if there exists a number $b$ such that $ab = 1$). Notice further that if both $a \mid b$ and $b \mid c$ hold, then $a \mid c$. We write down the following simple fact:

**Lemma 1.3.3.** *Let $a$ and $b$ be non zero integers. We have $a \mid b$ and $b \mid a$ if and only if $a$ and $b$ are associated, that is either $a = b$ or $a = -b$ holds.*

PROOF. By the hypothesis, there exist two integer numbers $n$, $m$ such that $b = na$ and $a = mb$. Then $b = nmb$ and so $nm = 1$. Therefore, either $n = m = 1$ or $n = m = -1$.                                             □

If $a > 1$ has only trivial divisors, it is said to be an *irreducible* or *prime* number. As we shall see, prime numbers are important, as they are the building blocks from which, by multiplication, all integers may be built. For the time being, however, we pass over this fundamental topic, delaying it until Chapter 4, to deal now with a simple and natural question: *given two integers a and b different from zero, which are their common divisors?* We shall show that, by repeatedly performing divisions, the problem reduces to computing the divisors of a single integer $d$.

### 1.3.2 The greatest common divisor

We begin with a trivial remark: the divisors of the integer $a$ are the same as those of the integer $-a$. Thus, in the problem we are studying, it is sufficient to consider the case in which $a$ and $b$ are both positive, and to look for their positive common divisors. So we shall study just this case.

We perform the following divisions; we suppose that in the first $n$ divisions the remainder is positive, while in the last one it is zero:

$$
\begin{array}{llll}
1. & a = b\,q_1 + r_1, & 0 < r_1 < b, \\[4pt]
2. & b = r_1\,q_2 + r_2, & 0 < r_2 < r_1, \\[4pt]
3. & r_1 = r_2\,q_3 + r_3, & 0 < r_3 < r_2, \\[4pt]
& \quad\vdots & \\[4pt]
i+2. & r_i = r_{i+1}\,q_{i+2} + r_{i+2}, & 0 < r_{i+2} < r_{i+1}, \\[4pt]
& \quad\vdots & \\[4pt]
n-1. & r_{n-3} = r_{n-2}\,q_{n-1} + r_{n-1}, & 0 < r_{n-1} < r_{n-2}, \\[4pt]
n. & r_{n-2} = r_{n-1}\,q_n + \boxed{r_n}, & 0 < r_n < r_{n-1}, \\[4pt]
n+1. & r_{n-1} = r_n\,q_{n+1} + 0.
\end{array}
$$

Notice that, when successively dividing in this way, in any case, *within at most b divisions*, we find a remainder equal to zero. Indeed, we have that $b > r_1 > r_2 > r_3 > \cdots$ is a strictly decreasing sequence of non-negative integers. We also remark that the common divisors of $a$ and $b$ are the same as the common divisors of $b$ and $r_1$: in fact, if an integer divides both $a$ and $b$, it divides each multiple of $b$, and the difference between $a$ and $q_1 b$, that is, $r_1$. On the other hand, by reasoning in the same way, if an integer divides $b$ and $r_1$, it also divides $a = bq_1 + r_1$. Using the second of the above divisions, we may see that the common divisors of $b$ and $r_1$ are the common divisors of $r_1$ and $r_2$. Going on like this, we find that the common divisors of $a$ and $b$ are the common divisors of $r_{n-1}$ and $r_n$. Clearly, as $r_{n-1}$ is a multiple of $r_n$, the common divisors of $r_{n-1}$ and $r_n$ coincide with the divisors of $r_n$.

Define $d = r_n$, the last remainder in the sequence of those divisions. We have seen that $d$ is a common divisor of $a$ and $b$. Furthermore, it is the greatest among the common divisors of $a$ and $b$: indeed, if $d'$ divides both $a$ and $b$ then, as we have seen, $d'$ divides $d$. Hence comes the name, for $d$, of *greatest common divisor* and the symbol $\mathrm{GCD}(a, b)$ to denote it. If $\mathrm{GCD}(a, b) = 1$, the numbers $a$ and $b$ have no non trivial common divisors: in this case we say that they are *coprime*, or *relatively prime*.

The algorithm we have just described is called *Euclidean algorithm* and yields a method to efficiently compute the greatest common divisor of two integers $a$ and $b$.

**Remark 1.3.4.** Given two positive integers $a$ and $b$, if we know all their divisors, clearly we can immediately find their greatest common divisor. In particular, let us announce in advance something we shall see in Chapter 4 but everyone knows since primary school: this holds if we know the prime factorisations of $a$ and $b$. In fact, as is well known, $\mathrm{GCD}(a, b)$ *is the product of the prime factors common to $a$ and $b$, taken each raised to the smallest exponent with which it appears in the factorisations.* Nevertheless, as we shall see in Chapter 4, finding the factorisation of an integer $n$ is a computationally hard problem, that is, in general *it requires a computation time that increases enormously as $n$ increases*, so much so that for a large enough $n$ this time becomes longer than the estimated life of the universe! So the method we learnt in school, requiring the prime factorisation of $a$ and $b$, is theoretically faultless, but is possibly less than useful in practice. The strong point of the Euclidean algorithm is that it enables us to find the greatest common divisor of two numbers $a$ and $b$ without having to know their prime factorisation. As we shall see in Chapter 2 this algorithm is more efficient, in a sense that will be made precise.

### 1.3.3 Bézout's identity

The Euclidean algorithm also provides a way of proving that a relation of the form

$$\mathrm{GCD}(a, b) = \alpha a + \beta b \tag{1.14}$$

holds, with $\alpha$ and $\beta$ suitable integers. Equation (1.14) is called *Bézout's identity* and turns out to be very useful. For instance, it is the starting point for the resolution of linear Diophantine equations, which we shall shortly deal with, and linear congruences, which will be covered in the next chapter. To prove this identity, it is sufficient to show that all the remainders of the successive divisions can be written as combinations of $a$ and $b$. In fact, notice that

$$r_1 = a - b\, q_1,$$
$$r_2 = b - r_1\, q_2,$$
$$\vdots$$
$$r_n = r_{n-2} - r_{n-1}\, q_n;$$

hence

$$r_2 = b - r_1 q_2 = b - (a - bq_1)q_2 = (-q_2)a + (1 + q_1 q_2)b,$$

that is, $r_1$ and $r_2$ may be written as combinations of $a$ and $b$. So $r_3$, being a combination with integer coefficients of $r_1$ and $r_2$, is a combination with integer coefficients of $a$ and $b$ too. In conclusion, $d = r_n$ is a combination with integer coefficients of $r_{n-1}$ and $r_{n-2}$, and so of $a$ and $b$.

Here follow some important consequences of Bézout's identity:

**Proposition 1.3.5.** *Let $a$ and $b$ be two positive integers. They are coprime if and only if there exist two integers $\alpha, \beta$ such that*

$$\alpha a + \beta b = 1. \tag{1.15}$$

PROOF. If $a$ and $b$ are coprime, we have $\mathrm{GCD}(a, b) = 1$ and the claim follows from Bézout's identity.

On the other hand, suppose equation (1.15) holds. Let $d$ be a common divisor of $a$ and $b$. Then clearly $d$ divides $\alpha a + \beta b$ too, and so divides 1. Thus either $d = 1$ or $d = -1$, and consequently $a$ and $b$ are relatively prime.  □

**Corollary 1.3.6.** *Let $a$ and $b$ be two positive integers and let $d = \mathrm{GCD}(a, b)$. If $a = dn$ e $b = dm$, we have $\mathrm{GCD}(n, m) = 1$.*

PROOF. Equation (1.14) may now be written as

$$d = dn\alpha + dm\beta.$$

By dividing by $d$ both sides we get

$$\alpha n + \beta m = 1$$

and by applying Proposition 1.3.5 we conclude that $n$ and $m$ are relatively prime.  □

**Corollary 1.3.7.** *Let $a$, $b_1$, and $b_2$ be integers such that $a$ and $b_1$ are coprime, and so are $a$ and $b_2$. Then $a$ is coprime with $b_1 b_2$.*

PROOF. The following relations hold:

$$1 = \alpha a + \beta_1 b_1, \qquad 1 = \alpha' a + \beta_2 b_2.$$

By multiplying them, we get

$$1 = (\alpha\alpha' a + \alpha\beta_2 b_2 + \alpha'\beta_1 b_1)a + (\beta_1\beta_2)(b_1 b_2),$$

proving the claim.  □

**Corollary 1.3.8.** *Let $a$, $b$, and $n$ be integers such that $a \mid n$, $b \mid n$ and $\mathrm{GCD}(a, b) = 1$. Then $ab \mid n$.*

PROOF. We have $n = n_1 a = n_2 b$. Moreover, a relation of the form (1.15) holds. Multiplying it by $n$ we get

$$n = \alpha na + \beta nb = \alpha n_2(ab) + \beta n_1(ab),$$

proving the claim.  □

**Corollary 1.3.9.** *Let $a$ and $b$ be two coprime positive integers, and let $n$ be any integer. If $a \mid bn$ then $a \mid n$.*

PROOF. By hypothesis there exists an integer $m$ such that

$$bn = am. \tag{1.16}$$

By Bézout's identity, there exist integers $\alpha$ and $\beta$ satisfying equation (1.15). Multiplying both sides of this equation by $n$ and keeping in mind equation (1.16), we get

$$n = \alpha an + \beta bn = \alpha an + \beta am;$$

hence, $a \mid n$.                                                                    □

Notice that the expression for $\mathrm{GCD}(a, b)$ yielded by equation (1.14) is not at all unique. For instance: $1 = 3 \cdot 7 + (-4) \cdot 5 = (-2) \cdot 7 + 3 \cdot 5$.

**Example 1.3.10.** We are now going to analyse an example to understand how to use Euclidean algorithm to find a Bézout relation. In doing so, we shall use a notation quite useful both for programming a computer to execute the algorithm and for applying it by hand.

We intend to find a Bézout's identity for $\mathrm{GCD}(1245, 56)$. Following the Euclidean algorithm, we proceed as follows:

$$\begin{aligned}
1245 &= 56 \cdot 22 + 13, \\
56 &= 13 \cdot 4 + 4, \\
13 &= 4 \cdot 3 + \boxed{1}, \\
4 &= 1 \cdot 4 + 0.
\end{aligned} \tag{1.17}$$

So we find $\mathrm{GCD}(1245, 56) = 1$. Now we want to express 1 in the form $\alpha a + \beta b$, with $a = 1245$, $b = 56$. In order to do this, it is convenient to use the following notation:

$$\alpha a + \beta b \equiv (\alpha, \beta).$$

In other words, we forget $a$ and $b$ and just write the coefficients of the linear combination as the elements of a pair. So we associate to $a$ the pair $(1, 0)$, and to $b$ the pair $(0, 1)$. Addition is defined on pairs by

$$(\alpha, \beta) + (\alpha', \beta') \stackrel{\text{def}}{=} (\alpha + \alpha', \beta + \beta');$$

moreover,

$$\gamma(\alpha, \beta) \stackrel{\text{def}}{=} (\gamma \cdot \alpha, \gamma \cdot \beta)$$

for all $\alpha, \beta, \gamma, \alpha', \beta' \in \mathbb{Z}$.

So we may rewrite the steps of Euclidean algorithm as follows:

$$\begin{aligned}
r_1 &= 13 = a + b \cdot (-22), \\
r_2 &= 4 = b + r_1 \cdot (-4), \\
r_3 &= 1 = r_1 + r_2 \cdot (-3),
\end{aligned}$$

which, in the new notation, become

$$r_1 = 13 = a + b \cdot (-22) \equiv (1,0) + (0,1)(-22) = (1,-22),$$
$$r_2 = 4 = b + r_1 \cdot (-4) \equiv (0,1) + (1,-22)(-4) = (-4,89),$$
$$r_3 = 1 = r_1 + r_2 \cdot (-3) \equiv (1,-22) + (-4,89)(-3) = (13,-289),$$

so a Bézout's identity for $\mathrm{GCD}(1245,56)$ is

$$1 = 13 \cdot 1245 + (-289) \cdot 56.$$

Notice that, as the algorithm puts in evidence, in determining the pair associated with a remainder $r_i$ we only use the two pairs associated with the two preceding remainders $r_{i-1}$ and $r_{i-2}$. So we may directly work with the pairs, without having to pass through the intermediate expressions.

### 1.3.4 Linear Diophantine equations

A first application of the material of this section concerns the study of so-called *linear Diophantine equations*. These are equations of the form

$$ax + by = c, \tag{1.18}$$

where $a, b, c$ are in $\mathbb{Z}$. The case when $a$ or $b$ is equal to zero is trivial, so we omit it. We want to ascertain whether the equation admits *integer solutions*, that is solutions $(x, y)$ with $x, y \in \mathbb{Z}$.

In a geometrical setting this equation represents, in a Cartesian plane, a line not parallel to either axis: we are interested in determining whether it passes through *integer points*, that is, points with integer numbers as coordinates.

The following proposition gives a necessary and sufficient condition for the equation $ax + by = c$ to admit integer solutions.

**Proposition 1.3.11.** *Equation* $ax + by = c$, *with* $a, b, c \in \mathbb{Z}$ *and* $a, b$ *different from zero, admits an integer solution* $(x, y)$ *if and only if* $\mathrm{GCD}(a, b)$ *divides* $c$.

PROOF. Let $(\bar{x}, \bar{y})$ be an integer solution of equation (1.18) and set $d = \mathrm{GCD}(a, b)$. Then $d$, being a divisor of both $a$ and $b$, divides the left-hand side of the equation and so divides $c$.

On the other hand, suppose that $d$ divides $c$, that is, $c$ may be written as $c = d \cdot h$. Write $d$ in the form $d = \alpha a + \beta b$. Multiplying both sides by $h$ we get

$$c = \alpha h a + \beta h b$$

and, setting $\bar{x} = \alpha h$ and $\bar{y} = \beta h$, we find that $(\bar{x}, \bar{y})$ is a solution of equation (1.18). □

For instance, equation

$$3x + 4y = -1 \tag{1.19}$$

has solutions in $\mathbb{Z}$, because $\text{GCD}(3,4) = 1$ divides $-1$. We may write $1 = \mathbf{3}(-1) + \mathbf{4}(1)$, and so we have $-1 = \mathbf{3}(1) + \mathbf{4}(-1)$. Thus a solution is $(1, -1)$. Notice that this solution is not unique: other solutions of equation (1.19) are $(-3, 2)$, $(-7, 5)$ e $(5, -4)$ (see figure 1.2).

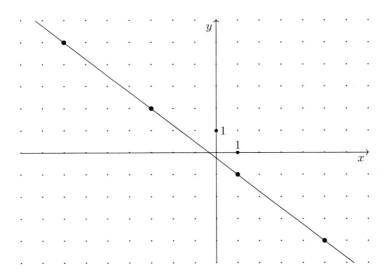

**Fig. 1.2.** The line $3x + 4y = -1$ in a Cartesian plane

### 1.3.5 Euclidean rings

It is useful to put what has been said about the division algorithm in $\mathbb{Z}$ in a wider, abstract context. For this purpose, let us recall some general notions about divisibility in a ring and in particular in Euclidean rings.

Consider a commutative ring with unity and no zero-divisors $A$. We may extend to $A$ most of the definitions about divisibility we have given regarding the ring $\mathbb{Z}$ of integer numbers.

First of all, an element $a \in A$ is said to be *invertible* if there exists an element $b \in A$ such that $ab = 1$. Clearly, 1 and $-1$ are invertible, all invertible elements are different from zero, and they form a group with respect to multiplication: this group is denoted by $A^*$ or $U(A)$.

Let $a$ and $b$ be elements of $A$ such that $b \neq 0$. We say that $b$ *divides* $a$, or that it is a *divisor* of $a$, or that $a$ is a *multiple* of $b$, and we write $b \mid a$, if there exists an element $x \in A$ such that $a = bx$. Notice that if $a \mid b$ and $b \mid c$ then $a \mid c$. Clearly, each invertible element divides each element of $A$. Two elements $a, b$ different from zero are said to be *associated* if $a = bx$ with $x$ an invertible element. A result analogous to Lemma 1.3.3 holds, that is $a \mid b$ and $b \mid a$ if and only if $a$ and $b$ are associated.

Consider an *integral domain* $A$, that is a commutative ring with unity $A$ with no zero-divisors. An integral domain $A$ is said to be *Euclidean* if there exists a map

$$v : A \setminus \{0\} \to \mathbb{N}$$

that satisfies the following properties:

(1) for each pair $(a, b)$ of elements different from zero we have $v(ab) \geq v(a)$;
(2) for each $a \in A$ and for each $b \in A \setminus \{0\}$, there exist $q, r \in A$ (respectively said *quotient* and *remainder* of the *division* of $a$ by $b$) such that $a = bq + r$ and either $r = 0$ or $v(r) < v(b)$.

Clearly, $\mathbb{Z}$ is a Euclidean ring, by taking $v(a) = |a|$, for each $a \in \mathbb{Z}$. So, $A$ is a Euclidean ring if *there exists in $A$ a division algorithm* analogous to the one in $\mathbb{Z}$.

Another trivial example of a Euclidean ring is given by any field $A$. It suffices to take as $v$ the constant map equal to 1 and, as quotient $q$ and remainder $r$ of the division of $a$ by $b$, respectively $q = a/b$ and $r = 0$.

Given an integral domain $A$ and given two elements $a, b$ different from zero, it is possible to consider the set $D(a, b)$ of common divisors of $a$ and $b$. Notice that for each $c \in D(a, b)$ and for each invertible $x$, we also have $cx \in D(a, b)$. We define $d \in D(a, b)$ to be a *greatest common divisor* of $a$ and $b$, and denote it by $\mathrm{GCD}(a, b)$, if for each $c \in D(a, b)$ we have $c \mid d$. Notice that if each of $d$ and $d'$ is a greatest common divisor of $a$ e $b$, they are associated, and conversely, if $d = \mathrm{GCD}(a, b)$ and if $d'$ is associated to $d$, then $d'$ is a $\mathrm{GCD}(a, b)$ too (see Exercise A1.44). If the greatest common divisor of $a$ e $b$ is invertible, and so we may assume that $\mathrm{GCD}(a, b) = 1$, then $a$ e $b$ are said to be *relatively prime*.

In an integral domain $A$ two elements may well not admit a greatest common divisor. However, if $A$ is Euclidean, greatest common divisors always exist, as it is always possible to apply the same procedure as in $\mathbb{Z}$. So we have the following theorem.

**Theorem 1.3.12.** *Let $A$ be a Euclidean ring. If $a, b \in A$ are elements different from zero, there exists a $\mathrm{GCD}(a, b)$, which can be determined by the Euclidean algorithm. Moreover, Bézout's identity holds, that is, there exist $\alpha, \beta \in A$ such that equation (1.14) holds.*

It is interesting to give a different interpretation of the greatest common divisor in a Euclidean ring. Recall that in a commutative ring with unity $A$ an ideal $I$ is a subset such that

(1) for each $x, y \in I$ we have $x + y \in I$;
(2) for each $x \in I$ and for each $y \in A$, we have $xy \in I$.

In general, $A$ and $\{0\}$ are ideals that are said to be *trivial*. The ideal $\{0\}$ is simply denoted by 0 and is called *zero ideal*.

Given $x_1, \ldots, x_n \in A$, consider the set $I$ of all the elements of $A$ of the form $x_1 y_1 + \cdots + x_n y_n$, with $y_1, \ldots, y_n$ elements of $A$. The set $I$ is an ideal said to be *(finitely) generated* by $x_1, \ldots, x_n$; it is denoted by the symbol $(x_1, \ldots, x_n)$. An ideal $(x)$ generated by a single element $x \in A$ is said to be *principal* and exactly consists of the multiples of $x$. For instance, $A = (1)$ and $0 = (0)$ are principal ideals. Notice that if $A$ is an integral domain, then $(x) = (y)$ if and only if $x$ and $y$ are associated (see Exercise A1.46). A commutative ring is said to be a *principal ideal ring* if every ideal of the ring is principal.

The following result is noteworthy.

**Proposition 1.3.13.** *If $A$ is a principal ideal integral domain then, for each pair of elements $a$, $b$ of $A$ different from zero, there exists the $\mathrm{GCD}(a,b)$, and every generator of the ideal $(a,b)$ is a $\mathrm{GCD}(a,b)$.*

PROOF. Let $d$ be a generator of the ideal $(a,b)$. As $a,b \in (a,b)$, clearly $d$ divides both $a$ and $b$. We know that Bézout's identity $d = \alpha a + \beta b$ holds; so, if $c \in D(a,b)$ is a common divisor of $a$ and $b$, clearly $c$ divides $d$. $\qquad\square$

For Euclidean rings the following remarkable theorem holds.

**Theorem 1.3.14.** *Every Euclidean ring is a principal ideal ring.*

PROOF. Let $A$ be a Euclidean ring, and let $I$ be an ideal of $A$. If $I = 0$ there is nothing to prove. If $I \neq 0$ let $b \in I$ be a non-zero element such that for each non-zero $a \in I$ inequality $v(b) \leq v(a)$ holds (here we use the well-ordering principle, see Exercise A1.2). Let $a \in I$ be any element. There exist $q, r$ such that $a = bq + r$ with $v(r) < v(b)$. Notice that $r \in I$ and, by the definition of $b$, we have $r = 0$. Thus $a$ is a multiple of $b$, and the theorem is proved. $\qquad\square$

As a consequence we get the following.

**Corollary 1.3.15.** *Let $A$ be a Euclidean ring and let $a,b$ be non-zero elements of $A$. Then $d = \mathrm{GCD}(a,b)$ if and only if $d$ is a generator of the ideal $(a,b)$.*

In particular, $\mathbb{Z}$ is a principal ideal ring. Another interesting example of a Euclidean ring will be shown in Exercise A1.49. Still another important example is described in the next section.

## 1.3.6 Polynomials

This is a good moment to recall some basics about polynomials, emphasising their similarities with integers, and giving an interpretation of some of their fundamental properties in terms of divisibility.

**Definition 1.3.16.** *A polynomial $p(x)$ with coefficients in a commutative ring with unity $A$ is a* formal *expression of the form*

$$p(x) = a_0 + a_1 x + a_2 x^2 + \cdots + a_n x^n = \sum_{i=0}^{n} a_i x^i, \qquad a_i \in A, \qquad (1.20)$$

*where $x$ is an* indeterminate *or* variable. *The elements $a_0, a_1, \ldots, a_n \in \mathbb{K}$ are said to be the* coefficients *of the polynomial. If $a_n \neq 0$, the integer $n$ is the* degree *of the polynomial and is denoted by $\deg(p(x))$ or by $\partial p(x)$. The polynomials of degree $0$ are called* constants *and may be identified with the elements of $A$. The polynomials of degree one are called* linear, *those of degree two* quadratic, *those of degree three* cubic, *and so on. The coefficient $a_n$ is called the* leading coefficient *of $p(x)$. If it is equal to $1$, the polynomial is said to be* monic.

Notice that the degree of a non-zero constant is zero. It is usual not to assign any degree to the zero polynomial, that is, the polynomial with all coefficients equal to zero.

In a more formal way, we may identify polynomial (1.20) with the sequence of elements of $A$

$$\{\, a_0,\, a_1,\, \ldots,\, a_n,\, 0,\, 0,\, \ldots \,\},$$

all terms of which are zero from a certain point onwards. We usually set $a_i = 0$ for each $i > n$; we may also write $p(x) = \sum_{i=0}^{\infty} a_i x^i$, keeping in mind that in any case it is a finite sum.

So, in general, if we use the "equal" sign to denote the above identification, we have

$$a = \{a, 0, 0, 0, \ldots\},$$
$$x = \{0, 1, 0, 0, \ldots\},$$
$$x^2 = \{0, 0, 1, 0, \ldots\},$$

$$\vdots$$

$$x^i = \{0, 0, \ldots, 0, \underbrace{1}_{(i+1)\text{-th position}}, 0, 0, \ldots\}).$$

The polynomial that corresponds to the sequence $\{0, 0, \ldots\}$, that is to the constant $0$, is the *zero polynomial*, which is denoted by the symbol $0$.

Let us look at an example:

$$x^3 + 2x - 1$$

is a degree 3 polynomial with integer, and so rational (or real, or complex) coefficients. Using the notation just introduced, the polynomial becomes

$$\{\, -1,\, 2,\, 0,\, 1,\, 0,\, 0,\, \ldots \,\}.$$

It follows from the definition that two polynomials $p(x) = \sum_{i=0}^{n} a_i x^i$ and $q(x) = \sum_{j=0}^{m} b_j x^j$, with $a_i, b_j \in \mathbb{K}$, are *equal* if and only if $a_i = b_i$, for each $i \in \mathbb{N}$. In particular, if $m > n$, then $b_{n+1} = b_{n+2} = \cdots = b_m = 0$.

The set of all polynomials with coefficients in $A$ is denoted by $A[x]$. In $A[x]$ two operation are defined: an *addition* and a *multiplication*, as follows. Let $p(x) = \sum_{i=0}^{n} a_i x^i$ and $q(x) = \sum_{j=0}^{m} b_j x^j$ in $A[x]$ be two polynomials, with $n \le m$. We put

$$p(x) + q(x) \stackrel{\text{def}}{=} \sum_{h=0}^{m} (a_h + b_h) x^h, \qquad p(x) q(x) \stackrel{\text{def}}{=} \sum_{h=0}^{n+m} \left( \sum_{i+j=h} a_i b_j \right) x^h.$$

The zero polynomials is the identity element for addition, and the opposite (or additive inverse) of the polynomial $p(x) = \sum_{i=0}^{n} a_i x^i$ is the polynomial having as its coefficients the opposite of the coefficients $a_i$, for each $i$.

Let us explicitly remark the following relations between the degrees of two polynomials with coefficients in a field, or in an integral domain, and the degrees of their sum and their product:

$$\partial(p(x) + q(x)) \le \max(\partial p(x), \partial q(x)), \quad \partial(p(x) q(x)) = \partial p(x) + \partial q(x). \tag{1.21}$$

With respect to the operations of addition and multiplication (1.21), polynomials form a ring. If $A$ is an integral domain, so is $A[x]$ (see Exercise A1.51). If this is the case, the invertible elements in $A[x]$ are the invertible elements of $A$. By iterating the construction of $A[x]$ it is possible to construct more generally the ring $A[x_1, \ldots, x_n]$ of polynomials in variables $x_1, \ldots, x_n$ on the ring $A$. The typical element of $A[x_1, \ldots, x_n]$ is a finite sum of *monomials* of the form $ax_1^{m_1} \cdots x_n^{m_n}$, with $a \in A$ and $m_1, \ldots, m_n$ non-negative integers. If $a = 0$, we get the zero monomial, otherwise the integer $m_1 + \cdots + m_n$ is called *degree* of the monomial, and the degree of a polynomial is the maximum of the degrees of its non-zero monomials.

In general we shall consider polynomials over fields.

The following remarkable result holds.

**Theorem 1.3.17.** *The ring $\mathbb{K}[x]$ of the polynomials over a field $\mathbb{K}$, endowed with the map*

$$\partial : \mathbb{K}[x] \setminus \{0\} \to \mathbb{N},$$

*is a Euclidean ring. In particular, let $f(x)$, $g(x) \in \mathbb{K}[x]$ be two polynomials, with $g(x) \neq 0$. Then there exist, and are uniquely determined, two polynomials $q(x)$ e $r(x)$ in $\mathbb{K}[x]$ such that*

$$f(x) = g(x) \cdot q(x) + r(x), \qquad \text{with } \partial r(x) < \partial g(x) \text{ or } r(x) = 0.$$

PROOF. First of all, it is easy to verify that the map $\partial$ satisfies property (1) of the definition of Euclidean rings.

As to the uniqueness of division, notice that if $f(x) = g(x) \cdot q_1(x) + r_1(x) = g(x) \cdot q_2(x) + r_2(x)$ such that $\partial r_i(x) < \partial g(x)$ or $r_i(x) = 0$ with $i = 1, 2$, then we would get

$$(q_1(x) - q_2(x))g(x) = r_2(x) - r_1(x).$$

If it is the case that $q_1(x) - q_2(x) \neq 0$, the left-hand side would have a greater degree than the second one, which is not possible. So we must have $q_1(x) = q_2(x)$, which also implies $r_2(x) = r_1(x)$.

As to the the existence of the algorithm for the division, if $f(x) = 0$ there is nothing to prove. So let $f(x) \neq 0$ and let $n, m$ be respectively the degrees of $f(x)$ and $g(x)$. If $m > n$ it suffices to take $q(x) = 0$ and $r(x) = f(x)$. So suppose $n \geq m$. As the theorem is trivially true for $n = 0$, we proceed by induction on $n$. If

$$f(x) = a_0 + a_1 x + \cdots + a_n x^n, \qquad g(x) = b_0 + b_1 x + \cdots + b_m x^m,$$

we put

$$h(x) = f(x) - \frac{a_n}{b_m} x^{n-m} g(x).$$

The degree of $h(x)$ is smaller than $n$, so we may apply to $h(x)$ the induction hypothesis. So there exist $q_1(x), r_1(x)$, with $r_1(x)$ either the zero polynomial or of degree smaller than $m$, such that

$$f(x) - \frac{a_n}{b_m} x^{n-m} g(x) = h(x) = q_1(x)g(x) + r_1(x).$$

Then it is enough to take $r(x) = r_1(x)$ and $q(x) = q_1(x) + (a_n/b_m)x^{n-m}$.     □

As a consequence of this result and of those given in the previous section, we conclude that it is possible to consider the greatest common divisor of non-zero polynomials over a field, and to compute it with the Euclidean algorithm. Notice that sometimes we identify the greatest common divisor of two polynomials with the unique *monic* polynomial having the same property (see Exercise A1.52).

**Example 1.3.18.** We shall compute the greatest common divisor of two polynomials

$$f(x) = x^3 + 1, \quad g(x) = x^2 + 1$$

on $\mathbb{Q}$. We have

$$x^3 + 1 = x(x^2 + 1) - x + 1,$$
$$x^2 + 1 = (-x - 1)(-x + 1) + 2,$$
$$-x + 1 = \frac{1}{2}(-x - 1) \cdot 2,$$

and so $\mathrm{GCD}(f(x), g(x)) = 2$. As 2 is invertible, we have that $f(x)$ and $g(x)$ are relatively prime. From the above divisions we get

$$1 = \frac{1}{2}(x^2 + 1) + \frac{1}{2}(x + 1)(1 - x) =$$
$$= \frac{1}{2}(x^2 + 1) + \frac{1}{2}(x + 1)[(x^3 + 1) - x(x^2 + 1)] =$$
$$= \frac{1}{2}(x^2 + 1)(-x^2 - x + 1) + \frac{1}{2}(x^3 + 1)(x + 1)$$

that is, a Bézout's identity.

If we consider the polynomials

$$p(x) = x^3 - 1, \quad q(x) = x^2 - 1$$

on $\mathbb{Q}$, we have

$$x^3 - 1 = x(x^2 - 1) + x - 1,$$
$$x^2 - 1 = (x + 1)(x - 1),$$

and so $\mathrm{GCD}(f(x), g(x)) = x - 1$.

We determine next the greatest common divisor of the real polynomials $f(x) = x^3 + 3x^2 - x - 3$ and $g(x) = x^2 + 3x + 2$, and find a Bézout's identity.

Applying the Euclidean algorithm we get

$$x^3 + 3x^2 - x - 3 = (x^2 + 3x + 2)x - 3(x + 1),$$
$$x^2 + 3x + 2 = -3(x + 1)\left(-\frac{1}{3}x - \frac{2}{3}\right) + 0.$$

So,

$$\mathrm{GCD}(x^3 + 3x^2 - x - 3, x^2 + 3x + 2) = x + 1.$$

Clearly, $-3(x+1)$ is *a* greatest common divisor too, but it is not monic. The greatest common divisor can be written down by generalising Equation (1.14) to polynomials as follows:

$$x + 1 = -\frac{1}{3}(x^3 + 3x^2 - x - 3) + \frac{1}{3}x(x^2 + 3x + 2)$$

with $h(x) = -1/3$ and $k(x) = (1/3)x$.

Remember that, given a polynomial $f(x) \in \mathbb{K}[x]$, an element $\alpha \in \mathbb{K}$ such that $f(\alpha) = 0$ is said to be a *root*, or a *zero*, of $f(x)$. A renowned theorem due to Ruffini, known as factor theorem, translates the question of the existence of roots into the context of divisibility.

**Theorem 1.3.19.** *If $f(x) \in \mathbb{K}[x]$ and $\alpha \in \mathbb{K}$ is an element such that $f(\alpha) = 0$, then $(x - \alpha) \mid f(x)$.*

PROOF. By dividing $f(x)$ by $x - \alpha$ we have

$$f(x) = q(x)(x - \alpha) + r(x) \tag{1.22}$$

where $r(x)$ has a smaller degree than $x - \alpha$; so, it is a constant $r$. By putting $x = \alpha$ in (1.22), we find $r = 0$. □

**Corollary 1.3.20.** *If $f(x) \in \mathbb{K}[x]$ is a non-zero polynomial of degree $n$, it has at most $n$ roots in $\mathbb{K}$.*

PROOF. If $\alpha_1, \ldots, \alpha_h$ are distinct roots of $f(x)$, Ruffini's theorem implies that $f(x)$ is divisible by the polynomial $(x - \alpha_1) \cdots (x - \alpha_h)$ of degree $h$ and so $n \geq h$. □

Given a polynomial $f(x) \in \mathbb{K}[x]$, we may consider the function

$$\varphi_f : x \in \mathbb{K} \to f(x) \in \mathbb{K},$$

called the *polynomial function* determined by $f(x)$.

**Proposition 1.3.21.** *If $\mathbb{K}$ is infinite, and if $f(x), g(x)$ are polynomials on $\mathbb{K}$, then $f(x) = g(x)$ if and only if $\varphi_f = \varphi_g$.*

PROOF. We have $\varphi_f = \varphi_g$ if and only if each $\alpha \in \mathbb{K}$ is a root of the polynomial $f(x) - g(x)$. By Corollary 1.3.20 this happens if and only if $f(x) - g(x) = 0$ and so if and only if $f(x) = g(x)$. □

**Definition 1.3.22.** *A root $\alpha \in \mathbb{K}$ of $f(x) \in \mathbb{K}[x]$ is said to be* simple *if $(x - \alpha) \mid f(x)$ but $(x - \alpha)^2 \nmid f(x)$. It is said to be of* multiplicity $m$ *if $(x - \alpha)^m \mid f(x)$, but $(x - \alpha)^{m+1} \nmid f(x)$. A root of multiplicity $m > 1$ is said to be a* multiple *root.*

For instance, $-1$ is a root of multiplicity 2 or, as it is also called, a *double root* of the polynomial $(x + 1)^2$.

Given a polynomial

$$f(x) = a_0 + a_1 x + a_2 x^2 + a_3 x^3 + \cdots + a_n x^n$$

over a field $\mathbb{K}$, the *derivative* polynomial $f'(x)$, also denoted by $D(f(x))$, is defined as

$$f'(x) = a_1 + 2a_2 x + 3a_3 x^2 + \cdots + n a_n x^{n-1}.$$

For instance, if

$$f(x) = 4x^4 + 6x^3 - 5x^2 + x - 1$$

is a polynomial with real coefficients, then

$$f'(x) = 16x^3 + 18x^2 - 10x + 1.$$

The derivative is a function

$$D : \mathbb{K}[x] \to \mathbb{K}[x] \tag{1.23}$$

satisfying the following properties (see Exercises A1.58 and A1.59):

- *linearity:* $D(f(x) + g(x)) = D(f(x)) + D(g(x))$, for each pair of polynomials $f(x), g(x) \in \mathbb{K}[x]$;
- *Leibniz's law:* $D(f(x) \cdot g(x)) = D(f(x)) \cdot g(x) + D(g(x)) \cdot f(x)$, for each pair of polynomials $f(x), g(x) \in \mathbb{K}[x]$.

The derivative composed with itself $h$ times, with $h \geq 2$, applied to a polynomial $f(x)$, is denoted by the symbol $f^{(h)}(x)$, or $D^{(h)}(f(x))$, and is called the *h-th derivative* of $f(x)$. It is usual to understand $D^{(1)}$ as meaning the same as $D$, while $D^{(0)}$ is the identity function in $\mathbb{K}[x]$.

Notice that $D^{(i)}(x^n) = 0$ if $i > n$, and so $D^{(i)}(f(x)) = 0$ if $i$ is greater than the degree of $f(x)$ (see Exercise A1.61). Moreover, it follows from Leibniz's law that, if $c \in \mathbb{K}$ and if $i \leq n$, then

$$D^{(i)}((x - c)^n) = n(n - 1) \cdots (n - i + 1)(x - c)^{n-i}; \tag{1.24}$$

in particular, we have

$$D^{(n)}((x - c)^n) = n! \tag{1.25}$$

(see Exercise A1.62).

The next proposition relates the concept of derivative of a polynomial with the multiplicity of its roots. Recall that the symbol $m!$ denotes the *factorial* of the integer number $m$ (see Exercise A1.12).

**Proposition 1.3.23.** *If $f(x)$ is a polynomial on the field $\mathbb{K}$ and if $\alpha$ is a root of $f$ of multiplicity $m$, we have $(D^{(i)}f)(\alpha) = 0$ for each $i \leq m - 1$. Moreover, if $m! \neq 0$ in $\mathbb{K}$, we have $(D^{(m)}f)(\alpha) \neq 0$.*

PROOF. We have $f(x) = (x - \alpha)^m g(x)$ with $g(\alpha) \neq 0$. By repeatedly applying Leibniz's law, we find

$$D^{(i)}(f(x)) = \sum_{j=0}^{i} D^{(j)}((x - \alpha)^m) D^{(i-j)}(g(x)).$$

Keeping in mind Equation (1.24), we get $(D^{(i)}f)(\alpha) = 0$ for each $i \leq m - 1$, while $(D^{(m)}f)(\alpha) = m! g(\alpha)$. So we get what was claimed.    □

This result gives a useful criterion to ascertain whether a polynomial admits multiple roots or not.

**Example 1.3.24.** Let us solve the following problem: *find a polynomial with real coefficients*

$$f(x) = x^3 - 3\lambda^2 x + 2$$

*admitting a double root.* We may proceed as follows. For $f(x)$ to admit a double root $\alpha$, it also has to be a root of its derivative

$$f'(x) = 3x^2 - 3\lambda^2,$$

so $\alpha = \pm \lambda$. Thus

$$(\pm \lambda)^3 - 3\lambda^2(\pm \lambda) + 2 = 0,$$

must hold; this implies $\lambda^3 = \pm 1$ and so $\lambda = \pm 1$. The polynomial we are looking for is

$$f(x) = x^3 - 3x + 2$$

and its unique double root is $\alpha = \pm \lambda = \pm(\pm 1) = 1$.

In conclusion we prove the following theorem, whose usefulness will become clear later on.

**Theorem 1.3.25 (Taylor's formula).** *Let*

$$f(x) = a_0 + a_1 x + \cdots + a_n x^n$$

*be a polynomial over a field* $\mathbb{K} = \mathbb{Q}$, $\mathbb{R}$ *or* $\mathbb{C}$. *If* $y$ *is another indeterminate on* $\mathbb{K}$, *we may consider the polynomial in two variables* $f(x + y)$ *on* $\mathbb{K}$. *For it the following formula holds, called* Taylor's formula:

$$f(x+y) = f(x) + f'(x)y + \frac{f^{(2)}(x)}{2}y^2 + \cdots + \frac{f^{(n)}(x)}{n!}y^n = \sum_{i=0}^{n} \frac{f^{(i)}(x)}{i!}y^i. \quad (1.26)$$

*Moreover, for each* $i = 1, \ldots, n$, *we have*

$$a_i = \frac{f^{(i)}(0)}{i!}. \quad (1.27)$$

PROOF. The last claim follows from the first one by exchanging $x$ and $y$ and putting $y = 0$. For the first claim, notice that Taylor's formula is *linear*, that is, if it is true for two polynomials $f(x)$ and $g(x)$ it is also true for every linear combination $\alpha f(x) + \beta g(x)$ with constant coefficients $\alpha, \beta$. Moreover, the formula is true for constants. So, to prove it for an arbitrary polynomial it suffices to prove it for monomials of the form $x^n$. In this case Taylor's formula is just the binomial theorem (see Exercise A1.18). □

Taylor's formula may also be written in a slightly different form. Choose a constant $k \in \mathbb{K}$ and substitute $k$ for $x$ and $x - k$ for $y$ in the formula. Then we get

$$f(x) = f(k) + f'(k)(x-k) + \frac{f^{(2)}(k)}{2}(x-k)^2 + \cdots + \frac{f^{(n)}(k)}{n!}(x-k)^n =$$

$$= \sum_{i=0}^{n} \frac{f^{(i)}(k)}{i!}(x-k)^i.$$

Written like this, the formula is called *Maclaurin's formula* around the point $k$.

For instance, the real polynomial

$$f(x) = x^3 + 2x - 1$$

may also be written, by applying Maclaurin's formula around 1, as

$$f(x) = 2 + 5(x - 1) + 6(x - 1)^2 + 6(x - 1)^3.$$

**Remark 1.3.26.** Let $\mathbb{A}$ be a ring with unity. We may consider the map

$$f : \mathbb{Z} \to \mathbb{A}$$

defined as follows: if $n$ is a positive number define $f(n) = \underbrace{1 + \cdots + 1}_{n}$, if $n$ is a negative number, define $f(n) = -f(-n)$, and put $f(0) = 0$. The value $f(n)$ is usually denoted by $n$. Clearly, we have

$$f(n + m) = f(n) + f(m), \quad f(nm) = f(n)f(m).$$

So $f(\mathbb{Z})$ is a subring of $A$, called the *prime subring* of $A$. When $f$ is injective, $A$ is said to have *characteristic zero* and the prime subring of $A$ can be identified with $\mathbb{Z}$.

If the characteristic of $A$ is different from zero, then there exists a least integer $p > 1$ such that $f(p) = 0$. It is said to be the *characteristic* of $A$. If $A$ is an integral domain, then $p$ is a prime number, that is, it is not a product of two positive integers smaller than $p$. In fact, if $p = nm$ with $n, m$ positive integers smaller than $p$, we would get $0 = f(p) = f(nm) = f(n)f(m)$ and so either $f(n) = 0$ or $f(m) = 0$, which is impossible.

Taylor's formula does not only hold for polynomials on $\mathbb{Q}$, $\mathbb{R}$ and $\mathbb{C}$, but more in general for polynomials over a field of characteristic zero. Also, the hypothesis $m! \neq 0$ in Proposition 1.3.23 is certainly satisfied if the characteristic of $\mathbb{K}$ is zero.

## 1.4 Counting in different bases

To use numbers it is necessary to *represent* them in a simple and efficient manner. There are several ways of doing so. In this section we shall expound the most well-known way, the one used since primary school by all of us, and by computers too, that is the representation of numbers *in a given base*. In the next section we shall however see another interesting and useful way of representing numbers.

### 1.4.1 Positional notation of numbers

When we write the number 3013 in base 10, we mean the following expression:

$$3013 = 3 \cdot 10^3 + 0 \cdot 10^2 + 1 \cdot 10^1 + 3 \cdot 10^0.$$

In this example it is clear that equal digits, for instance 3, represent different numbers, depending on the position it occupies within the number. Indeed, it represents number 3 if 3 is the coefficient of $10^0$, and number 3000 if it multiplies $10^3$. So it is a *positional* notation in base 10, in which the ten *digits* from 0 to 9 are used.

Examining further this example, notice that by dividing 3013 by 10 we get

$$3013 = 10 \cdot 301 + 3,$$

that is, 3, the rightmost digit, gives the remainder of the division by 10 of the original number. Going on, we divide the quotient we found by 10 again. We get

$$301 = 10 \cdot 30 + 1.$$

So the second digit from the right, that is 1, once more gives the remainder of a division, and is so uniquely determined. In conclusion, the digits appearing in the decimal representation of the number are uniquely determined by successive divisions.

The choice of 10 as a notational base is purely conventional: in fact, along the centuries different cultures used different bases in their numeral systems: Babylonians used the base 60, Mayans the base 20 and so on. Computers use the base 2, that is they use just two digits, 0 and 1, to represent a number. In fact, in the *binary system*, that is to say in base 2, each digit conveys one *bit* of information: the symbol 0 is interpreted by the computer as the command *off* and the symbol 1 as the command *on*. Other bases used in computer science are $\beta = 8$ and $\beta = 16$.

So the role played in numeral systems by the number 10 may be played by any other integer greater than 1. For instance, if we choose 9 as a base, the number $3 \cdot 9^3 + 0 \cdot 9^2 + 1 \cdot 9^1 + 3 \cdot 9^0$ is represented in base 10 as 2199. On the other hand, the number 3013 is represented in base 9 by 4117.

The result on which the possibility of counting in any base depends is the following theorem.

**Theorem 1.4.1.** *Let $\beta$ be an integer greater than or equal to 2. Then, for each $n \in \mathbb{N}$, there exist a non-negative integer $k$ and $k + 1$ integers $a_0, a_1, \ldots, a_k$ such that $0 \le a_i < \beta$, for each $i = 0, \ldots, k$, these being the only such integers satisfying:*

$$n = a_k\beta^k + a_{k-1}\beta^{k-1} + \cdots + a_2\beta^2 + a_1\beta + a_0. \tag{1.28}$$

PROOF. Apply the Euclidean algorithm in the following way: divide $n$ by $\beta$ obtaining

$$n = \beta \cdot q_0 + a_0, \qquad\qquad 0 \le a_0 < \beta.$$

Now, if $q_0 \ne 0$, divide again $q_0$ by $\beta$ obtaining

$$q_0 = \beta \cdot q_1 + a_1, \qquad\qquad 0 \le a_1 < \beta.$$

Going on in the same way we get

$$q_1 = \beta \cdot q_2 + a_2, \qquad\qquad 0 \le a_2 < \beta,$$

$$\vdots$$

$$q_{s-2} = \beta \cdot q_{s-1} + a_{s-1}, \qquad\qquad 0 \le a_{s-1} < \beta,$$
$$q_{s-1} = \beta \cdot q_s + a_s, \qquad\qquad 0 \le a_s < \beta.$$

As $n > q_0 > q_1 > \cdots > q_s \ge 0$ is a strictly decreasing sequence of non-negative integers, it necessarily reaches zero. Let $k$ be the first integer such that $q_k = 0$.

Rewrite the list of relations obtained above:

$$n = \beta \cdot q_0 + a_0, \qquad 0 \le a_0 < \beta,$$
$$q_0 = \beta \cdot q_1 + a_1, \qquad 0 \le a_1 < \beta,$$
$$q_1 = \beta \cdot q_2 + a_2, \qquad 0 \le a_2 < \beta,$$
$$\vdots$$
$$q_{k-3} = \beta \cdot q_{k-2} + a_{k-2}, \qquad 0 \le a_{k-2} < \beta,$$
$$q_{k-2} = \beta \cdot q_{k-1} + a_{k-1}, \qquad 0 \le a_{k-1} < \beta,$$
$$q_{k-1} = \beta \cdot 0 + a_k, \qquad 0 \le a_k < \beta.$$

Proceed backwards by substituting the value $q_{k-1} = a_k$, obtained from last equation, in the previous one. We find

$$q_{k-2} = \beta \cdot a_k + a_{k-1}.$$

Substituting in the previous equation we find

$$q_{k-3} = \beta \cdot (\beta \cdot a_k + a_{k-1}) + a_{k-2} = \beta^2 \cdot a_k + \beta \cdot a_{k-1} + a_{k-2}.$$

Going on like this up to the first equation we finally get expression (1.28). The uniqueness of this expression is clear, as the digits $a_i$ appearing in it are uniquely determined as the remainders of the successive divisions.     □

In conclusion, each number stricly smaller than the base $\beta$ is represented by a unique *symbol* or *digit*. To represent a number in base $\beta$, $\beta$ different symbols are necessary. The number having as its expression (1.28) is usually denoted by the symbol $(a_k a_{k-1} \ldots a_1 a_0)_\beta$ or simply by the symbol $a_k a_{k-1} \ldots a_1 a_0$, without explicitly mentioning the base, when no confusion can possibly arise.

### 1.4.2 Base 2

How are the representations of a number in different bases related? To understand this, it suffices to keep in mind the proof of Theorem 1.4.1, which gives an algorithm based on division to determine the expression of a number in a base $\beta$. For instance, to go from the base 10 to the base 2, it suffices to divide the given number by 2, and to divide successively each quotient by 2, till we arrive to a *zero quotient*. The remainders we obtain, read from the bottom up, give the representation in base 2 of the given number.

Suppose for instance we want to write the number 8112 in base 2:

$$8112 = 2 \cdot 4056 + 0, \qquad r_0 = 0,$$
$$4056 = 2 \cdot 2028 + 0, \qquad r_1 = 0,$$
$$2028 = 2 \cdot 1014 + 0, \qquad r_2 = 0,$$
$$1014 = 2 \cdot 507 + 0, \qquad r_3 = 0,$$
$$507 = 2 \cdot 253 + 1, \qquad r_4 = 1,$$

$$253 = 2 \cdot 126 + 1, \qquad\qquad r_5 = 1,$$
$$126 = 2 \cdot 63 + 0, \qquad\qquad r_6 = 0,$$
$$63 = 2 \cdot 31 + 1, \qquad\qquad r_7 = 1,$$
$$31 = 2 \cdot 15 + 1, \qquad\qquad r_8 = 1,$$
$$15 = 2 \cdot 7 + 1, \qquad\qquad r_9 = 1,$$
$$7 = 2 \cdot 3 + 1, \qquad\qquad r_{10} = 1,$$
$$3 = 2 \cdot 1 + 1, \qquad\qquad r_{11} = 1,$$
$$1 = 2 \cdot 0 + 1, \qquad\qquad r_{12} = 1.$$

Thus, the number which is written as 8112 in base 10, in base 2 is written as

$$(8112)_{10} = (r_{12}r_{11}r_{10}r_9r_8r_7r_6r_5r_4r_3r_2r_1r_0)_2 = (1111110110000)_2.$$

To reverse this operation, that is, to go from base 2 to base 10, it suffices to sum up the right powers of two. For instance, to go back from the number $(1111110110000)_2$ to its decimal representation, we just write

$$(1111110110000)_2 = 1 \cdot 2^{12} + 1 \cdot 2^{11} + 1 \cdot 2^{10} + 1 \cdot 2^9 + 1 \cdot 2^8 + 1 \cdot 2^7 +$$
$$+ 0 \cdot 2^6 + 1 \cdot 2^5 + 1 \cdot 2^4 + 0 \cdot 2^3 + 0 \cdot 2^2 + 0 \cdot 2^1 + 0 \cdot 2^0 =$$
$$= 2^{12} + 2^{11} + 2^{10} + 2^9 + 2^8 + 2^7 + 2^5 + 2^4 = 8112$$

using, when developing the powers of two, the 10 digits from 0 to 9.

Notice that in every base the expression "10" denotes the base itself, as

$$(1\,0)_\beta = 1 \cdot \beta + 0 \cdot \beta^0 = \beta.$$

### 1.4.3 The four operations in base 2

Positional notation in base 10 represented a substantial advancement with respect to non-positional notations, for instance the Roman one, where performing calculations was far from easy. The binary system brings about a further simplification in this direction. Let us see why.

We describe now the rules to perform the four operations when we represent the numbers in base 2. The reader will notice that they are not too different from the calculation in base 10, well known since primary school. On the other hand, using base 2 is particularly simple and, moreover, useful and of practical use because, as we already recalled, this is the base used by computers. Notice that $1 + 1 = 2$ is written $1 + 1 = 10$ in base 2. So the addition tables we must learn in order to perform additions in base 2 are very simple:

$$0 + 0 = 0,$$
$$0 + 1 = 1,$$
$$1 + 0 = 1,$$
$$1 + 1 = 1\,0.$$

Let $a = \sum_{i=0}^{n} a_i \cdot 2^i$ and $b = \sum_{i=0}^{n} b_i \cdot 2^i$ be two positive integers. Then

$$a + b = \sum_{i=0}^{n} (a_i + b_i) \cdot 2^i.$$

If we want to write this in base 2, we must act as follows. Suppose $a_i + b_i$ is the lowest-indexed coefficient different from 0 and 1. This means that $a_i = b_i = 1$.

Now use the addition table and remember that $1 + 1 = 10$. This means that the term $(a_i + b_i) \cdot 2^i = 2 \cdot 2^i = 2^{i+1}$, and so we must *carry* 1 to the next coefficient, so the coefficient of $2^{i+1}$ becomes $a_{i+1} + b_{i+1} + 1$, while the coefficient of $2^i$ is 0. We go on like this for all indices.

Let us see an example to demonstrate this procedure. Suppose we have to sum the numbers 10111111 e 1011:

$$
\begin{array}{r}
\text{carries} \quad 1\,1\,1\,1\,1\,1 \\
1\,0\,1\,1\,1\,1\,1\,1 + \\
1\,0\,1\,1 = \\
\hline
1\,1\,0\,0\,1\,0\,1\,0
\end{array}
$$

Recall that the terms on the columns represent the coefficients of powers of 2, increasing from right and beginning with $2^0$. In this case $a_0 = b_0 = 1$ and $a_1 = b_1 = 1$, so the coefficient of $2^0$ becomes 0 and the coefficient of 2 becomes $(1+1)+1 = 2^1 + 2^0$, so that the coefficient of $2^1$ is 1 while that of $2^2$ becomes $a_2 + b_2 + 1 = (1 + 0) + 1$, as shown in the table. Iterating this reasoning, *write zero* for the coefficient of $2^2$, and *carry one*, that is, add one to the next coefficient, and so on. For the curious reader: the numbers we are adding up are respectively, in base 10, 191 e 11 and their sum is 202, corresponding to the base 2 result as shown.

In the same way, if

$$a = \sum_{i=0}^{n} a_i \cdot 2^i, \qquad b = \sum_{i=0}^{n} b_i \cdot 2^i$$

and $a \geq b$, then

$$a - b = \sum_{i=0}^{n} (a_i - b_i) \cdot 2^i.$$

Here we face the problem of *having all coefficient in this operation become 0 or 1*. The problem obviously only arises if for some $i$ we have $a_i = 0$ and $b_i = 1$. In this case, in the row representing $a$, we *borrow* a 1 from the first 1 appearing when we move leftwards starting from $a_i$.

Let us see an example, with

$$a = 33 = 2^5 + 1 \cdot 2^0 = (1\,0\,0\,0\,0\,1)_2,$$
$$b = 10 = 2^3 + 1 \cdot 2^1 = (1\,0\,1\,0)_2.$$

Then

$$a - b = 2^5 - 2^3 - 2^1 + 2^0 = 2^5 - 2^3 - 2^2 + (2-1) \cdot 2^1 + 2^0 =$$
$$= 2^5 - 2^3 - 2^2 + 1 \cdot 2^1 + 2^0 =$$
$$= 2^5 - 2^3 - 2^3 + 2^3 - 2^2 + 2^1 + 2^0 =$$
$$= 2^5 - 2^3 - 2^3 + (2-1) \cdot 2^2 + 2^1 + 2^0 =$$
$$= 2^5 - 2 \cdot 2^3 + 1 \cdot 2^2 + 2^1 + 2^0 = 2^5 - 2^4 + 2^2 + 2^1 + 2^0 =$$
$$= (2-1) \cdot 2^4 + 2^2 + 2^1 + 2^0 = 2^4 + 2^2 + 2^1 + 2^0 =$$
$$= (1\ 0\ 1\ 1\ 1)_2$$

which represents in base 2 the number 23.

Let us see how we actually perform the subtraction of the same two numbers:

- $a_0 - b_0 = 1 - 0 = 1$, so we may write 1 under the rightmost column:

$$
\begin{array}{r}
1\,0\,0\,0\,0\,1\ - \\
1\,0\,1\,0\ = \\
\hline
1
\end{array}
$$

- notice now that $a_1 - b_1 = 0 - 1$ is negative: so we must begin "borrowing". The first 1 we meet going leftwards is the coefficient of $2^5$. We must *move* the coefficient 1 to the right, until it is over the second column from right, keeping in mind that for each $i$ we have $2^i = 2^{i-1} + 2^{i-1}$. So we perform the following steps:

1.      borrowings   1
$$
\begin{array}{r}
\not{1}\,1\,0\,0\,0\,1\ - \\
1\,0\,1\,0\ = \\
\hline
1
\end{array}
$$

2.      borrowings  $\not{1}$ 1
$$
\begin{array}{r}
0\,1\,1\,0\,0\,1\ - \\
1\,0\,1\,0\ = \\
\hline
1
\end{array}
$$

3.      borrowings    $\not{1}$ 1
$$
\begin{array}{r}
0\,1\,1\,1\,0\,1\ - \\
1\,0\,1\,0\ = \\
\hline
1
\end{array}
$$

4.      borrowings      $\not{1}$ 1
$$
\begin{array}{r}
0\,1\,1\,1\,1\,1\ - \\
1\,0\,1\,0\ = \\
\hline
1
\end{array}
$$

Finally,

$$
\begin{array}{r}
\text{borrowings} \qquad 1 \\
0\ 1\ 1\ 1\ 1\ 1\ - \\
1\ 0\ 1\ 0\ = \\
\hline
\mathbf{1\ 0\ 1\ 1\ 1}
\end{array}
$$

and so the result of the subtraction is

$$\boxed{1\ 0\ 1\ 1\ 1}.$$

**Remark 1.4.2.** Computers actually transform each subtraction operation into an addition by a method called *two's complement*. To demonstrate this method, let us take a step backward. We write down in the following table the integer numbers from 0 to 15, written in base 2.

| base 10 | base 2 | base 10 | base 2 | |
|---------|--------|---------|--------|---|
| 0 | 0000 | 8 | 1000 | |
| 1 | 0001 | 9 | 1001 | |
| 2 | 0010 | 10 | 1010 | |
| 3 | 0011 | 11 | 1011 | (1.29) |
| 4 | 0100 | 12 | 1100 | |
| 5 | 0101 | 13 | 1101 | |
| 6 | 0110 | 14 | 1110 | |
| 7 | 0111 | 15 | 1111 | |

Notice that to represent all of them we need 4 *bits*, where each bit represents a symbol: 0 or 1. In the above table we added zeros on the left in order to fill up all 4 available positions. With five bits we may represent numbers up to $31 = 32 - 1 = 2^5 - 1$.

In general, with $n$ bits we may represent all positive integers $a$ with $0 \le a < 2^n$. In a computer, information is stored in 8-bit units called *bytes*; so a machine with a 1-byte memory unit can represent in a single cell any integer number between 0 and $2^8 - 1 = 255$, while a machine with 2-byte cells can represent in a single cell numbers up to $2^{16} - 1 = 65535$. Finally, a machine with 4-byte cells can store in a single cell numbers up to $2^{32} - 1 = 4{,}294{,}967{,}295$.

In $n$-bit representation we add to the left all the zeros necessary to fill up the available positions.

To get back to how computers actually perform subtractions by transforming them into additions, the critical point is the maximum length of the strings computers use to represent integers. In other words, if the maximum length is 4, the number 10001 is read as 0001, that is, the computer *forgets* all the coefficients of the terms of the form $2^h$, with $h \ge 4$ and, for instance, read the number $17 = (10001)_2$ as 1.

Suppose the computer has 4-bit cells and we want to subtract $b = 0111$ from $a = 1001$.

The following steps are performed:

- *ones' complement:* exchange 1 and 0 in the digits of the binary representation of $b$, obtaining $b' = 1000$;

- *two's complement:* add the number 0001 to $b'$, obtaining $b'' = 1001$;
- *sum:* add $b'' = 1001$ to $a = 1001$.

We shall call $b''$ the *opposite* of $b$. But care is needed! This is not the true opposite of $b$, that is $-b$. However, for the computer it works as such, because the computer only reads digits up to the fourth one. So $b'' + b = 1001 + 0111 = 1|0000$ which the computer *reads* as 0000. In the example, subtracting $b$ from $a$ as above we obtain

$$
\begin{array}{r}
1\ 0\ 0\ 1\ - \\
0\ 1\ 1\ 1\ = \\
\hline
0\ 0\ 1\ 0
\end{array}
$$

By using the computer's method, that is two's complement, we have to compute $a + b''$ obtaining

$$
\begin{array}{r}
1\ 0\ 0\ 1\ + \\
1\ 0\ 0\ 1\ = \\
\hline
\cancel{1}\ 0\ 0\ 1\ 0
\end{array}
$$

which is the *right* result because, as already remarked, the initial 1, crossed out here, does not exist for the computer!

Let us now explain theoretically why things work. Let $a, b$ be positive integers with $a \geq b$ and $a, b$ both smaller than $2^n$. We have

$$
a - b + 2^n = a + ((2^n - 1) - b + 1).
$$

As $2^n = (1\,0\ldots0)$, a string of length $n + 1$, and $2^n - 1 = (1\,1\ldots1\,1)$, a string of length $n$, then $b'$, the complement to 1 of $b$, is

$$
(2^n - 1) - b = (1\,1\ldots1\,1) - b
$$

and so $b''$, the opposite of $b$, is $(2^n - 1) - b + 1 = 2^n - b$. Thus, we have

$$
a + b'' = 2^n + a - b,
$$

which is read by the computer as $a - b$, if it uses an $n$-bit cell.

Let us discuss now *multiplication*. It easily reduces to addition, due to the following simple remark. If we are considering the number

$$
a = \sum_{i=0}^{n} a_i \cdot 2^i,
$$

then

$$
2^j \cdot a = \sum_{i=0}^{n} a_i \cdot 2^{i+j},
$$

that is, $2^j \cdot a$ may be written simply moving leftwards by $j$ positions the digits of $a$ and putting on their right the same number of zeros. On the other hand, suppose we have the numbers

$$
a = \sum_{i=0}^{n} a_i \cdot 2^i, \qquad b = \sum_{j=0}^{m} b_j \cdot 2^j.
$$

Suppose the non-zero (and so equal to 1) digits of $b$ are exactly the $h$ digits $b_{j_1}, \cdots, b_{j_h}$, that is, let

$$b = \sum_{l=1}^{h} 2^{j_l}.$$

In this case we have

$$a \cdot b = \sum_{l=1}^{h} \sum_{i=0}^{n} (a_i \cdot 2^{i+j_l}).$$

In conclusion, multiplying $a$ and $b$ reduces to summing up the $h$ numbers obtained from $a$ by moving leftwards its digits by $b_{j_1}, \ldots, b_{j_h}$ positions, and putting on the right the same number of zeros. These zeros are usually omitted: we may simply draw up in a column and sum up the numbers obtained by suitably moving leftwards the digits of $a$. For instance,

$$
\begin{array}{r}
1\,1\,1\,0\,1\,\times \\
1\,1\,0\,1\,= \\
\hline
1\,1\,1\,0\,1 \\
1\,1\,1\,0\,1\text{-\,-} \\
1\,1\,1\,0\,1\text{-\,-\,-} \\
\hline
1\,0\,1\,1\,1\,1\,0\,0\,1
\end{array}
$$

(1.30)

The reader is encouraged to verify the correctness of this calculation by performing it in base 10 too (see Exercise B1.56).

Finally, as regards *division*, we may proceed in the same way as in base 10, with thing made easier by the successive divisions having quotients equal to 1 or 0 only, while the remainders are computed by subtractions, as described above. For instance, consider $a = 30 = (11110)_2$ and $b = 4 = (100)_2$. Let us see the steps to compute the quotient and the remainder of the division of $a$ by $b$:

$$
\begin{array}{r|l}
1\,1\,1\,1\,0 & 1\,0\,0 \\
1\,0\,0 & \overline{1\,1\,1} \\
\cline{1-1}
1\,1\,1 & \\
1\,0\,0 & \\
\cline{1-1}
1\,1\,0 & \\
1\,0\,0 & \\
\cline{1-1}
1\,0 &
\end{array}
$$

Analogously, let $a = 26 = (11010)_2$ and $b = 5 = (101)_2$. The division of $a$ by $b$ is

$$
\begin{array}{r|l}
1\,1\,0\,1\,0 & 1\,0\,1 \\
1\,0\,1 & \overline{1\,0\,1} \\
\cline{1-1}
1\,1\,0 & \\
1\,0\,1 & \\
\cline{1-1}
1 &
\end{array}
$$

**Remark 1.4.3.** Writing a number in base 2, the ease of performing operations notwithstanding, requires on the other hand more space than writing it in a larger base, for instance in the usual base 10. A remedy to this inconvenient situation is given by proceeding as follows, using table (1.29): consider the 16-element set of all 4-digit strings formed by the digits in $\{0, 1\}$. Each of these strings represents a number in base 2, as shown in table (1.29).

By putting $1010 = A$, $1011 = B$, $1100 = C$, $1101 = D$, $1110 = E$ and $1111 = F$, we have 16 symbols $0, \ldots, 9$, $A$, $B$, $C$, $D$, $E$, $F$ which we may take as the digits of a *hexadecimal system*. It is very easy to obtain a number written in base 16 from a number written in base 2: it is sufficient to subdivide the binary number, from right leftwards, in 4-digit groups (four bits = half a byte, one byte consisting of 8 bits). Here is an example:

$$(\underbrace{1000}_{8}\ \underbrace{1110}_{E}\ \underbrace{1101}_{D}\ \underbrace{0110}_{6}\ \underbrace{0001}_{1})_2 = (8ED61)_{16} =$$

$$= 8 \cdot (16)^4 + 14 \cdot (16)^3 + 13 \cdot (16)^2 + 6 \cdot 16 + 1 = (585057)_{10}.$$

To go from a binary system to a base 8 system, it suffices to group digits in 3-digit groups, starting from the right.

### 1.4.4 Integer numbers in an arbitrary base

Coming back to arbitrary bases we point out, without delving too much into concepts that are by now quite clear and are studied further in the exercises, that here too the four fundamental operations on numbers may be performed. It is enough to know the addition and multiplication tables of 1-digit numbers, and to follow the ordinary rules for *carries*, as expounded in the case of base 2. Here are the addition and multiplication tables in base 3:

| + | 0 | 1 | 2 |
|---|---|---|---|
| 0 | 0 | 1 | 2 |
| 1 | 1 | 2 | 10 |
| 2 | 2 | 10 | 11 |

| · | 0 | 1 | 2 |
|---|---|---|---|
| 0 | 0 | 0 | 0 |
| 1 | 0 | 1 | 2 |
| 2 | 0 | 2 | 11 |

For instance, to calculate $(221)_3 + (12)_3$ it is enough to proceed as follows. Write $1 + 2 = 10$, then write 0 and carry 1. Then compute $1 + 2 + 1 = 11$. Write 1 and carry 1. Finally, $1 + 2 = 10$. In conclusion, we have

$$
\begin{array}{r}
1\,1 \\
2\,2\,1\,+ \\
1\,2 \\
\hline
1\,0\,1\,0
\end{array}
$$

We proceed analogously for multiplication. For instance, the reader will be able to verify that $(221)_3 \cdot (12)_3 = (11122)_3$.

### 1.4.5 Representation of real numbers in an arbitrary base

Not just integer numbers but more in general real numbers may be represented in a given base $\beta$. Indeed, the following result holds.

**Theorem 1.4.4.** *Let $a$ be a real number such that $0 \le a < 1$ and let $\beta > 1$ be an integer. Then $a$ may be written as the sum of a series*

$$a = \sum_{i=1}^{\infty} \frac{c_i}{\beta^i} \qquad (1.31)$$

*where the coefficients are integers $c_i$ such that $0 \le c_i \le \beta - 1$. This series is unique if we require that for each positive integer $m$ there is an integer $n \ge m$ such that $c_n \ne \beta - 1$.*

PROOF. When the interval $[0, 1]$ is partitioned into $\beta$ equal parts there exists a unique integer $c_1$ such that

$$\frac{c_1}{\beta} \le a < \frac{c_1 + 1}{\beta}.$$

Partitioning further the interval $[c_1/\beta, (c_1 + 1)/\beta]$ into $\beta$ equal parts, there exists a unique integer $c_2$ such that

$$\frac{c_1}{\beta} + \frac{c_2}{\beta^2} \le a < \frac{c_1}{\beta} + \frac{c_2 + 1}{\beta^2}.$$

By iterating this procedure, we prove the existence of the series in (1.31). As to the uniqueness, the procedure just described is obviously unique and cannot give a series such that for some positive integer $m$ one has $c_{m-1} < \beta - 1$ and for each $n \ge m$ one has $c_n = \beta - 1$. Indeed, keeping in mind that

$$\sum_{i=0}^{\infty} \frac{1}{\beta^i} = \frac{\beta}{\beta - 1},$$

in this case we would get

$$a = \sum_{i=1}^{\infty} \frac{c_i}{\beta^i} = \sum_{i=1}^{m-1} \frac{c_i}{\beta^i} + \sum_{i=m}^{\infty} \frac{\beta - 1}{\beta^i} = \sum_{i=1}^{m-1} \frac{c_i}{\beta^i} + \frac{\beta - 1}{\beta^m} \sum_{i=0}^{\infty} \frac{1}{\beta^i} =$$

$$= \sum_{i=1}^{m-1} \frac{c_i}{\beta^i} + \frac{1}{\beta^{m-1}}.$$

Thus, the above procedure would yield the series

$$a = \sum_{i=1}^{m-2} \frac{c_i}{\beta^i} + \frac{c_{m-1} + 1}{\beta^{m-1}}$$

rather than

$$a = \sum_{i=1}^{m-1} \frac{c_i}{\beta^i} + \sum_{i=m}^{\infty} \frac{\beta - 1}{\beta^i}. \qquad \square$$

By this theorem it is possible to write real numbers in base $\beta$. Indeed, if $a$ is a positive real number, its *integral part* $[a]$, defined as the unique integer $x$ such that $x \leq a < x + 1$, may be written in base $\beta$ as $(a_k a_{k-1} \ldots a_1 a_0)_\beta$. Moreover, we may apply the above theorem to the real number $a - [a]$, which belongs to the interval $[0, 1)$ and is usually written

$$a - [a] = (0.c_1 c_2 c_3 \ldots)_\beta.$$

In conclusion, we write

$$a = (a_k a_{k-1} \ldots a_1 a_0 . c_1 c_2 c_3 \ldots)_\beta,$$

or simply

$$a = a_k a_{k-1} \ldots a_1 a_0 . c_1 c_2 c_3 \ldots,$$

without mentioning base $\beta$ if no confusion can possibly arise. This expression is unique if it satisfies the conditions of Theorem 1.4.4. As regards the non-uniqueness, in general, of expression (1.31), notice that, for instance, in base 10 we have $1 = 0.9999 \ldots$

The above remarks apply in particular to *rational* numbers, that is to the elements of field $\mathbb{Q}$. We recall that the elements of $\mathbb{Q}$ are the *fractions*, that is they may be written as $n/m$ with $n, m$ integers, $m \neq 0$. Clearly it is always possible to assume that $\text{GCD}(n, m) = 1$, and in this case the fraction $n/m$ is said to be *reduced*. Notice the following fact, an immediate consequence of Corollary 1.3.9 which is left to the reader as an exercise (see Exercise A1.70).

**Lemma 1.4.5.** *If $n/m$ and $n'/m'$ are equal reduced fractions, then $n$ is associated with $n'$ and $m$ to $m'$.*

Some quick remarks about the representation of a rational number in a given base $\beta$ follow. The watchful readers will recognise notions learnt in school about the base 10.

As a first step, we call $\beta$–*defined*, or just *defined* a number $a$ written as $a_k a_{k-1} \ldots a_1 a_0 . c_1 c_2 c_3 \ldots$ in base $\beta$, if there exist a positive integer $m$ such that $c_i = 0$ for each $i \geq m$. In this case we simply write

$$a = a_k a_{k-1} \ldots a_1 a_0 . c_1 c_2 c_3 \ldots c_m.$$

The digits $c_1 c_2 c_3 \ldots c_m$ are said to be the $\beta$–*mantissa*, or simply the *mantissa*, of $a$. It is clear that a defined number is rational. It is not true that every rational number is defined. Indeed, the following proposition holds, and its easy proof is left to the reader (see Exercise A1.71).

**Proposition 1.4.6.** *A rational number $a$ is $\beta$–defined if and only if it may be written as $a = n/m$, with $m$ a divisor of a power of $\beta$ with non-negative integer exponent.*

So we are left with the question of understanding how to represent in base $\beta$, in general, rational numbers, and in particular those which are not $\beta$–defined or, as they are called, $\beta$–*undefined*, or simply *undefined*.

We shall need some definitions. Given $a = a_k a_{k-1} \ldots a_1 a_0.c_1 c_2 c_3 \ldots$, $a$ is said to be $\beta$–*recurring*, or simply *recurring*, if there exist integers $m \geq 0$ and $h \geq 1$ such that for each positive integer $n$ and for each $i = 1, \ldots, h$ one has $c_{m+nh+i} = c_{m+i}$. In other words, after the digits $c_1, \ldots, c_m$, we find the digits $c_{m+1}, \ldots, c_{m+h}$, and they *are periodically repeated*:

$$a - [a] = 0.c_1 \ldots c_m c_{m+1} \ldots c_{m+h} c_{m+1} \ldots c_{m+h} \cdots c_{m+1} \ldots c_{m+h} \cdots;$$

this may be written more concisely as

$$a - [a] = 0.c_1 \ldots c_m \overline{c_{m+1} \ldots c_{m+h}}.$$

The digits $c_{m+1} \ldots c_{m+h}$ are called *recurring digits* of $a = a_k a_{k-1} \ldots a_1 a_0.c_1 c_2 c_3 \ldots$, the digits $c_1 \ldots c_m$ are the *non-recurring digits* (after the decimal separator). If $m = 0$, that is if there are no non-recurring digits, $a$ is said to be a *simple recurring* number, otherwise it is said to be *mixed recurring*.

Coming back to the problem of understanding the representation in base $\beta$ of rational numbers, we may clearly reduce to considering numbers in the interval $(0, 1)$. The problem is completely settled by the following result:

**Proposition 1.4.7.** *Let $a$ be a real number such that $0 < a < 1$. The number $a$ is rational if and only if it is $\beta$–recurring. More precisely, we have*

$$a = 0.c_1 \ldots c_m \overline{c_{m+1} \ldots c_{m+h}}$$

*if and only if*

$$a = \frac{(c_1 \ldots c_m c_{m+1} \ldots c_{m+h})_\beta - (c_1 \ldots c_m)_\beta}{(\underbrace{\beta - 1, \ldots, \beta - 1}_{h}, \underbrace{0, \ldots, 0}_{m})_\beta} :$$

*the digit $\beta - 1$ is repeated $h$ times, that is a number of times equal to the number of repeating digits, and $0$ is repeated $m$ times, that is a number of times equal to the number of non-repeating digits after the decimal separator.*

In general, if

$$a = a_k a_{k-1} \ldots a_0.c_1 \ldots c_m \overline{c_{m+1} \ldots c_{m+h}},$$

we have

$$a = a_k a_{k-1} \ldots a_0 + \frac{(c_1 \ldots c_m c_{m+1} \ldots c_{m+h})_\beta - (c_1 \ldots c_m)_\beta}{(\underbrace{\beta - 1, \ldots, \beta - 1}_{h}, \underbrace{0, \ldots, 0}_{m})_\beta}$$

and this fraction is sometimes called *generating fraction* of $a$.

For instance, the recurring number written in base 10 as

$$107.455\overline{10}$$

is the rational number

$$107 + \frac{45510 - 455}{99000} = 107 + \frac{45055}{99000}.$$

We postpone the proof of this proposition to Chapter 4 (see § 4.3), where we shall give further information about the structure of the recurring digits of a rational number in a given base.

## 1.5 Continued fractions

A very important application of the Euclidean algorithm lies in the continued fractions, which also gives an alternative way of representing real numbers. In this section we shall examine this subject and shall see its link with the solution of linear Diophantine equations. We shall again find them when studying the problem of factoring.

Let us begin with the numbers $a = 214$ and $b = 35$. By applying the Euclidean algorithm to these numbers we find

$$214 = 35 \cdot 6 + 4, \tag{1.32}$$
$$35 = 4 \cdot 8 + 3, \tag{1.33}$$
$$4 = 3 \cdot 1 + 1, \tag{1.34}$$
$$3 = 1 \cdot 3 + 0. \tag{1.35}$$

We now divide both sides of Equation (1.32) by 35, obtaining

$$\frac{214}{35} = 6 + \frac{4}{35}. \tag{1.36}$$

So we have obtained a first piece of information: the rational number $214/35$ lies between 6 and 7, as $0 < 4/35 < 1$. By writing $4/35$ as the inverse of a number greater than 1, formula (1.36) becomes

$$\frac{214}{35} = 6 + \frac{1}{\dfrac{35}{4}}. \tag{1.37}$$

Divide the next formula (1.33) by 4:

$$\frac{35}{4} = 8 + \frac{3}{4} \tag{1.38}$$

and rewrite it in the form

$$\frac{35}{4} = 8 + \frac{1}{\dfrac{4}{3}}. \tag{1.39}$$

Finally, dividing Equation (1.34) by 3 we get

$$\frac{4}{3} = 1 + \frac{1}{3}.$$

In this way we have found the following expression for $214/35$:

$$\frac{214}{35} = 6 + \cfrac{1}{8 + \cfrac{1}{1 + \cfrac{1}{3}}}.$$

It is called the *continued fraction* representation of $214/35$.

Representing a rational number by a continued fraction might seem just a useless curiosity: actually, it yields in a natural way simple and effective approximations of the rational number we are considering. For instance, in the present case the two fractions

$$6, \qquad 6 + \frac{1}{8} = \frac{49}{8} = 6.125,$$

approximate quite well the rational number $214/35 = 6.1\overline{142857}$. The meaning of "quite well" will be specified later.

Let us see one more example.

**Example 1.5.1.** Find the expression of $17/7$ as a continued fraction.

The Euclidean algorithm is as follows.

$$\begin{aligned}
17 &= 7 \cdot 2 + 3, \\
7 &= 3 \cdot 2 + 1, \\
3 &= 1 \cdot 3 + 0;
\end{aligned}$$

from this we get

$$\frac{17}{7} = 2 + \frac{3}{7}, \qquad \frac{7}{3} = 2 + \frac{1}{3},$$

so the expression of $17/7$ as a continued fraction is

$$\frac{17}{7} = 2 + \cfrac{1}{2 + \cfrac{1}{3}}.$$

The fractions approximating $17/7$ are the following:

$$2, \qquad 2 + \frac{1}{2} = \frac{5}{2} = 2.5,$$

while the exact value of $17/7$ is $2.\overline{428571}$.

### 1.5.1 Finite simple continued fractions and rational numbers

We have just seen an example of a continued fraction. In order to formalise what we have seen, we give the following definition.

**Definition 1.5.2.** *A finite continued fraction is a fraction of the form*

$$a_1 + \cfrac{1}{a_2 + \cfrac{1}{a_3 + \cfrac{1}{a_4 + \cfrac{\ddots}{\phantom{a} a_{n-1} + \cfrac{1}{a_n}}}}} \tag{1.40}$$

with $a_1, a_2, \ldots, a_n$ *real numbers, all positive with the possible exception of* $a_1$. *The numbers* $a_2, \ldots, a_n$ *are called* partial denominators, *or* partial quotients, *of the fraction.*

*A finite continued fraction is said to be* simple *if all of its partial quotients are integer.*

We shall mostly deal with simple continued fractions.

Clearly, every simple finite continued fraction is a rational number. On the other hand, as the Euclidean algorithm can be applied to arbitrary integers $a$ and $b$, with $b \neq 0$, it is easily understood, keeping also in mind the above example, that any rational number can be expanded as a finite simple continued fraction. This is the gist of the following proposition.

**Proposition 1.5.3.** *Every finite simple continued fraction is equal to a rational number, and every rational number can be written as a finite simple continued fraction.*

PROOF. The first part is trivial. For the second one, let $a/b$ be the rational number, $b > 0$. Apply the Euclidean algorithm to find the GCD of $a$ and $b$:

$$a = ba_1 + r_1, \qquad\qquad 0 < r_1 < b,$$
$$b = r_1 a_2 + r_2, \qquad\qquad 0 < r_2 < r_1,$$
$$r_1 = r_2 a_3 + r_3, \qquad\qquad 0 < r_3 < r_2,$$
$$\vdots$$
$$r_i = r_{i+1} a_{i+2} + r_{i+2}, \qquad\qquad 0 < r_{i+2} < r_{i+1},$$
$$\vdots$$
$$r_{n-3} = r_{n-2} a_{n-1} + r_{n-1}, \qquad\qquad 0 < r_{n-1} < r_{n-2},$$
$$r_{n-2} = r_{n-1} a_n + 0.$$

As all the remainders are positive, so are all the quotients $a_i$, with the possibe exception of the first one. Rewrite the equations given by the Euclidean algorithm dividing the first one by $b$, the second one by $r_1$, the third one by $r_2$ and so on, till the last one, to be divided by $r_n$. So we obtain

$$\frac{a}{b} = a_1 + \frac{r_1}{b} = a_1 + \cfrac{1}{\cfrac{b}{r_1}},$$

$$\frac{b}{r_1} = a_2 + \frac{r_2}{r_1} = a_2 + \cfrac{1}{\cfrac{r_1}{r_2}},$$

$$\frac{r_1}{r_2} = a_3 + \frac{r_3}{r_2} = a_3 + \cfrac{1}{\cfrac{r_2}{r_3}},$$

$$\frac{r_2}{r_3} = a_4 + \frac{r_4}{r_3} = a_4 + \frac{1}{\dfrac{r_3}{r_4}} ,$$

$$\vdots$$

$$\frac{r_{n-2}}{r_{n-1}} = a_n.$$

The left-hand sides of these equations are rational numbers, which are rewritten as the sum of an integer and a fraction with numerator equal to 1. By successive eliminations, we get

$$\frac{a}{b} = a_1 + \frac{1}{\dfrac{b}{r_1}} = a_1 + \frac{1}{a_2 + \dfrac{1}{\dfrac{r_1}{r_2}}} ;$$

hence

$$\frac{a}{b} = a_1 + \frac{1}{\dfrac{b}{r_1}} = a_1 + \frac{1}{a_2 + \dfrac{1}{\dfrac{r_1}{r_2}}} = a_1 + \frac{1}{a_2 + \dfrac{1}{a_3 + \dfrac{1}{\dfrac{r_2}{r_3}}}} ,$$

until we obtain the expression

$$\frac{a}{b} = a_1 + \cfrac{1}{a_2 + \cfrac{1}{a_3 + \cfrac{1}{a_4 + \cfrac{\ddots}{\quad a_{n-1} + \cfrac{1}{a_n}}}}} .$$

So we have represented the rational number $a/b$ as a finite simple continued fraction.  □

The expression (1.40) is visually quite bulky, so it is more convenient to denote the same continued fraction as follows.

$$[a_1; a_2, a_3, \dots, a_n],$$

that is, as the finite sequence of its partial quotients. For instance, the fraction in the above example can be written as

$$\frac{214}{35} = [6; 8, 1, 3].$$

Notice that the initial integer $a_1$ is equal to zero if and only if the fraction is positive and smaller than 1. Moreover, notice that $a_1$ is the integer value that approximates $a/b$ from below, that is $a_1 = [a/b]$; $a_2$ is the value approximating $b/r_1$ from below, that is $a_2 = [b/r_1]$, and in general

$$a_i = \left[\frac{r_{i-2}}{r_{i-1}}\right].$$

**Remark 1.5.4.** The representation of a rational number as a finite simple continued fraction is not unique. Indeed, having found the representation as shown, we always may modify its last term. In fact, if the last term $a_n$ is greater than 1, we may write

$$a_n = (a_n - 1) + 1 = (a_n - 1) + \frac{1}{1}$$

where $a_n - 1$ is positive, and so

$$[a_1; a_2, a_3, \ldots, a_{n-1}, a_n] = [a_1; a_2, \ldots, a_{n-1}, a_n - 1, 1].$$

If, on the other hand, the last term $a_n$ is equal to 1, then

$$[a_1; a_2, a_3, \ldots, a_n] = [a_1; a_2, a_3, \ldots, a_{n-2}, a_{n-1} + 1].$$

It is easy to verify that this is the only indeterminacy in the possible expression of a continued fraction (see Exercise A1.73). So, every rational number may be written as a finite simple continued fraction in exactly two ways: one with the last partial quotient equal to 1, and one with the last partial quotient greater than 1.

**Example 1.5.5.** Write the ratio $f_n/f_{n-1}$ of two consecutive Fibonacci numbers as a continued fraction.

From the relations

$$f_n = f_{n-1} \cdot 1 + f_{n-2}, \qquad\qquad 0 < f_{n-2} < f_{n-1},$$
$$f_{n-1} = f_{n-2} \cdot 1 + f_{n-3}, \qquad\qquad 0 < f_{n-3} < f_{n-2},$$

$$\vdots$$

$$f_3 = f_2 \cdot 2 + 0,$$

we get

$$\frac{f_n}{f_{n-1}} = 1 + \frac{f_{n-2}}{f_{n-1}} = 1 + \frac{1}{\dfrac{f_{n-1}}{f_{n-2}}},$$

$$\frac{f_{n-1}}{f_{n-2}} = 1 + \frac{f_{n-3}}{f_{n-2}} = 1 + \frac{1}{\dfrac{f_{n-2}}{f_{n-3}}},$$

$$\vdots$$

$$\frac{f_3}{f_2} = 2 + 0,$$

and hence

$$\frac{f_n}{f_{n-1}} = 1 + \cfrac{1}{1 + \cfrac{1}{1 + \cfrac{1}{1 + \cfrac{\ddots}{\cfrac{1}{2}}}}},$$

or

$$\frac{f_n}{f_{n-1}} = \underbrace{[1; 1, 1, 1, \ldots, 1, 2]}_{n-2} = \underbrace{[1; 1, 1, 1, \ldots, 1]}_{n-1}.$$

### 1.5.2 Infinite simple continued fractions and irrational numbers

We have seen that all rational numbers, and no other number, can be represented as finite simple continued fractions.

The main reason of interest of continued fractions, however, is in their application to the representation of *irrational numbers*. To that end we shall need *infinite* simple continued fractions.

As a quick historical aside, recall that studies about continued fractions are found in Indian mathematics of 6th and 12th century, where they were used to study linear equations. Fibonacci in his *Liber Abaci* attempted to give a general definition of a continued fraction. The first rigorous investigations into continued fraction appeared in a 1572 book by Rafael Bombelli, the inventor, among other things, of complex numbers. He wrote that *"methods to form fractions can be found in other authors' works which attack and accuse each other, in my opinion without reason because all of them are bent on the same goal"*. This *same goal* was, as Bolognese mathematician P.A. Cataldi wrote in 1613 in his *Trattato del modo brevissimo di trovare la radice quadrata dei numeri* (*Treatise about the shortest way of finding square roots of numbers*), the problem of approximating the square roots of a non-square integer by a rational expression. The phrase *continued fraction* first appeared in the 1653 edition of J. Wallis's *Arithmetica infinitorum*. Other great mathematicians studied infinite continued fractions: among them, Euler in his *De fractionibus continuis*, Lagrange, Gauss, and Liouville, who used them in his famous proof of the existence of transcendental numbers.

In order to approach the idea of infinite continued fractions, consider, as a simple example, the number $\sqrt{2}$. As it is greater than 1 and smaller than 2, we may write it in the form

$$\sqrt{2} = 1 + \frac{1}{x}$$

for some *real* number $x > 1$. Alternatively, we may write $x = \sqrt{2} + 1$ and hence

$$x = \sqrt{2} + 1 = \left(1 + \frac{1}{x}\right) + 1 = 2 + \frac{1}{x}.$$

From this equation the following ones may be deduced.

$$\left. \frac{1}{x} = \cfrac{1}{2 + \cfrac{1}{x}} = \cfrac{1}{2 + \cfrac{1}{2 + \cfrac{1}{x}}} = \cdots = \cfrac{1}{2 + \cfrac{1}{2 + \cfrac{1}{2 + \cfrac{1}{\ddots \cfrac{1}{2 + \cfrac{1}{2 + \cfrac{1}{x}}}}}}} \right\} n$$

As $n$ approaches infinity we obtain

$$\sqrt{2} = 1 + \cfrac{1}{2 + \cfrac{1}{2 + \cfrac{1}{2 + \cfrac{1}{2 + \cfrac{1}{2 + \cdots}}}}}.$$

This is, intuitively, the way of writing $\sqrt{2}$ as a continued fraction. We have proved that *finite* simple continued fractions are rational numbers, while $\sqrt{2}$ is not (see Exercise A1.72), so it is to be expected that the continued fraction we get for $\sqrt{2}$ *is not finite*, even if we have not yet defined the concept of an infinite continued fraction.

The expression of $\sqrt{2}$ as a continued fraction uncovers a remarkable elegance and regularity, as opposed to its decimal representation, which does not show any regularity.

By using a generalised form of continued fraction, in which the numerators are allowed not to be equal to 1, we may write the real number $\sqrt{a^2 + b}$, where $a, b$ are integer numbers, in the following form:

$$\sqrt{a^2 + b} = \cfrac{b}{\cfrac{2a}{b} + \cfrac{b}{2a + \cfrac{b}{2a + \cfrac{b}{2a + \cfrac{b}{2a + \cdots}}}}}.$$

In particular, we get

$$\sqrt{a^2 + 1} = [a; 2a, 2a, \ldots] = [a, \overline{2a}],$$

where $\overline{2a}$ means that the continued fraction is periodic, that is to say, $2a$ is repeated infinitely many times. The case of $\sqrt{2}$ corresponds to putting $a = 1$ and $b = 1$.

Here follow some more examples.

$$\sqrt{3} = [1; \overline{1,2}], \qquad \sqrt{5} = [2; \overline{4}], \qquad \sqrt{10} = [3; \overline{6}],$$

while the golden ratio has a truly perfect representation as a continued fraction:

$$\frac{\sqrt{5} + 1}{2} = [1; 1, 1, 1, \ldots].$$

Before going on, we must attribute a meaning to the concept of continued fraction with infinitely many terms. We begin by giving the following definition.

**Definition 1.5.6.** *Let $[a_1; a_2, a_3, \ldots, a_n]$ be a finite simple continued fraction. The continued fraction obtained by* truncating *this continued fraction after the $k$-th partial quotient is called $k$-th convergent and is denoted as follows:*

$$C_k = [a_1; a_2, a_3, \ldots, a_k], \qquad \text{for each } 1 \le k \le n.$$

Notice that $C_{k+1}$ may be obtained from $C_k$ by substituting $a_k + 1/a_{k+1}$ for $a_k$. Clearly, for $k = n$ we get the complete original continued fraction.

Every $C_k = [a_1; a_2, \ldots, a_k]$ is a rational number which will be denoted by $p_k/q_k$, where $\text{GCD}(p_k, q_k) = 1$. It is important to find formulas for the numerator $p_k$ and the denominator $q_k$ when the $a_i$s are known. Clearly, if $C_1 = [a_1] = p_1/q_1$, then $p_1 = a_1$ and $q_1 = 1$. Moreover, if

$$C_2 = [a_1; a_2] = a_1 + \frac{1}{a_2} = \frac{a_1 a_2 + 1}{a_2} = \frac{p_2}{q_2},$$

then $p_2 = a_1 a_2 + 1$ and $q_2 = a_2$. Analogously, if

$$C_3 = [a_1; a_2, a_3] = a_1 + \cfrac{1}{a_2 + \cfrac{1}{a_3}} = \frac{a_3(a_2 a_1 + 1) + a_1}{a_3 a_2 + 1},$$

then $p_3 = a_3(a_2 a_1 + 1) + a_1 = a_3 p_2 + a_1$ and $q_3 = a_3 a_2 + 1 = a_3 q_2 + q_1$.

It can be shown (see Exercise A1.75) that in general the following recurrence relations hold:

$$\boxed{p_k = a_k p_{k-1} + p_{k-2}, \qquad q_k = a_k q_{k-1} + q_{k-2}.} \qquad (1.41)$$

From these relations it is straightforward to obtain

$$p_k q_{k-1} - q_k p_{k-1} = -(p_{k-1} q_{k-2} - q_{k-1} p_{k-2}).$$

As

$$p_2 q_1 - q_2 p_1 = (a_1 a_2 + 1) \cdot 1 - a_2 a_1 = 1,$$

it follows that

$$\boxed{p_k q_{k-1} - q_k p_{k-1} = (-1)^k :} \qquad (1.42)$$

hence, in particular, for each $k = 1, \ldots, n$, the numbers $p_k$ and $q_k$ are relatively prime. By dividing the last relation by $q_k q_{k-1}$ we get

$$\frac{p_k}{q_k} - \frac{p_{k-1}}{q_{k-1}} = \frac{(-1)^k}{q_k q_{k-1}},$$

that is,

$$\boxed{C_k - C_{k-1} = \frac{(-1)^k}{q_k q_{k-1}},} \qquad (1.43)$$

which holds for each $k \geq 1$. It can be shown in a completely analogous way that the following relation

$$\boxed{C_k - C_{k-2} = \frac{(-1)^{k+1} a_k}{q_k q_{k-2}}} \qquad (1.44)$$

holds for each $k \geq 2$ (see Exercise A1.76). We shall shortly use these equations.

The following formula is a direct consequence of Equation (1.41), and it is more convenient in practice. The convergents $C_k = p_k/q_k$ of the continued fraction $[a_1; a_2, a_3, \ldots, a_n]$, for $2 \leq k \leq n$, may be obtained with the following formula (see Exercise A1.77).

$$\boxed{\begin{pmatrix} a_1 & 1 \\ 1 & 0 \end{pmatrix} \cdot \begin{pmatrix} a_2 & 1 \\ 1 & 0 \end{pmatrix} \cdot \begin{pmatrix} a_3 & 1 \\ 1 & 0 \end{pmatrix} \cdots \begin{pmatrix} a_k & 1 \\ 1 & 0 \end{pmatrix} = \begin{pmatrix} p_k & p_{k-1} \\ q_k & q_{k-1} \end{pmatrix}.} \qquad (1.45)$$

**Example 1.5.7.** Determine with the above method the convergents of the continued fraction $17/7 = [2; 2, 3]$.

We have

$$\begin{pmatrix} 2 & 1 \\ 1 & 0 \end{pmatrix} \begin{pmatrix} 2 & 1 \\ 1 & 0 \end{pmatrix} = \begin{pmatrix} 5 & 2 \\ 2 & 1 \end{pmatrix} = \begin{pmatrix} p_2 & p_1 \\ q_2 & q_1 \end{pmatrix},$$

hence $p_2/q_2 = 5/2$, and $p_1/q_1 = 2/1$, as we saw above. Going on, we find

$$\begin{pmatrix} 5 & 2 \\ 2 & 1 \end{pmatrix} \begin{pmatrix} 3 & 1 \\ 1 & 0 \end{pmatrix} = \begin{pmatrix} 17 & 5 \\ 7 & 2 \end{pmatrix} = \begin{pmatrix} p_3 & p_2 \\ q_3 & q_2 \end{pmatrix},$$

or, as it is supposed to be, $p_3/q_3 = 17/7$.

**Example 1.5.8.** Find the convergents of $214/35 = [6; 8, 1, 3]$.

Applying the same method as above, we have

$$\begin{pmatrix} 6 & 1 \\ 1 & 0 \end{pmatrix} \begin{pmatrix} 8 & 1 \\ 1 & 0 \end{pmatrix} = \begin{pmatrix} 49 & 6 \\ 8 & 1 \end{pmatrix} = \begin{pmatrix} p_2 & p_1 \\ q_2 & q_1 \end{pmatrix},$$

hence $p_2/q_2 = 49/8$, and $p_1/q_1 = 6/1$. Moreover,

$$\begin{pmatrix} 49 & 6 \\ 8 & 1 \end{pmatrix} \begin{pmatrix} 1 & 1 \\ 1 & 0 \end{pmatrix} = \begin{pmatrix} 55 & 49 \\ 9 & 8 \end{pmatrix} = \begin{pmatrix} p_3 & p_2 \\ q_3 & q_2 \end{pmatrix},$$

hence $p_3/q_3 = 55/9$. It is obvious that $p_4/q_4 = 214/35$ and it is not necessary to compute it.

**Example 1.5.9.** Which rational number is $[1; 2, 3]$?

We have

$$\begin{pmatrix} p_3 & p_2 \\ q_3 & q_2 \end{pmatrix} = \begin{pmatrix} 1 & 1 \\ 1 & 0 \end{pmatrix} \cdot \begin{pmatrix} 2 & 1 \\ 1 & 0 \end{pmatrix} \cdot \begin{pmatrix} 3 & 1 \\ 1 & 0 \end{pmatrix} = \begin{pmatrix} 10 & 3 \\ 7 & 2 \end{pmatrix},$$

which means

$$[1; 2, 3] = C_3 = \frac{p_3}{q_3} = \frac{10}{7}.$$

**Example 1.5.10.** The orbital period of Saturn, that is to say the time taken for it to complete one orbit around the Sun, is 29.46 years. In Huygens's times, it was believed to be 29.43 years. In order to simulate Saturn's trajectory around the Sun, Huygens had to build two gears, one with $p$ teeth and the other with $q$ teeth, such that $p/q$ were approximately 29.43. Which values did Huygens choose for $p$ and $q$? To be useful, $p$ and $q$ had to be quite small: he could not use a gear with 2943 teeth and one with 100 teeth. So he computed the convergents of $29.43 = 2943/100$. Using the Euclidean algorithm, one gets

$$2943 = 100 \cdot 29 + 43,$$
$$100 = 43 \cdot 2 + 14,$$
$$43 = 14 \cdot 3 + 1,$$
$$14 = 1 \cdot 14 = 0;$$

hence

$$\frac{2943}{100} = 29 + \frac{43}{100}, \qquad \frac{100}{43} = 2 + \frac{14}{43}, \qquad \frac{43}{14} = 3 + \frac{1}{14},$$

so the expression $2943/100$ as a continued fraction is

$$\frac{2943}{100} = [29; 2, 3, 14].$$

Thus, the first three convergents are

$$C_1 = \frac{29}{1},$$

$$C_2 = \frac{a_1 a_2 + 1}{a_2} = \frac{59}{2},$$

$$C_3 = a_3(a_2 a_1 + 1) + a_1 a_3 a_2 + 1 = \frac{3 \cdot 59 + 29}{7} = \frac{206}{7}.$$

A quite good approximation of 29.43 is given by the fraction $206/7 = 29.4285$. So, in order to simulate the motion of Saturn with respect to Earth's motion, Huygens built two gears, one with 7 teeth and one with 206 teeth.

Notice that the convergents $C_k$ of a finite simple continued fraction oscillate. Indeed, the following lemma holds, the easy proof of which is left as an exercise to the reader (see Exercise A1.78).

**Lemma 1.5.11.** *Let* $a/b = [a_1; a_2, \dots, a_n]$ *be a simple continued fraction. Then its convergents satisfy the following properties:*

- $C_1 < C_3 < C_5 < \cdots$,
- $C_2 > C_4 > C_6 > \cdots$,
- $C_{2k} > C_{2j-1}$, *for each* $j, k \geq 1$.

Hence we deduce that

$$C_1 < C_3 < C_5 < \cdots \leq \frac{a}{b} \leq \cdots < C_6 < C_4 < C_2, \tag{1.46}$$

that is to say that the convergents approximate the continued fraction, but *oscillating*: the odd-indexed ones increase and approximate it by defect, while the even-indexed ones decrease and approximate it by excess, and each even-indexed convergent is greater than all odd-indexed convergent.

We may finally attribute a precise meaning to an expression of the form $[a_1; a_2, a_3, \dots]$, with $a_1, a_2, a_3, \dots$ an infinite sequence of integers such that $a_i > 0$ for $i > 1$: it will be called *infinite simple continued fraction* with *partial quotients* $a_1; a_2, a_3, \dots, a_n, \dots$ Having put $C_n = [a_1; a_2, a_3, \dots, a_n]$, which shall be called *convergents* of the infinite continued fraction, define

$$\boxed{[a_1; a_2, a_3, \dots] = \lim_{n \to \infty} C_n.}$$

The existence of this limit is ensured by the following theorem.

**Theorem 1.5.12.** *The concept of an infinite simple continued fraction is well-defined, that is to say the following limit exists and it is an irrational number:*

$$\lim_{n \to \infty} C_n = \lim_{n \to \infty} [a_1; a_2, a_3, \dots, a_n].$$

PROOF. As we have seen, the sequence of odd-indexed convergents, $C_1$, $C_3$, $C_5$, ..., is increasing, while the sequence of even-indexed ones, $C_2$, $C_4$, $C_6$, ..., is decreasing. Moreover, the first sequence is bounded above and the second is bounded below. It is well-known that

$$\lim_{n\to\infty} C_{2n-1} = \sup\{C_{2n-1}\}, \qquad \lim_{n\to\infty} C_{2n} = \inf\{C_{2n}\};$$

we shall denote these two numbers by $\alpha_1$ and $\alpha_2$. We have next, by using Equation (1.43),

$$\alpha_2 - \alpha_1 = \lim_{n\to\infty} (C_{2n} - C_{2n-1}) = \lim_{n\to\infty} \frac{1}{q_{2n}q_{2n-1}} = 0,$$

as $\lim_{n\to\infty} q_n = +\infty$. Consequently, $\alpha_1 = \alpha_2$. If we put $\alpha = \alpha_1 = \alpha_2$, we get

$$\alpha = \lim_{n\to\infty} C_n.$$

As to the irrationality of $\alpha$, assume by contradiction $\alpha = p/q$, for integer $p, q$, $q \neq 0$. For each $n \geq 1$ we have

$$C_{2n-1} < \alpha < C_{2n},$$

and so we also have

$$0 < \alpha - C_{2n-1} < C_{2n} - C_{2n-1},$$

or

$$0 < \alpha - \frac{p_{2n-1}}{q_{2n-1}} < \frac{1}{q_{2n}q_{2n-1}},$$

that is

$$0 < \alpha q_{2n-1} - p_{2n-1} < \frac{1}{q_{2n}},$$

and finally

$$0 < pq_{2n-1} - qp_{2n-1} < \frac{q}{q_{2n}}.$$

This last relation immediately leads to a contradiction. Indeed, the central term is a non-zero *integer* because $\alpha > C_{2n-1}$, while the right-hand term has limit zero as $n$ approaches infinity. ☐

We have so clarified the concept of an infinite simple continued fraction and its convergents. For instance, in the case of $\sqrt{2}$, we have $C_1 = 1$, $C_2 = 3/2 = 1.5$, $C_3 = 1.4$ and then

$$C_4 = 1 + \frac{5}{12} = 1.41\bar{6},$$

$$C_5 = 1 + \frac{12}{29} = 1.\overline{4137931034482758620689655172},$$

$$C_6 = 1 + \frac{29}{70} = 1.4\overline{142857},$$

$$\vdots$$

The convergents are oscillating for infinite continued fractions too. In the case of $\sqrt{2}$, all the $C_{2k}$ are greater than $\sqrt{2}$, while all the $C_{2k+1}$ are smaller than it. In other words, the sequence $\{C_{2k+1}\}$ is monotonically increasing, while $\{C_{2k}\}$ is monotonically decreasing, and both sequences have limit $\sqrt{2}$.

We may complete the last theorem as follows.

**Theorem 1.5.13.** *Every positive irrational number $\alpha$ may be expressed as an infinite simple continued fraction in a unique way.*

PROOF. We give a sketch of the proof, leaving the details to the reader (see Exercise A1.79). We define recursively the partial quotients $a_1, a_2, \ldots, a_n, \ldots$ of the continued fraction expressing $\alpha$. We may do this as follows.

- put $\alpha_1 = \alpha$;
- suppose $\alpha_1, \ldots, \alpha_n$ have been defined; put $a_n = [\alpha_n]$ and define recursively $\alpha_{n+1} = 1/(\alpha_n - a_n)$.

It is possible to verify that $\alpha = [a_1; a_2, \ldots, a_n, \ldots]$.
Suppose now that $\alpha = [a_1; a_2, \ldots, a_n, \ldots] = [b_1; b_2, \ldots, b_n, \ldots]$. We have

$$C_1 = a_1 < \alpha < a_1 + \frac{1}{a_2} = C_2,$$

and so $a_1 = [\alpha]$. Analogously $b_1 = [\alpha]$ must hold, and so $a_1 = b_1$. Notice now that

$$\alpha = [a_1; a_2, \ldots, a_n, \ldots] = a_1 + \frac{1}{[a_2; a_3, \ldots, a_n, \ldots]},$$

$$\alpha = [b_1; b_2, \ldots, b_n, \ldots] = a_1 + \frac{1}{[b_2; b_3, \ldots, b_n, \ldots]};$$

hence $[a_2; a_3, \ldots, a_n, \ldots] = [b_2; b_3, \ldots, b_n, \ldots]$, and so $a_2 = b_2$. By induction one has $a_n = b_n$ for each $n > 0$.  □

**Example 1.5.14.** Write $\sqrt{6}$ as a continued fraction.

Put $a_1 = [\sqrt{6}] = 2$. Consequently, define

$$\alpha_2 = \frac{1}{\sqrt{6} - 2} = \frac{\sqrt{6} + 2}{2}, \qquad a_2 = [\alpha_2] = 2,$$

and successively

$$\alpha_3 = \frac{1}{\dfrac{\sqrt{6} + 2}{2} - 2} = \sqrt{6} + 2, \qquad a_3 = [\alpha_3] = 4,$$

$$\alpha_4 = \frac{1}{\sqrt{6} - 2} = \frac{\sqrt{6} + 2}{2}, \qquad a_4 = [\alpha_3] = 2,$$

and so forth. Hence,

$$\sqrt{6} = [2; 2, 4, 2, 4, \ldots] = [2, \overline{2, 4}].$$

### 1.5.3 Periodic continued fractions

It is not by chance that the continued fraction of $\sqrt{6}$, as well as that of $\sqrt{2}$ studied above, are periodic. First of all, in order to better define and extend the concept of a periodic continued fraction, which has already been partially shown, we shall say that a continued fraction $[a_1; a_2, a_3, a_4, \ldots]$ is *periodic* if there exist positive integers $m, t \in \mathbb{N}$ such that $a_n = a_{n+t}$ for each $n > m$. In this case the following notation is used

$$[a_1; a_2, a_3, a_4, \ldots] = [a_1; a_2, \ldots, a_m, \overline{a_{m+1}, \ldots, a_{m+t}}]$$

and $a_1$, ..., $a_m$ are said to be the *non-repeating quotients* of the fraction; while $a_{m+1}$, ..., $a_{m+t}$ are said to be the *repeating quotients* of the fraction. If there are no non-repeating quotients, that is to say if the fraction is of the form $[\overline{a_1; a_2, \ldots, a_t}]$, it is said to be *purely periodic*.

Let us now determine the irrational numbers such that their expression as continued fractions is periodic. We need a definition first. A real number $\alpha$ is said to be *quadratic* if $\alpha$ is a root of a quadratic equation with integer coefficients. In this case $\alpha$ is of the form

$$\alpha = a + b\sqrt{d}, \qquad a, b \in \mathbb{Q},$$

with $d$ a positive integer (see Exercise A1.81). The number

$$\alpha' = a - b\sqrt{d},$$

which is the other root of the same quadratic equation, is said to be the *conjugate* of $\alpha$. For instance,

$$1 + \frac{\sqrt{5}}{4}, \qquad 1 - \frac{\sqrt{5}}{4}$$

are irrational conjugate quadratic numbers.

A quadratic number $\alpha$ is said to be *reduced* if $\alpha > 0$ while $-1 < \alpha' < 0$. For instance, if $n$ is any non-square positive integer, it is clear that the number $\sqrt{n} + [\sqrt{n}]$ is reduced (see Exercise A1.85). For instance, $2 + \sqrt{5}$ is reduced.

The proof of the following theorem will be given to the reader through a series of exercises (see Exercises A1.87 to A1.94).

**Theorem 1.5.15 (Lagrange).** *A continued fraction* $\alpha = [a_1; a_2, a_3, a_4, \ldots]$ *is periodic if and only if* $\alpha$ *is an irrational quadratic number. Moreover,* $[a_1; a_2, a_3, a_4, \ldots]$ *is purely periodic if and only if* $\alpha$ *is reduced. In this case we have*

$$\alpha = [\overline{a_1; a_2, \ldots, a_t}], \qquad -\frac{1}{\alpha'} = [\overline{a_t; a_{t-1}, \ldots, a_1}].$$

**Example 1.5.16.** A positive integer $n$ is said to be a *perfect square* or simply a *square* if there exists an integer $a$ such that $n = a^2$. If $n$ is a positive integer that is not a perfect square, then $\sqrt{n}$ is an irrational number (see Exercise A1.80). Let us see now the structure of the expression of $\sqrt{n}$ as a continued fraction. Put $a_0 = \sqrt{n}$.

We have already seen that $\sqrt{n} + [\sqrt{n}]$ is reduced. By applying Theorem 1.5.15 and noticing that $[\sqrt{n} + [\sqrt{n}]] = 2a_0$, we obtain an expression of the form

$$\sqrt{n} + [\sqrt{n}] = [\overline{2a_0; a_1, a_2, \ldots, a_n}] = [2a_0; a_1, a_2, \ldots, a_n, \overline{2a_0, a_1, a_2, \ldots, a_n}];$$

hence,

$$\sqrt{n} = [a_0; a_1, a_2, \ldots, a_n, \overline{2a_0, a_1, a_2, \ldots, a_n}] = [a_0; \overline{a_1, a_2, \ldots, a_n, 2a_0}].$$

Thus

$$\sqrt{n} - [\sqrt{n}] = [0; \overline{a_1, a_2, \ldots, a_n, 2a_0}],$$

and

$$\frac{1}{\sqrt{n} + [\sqrt{n}]} = [\overline{a_1; a_2, \ldots, a_n, 2a_0}].$$

On the other hand, by applying Theorem 1.5.15 again, we get

$$\frac{1}{\sqrt{n} + [\sqrt{n}]} = [\overline{a_n; a_{n-1}, a_{n-2}, \ldots, a_1, 2a_0}].$$

So we obtain

$$a_1 = a_n, \quad a_2 = a_{n-1}, \quad \ldots, \quad a_n = a_1,$$

that is to say, the expression of $\sqrt{n}$ as a continued fraction has an interesting symmetry:

$$\sqrt{n} = [a_0; \overline{a_1, a_2, \ldots, a_2, a_1, 2a_0}].$$

We have already seen several instances of this state of things. The curious reader might wish to find a further confirmation of what we have seen here by having a look at the table of the expressions as continued fractions of all irrational numbers of the form $\sqrt{n}$, with $n$ a positive integer not greater than 100, in [49] (Table 5).

### 1.5.4 A geometrical model for continued fractions

We quickly mention a very simple and quite interesting geometrical model for continued fractions and their convergents. This interpretation was found by F. Klein in 1895 (see [16]).

Let $\alpha$ be an irrational positive number. In a given Cartesian coordinate system consider the line $y = \alpha x$. As $\alpha$ is irrational, this line does not pass through any point with integer coordinates (see figure 1.3).

Imagine now nails driven in all points with positive integer coordinates, that is in all points with integer coordinates which lie in the first quadrant of the plane, and imagine to lay along the line $y = \alpha x$ a string with one end fixed in an infinitely far point and the other end fixed in the origin. If we move downwards the end of the string that was placed in the origin, the whole string will be moved and it will run into some of the nails, that is some of the points with positive integer coordinates. Analogously, by moving upwards the same end of the string, it will run into some nails, that is some points with integer coordinates: so the string will form two polygonal curves, one lying completely under the line $y = \alpha x$ and one above.

If we denote by $(n_1, m_1)$, $(n_3, m_3)$, $(n_5, m_5)$, $\ldots$, the nails hit by string in its movement downwards, that is to say the vertices of the first polygonal curve, and by $(n_2, m_2)$, $(n_4, m_4)$, $(n_6, m_6)$, $\ldots$, the nails hit in its movement upwards, that is to say the vertices of the second polygonal curve, the quotients $m_i / n_i$ are exactly the convergents $C_i = p_i / q_i$ of the continued fraction $[a_1; a_2, a_3, \ldots]$, which represents the irrational number $\alpha$ (see figure 1.3), that is

$$\boxed{\frac{m_i}{n_i} = C_i = \frac{p_i}{q_i} \qquad \text{for all } i.}$$

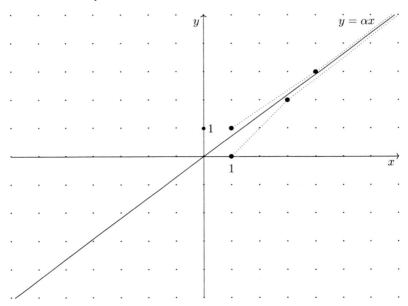

**Fig. 1.3.** Approximating polygonal curves for the line $y = \alpha x$ in the Cartesian Plane

### 1.5.5 The approximation of irrational numbers by convergents

Given an infinite simple continued fraction $[a_1; a_2, a_3, \ldots]$ it is possible to approximate the corresponding irrational number $\alpha$ by rational numbers with arbitrary precision, by computing the convergents $C_n$. Formulas (1.43) together with (1.46) tell us that by using $C_{n-1}$ to approximate $\alpha$ the error is smaller than

$$\frac{1}{q_n q_{n-1}}.$$

This fact may also be verified keeping in mind the geometrical model we have described above.

As already mentioned, one of the reasons for the importance of the representation of numbers by continued fractions lies in the fact that the convergents yield, in a sense we are now making precise, *the best possible approximation of an irrational number by rational numbers*. We prove some results in this spirit.

**Theorem 1.5.17.** *Let $\alpha$ be an irrational number and let $\alpha = [a_1; a_2, a_3, \ldots, a_n, \ldots]$ be its expression as a continued fraction, and let $C_n = p_n/q_n$ be its convergents. If $p, q$ are integers with $q > 0$ and if $n$ is a positive integer such that*

$$|q\alpha - p| < |q_n\alpha - p_n|, \tag{1.47}$$

*then $q \geq q_{n+1}$. Moreover, if*

$$\left|\alpha - \frac{p}{q}\right| < |\alpha - C_n|, \tag{1.48}$$

*then $q > q_n$. In other words, every convergent $C_n = p_n/q_n$ approximates the value of $\alpha$ better than any fraction whose denominator is smaller than or equal to $q_n$.*

PROOF. Assume that Equation (1.47) hold and assume by contradiction that $q < q_{n+1}$. Consider the system

$$\begin{cases} p_n x + p_{n+1} y = p, \\ q_n x + q_{n+1} y = q. \end{cases}$$

By using Equation (1.42), we find

$$y = (-1)^{n+1}(pq_n - qp_n), \qquad x = (-1)^{n+1}(qp_{n+1} - pq_{n+1}).$$

Notice that $x \neq 0$, or else we would have $q = q_{n+1} y \geq q_{n+1}$, contradicting the hypothesis. Analogously, we have $y \neq 0$, or we would have $p = p_n x$, $q = q_n x$ and then

$$|q\alpha - p| = |x||q_n\alpha - p_n| \geq |q_n\alpha - p_n|$$

contradicting Equation (1.47).

We verify now that $x$ and $y$ have opposite signs. Let $y < 0$. Then $q_n x = q - q_{n+1} y > 0$ and so $x > 0$ because $q_n > 0$. Let $y > 0$. As $q_{n+1} y \geq q_{n+1} > q$, we have $q_n x = q - q_{n+1} y < 0$, and so $x < 0$.

It immediately follows from the fact that the convergents oscillate that $q_n\alpha - p_n$ and $q_{n+1}\alpha - p_{n+1}$ have opposite signs. Thus, $x(q_n\alpha - p_n)$ and $y(q_{n+1}\alpha - p_{n+1})$ have equal signs. As

$$|q\alpha - p| = |(q_n x + q_{n+1} y)\alpha - (p_n x + p_{n+1} y)|,$$

we have

$$|q\alpha - p| = |x||q_n\alpha - p_n| + |y||q_{n+1}\alpha - p_{n+1}| \geq |q_n\alpha - p_n|$$

contradicting Equation (1.47). This proves the first part of the theorem.

Assume now that Equation (1.48) holds and, by contradiction, that $q \leq q_n$. Then we have

$$q\left|\alpha - \frac{p}{q}\right| < q_n|\alpha - C_n|,$$

that is, Equation (1.47). But in this case it follows that $q \geq q_{n+1}$, and so we would have $q_n \geq q_{n+1}$, contradicting Equation (1.41).                                    □

**Example 1.5.18.** We have seen that the convergent $C_3$ of the continued fraction $\sqrt{2} = [1, \overline{2}]$ equals $7/5 = 1.4$. By approximating $\sqrt{2}$ with $1.4$ the error is $\sqrt{2} - 1.4 < C_4 - C_3 = 1.41\overline{6} - 1.4 = 0.01\overline{6}$. We show now that every fraction having, for instance, denominator equal to 4 approximates $\sqrt{2}$ worse than 1.4. The only fractions to be considered are $5/4 = 1.25$, $6/4 = 3/2 = 1.5$, and $7/4 = 1.75$, as $\sqrt{2}$ is greater than 1 and smaller than 2. We have $1.25 < 1.4$, and so it is clear that 1.25 approximates $\sqrt{2}$ worse than 1.4. On the other hand,

$$1.5 - \sqrt{2} > 1.5 - C_4 = 1.5 - 1.41\overline{6} = 0.18\overline{3} > 0.01\overline{6}.$$

Thus 1.5 approximates $\sqrt{2}$ worse than 1.4 too. This holds a fortiori for $1.75 > 1.5$.

**Theorem 1.5.19.** *Let $\alpha$ be an irrational number, let $\alpha = [a_1; a_2, a_3, \ldots, a_n, \ldots]$ be its expression as a continued fraction, and let $C_n = p_n/q_n$ be its convergents. If $p, q$ are relatively prime integers with $q > 0$ and such that*

$$\left| \alpha - \frac{p}{q} \right| < \frac{1}{2q^2},$$

*then $p/q$ is a convergent of the continued fraction $[a_1; a_2, a_3, \ldots, a_n, \ldots]$.*

PROOF. Suppose that $p/q$ is not a convergent. We may find two consecutive convergents, $C_n, C_{n+1}$ such that $q_n \leq q < q_{n+1}$. By Theorem 1.5.17 we have

$$|q_n \alpha - p_n| \leq |q\alpha - p| = q \left| \alpha - \frac{p}{q} \right| < \frac{1}{2q},$$

hence

$$|\alpha - C_n| < \frac{1}{2qq_n}.$$

As $p/q \neq C_n$, we have $|qp_n - pq_n| \geq 1$, and so

$$\frac{1}{qq_n} \leq \frac{|qp_n - pq_n|}{qq_n} = \left| C_n - \frac{p}{q} \right| \leq |\alpha - C_n| + \left| \alpha - \frac{p}{q} \right| < \frac{1}{2qq_n} + \frac{1}{2q^2},$$

which yields

$$\frac{1}{2qq_n} < \frac{1}{2q^2},$$

that is to say, $q_n > q$, contradicting the hypothesis.           □

**Proposition 1.5.20.** *Let $\alpha$ be an irrational number, let $\alpha = [a_1; a_2, a_3, \ldots, a_n, \ldots]$ be its expression as a continued fraction, and let $C_n = p_n/q_n$ be its convergents. For each $n \geq 1$, we have*

$$|p_n^2 - \alpha^2 q_n^2| < 2\alpha.$$

PROOF. Keeping in mind Equations (1.43) and (1.46), we have

$$|p_k^2 - \alpha^2 q_k^2| = q_k^2 |\alpha^2 - C_k^2| = q_k^2 |\alpha - C_k| \cdot |\alpha + C_k| <$$

$$< \frac{q_k}{q_{k+1}} \left( \alpha + \alpha + \frac{1}{q_k q_{k+1}} \right)$$

and so

$$|p_k^2 - \alpha^2 q_k^2| - 2\alpha < 2\alpha \left( -1 + \frac{q_k}{q_{k+1}} + \frac{1}{2\alpha q_{k+1}^2} \right) <$$

$$< 2\alpha \left( -1 + \frac{q_k}{q_{k+1}} + \frac{1}{q_{k+1}} \right) < 2\alpha \left( -1 + \frac{q_{k+1}}{q_{k+1}} \right) = 0,$$

as was to be proved.           □

We may apply the above result to obtain an information that will be useful later on (see Chapter 6). Let $n$ and $a$ be positive integers and let $r$ be the remainder of the division of $a$ by $n$. So $0 \leq r < n$. Define *least absolute residue* of $a$ with respect to $n$, or *modulo $n$*, the integer $r$, if $0 \leq r \leq n/2$, or the integer $r - n$ if $n/2 < r < n$. In this last case, we have $0 > r - n > -n/2$. Denote the least absolute residue of $a$

with respect to $n$ by the symbol $\mathrm{LAR}(a, n)$. As a consequence of the definition, we always have

$$-\frac{n}{2} < \mathrm{LAR}(a, n) \le \frac{n}{2} \qquad (1.49)$$

and $\mathrm{LAR}(a, n)$ is the only integer satisfying Equation (1.49) such that $a - \mathrm{LAR}(a, n)$ is divisible by $n$.

For instance, $\mathrm{LAR}(35, 9) = -1$.

**Proposition 1.5.21.** *Let $n \ge 17$ be a non-square positive integer and let $C_k = p_k/q_k$ be the $k$-th convergent of the continued fraction expressing $\sqrt{n}$. Then for each $k \ge 1$ we have*

$$|\mathrm{LAR}(p_k^2, n)| < 2\sqrt{n}.$$

PROOF. Last proposition implies that

$$|p_k^2 - nq_k^2| < 2\sqrt{n}$$

for each $k \ge 1$. As $n \ge 17$, we have $\sqrt{n} < n/4$ and so

$$|p_k^2 - nq_k^2| < \frac{n}{2}.$$

It follows that $\mathrm{LAR}(p_k^2, n) = p_k^2 - nq_k^2$, hence the the claim.    $\square$

### 1.5.6 Continued fractions and Diophantine equations

We mention now an interesting relation between the solution of linear Diophantine equations and continued fractions, given by the following theorem.

**Theorem 1.5.22.** *If $C_{n-1} = p_{n-1}/q_{n-1}$ is the last but one convergent of $a/b$, with $a$ and $b$ relatively prime positive integers, then*

$$x = (-1)^n q_{n-1}, \quad y = (-1)^n p_{n-1}$$

*is a solution of the equation $ax - by = 1$.*

The proof is not difficult and is left as an exercise to the interested reader (see Exercise A1.84); we give just an example of its application.

**Example 1.5.23.** Solve the Diophantine equation $214x - 35y = 1$.

It has already been shown that

$$\frac{214}{35} = [6; 8, 1, 3],$$

its convergents being

$$C_1 = \frac{6}{1}, \quad C_2 = \frac{49}{8}, \quad C_3 = \frac{55}{9}, \quad C_4 = \frac{214}{35},$$

so a solution of the equation is $x = 9, y = 55$.

# Appendix to Chapter 1

In this appendix we give several exercises, as we shall do for each chapter: first those of a theoretical nature, then more practical ones and finally a set of programming exercises. Some exercises consist in a guide to the proof of some of the results which have not been proved in the text. The exercises we regard as more difficult are marked by an asterisk.

Sometimes we have given multiple choice questions. We urge the reader to attempt their solution without trying to exclude the less likely answers by following a process of elimination or (even worse) by guessing at the right answer. Indeed, to randomly find the correct solution is quite useless and by far less worthwhile than finding a wrong solution that has been reasoned through.

# A1 Theoretical exercises

**A1.1.** Prove that complete induction (or strong induction) (CI) implies mathematical induction.

**A1.2.**\* Prove that mathematical induction implies the following result.

**Well-ordering principle** (WOP). Any non-empty subset $A$ of $\mathbb{N}$ contains a least element, that is to say, there exists an element $m \in A$ such that for each $a \in A$ we have $m \leq a$.

**A1.3.**\* Prove that the well-ordering principle (WOP) implies complete induction (or strong induction) (CI).

**A1.4.** Exactly one of the following claims is true. Which one?

(a) Complete induction implies mathematical induction, but not the other way around.
(b) Mathematical induction is equivalent to complete induction, which in turn is equivalent to the well-ordering principle.
(c) Mathematical induction is equivalent to complete induction which is a consequence of, but is not equivalent to, the well-ordering principle.
(d) None of the above.

**A1.5.** We prove that every city is *small*, by induction on the number of its inhabitants. Clearly, if a city has just one inhabitant, it is small. Assume a city has $n$ inhabitants. If this city is small, it is still so even if we add one inhabitant. So we deduce that all cities are small. Is everything right or is something amiss?

(a) Everything is right and it proves that mathematics cannot possibly be applied to concrete questions.
(b) The basis of the induction is not right: if there is a single inhabitant, it is not a city.
(c) The basis of the induction is right, but the proof of the inductive step depends upon the definition of a small city.
(d) None of the above.

**A1.6.** Assume we have defined a city to be *small* if it has less than 50,000 inhabitants. Which one of the statements (a)–(d) of Exercise A1.5 is correct? (N.B. It is not the same one as in Exercise A1.5!)

**A1.7.** We are going to show that all students in a course get the same mark in a given written test, by proving it by induction on the number $n$ of students. The basis of the induction, for $n = 1$, is trivial. Assume each of the $n$ students has completed his test: we mark all the tests and put them in a stack. By induction hypothesis, all the tests but the last one get the same mark. For the same reason, all the tests but the first one get the same mark. Then, all the tests under the first one and on the last one get the same mark, which is then the same for all the tests, including the first and last ones. We deduce from this that it is a waste of energy to correct all the tests, it being enough to correct just one of them. What is wrong with this reasoning?

(a) The basis of the induction is false.
(b) The inductive step is not true for $n = 2$.
(c) The final deduction is not right: correcting the tests would be a waste of energy even if the marks are different.
(d) None of the above.

**A1.8.** By using the well-ordering principle, prove that there exist two integers $qb$ and $(q+1)b$ as needed in the proof of Proposition 1.3.1.

**A1.9.** Assume we know the initial values $a_0, a_1, a_2, \ldots, a_{n_0-1}, a_{n_0}$ of a sequence $\{a_n\}$ of elements from a set $X$; assume further that we are able to determine $a_n$, for each $n > n_0$, if the preceding terms $a_m$, with $m < n$, are known. Prove that it is possible to determine $a_n$ for each $n \in \mathbb{N}$.

**A1.10.** Assume we know the initial values $a_0, a_1, a_2, \ldots, a_{n_0-1}, a_{n_0}$ of a sequence $\{a_n\}$ of elements from a set $X$; assume further that we are able to determine $a_n$, for each $n > n_0$, if the term $a_{n-1}$ is known. Prove that it is possible to determine $a_n$ for each $n \in \mathbb{N}$.

The goal of the exercises from A1.11 to A1.25 is to recall some notions of *combinatorial analysis*.

**A1.11.** Given a non-empty set $X$, a *permutation* of $X$ is defined to be any bijection of $X$ in itself. Denote by $I_n$ the set $\{1, \ldots, n\}$ of the first $n$ natural numbers. The set of all bijections of $I_n, n \geq 1$, in itself is denoted by the symbol $\mathfrak{S}_n$.

Given an element $f$ of $\mathfrak{S}_n$, put $f(1) = i_1, \ldots, f(n) = i_n$. So we associate to $f$ the ordered $n$-tuple $(i_1, \ldots, i_n)$ of elements of $I_n$, which is said to be the *symbol* of the permutation $f$ and is denoted by $\sigma(f)$. In particular $(1, \ldots, n)$ is the symbol of the identity map of $I_n$ in itself, also called *identity permutation*.

Prove that the mapping

$$\sigma : f \in \mathfrak{S}_n \to \sigma(f) \in \underbrace{I_n \times \cdots \times I_n}_{n}$$

is injective; hence a bijection is uniquely determined by its symbol. So we may identify a permutation with its symbol, as we shall do in what follows.

**A1.12.** Let $n \in \mathbb{N}$ be a natural number. Define the *factorial* of $n$ as the number, denoted by $n!$, equal to 1 if $n = 0$ or 1 and to $n! = (n-1)! \cdot n = n(n-1)(n-2) \cdots 2 \cdot 1$

if $n > 1$. Prove by induction that, for each integer $n \geq 1$, the number of elements of $\mathfrak{S}_n$ is $n!$.

**A1.13.*** How many injective mappings $A \hookrightarrow B$ are there between two sets $A$ and $B$ having $n$ elements each? How many surjective mappings?

(a) The injective mappings are $2^n$, while the surjective ones are $2^n + 1$.
(b) All injective maps are surjective as well, and vice versa, and there are $n^n$ of them.
(c) There are as many surjective mappings as injective ones, that is to say $n!$.
(d) None of the above.

**A1.14.** Let $n$ and $k$ be two positive integer numbers such that $n \geq k$. Recall the definition of *binomial coefficient*:

$$\binom{n}{k} = \frac{n!}{k!(n-k)!} = \frac{n(n-1)(n-2)\cdots(n-k+1)}{k!}. \tag{1.50}$$

Notice that

$$\binom{n}{k} = \binom{n}{n-k}.$$

Prove that

$$\binom{n+1}{k} = \binom{n}{k} + \binom{n}{k-1}. \tag{1.51}$$

From this, deduce that

$$\binom{n+1}{k+1} = \binom{n}{k+1} + \binom{n}{k} = \binom{n}{k+1} + \binom{n-1}{k} + \binom{n-1}{k-1}$$
$$= \binom{n}{k+1}\binom{n-1}{k}\binom{n-2}{k-1}\binom{n-2}{k-2} \tag{1.52}$$

and so on.

**A1.15.*** Prove by induction on $n$ that the binomial coefficient $\binom{n}{m}$ is the number of subsets of $I_n$ having size equal to $m$.

**A1.16.** Prove the inequality

$$\binom{n+m-1}{m} \geq \left(\frac{n}{m}\right)^m.$$

**A1.17.** Prove that the subset of $I_n \times I_n$ consisting of the pairs $(i,j)$ with $1 \leq i < j < n$ has the same size as the set of subsets of $I_n$ of size 2, that is to say, has size $\binom{n}{2} = n(n-1)/2$.
    Prove that, on the other hand, the subset of $I_n \times I_n$ consisting of the pairs $(i,j)$ with $1 \leq i \leq j < n$, has size $\binom{n+1}{2} = n(n+1)/2$.

**A1.18.** Let $a, b$ be two elements of a field. Prove the *binomial theorem*:

$$(a+b)^n = a^n + na^{n-1}b + \binom{n}{2}a^{n-2}b^2 + \binom{n}{3}a^{n-3}b^3 + \cdots + b^n = \sum_{i=0}^{n}\binom{n}{i}a^{n-i}b^i.$$

(Hint: prove it by induction on $n$; it is convenient to use formula (1.51) for the inductive step.)

**A1.19.*** Let $A$ and $B$ be two sets of size, respectively, $n$ and $m$, and assume $n \le m$. How many injective functions $A \hookrightarrow B$ are there?

(a) $m! - n!$.
(b) $(n+m)(n+m-1)(n+m-2)\cdots(n+1)$.
(c) $m(m-1)(m-2)\cdots(m-n+1)$.
(d) None of the above.

Why did we assume $n \le m$?

**A1.20.*** Let $A$ and $B$ be two sets of size, respectively, $n$ and $m$. How many functions $A \to B$ are there?

(a) $m^n$.
(b) $n \cdot m$.
(c) $n^m$.
(d) None of the above.

**A1.21.** Let $A$ be a set of size $n$ and let $m$ be a positive integer. Prove that the set $\underbrace{A \times \cdots \times A}_{m}$ consisting of the ordered $m$-tuples of elements of $A$ has size $n^m$.

**A1.22.*** Recall that the power set $\mathcal{P}(X)$ of a set $X$ is the set whose elements are all the subsets of $X$, including $X$ itself and the empty set. If $X$ is a finite set of size $n$, which is the size of $\mathcal{P}(X)$?

(a) $2^n - 1$.
(b) $2^n$.
(c) $2(2^{n-1} - 1)$.
(d) None of the above.

**A1.23.** Prove that for each integer $n \ge 0$ the following holds:

$$2^n = \binom{n}{n} + \binom{n}{n-1} + \cdots + \binom{n}{1} + \binom{n}{0}.$$

**A1.24.*** Let $\mathbb{K}$ be a field and let $x_1, \ldots, x_n$ be variables over $\mathbb{K}$. We call *monic monomial* of degree $h$ in $x_1, \ldots, x_n$ an expression of the form $x_1^{i_1} \cdots x_n^{i_n}$, with $i_1, \ldots, i_n$ non-negative integers such that $i_1 + \cdots + i_n = h$. Denote by $s(n, h)$ the number of such monomials.
    Prove by induction on $n$ that

$$s(n, h) = \binom{n+h-1}{h}.$$

**A1.25.*** Let $a_1, \ldots, a_r$ be distinct elements of a field. Prove by induction the following generalisation of the binomial theorem

$$(a_1 + \cdots + a_r)^n = \sum_{s=1}^{r} \sum_{\substack{\{a_{i_1},\ldots,a_{i_s}\} \\ n_{i_1}+\cdots+n_{i_s}=n}} \frac{n!}{n_{i_1}!\cdots n_{i_s}!} a_{i_1}^{n_{i_1}} \cdots a_{i_s}^{n_{i_s}},$$

where the internal sum is over all subsets $\{a_{i_1},\ldots,a_{i_s}\}$ of size $s$ of $\{a_1,\ldots,a_r\}$ and over all $s$-tuples consisting of positive integers $n_{i_1},\ldots,n_{i_s}$ and such that $n_{i_1} + \cdots + n_{i_s} = n$.

**A1.26.** Compare Equation (1.4) with the *homogeneous* linear recurrence relation

$$a'_n = b_{k-1}a'_{n-1} + b_{k-2}a'_{n-2} + \cdots + b_0 a'_{n-k}. \tag{1.53}$$

Prove that if $\{y_n\}$ is a solution of the linear recurrence relation (1.4), then every solution is of the form $\{y_n + x_n\}$, where $\{x_n\}$ is a solution of the associated homogeneous linear recurrence relation (1.53).

**A1.27.** Let $\{a_n\}_{n \in \mathbb{N}}$ be a sequence of elements of the field $\mathbb{K}$ satisfying a linear recurrence relation of the form

$$a_{n+1} = ba_n + c,$$

with $b, c \in \mathbb{K}$. Prove that if $a_0, a_1, a_2$ are known, it is possible to determine all the elements of the sequence.

**A1.28.** Let $A$ be the matrix (1.6). Prove Proposition 1.2.3 by induction on $n$.

**A1.29.** Given a line segment of length $a$, construct using ruler and compasses a line segment of length $b$ such that $a/b$ is the golden ratio.

(Hint: if $M$ is the midpoint of a side $AB$ of a square having side-length $a$, the the line segment joining $M$ and a vertex of the square different from $A$ and $B$ has length $r = (\sqrt{5}/2)a$; the line segment of length $b$ we are looking for has endpoints $B$ and the intersection of the circumference with centre $M$ and radius $r$ with the line on which $AB$ lies, on $B$'s side.)

The following exercises, from A1.30 to A1.35, are about Fibonacci numbers and some of their properties.

**A1.30.** Prove that for each $n$ and each $k$ in $\mathbb{N}$ the following holds

$$f_{n+k} = f_k f_{n+1} + f_{k-1}f_n. \tag{1.54}$$

Deduce that $f_{kn}$ is a multiple of $f_n$.

**A1.31.*** Prove by induction that each positive integer can be written as a sum of finitely many distinct Fibonacci numbers.

**A1.32.** Prove, using Formula (1.10), that is to say, assuming not to know the recurrence relation (1.5), that Fibonacci numbers are integers.

**A1.33.** Prove by induction that for each $n \geq 1$ the following holds:

$$f_n \geq \left(\frac{1+\sqrt{5}}{2}\right)^{n-2} \tag{1.55}$$

and the inequality is strict if $n \geq 3$.

**A1.34.** Prove that for each $n$ we have

$$f_{2n} = f_1 + f_3 + f_5 + \cdots + f_{2n-1}.$$

**A1.35.** Prove that for each $n$ we have

$$f_{n+1} = \sum_{\substack{h+k=n \\ h \geq k \geq 0}} \binom{h}{k}.$$

(Hint: prove it by induction on $n$, using the recurrence relation (1.5) and Formula (1.51).)

**A1.36.** Study the following variant of the division algorithm. Let $a, b$ be integer numbers, with $b \neq 0$. Prove that there exist and are uniquely determined two integers $q, r$ with $-|b|/2 < r \leq |b|/2$ such that $a = qb + r$.

**A1.37.** Let $a, b, c \in \mathbb{Z}$ be different from zero. Prove that if $\mathrm{GCD}(a, b) = 1$ and if $c \mid a$ then $\mathrm{GCD}(c, b) = 1$ too.

**A1.38.** Let $a, b \in \mathbb{Z}$ be different from zero. Prove that if $\mathrm{GCD}(a, b) = 1$ and if $b \mid a$ then $b = \pm 1$.

**A1.39.*** This exercise leads to a theoretical but non-constructive proof of the existence in $\mathbb{Z}$ of the greatest common divisor of two integers $a$ and $b$ at least one of which is different from zero. Let $S = \{\, xa + yb \mid x, y \in \mathbb{Z}, \ xa + yb > 0 \,\} \subseteq \mathbb{N}$. Prove that this set has a least element $d = x_0 a + y_0 b$. Prove that $d$ is the greatest common divisor of $a$ and $b$.

**A1.40.** Work out an algorithm to determine the greatest common divisor of two integer numbers that uses the alternative division algorithm described in Exercise A1.36 rather than the usual division algorithm.

**A1.41.** Let $(\bar{x}, \bar{y})$ be an integer solution of the Equation (1.18). Prove that all the pairs obtained by adding to $(\bar{x}, \bar{y})$ an integer solution $(x_0, y_0)$ of the associated homogeneous equation $ax + by = 0$ are integer solutions of (1.18), and all of them can be obtained in this way.

**A1.42.** Suppose we have to divide 123456 by 365, that is to say we have to find the integers $q$ and $r$ such that $123456 = q \cdot 365 + r$, with $0 \leq r < 365$. We do not feel like calculating them by hand, so we take a calculator and find that the result of the division of 123456 by 365 is 338.23561644. How may we determine $q$ and $r$ as requested?

**A1.43.** Prove that in a Euclidean ring results analogous to those in Proposition 1.3.5 and in its Corollaries 1.3.6, 1.3.7, 1.3.8, and 1.3.9 hold.

**A1.44.** Let $A$ be an integral domain and let $a$ and $b$ be non-zero elements of $A$. Prove that if each of $d$ and $d'$ is a greatest common divisor of $a$ and $b$, they are associated. Conversely, if $d = \mathrm{GCD}(a, b)$ and if $d'$ is associated to $d$, then $d'$ is a $\mathrm{GCD}(a, b)$ too.

**A1.45.** Let $A$ be an integral domain. Prove that the relation in $A$ defined by $x \sim y$ if and only if $x$ is associated with $y$ is an equivalence relation.

**A1.46.** Let $A$ be an integral domain. Prove that in $A$ the ideals $(x)$ and $(y)$ coincide if and only if $x$ and $y$ are associated. Deduce that for each ideal $I$ of $\mathbb{Z}$ there exists a unique positive integer $x$ such that $I = (x)$.

**A1.47.*** Let $n, m$ be positive integers. Prove that the subset $(n) \cap (m)$ of $\mathbb{Z}$ is an ideal. Let $D$ be the positive integer such that $(D) = (n) \cap (m)$. Prove that $D$ is the unique positive integer that is a multiple both of $n$ and $m$ and such that, for each positive multiple $N$ of $n$ and $m$, we have $D \mid N$. The integer $D$ is said to be the *least common multiple* of $n$ and $m$ and is denoted by $\operatorname{lcm}(n, m)$.

**A1.48.*** Let $A$ be an integral domain. Notice that for each $a \neq 0$, the map $x \in A \rightarrow ax \in A$ is injective. Deduce that in a finite integral domain each non-zero element has an inverse; hence, every finite integral domain is a field.

Finally, deduce that a finite ring with unity is a field if and only if it is an integral domain.

**A1.49.*** Consider the subset $\mathbb{G}$ of the set $\mathbb{C}$ of complex numbers consisting of the numbers, called *Gaussian integers*, of the form $x + iy$, with $x, y \in \mathbb{Z}$. Prove that $\mathbb{G}$ is a subring of $\mathbb{C}$.

Define in $\mathbb{G}$ the map

$$v : x + iy \in \mathbb{G} \rightarrow x^2 + y^2 \in \mathbb{N}.$$

In other words, $v(x+iy)$ is the *norm* $\|x+iy\|$ of the complex number $x + iy$. Prove that it satisfies property (1) of the definition of a Euclidean ring.

Let $a, b \in \mathbb{G}$, with $b \neq 0$. Consider the complex number $x + iy = a/b$. Prove that there exist integers $x_1, y_1$ such that $|x - x_1| \leq 1/2$ and $|y - y_1| \leq 1/2$.

Put $x_2 = x - x_1$, $y_2 = y - y_1$; hence

$$a = b(x_1 + iy_1) + b(x_2 + iy_2).$$

Notice that $x_1 + iy_1$ and $b(x_2 + iy_2)$ are Gaussian integers. Moreover, notice that

$$v(b(x_2 + iy_2)) = v(b)\|x_2 + iy_2\| \leq \frac{v(b)}{2} < v(b).$$

Deduce that $\mathbb{G}$ is a Euclidean ring.

**A1.50.*** If $\alpha$ is a complex number, denote by $\mathbb{Z}[\alpha]$ the smallest subring of $\mathbb{C}$ containing both $\mathbb{Z}$ and $\alpha$. Prove that if $\alpha^2 \in \mathbb{Z}$, then $\mathbb{Z}[\alpha]$ consists of all complex numbers of the form $x + \alpha y$, with $x, y \in \mathbb{Z}$.

**A1.51.** Prove that the ring $A[x]$ of polynomials with coefficients in an integral domain $A$ is an integral domain.

**A1.52.** Let $f(x)$ be a polynomial with coefficients in a field $\mathbb{K}$. Prove that there exists a unique monic polynomial associated with $f(x)$.

(Hint: divide $f(x)$ by its leading coefficient.)

**A1.53.** Let $f_1(x), f_2(x)$ be polynomials with coefficients in a field $\mathbb{K}$ and let $\alpha \in \mathbb{K}$. Prove that if $\alpha$ is a root both of $f_1(x)$ and of $f_2(x)$ with multiplicity respectively $k_1$ and $k_2$, then $\alpha$ is a root of the polynomial $f_1(x)f_2(x)$ with multiplicity $k_1 + k_2$. Let $k$ the multiplicity of $\alpha$ as a root of the polynomial $f_1(x) + f_2(x)$. Prove that $k \geq \min\{k_1, k_2\}$ and that if $k_1 \neq k_2$, then $k = \min\{k_1, k_2\}$.

(Hint: use the factor theorem.)

**A1.54.**\* Let $\mathbb{K}$ be a field, $c$ an element of $\mathbb{K}$, and $f(x) \in \mathbb{K}[x]$ a polynomial of degree $t > 0$ such that $f(c) \neq 0$. Prove that there exists exactly one polynomial $g(x) \in \mathbb{K}[x]$ of degree smaller than $t$ such that $(x - c) \cdot g(x) - 1$ is divisible by $f(x)$.

**A1.55.** Let $f(x)$ be a polynomial with coefficients in a field $\mathbb{K}$. If $f(x)$ has degree $n$, then $f(x)$ admits at most $n$ roots, counting each with its multiplicity.

(Hint: use the factor theorem and reason by induction.)

**A1.56.** Recall the following theorem, which shall be proved in Chapter 4 (see Theorem 4.5.21):

**Theorem A1.1 (Fundamental theorem of algebra).** *Every non-zero polynomial with coefficients in $\mathbb{C}$ has at least one root in $\mathbb{C}$.*

Using this theorem, prove that a polynomial $f(x) \in \mathbb{C}[x]$ of degree $n$ has exactly $n$ roots, counted with multiplicity.

(Hint: use the factor theorem and reason by induction on $n$.)

**A1.57.** Let $f(x)$ be a polynomial in $\mathbb{R}[x]$, of degree greater than 1. Let $\alpha \in \mathbb{C}$ be a root of $f(x)$ considered as a polynomial with complex coefficients. Prove that among the roots $f(x)$ there is also $\bar{\alpha}$, the complex conjugate of $\alpha$.

**A1.58.** Prove that the derivative mapping $D : \mathbb{K}[x] \to \mathbb{K}[x]$, defined on page 27, is linear, that is to say, $D(f(x)+g(x)) = D(f(x))+D(g(x))$, for all pairs of polynomials $f(x), g(x) \in \mathbb{K}[x]$.

**A1.59.**\* Prove that Leibniz's law for the derivative of polynomials: $D(f(x) \cdot g(x)) = D(f(x)) \cdot g(x) + D(g(x)) \cdot f(x)$ holds for all pairs of polynomials $f(x), g(x) \in \mathbb{K}[x]$.

**A1.60.**\* Prove that the derivative mapping is uniquely determined by its being linear, by Leibniz's law and by the fact that $D(x) = 1$.

**A1.61.** Prove that if $f(x)$ is a polynomials of degree $n$, then the $i$th derivative $f^{(i)}(x)$ of $f$ is 0 for all $i > n$.

(Hint: prove the claim for $f(x) = cx^n$, by induction on $n$).

**A1.62.** Prove Formula (1.25) on page 28.

**A1.63.** Let $f(x)$ be a polynomial with coefficients in $\mathbb{Z}$ and let $s \in \mathbb{Z}$. Prove that, for each positive integer $n$, $n!$ divides $f^{(n)}(s)$.

(Hint: prove the claim for monomials.)

The following exercises, from A1.64 to A1.69, will underline the importance of Fibonacci numbers with respect to the Euclidean algorithm.

**A1.64.** Prove by induction that two consecutive Fibonacci numbers are relatively prime.

**A1.65.**\* Prove that the greatest common divisor of two Fibonacci numbers is again a Fibonacci number. Precisely, $\mathrm{GCD}(f_n, f_m) = f_d$, where $d = \mathrm{GCD}(m, n)$.

**A1.66.**\* Prove that $f_k$ divides $f_n$ if and only if $k$ divides $n$.

**A1.67.** Which is the value of $\lim_{n \to \infty} (f_{n+1}/f_n)$ ?

(a) 1.

(b) $(1 + \sqrt{5})/2$.

(c) $\sqrt{5}/2$.

(d) None of the above.

**A1.68.*** Prove by induction on $k$ that if $r_n$ is the first remainder equal to zero in the Euclidean algorithm applied to two numbers $a$ and $b$, then

$$r_{n-k} \geq f_k,$$

where $f_k$ is the $k$th Fibonacci number.

**A1.69.*** Use the result found in Exercise A1.68 to prove that if $b < f_n$, then, for each $a \geq b$, the number $D(a, b)$ of divisions that are necessary to get a zero remainder in the Euclidean algorithm is smaller than $n$.

**A1.70.** Prove Lemma 1.4.5.

**A1.71.** Prove Proposition 1.4.6.

**A1.72.** Prove that $\sqrt{2}$ is not a rational number.

**A1.73.** Prove that the expression of a rational number as a continued fraction is unique, if we assume that the last partial quotient is greater than 1.

**A1.74.** Let $\alpha > 1$ be a real number. Prove that the $n$th convergent of the expression of $1/\alpha$ as a continued fraction is equal to the reciprocal of the $(n - 1)$th convergent of the expression of $\alpha$ as a continued fraction.

**A1.75.** Prove Formula (1.41) of page 50.

(Hint: reason by induction on $k$.)

**A1.76.** Prove Formula (1.44) of page 51.

**A1.77.** Prove by induction Formula (1.45) of page 51.

**A1.78.** Prove Lemma 1.5.11 of page 53.

**A1.79.** Fill in the missing details of the proof of Theorem 1.5.13 of page 55.

(Hint: denoting, as usual, by $C_n = p_n/q_n$ the convergents of the continued fraction $[a_0; a_1, \ldots, a_n, \ldots]$, prove that, for each positive integer $n$, the following holds:

$$\alpha = \frac{\alpha_{n+1}p_n + p_{n-1}}{\alpha_{n+1}q_n + q_{n-1}}. \tag{1.56}$$

Using Equation (1.42), deduce that

$$\alpha - C_n = \frac{(-1)^{n-1}}{q_n(\alpha_{n+1}q_n + q_{n-1})}$$

and so that $\lim_{n \to \infty} C_n = \alpha$.)

**A1.80.** Prove that, if $n$ is a non-square positive integer, then $\sqrt{n}$ is not a rational number.

**A1.81.** Prove that each root $\alpha$ of a quadratic equation with integer coefficients is of the form $\alpha = a + b\sqrt{d}$, where $d$ is a positive integer and $a, b$ are rational. Determine $a$, $b$ and $d$ as functions of the coefficients of the quadratic equation.

(Hint: write down the quadratic formula for the equation.)

Conversely, prove that if $\alpha = a + b\sqrt{d}$, where $d$ is a positive integer and $a, b$ are rational, then $\alpha$ is a quadratic number.

**A1.82.** Prove that, if $\alpha$ is a quadratic number and $a, b, c, d$ are rational numbers such that $c\alpha + d \neq 0$, then $(a\alpha + b)/(c\alpha + d)$ is a quadratic number too.

**A1.83.** Recall that if $\alpha$ is a quadratic number, we can consider its conjugate $\alpha'$. Prove that $\alpha = \alpha'$ if and only if $\alpha$ is a rational number. Moreover, if $\alpha$ and $\beta$ are quadratic numbers, we have

(1) $(\alpha + \beta)' = \alpha' + \beta'$;
(2) $(\alpha - \beta)' = \alpha' - \beta'$;
(3) $(\alpha\beta)' = \alpha'\beta'$;
(4) $(1/\alpha)' = 1/\alpha'$.

**A1.84.** Prove Theorem 1.5.22 of page 61.

**A1.85.** Prove that, for each non-square positive integer $n$, the number $\sqrt{n} + [\sqrt{n}]$ is reduced.

**A1.86.*** Let $\alpha$ be an irrational number and let $\alpha = [a_1; a_2, a_3, \ldots, a_n, \ldots]$ be its expression as a continued fraction, and let $C_n = p_n/q_n$ be its convergents. If we take two consecutive convergents $C_n = p_n/q_n$ and $C_{n+1} = p_{n+1}/q_{n+1}$, at least one, say $C_i$, satisfies the relation

$$|\alpha - C_i| < \frac{1}{2q_i^2}.$$

The next exercises, from A1.87 to A1.94, lead to the proof of the theorem of Lagrange 1.5.15 and to the description of an algorithm to compute the expression of an irrational number, and in particular of a quadratic irrational number, as a continued fraction.

**A1.87.*** Let $\alpha = [\overline{a_1; a_2, \ldots, a_t}]$. Notice that $\alpha = [a_1; a_2, \ldots, a_t, \alpha]$. Using Equations (1.41) deduce that

$$\alpha = \frac{\alpha p_t + p_{t-1}}{\alpha q_t + q_{t-1}}$$

where $C_n = p_n/q_n$ are the convergents of $[a_1; a_2, \ldots, a_t]$. Deduce that $\alpha$ is a quadratic number satisfying the equation

$$x^2 q_t + x(q_{t-1} - p_t) - p_{t-1} = 0.$$

**A1.88.*** Let $\alpha = [a_1; a_2, \ldots, a_m, \overline{a_{m+1}, \ldots, a_{m+t}}]$ be a periodic continued fraction. Put $\beta = [\overline{a_{m+1}; a_{m+2}, \ldots, a_{m+t}}]$, which is a quadratic number by the result of the previous exercise. Notice that $\alpha = [a_1; a_2, \ldots, a_m, \beta]$. Mimicking the argument of the previous exercise and keeping in mind Exercise A1.82 prove that $\alpha$ is a quadratic number.

**A1.89.*** Prove that an irrational quadratic number $\alpha$ may be written as $\alpha = (p + \sqrt{d})/q$, where $p, q, d$ are integers, $q \neq 0$, $d > 0$ is not a perfect square, and $q$ divides $d - p^2$.

**A1.90.*** Let $\alpha$ be an irrational quadratic number; so, by the previous exercise, we may write

$$\alpha = \frac{p_0 + \sqrt{d}}{q_0},$$

where $p_0, q_0, d$ are integers, $q_0 \neq 0$, $d > 0$ is not a perfect square, and $q_0$ divides $d - p_0^2$.

Define recursively

$$\alpha_n = \frac{p_n + \sqrt{d}}{q_n}, \qquad a_n = [\alpha_n], \qquad p_{n+1} = a_n q_n - p_n, \qquad q_{n+1} = \frac{d - p_{n+1}^2}{q_n}.$$

Prove that for each positive integer $n$ the numbers $p_n$, $q_n$ are integers, $q_n \neq 0$ and $q_n$ divides $d - p_n^2$.

**A1.91.** * Continuing the previous exercise, consider the continued fraction $[a_0; a_1, \ldots, a_n, \ldots]$, and denote as usual by $C_n = p_n/q_n$ the convergents. Using Theorem 1.5.13, prove that $\alpha = [a_0; a_1, \ldots, a_n, \ldots]$.

**A1.92.** * Continuing the previous exercises, notice that, for each positive integer $n$ we have $\alpha = [a_0; a_1, \ldots, a_n, \alpha_{n+1}]$, and so Equation (1.56). Conjugating all the terms in it (use here Exercise A1.83), verify that

$$\alpha_n' = -\frac{q_{n-1}}{q_n} \frac{\alpha' - C_{n-1}}{\alpha' - C_n}.$$

Keeping in mind that $\lim_{n \to \infty} C_n = \alpha$, deduce that there exists an integer number $m$ such that, for $n > m$, we have $\alpha_n' < 0$.

**A1.93.** * Still continuing the previous exercises, prove that there exists an integer $m$ such that, for $n > m$, we have $q_n > 0$. Keeping in mind that $q_n q_{n+1} = d - p_{n+1}^2 < d$, deduce that for each $n > m$ we have $q_n < d$.

Analogously, notice that for each $n > m$ one has $p_{n+1}^2 = d - q_n q_{n+1} < d$. Deduce that there are infinitely many integers $i, j$ such that $(p_i, q_i) = (p_j, q_j)$; hence it follows that $[a_0; a_1, \ldots, a_n, \alpha_{n+1}]$ is periodic.

**A1.94.** * Assume now that $\alpha$ is an irrational reduced quadratic number. Using again the notation of the previous exercises, prove by induction that for each positive integer $n$ one has $-1 < \alpha_n' < 0$. Deduce that $a_n = [-1/\alpha_{n+1}']$.

Using what has been proved in the previous exercises, we know that there exist integer numbers $i, j$, with $j > i$, such that $\alpha_i = \alpha_j$, and so also $a_{i-1} = a_{j-1}$. Deduce that $\alpha_{i-1} = \alpha_{j-1}$, and so $\alpha = \alpha_0 = \alpha_{j-i}$. Noticing that

$$\alpha = [a_0; a_1, \ldots, a_{j-i-1}, \alpha_{j-i}] = [a_0; a_1, \ldots, a_{j-i-1}, \alpha],$$

deduce that $[a_0; a_1, \ldots, a_n, \ldots]$ is purely periodic.

**A1.95.** * Assume now that $\alpha = \overline{[a_1; a_2, \ldots, a_t]}$ with $C_n = p_n/q_n$ convergents of the continued fraction. Put $\beta = \overline{[a_t; a_{t-1}, \ldots, a_1]}$, and denote by $C_n' = p_n'/q_n'$ the convergents of the latter continued fraction. Using Equation (1.41), prove that

$$C_t' = \frac{p_t}{p_{t-1}}, \qquad C_{t-1}' = \frac{q_t}{q_{t-1}}.$$

Deduce that

$$p_t = p_t', \quad q_t' = p_{t-1}, \quad p_{t-1}' = q_t, \quad q_{t-1}' = q_{t-1}.$$

Keeping in mind Exercise A1.87, prove that $\alpha$ and $-1/\beta$ are solutions of the same equation with integer coefficients, and so they are conjugate numbers. Notice that, as $a_t > 0$, one has $\beta > 1$. Deduce that $\alpha$ is reduced.

# B1 Computational exercises

**B1.1.** For which natural numbers $n$ does the inequality $n < 2^n$ hold?

(a) Every even $n$.
(b) Every $n > 0$.
(c) Every $n > 1$.
(d) None of the above.

Prove by induction the correct claim.

**B1.2.** Let $S(n)$ be the sum of the first $n$ odd natural numbers. Which of the following is a recurrence relation satisfied by $S(n)$?

(a) $S(n + 1) = S(n) + 2n + 1$.
(b) $S(n + 1) = S(n) + n + 1$.
(c) $S(n + 1) = S(n) + 2n - 1$.
(d) None of the above.

**B1.3.** Let $S(n)$ be the sum of the first $n$ odd natural numbers. Which is the closed formula to compute $S(n)$?

(a) $S(n) = n(n + 1)/2$.
(b) $S(n) = n^2 - n + 1$.
(c) $S(n) = n^2$.
(d) None of the above.

Prove by induction the correct formula.

**B1.4.** Let $S(n)$ be the sum of the first $n$ even positive numbers. Which of the following is a recurrence relation satisfied by $S(n)$?

(a) $S(n + 1) = S(n) + n + 2$.
(b) $S(n + 1) = S(n) + 2n$.
(c) $S(n + 1) = S(n) + 2n + 2$.
(d) None of the above.

**B1.5.** Let $S(n)$ be the sum of the first $n$ even positive numbers. Which is the closed formula to compute $S(n)$?

(a) $S(n) = n(n + 1)/2$.
(b) $S(n) = n^2 + 1$.
(c) $S(n) = n(n + 1)$.
(d) None of the above.

It is recommended to prove it by induction on $n$.

**B1.6.** For each natural number $n$, which is the value of

$$\sum_{k=0}^{n} (4k + 1) \ ?$$

(a) $n^2 + 4n + 1$.
(b) $(2n + 1)(n + 1)$.
(c) $3n^2 + 2n + 1$.

(d) None of the above.

Prove by induction the correct formula.

**B1.7.** For each natural number $n$, which is the value of the sum of the first $n$ squared numbers

$$\sum_{k=1}^{n} k^2 = 1^2 + 2^2 + 3^2 + \cdots + n^2 \ ?$$

(a) $(n+1)(2n+1)/2$.
(b) $(n+1)(n+2)(n+3)/6$.
(c) $n(n+1)(2n+1)/6$.
(d) None of the above.

Prove by induction the correct formula.

**B1.8.** For each natural number $n$, which is the value of the sum of the first $n$ cubed numbers

$$\sum_{k=1}^{n} k^3 = 1^3 + 2^3 + 3^3 + \cdots + n^3 \ ?$$

(a) $(n^4 + 2n^3 + n^2)/4$.
(b) $(n^4 + 18n^3 - 19n^2 + 12n)/12$.
(c) $(6n^3 - 17n^2 + 25n - 12)/2$.
(d) None of the above.

**B1.9.** For each natural number $n$, which is the value of the sum of the first $n$ cubed even numbers

$$\sum_{k=1}^{n} (2k)^3 = 2^3 + 4^3 + \cdots + (2n)^3 \ ?$$

(a) $2n(n+1)(n+2)(n+3)$.
(b) $n(n+1)(n+2)(n+3)/3$.
(c) $2n^2(n+1)^2$.
(d) None of the above.

Prove by induction the correct formula.

**B1.10.** Which is the value of

$$\sum_{k=0}^{n-1} 2 \cdot 3^k \ ?$$

(a) $3^n - 1$.
(b) $3(3^n - 1)/2$.
(c) $3^{n+1} - 1$.
(d) None of the above.

Prove by induction the correct formula.

**B1.11.** Which is the value of

$$\sum_{k=0}^{n} 5 \cdot 9^k \ ?$$

(a) $4(9^{n+1} - 1)/5$.

(b) $5(9^{n+1} - 1)/8$.
(c) $5(9^n - 1)/4$.
(d) None of the above.

Prove by induction the correct formula.

**B1.12.** Find the eigenvectors of the matrix $A$ defined in Equation (1.6). In particular, verify that $(1, \lambda_1)$ and $(1, \lambda_2)$ are eigenvectors with corresponding eigenvalues $\lambda_1$ and $\lambda_2$, as defined by formula (1.7) on page 9.

**B1.13.** Let $C$ and $C^{-1}$ be the matrices of formula (1.9), $D$ the matrix (1.8) and $A$ the matrix (1.6). Verify that $C \cdot C^{-1} = C^{-1} \cdot C = I$, and that $C \cdot D \cdot C^{-1} = A$.

**B1.14.** Which is the solution of the *geometric progression*

$$a_n = r \cdot a_{n-1}, \qquad a_0 = k,$$

where $r$ and $k$ are fixed integers?

(a) $a_n = kr^n$, for each $n > 0$.
(b) $a_n = rk^n$, for each $n > 0$.
(c) $a_n = nk^r$, for each $n > 0$.
(d) None of the above.

**B1.15.** Assume the world population increases by 3% each year. Which is the recurrence relation giving world population $P(n)$?

(a) $P(n) = P(n-1) \cdot 3/100$.
(b) $P(n) = P(n-1) \cdot 103/100$.
(c) $P(n) = P(n-1) + 3/100$.
(d) None of the above.

**B1.16.** Assume that in 2000 the world population were of 5 billion people, and that it evolves according to the recurrence relation given in Exercise B1.15. How many billions people will live in the world in 2024?

(a) About 50 billion.
(b) About 20 billion.
(c) About 10 billion.
(d) None of the above.

**B1.17.** Give a closed formula for the beetle population of Example 1.2.8.

**B1.18.** Consider Example 1.2.8 on page 11. Assume that in a given year there were 120 newborn beetles and 20 two-year beetles. How many beetles will live 10 years later?

(a) 180.
(b) 140.
(c) 120 (60 newborn and 60 one-year beetles).
(d) 120 (all newborn).

**B1.19.** Give a closed formula for the population of Liguria, as described in Example 1.2.9.

**B1.20.** Consider Example 1.2.9. Assume that in a given year 5 million people live in Liguria and 10 million outside. How many inhabitants will Liguria have 30 years later?

(a) 20 million.
(b) 10 million.
(c) 7 million.
(d) None of the above.

**B1.21.** Consider the regions a plane is divided into by $n$ lines in generic position (that is to say, no two lines are parallel and no three lines meet in a point). How many regions are there?

(a) $n^2 + 1$.
(b) $n^2 + n - 1$.
(c) $n(n + 1)/2 + 1$.
(d) None of the above.

**B1.22.** Consider the regions a plane is divided into by $n$ circles in generic position (that is to say, no three circles meet in a point and any two circles meet in exactly two points). How many regions are there?

(a) $(n^3 - 5n^2 + 14n - 8)/2$.
(b) $(n^2 + 3n - 2)/2$.
(c) $(n^3 - 3n^2 + 20n - 4)/6$.
(d) None of the above.

**B1.23.** Prove that it is possible to colour the regions into which an arbitrary number of lines divides the plane (even when some of the lines are not in generic position) with just two colours. Notice that by *colouring* we mean to assign colours to the regions in such a way that *neighbouring* regions (that is, with a *side* in common) always have distinct colours.

**B1.24.** Mark has opened a bank account for which there are no charges and yielding a yearly 4% interest which is computed and paid to his account every third month. Suppose Mark deposited a certain amount of money when he opened the account and after that he neither withdrew nor deposited money from the account. Which is the recurrence relation determining the amount of money $S(n)$ Mark has in his account after $n$ years?

(a) $S(n) = S(n - 1) \cdot (101/100)^4$.
(b) $S(n) = S(n - 1) \cdot 104/100$.
(c) $S(n) = S(n - 1) + 4/100$.
(d) None of the above.

**B1.25.** Consider Mark's bank account as in Exercise B1.24. Suppose now that Mark deposited 5,000 euros when he opened the account. How much money will Mark have in his account 20 years later?

(a) About 15,000 euros.

(b) About 11,000 euros.
(c) About 9,000 euros.
(d) None of the above.

**B1.26.** Compute the first 10 terms $a_1, \ldots, a_{10}$ of each of the following recurrence relations, always assuming the initial value is $a_0 = 1$:

$$a_{n+1} = a_n + 3; \qquad a_{n+1} = 3a_n + 1; \qquad a_{n+1} = 2a_n - n; \qquad a_{n+1} = a_n^2 - a_n + 2.$$

**B1.27.** Compute the terms $a_2, \ldots, a_{10}$ of each of the following recurrence relations, always assuming the initial values are $a_0 = a_1 = 1$:

$$a_{n+2} = 3a_{n+1} - a_n; \qquad\qquad a_{n+2} = a_{n+1} + a_n + 1;$$

$$a_{n+2} = -a_{n+1} + 2a_n + n^2; \qquad\qquad a_{n+2} = 2a_{n+1} \cdot a_n.$$

**B1.28.** Which is the solution of the recurrence relation

$$a_{n+2} = a_{n+1} + 2a_n$$

with initial values $a_0 = 2$ and $a_1 = 7$?

(a) $a_n = 3 \cdot 2^n - (-1)^n$.
(b) $a_n = 2^n - 1$.
(c) $a_n = 2^n - 3 \cdot (-1)^n$.
(d) None of the above.

**B1.29.** Which is the solution of the recurrence relation

$$a_n = 2a_{n-1} + n + 5$$

with initial value $a_0 = 4$?

(a) $a_n = 11 \cdot 2^n - n - 7$.
(b) $a_n = 9 \cdot 2^n - n - 5$.
(c) $a_n = 11 \cdot 2^n - n + 7$.
(d) None of the above.

(Hint: first, find a solution of the given recurrence relation, ignoring the initial condition, in the form $\alpha n + \beta$, where $\alpha$ and $\beta$ are constants, then try to solve the associated homogeneous recurrence relation.)

**B1.30.** Which is the solution of the recurrence relation

$$a_{n+1} = 2a_n$$

with initial value $a_0 = 3$?

(a) $a_n = 3 \cdot 2^n$.
(b) $a_n = 2^n$.
(c) $a_n = 2^n + 3$.
(d) None of the above.

**B1.31.** Which is the solution of the recurrence relation

$$a_n = 2a_{n-1} + 2^n$$

with initial value $a_0 = 2$?

(a) $a_n = n2^n$.
(b) $a_n = 2^n + 3^n$.
(c) $a_n = (1 + n)2^n$.
(d) None of the above.

(Hint: first, find a solution of the given recurrence relation, ignoring the initial condition, in the form $n\alpha^n$, where $\alpha$ is a constant, then solve the associated homogeneous recurrence relation.)

**B1.32.** How many moves must the priests of Brahma's temple perform in order to transfer the tower of Hanoi consisting of 64 discs (see Example 1.2.10)?

(a) 18,446,744,073,709,551,615.
(b) 18,446,744,073,709,541,615.
(c) 18,446,744,073,709,541,613.
(d) None of the above.

**B1.33.** Assuming that it takes to the priests of Brahma's temple 30 seconds to perform a move (the discs are very heavy!) and that they started in 1999 BCE, in how many years should the end of the world arrive?

(a) About $1 \cdot 10^{12}$.
(b) About $1 \cdot 10^{13}$.
(c) About $2 \cdot 10^{13}$.
(d) None of the above.

**B1.34.** Which is the greatest common divisor of 491 and 245?

(a) 1.
(b) 7.
(c) 13.
(d) None of the above.

**B1.35.** How many steps does the Euclidean algorithm require to compute GCD $(491, 245)$?

(a) 1.
(b) 2.
(c) 3.
(d) None of the above.

**B1.36.** How many steps does the Euclidean algorithm require to compute GCD $(3072, 165)$?

(a) 7.
(b) 8.
(c) 9.
(d) None of the above.

**B1.37.** Compute GCD(34567, 457) using the algorithm described in Exercise A1.40 and verify that it requires just five steps, while the one based on the usual division algorithm requires eight steps (see Example 2.5.5 on page 100).

**B1.38.** Which are the coefficients of 491 and of 245 in Bézout's identity for GCD(491, 245), obtained using the Euclidean algorithm?

(a) 1 e −2.
(b) −1 e 2.
(c) −244 e 489.
(d) None of the above.

**B1.39.** Which is the greatest common divisor of 28762 and 1515?

(a) 3.
(b) 7.
(c) 1.
(d) None of the above.

**B1.40.** How many steps of the Euclidean algorithm are required to compute GCD(28762, 1515)?

(a) 3.
(b) 5.
(c) 7.
(d) None of the above.

**B1.41.** Which are the coefficients of 28762 and of 1515 in Bézout's identity for GCD(28762, 1515), obtained using the Euclidean algorithm?

(a) −527 e 10005.
(b) −1 e 2.
(c) −244 e 489.
(d) None of the above.

**B1.42.** Determine all integers $x, y$ satisfying the equation

$$92x + 28y = 180.$$

(a) The equation has no integer solutions.
(b) $(-135 - 7t, 450 + 23t)$ for each integer $t$.
(c) $(-135 + 7t, 450 + 23t)$ for each integer $t$.
(d) None of the above.

**B1.43.** Determine all integers $x, y$ satisfying the equation

$$482x + 20y = 35.$$

(a) The equation has no integer solutions.
(b) $(12 - 7t, -33 + 5t)$ for each integer $t$.
(c) $(15 - 7t, -31 + 5t)$ for each integer $t$.
(d) None of the above.

**B1.44.** Determine all integers $x, y$ satisfying the equation

$$1859x + 2057y = 143.$$

(a) The equation has no integer solutions.
(b) $(37 + 187t, -41 - 169t)$ for each integer $t$.
(c) $(-676 + 187t, 611 - 169t)$ for each integer $t$.
(d) None of the above.

**B1.45.** Which sequence does the polynomial $x^7 - 2x^3 + x - 1$ correspond to?

(a) $\{1, 0, 0, -2, 0, 1, -1, 0, \dots\}$.
(b) $\{1, 0, 0, 0, -2, 0, 1, -1, 0, \dots\}$.
(c) $\{-1, 1, 0, -2, 0, 0, 0, 1, 0, \dots\}$.
(d) None of the above.

**B1.46.** Which sequence does the product of the polynomials $x^2 - 1$ and $x^2 + 1$ correspond to?

(a) $\{1, 0, 2, 0, -1, 0, \dots\}$.
(b) $\{1, 0, 1, 0, -1, 0, \dots\}$.
(c) $\{-1, 0, 0, 0, 1, 0, \dots\}$.
(d) None of the above.

**B1.47.** Is it true that the polynomial $x^2 - 3x^4 - 2x + 1 + x^3 \in \mathbb{Q}[x]$ is monic?

(a) Yes, because its first coefficient is 1.
(b) Yes, because its leading coefficient is 1.
(c) No.
(d) None of the above.

**B1.48.** Which are the quotient and the remainder of the division in $\mathbb{Q}[x]$ of the polynomial $x^4 - 2x^3 - 2x - 1$ by $2x^3 - x - 1$?

(a) The quotient is $x/2$ and the remainder is $x^2/2 - 5x/2 - 2$.
(b) The quotient is $x/2 - 1$ and the remainder is $x^2/2 - 5x/2 - 2$.
(c) The quotient is $x/2 - 1$ and the remainder is $x^2/2 - 3x/2 - 1$.
(d) The quotient is $x/2$ and the remainder is $x^2/2 - 3x/2 - 1$.

**B1.49.** Is it true that $-x - 1$ is the greatest common divisor of the polynomials $2x^3 - x + 1$ and $x^2 + 2x + 1$ in $\mathbb{Q}[x]$?

(a) No, because the greatest common divisor of those polynomials is 1.
(b) No, because $-x - 1$ does not divide $2x^3 - x + 1$.
(c) Yes.
(d) None of the above.

**B1.50.** Is it true that $2x + 2$ is a greatest common divisor of the polynomials $2x^4 + x - 1$ and $x^2 - 1$ in $\mathbb{Q}[x]$?

(a) No, because the greatest common divisor of those polynomials is 1.
(b) No, because it is not a monic polynomial.
(c) Yes.
(d) None of the above.

**B1.51.** Using the Euclidean algorithm, how many steps are necessary to find the greatest common divisor of the polynomials $2x^3 - x + 1$ and $x^2 + 2x + 1$ in $\mathbb{Q}[x]$? (Remember that the last step is the one in which the zero remainder is found.)

(a) 1.
(b) 2.
(c) 3.
(d) 4.

**B1.52.** Do two polynomials $f(x)$ and $g(x)$ in $\mathbb{Q}[x]$ such that

$$(x^2 + 2x + 1)f(x) + (2x^3 - x + 1)g(x) = 4x + 4$$

exist?

(a) No, they do not exist, because the GCD of $x^2 + 2x + 1$ and $2x^3 - x + 1$ is $-x - 1$.
(b) No, because the GCD of $x^2 + 2x + 1$ and $2x^3 - x + 1$ is 1.
(c) Yes, they have to exist, but there is no way to compute them.
(d) None of the above.

**B1.53.** Which is the derivative of $3x^3 - 5x^5 + 123 - 2x^4$ in $\mathbb{Q}[x]$?

(a) $9x^2 + 5x^4 - 8x^3$.
(b) $9x^2 - 5x^4 - 8x^3$.
(c) $3x^2 - 5x^4 - 8x^3$.
(d) None of the above.

**B1.54.** Let $f(x) = 15/7 + 3x - 24/13x^3$. Which polynomial is a primitive polynomial with integer coefficients, associated to $f(x)$?

(a) $195 + 293x - 168x^3$.
(b) $195 + 271x - 168x^3$.
(c) $56x^3 - 91x - 65$.
(d) None of the above.

**B1.55.*** Consider the Euclidean ring $\mathbb{G}$ described in Exercise A1.49. Find a greatest common divisor of 2 and $1 - 5i$ in $\mathbb{G}$.

**B1.56.** Verify formula (1.30) carrying out the same multiplication in base 10.

**B1.57.** Consider the numbers 100100101110111 and 1101 in base 2. Carry out the division in base 2 step by step and find the quotient and the remainder.

**B1.58.** Consider a number system in base $b$ where $b = n^2 + 1$, with $n \in \mathbb{N}$. Write in base $b$ the following numbers:

$$n^2 + 2, \qquad n^2 + 2n, \qquad (n^2 + 2)^2, \qquad n^4, \qquad n^2(n^2 + 2)^2.$$

**B1.59.** Convert the number $(2345)_{10}$ to base 3.

(a) 10012212.
(b) 10111212.
(c) 10102212.
(d) None of the above.

**B1.60.** Convert $(234)_{10}$ to base 8.

(a) 352.
(b) 344.
(c) 350.
(d) None of the above.

**B1.61.** Convert $(456)_8$ to base 2.

(a) 1001101110.
(b) 1001010110.
(c) 100101110.
(d) None of the above.

**B1.62.** Which are the sum and the product of 10010111 e 11101 in base 2?

(a) 10110100, 1001000011011.
(b) 11110100, 1010100011011.
(c) 11110100, 1000100011011.
(d) None of the above.

**B1.63.** Determine the sum and the product of 12323 e 321 in base 4.

(a) 12310, 12023203.
(b) 13310, 12023203.
(c) 13310, 12022203.
(d) None of the above.

**B1.64.** Consider base 21 integer numbers, where the digits are denoted by the 21 alphabet letters $A, B, C, D, E, F, G, H, I, L, M, N, O, P, Q, R, S, T, U, V, Z$, with $A = 1$, $B = 2$, ..., $V = 20$ e $Z = 0$. Which are the sum and the product of $QUI$ and $QUO$?

(a) $ANTA$, $LAEQUA$.
(b) $ALTA$, $NAQTCN$.
(c) $ALTA$, $NAPTCN$.
(d) $ANTA$, $NAQTCN$.

**B1.65.** Consider base 21 numbers, where the digits are denoted by the 21 alphabet letters $A, B, C, D, E, F, G, H, I, L, M, N, O, P, Q, R, S, T, U, V, Z$, with $A = 1$, $B = 2$, ..., $V = 20$ e $Z = 0$. Which are the sum and the difference of $CASA$ e $CANE$?

(a) $FCHF$, $ES$.
(b) $FBHF$, $DS$.
(c) $FCHF$, $DS$.
(d) $FBHF$, $ES$.

It is advisable to write down a table with correspondence between alphabet letters and integer numbers in $\{0, 1, \ldots, 20\}$.

**B1.66.** Determine the base 10 fraction corresponding to the number $0.\overline{40}$.

**B1.67.** Determine the base 10 fraction corresponding to the number $55.3\overline{5}$.

**B1.68.** Determine the base 2 fraction corresponding to the number $0.\overline{101}$.

**B1.69.** Determine the base 2 fraction corresponding to the number $101.1011\overline{01}$.

**B1.70.** Determine the base 7 fraction corresponding to the number $0.\overline{40}$.

**B1.71.** Determine the base 7 fraction corresponding to the number $55.3\overline{5}$.

**B1.72.** Write the fraction $7/3$ as a recurring number in base 10.

**B1.73.** Write the fraction $31/5$ as a recurring number in base 7.

**B1.74.** Write the fraction $1011/11$ as a recurring number in base 2.

**B1.75.** Which is the expression of $113/90$ as a continued fraction?

(a) $[1; 3, 1, 2]$.
(b) $[1; 3, 1, 10]$.
(c) $[1; 3, 10, 2]$.
(d) None of the above.

**B1.76.** Which rational number does the continued fraction $[1; 3, 1, 4, 1, 1, 2]$ represent?

(a) $135/107$.
(b) $133/107$.
(c) $133/105$.
(d) None of the above.

**B1.77.** How many terms are there in the expression of $350/211$ as a continued fraction? (Assume that the last term is not 1.)

(a) 6 terms.
(b) 5 terms.
(c) 4 terms.
(d) None of the above.

**B1.78.** Prove that $\sqrt{3} = [1; \overline{1, 2}]$.

**B1.79.** Prove that $\sqrt{5} = [2, \overline{4}]$.

**B1.80.** Prove that the expression of the golden ratio as a continued fraction is $[1; \overline{1}]$.

**B1.81.** Find the positive number corresponding to the periodic simple continued fraction $[\overline{4; 1, 3}]$.

**B1.82.** Which is the expression of $\sqrt{27}$ as a continued fraction?

(a) $[5; \overline{5}]$.
(b) $[5; \overline{6}]$.
(c) $[5; \overline{10}]$.
(d) None of the above.

# C1 Programming exercises

**C1.1.** Write a program that computes the factorial of a positive integer number.

**C1.2.** Write a program that computes the binomial coefficients.

**C1.3.** Consider the following recurrence relation:

$$a_{n+k} = b_1 a_{n+k-1} + b_2 a_{n+k-2} + \cdots + b_k a_n + d(n),$$

where $b_i$s are constants, $d(n)$ is an arbitrary function of $n$, and the initial values are

$$a_1 = \alpha_1, \ a_2 = \alpha_2, \ \ldots, a_r = \alpha_r.$$

Write a program that computes the term $a_n$.

**C1.4.** Write a program that computes Fibonacci numbers.

**C1.5.** Write a program that determines whether a given integer $n$ is a Fibonacci number. Use this program to verify Exercises A1.30, A1.66, and A1.33.

**C1.6.** Write a program that computes the greatest common divisor of two integers $a$ and $b$ using the Euclidean algorithm.

**C1.7.** Write a program that computes the greatest common divisor of two integers $a$ and $b$ using the algorithm described in Exercise A1.40.

**C1.8.** Write a program that computes a Bézout's identity for two integer numbers $a$ and $b$, that is to say, that expresses the greatest common divisor of two numbers $a$ and $b$ in the form $\alpha a + \beta b$ for suitable coefficients $\alpha$ and $\beta$ in $\mathbb{Z}$, by using the Euclidean algorithm.

**C1.9.** Write a program that computes the greatest common divisor of two Fibonacci numbers, and use this program to verify Exercises A1.64 and A1.65.

**C1.10.** Write a program that expresses each integer $n$ as a sum of distinct Fibonacci numbers (see Exercise A1.31).

**C1.11.** Write a program that computes, given two integers $a$ and $b$, with $a \geq b$, the number of divisions occurring during the execution of the Euclidean algorithm (see Exercise A1.69).

**C1.12.** Write a program that carries out the division of two given polynomials, computing the quotient polynomial and the remainder,

**C1.13.** Write a program that computes the greatest common divisor of two polynomials $f(x)$ and $g(x)$ over a field using the Euclidean algorithm.

**C1.14.** Write a program that computes a Bézout's identity $d(x) = f(x)h(x) + g(x)k(x)$ for two polynomials $f(x)$ and $g(x)$, where $d(x)$ is their greatest common divisor.

**C1.15.** Write a program that computes the $n$th derivative of a given polynomial.

**C1.16.** Write a program that converts the base 10 expression of integer numbers to an arbitrary base $b > 1$ and vice versa.

**C1.17.** Write a program that, for any fraction, determines its decimal representation, finding its recurring and non-recurring digits.

**C1.18.** Write a program that computes the expression of an arbitrary rational number as a continued fraction.

**C1.19.** Write a program that computes the convergents of an infinite continued fraction, up to an assigned approximation of the irrational number represented by the continued fraction.

**C1.20.** Write a program that computes the expression of $\sqrt{n}$ as a continued fraction, for a positive integer $n$.

# 2

## Computational complexity

### 2.1 The idea of computational complexity

In Chapter 1 we have recalled some basic notions about natural, integer, rational numbers, and about the operations on numbers. The need to solve concrete problems leads, as we have discussed in § 1.2, to devise *mathematical models*; they in turn pose the question of developing computing procedures, which may be very complex, but that in any case result in the solution of the studied problem in finitely many steps. Such a procedure is called an *algorithm*, from the name of the Arab-language mathematician and astronomer al-Khowarizmi, who lived in the 9th century. This is a good moment to pose the main question: which is the cost of an algorithm, in terms of effort and time needed to execute its steps from the beginning to the end of the necessary operations? For instance, if we use a computer to perform the computations, which is the cost of a given algorithm, in terms of computation time and programming difficulty? Is there a mathematically sound way to distinguish *good algorithms*, those requiring little *time* and little *programming effort*, from those requiring *too much time*, in relative or absolute terms, for their execution?

As we have said, an algorithm is just a finite sequence of *elementary operations*. Thus, in order to answer the above questions, that is to say, to determine the *computational complexity* of a given algorithm, it is necessary to calculate with precision the number of basic operations composing the algorithm. Multiplying this number by the time taken to program or perform a single basic operation with the available computing devices, we obtain a measure of the total time needed to perform the calculations the algorithm consists of. Notice that, while this measure of complexity is crucially affected by the characteristics of the computing devices we are using, the computational complexity is an intrinsic property of the algorithm. In any case, the computation time of an algorithm is proportional to its computational complexity, and the ratio is determined by the characteristics of the computing devices.

We shall only deal with computational complexity rather than with actual computation time, because the former only requires mathematical notions, while the latter concerns computer science. So in what follows when we shall refer to *time*, we shall only mean *computational time*.

The importance of the notion of computational complexity is obvious. For instance, it allows us to distinguish between:

- algorithms than can be executed in a reasonable time with the available computing devices;
- algorithms that cannot be possibly executed in a reasonable time with the available computing devices, for instance by hand, but than can be executed in a reasonable time using more advanced instruments, of which we may estimate the power, i.e. it is possible to evaluate the time they need to perform a single basic operation;
- algorithms only theoretically feasible, because in practice their execution requires a time either unconditionally too long or exceeding the time at our disposal, *independently of the available computing resources*.

The unit which will be used to measure the complexity of an algorithm is the *bit operation* which, basically, corresponds to one of the elementary operations the algorithm consists of. Let us give this notion in detail.

We shall limit our treatment to algorithms operating on integer numbers. To fix the ideas, when performing the operations prescribed by an algorithm we shall use base 2 integer numbers. The reason for doing so lies in the fact that this is what happens when operates with a computer, and we assume that most of the calculations that follow will be performed using a computer. Moreover, we shall soon see that this choice does not modify the complexity of the algorithms, in a sense that will be made precise.

We shall write a number $n$ as $n = (a_{k-1} \ldots a_0)_2$, or simply as

$$n = a_{k-1} a_{k-2} \ldots a_0$$

if $n = \sum_{i=0}^{k-1} 2^i a_i$, where $a_i = 0, 1$, and $i = 0, \ldots, k-1$. The terms $a_i$ are called *binary digits*.

By a *bit operation* we mean one of the following elementary operations:

(1) addition of two binary digits (e.g. $0 + 1$);
(2) subtraction of two binary digits (e.g. $1 - 0$);
(3) multiplication of two binary digits (e.g. $1 \cdot 1$);
(4) division of a *two* binary digit integer by 1;
(5) *left* translation by one place, that is multiplication by 2 (for instance 111 becomes 1110), and *right* translation by one place, that is division by 2.

Clearly, each of the four basic arithmetic operations on integers can be expressed by means of a sequence of bit operations. Thus, every algorithm can be expressed by means of a sequence of bit operations.

**Definition 2.1.1.** *The computational complexity of an algorithm operating on integer numbers is given by the number of bit operations necessary to execute it.*

Notice that the computational complexity of an algorithm *is not a number, but a function*. Indeed, it depends in general on the integer numbers on which the algorithm operates. In other words, it is to be expected that the larger the integers on which the algorithm operates, the larger the number of bit operations necessary to execute the algorithm, and so the larger its complexity. We shall come back shortly, in § 2.3, on this fundamental notion of the dependence of an algorithm on the numbers on which it operates.

We are not usually interested in a precise determination of the complexity of an algorithm, but in a *reasonable* estimate from above. This is, in fact, what we need in order to evaluate the time necessary for the execution of the algorithm. To express this estimate in an efficient way, that is to say without unnecessary details or redundancy, it is common to use the $\mathcal{O}$ notation, commonly said *Big-Oh* notation, introduced by Paul Bachmann in 1894. It is described in next section.

One further note of caution: so far we defined the notion of computational complexity just for algorithms operating on integer numbers. We shall later extend this notion to algorithms operating on other objects, for instance on polynomials or on elements of a field (see § 2.5.3, § 5.1.14).

## 2.2 The symbol $\mathcal{O}$

The importance of the symbol $\mathcal{O}$ lies in the fact that it describes in a very simple and explicit way, as we shall soon make clear, the asymptotic behaviour of a given function.

**Definition 2.2.1.** *Let $f$ and $g$ be two* positive-valued *functions, defined on a subset $S$ of $\mathbb{R}$ containing $\mathbb{N}$. We shall say that $f$ is* dominated *by $g$ (or that $g$ dominates $f$, or that $g$ has* higher order *than $f$) if there exist constants $k$ and $c$ in $\mathbb{R}$ such that*

$$f(x) \leq k \cdot g(x), \qquad \text{for all } x \in S, \text{ with } x > c. \qquad (2.1)$$

We shall denote by $\mathcal{O}(g)$ the set of the functions dominated by $g$, that is those *increasing not faster than $g$*. If $f \in \mathcal{O}(g)$, we say that the function $f$ is *in the class $\mathcal{O}(g)$* and that we have given a $\mathcal{O}$-*estimate* for the function $f$ by means of the *reference function $g$*. Notice that, if $f \in \mathcal{O}(g)$ and $g \in \mathcal{O}(h)$, then $f \in \mathcal{O}(h)$ (see Exercise A2.1). If $f \in \mathcal{O}(g)$ and $g \in \mathcal{O}(f)$, we shall say that $f$ and $g$ have the *same order*.

When we write $f \in \mathcal{O}(g)$, one does not usually try to determine the least value of the constant $k$ for which Equation (2.1) holds, as we are generally interested only in giving an estimate for the function $f(x)$. It is only when dealing with subtler questions that we take in account the problem of *improving the constants*.

Let us see some examples with $S = \mathbb{N}$.

**Example 2.2.2.** Let $f = n^4$ e $g = n^4 + 8n^2 + 5n + 10$. Then $f \in \mathcal{O}(g)$, with $k = 1$ and $c = 0$. Moreover, $g \in \mathcal{O}(f)$ too, with $k = 24$ and $c = 0$.

Indeed, we have that $8n^2 + 5n + 10 > 0$ for all $n \in \mathbb{N}$, and so

$$f(n) = n^4 < 1 \cdot (n^4 + 8n^2 + 5n + 10) = 1 \cdot g(n) \quad \forall n \in \mathbb{N}.$$

Moreover it is straightforward to check that, for all $n \in \mathbb{N}$,

$$g(n) = n^4 + 8n^2 + 5n + 10 < n^4 + 8n^4 + 5n^4 + 10n^4 = 24n^4;$$

hence $g(n) < 24 \cdot f(n)$ for all $n$.

**Example 2.2.3.** Let $f(n) = n^2 \log n$. Then $f \in \mathcal{O}(n^3)$.

Indeed, it is well known that if $\alpha$ is a positive real number, then

$$\lim_{x \to \infty} \frac{\log x}{x^\alpha} = 0.$$

Hence, there exists $c$ such that $\log n \leq n$ for all $n > c$.

The symbol $\mathcal{O}$ is suitable to express properties regarding the *asymptotic behaviour* of functions. In what follows, we shall assume that all the functions studied are defined in the same domain $S$, implying, when necessary, that we are redefining them taking as domain the intersection of their domains, so as to be able to talk about sum and product of functions, and so on.

Consider two functions $f$ and $g$; the following table summarises the possible cases (see Exercises A2.3, A2.2, A2.4):

| | |
|---|---|
| $\exists$ a finite $\displaystyle\lim_{n \to \infty} \frac{f(n)}{g(n)} = a \neq 0$ | $f \in \mathcal{O}(g)$ e $g \in \mathcal{O}(f)$, that is, $f$ and $g$ are *of the same order;* |
| $\displaystyle\lim_{n \to \infty} \frac{f(n)}{g(n)} = \infty$ | $g \in \mathcal{O}(f)$, that is, $f$ dominates $g$, and we say that $f$ is of *higher order* than $g$; |
| $\displaystyle\lim_{n \to \infty} \frac{f(n)}{g(n)} = 0$ | $f \in \mathcal{O}(g)$, that is, $g$ dominates $f$, and we say that $g$ is of *higher order* than $f$; |
| $\nexists \displaystyle\lim_{n \to \infty} \frac{f(n)}{g(n)}$ | we cannot say anything in general, neither that $g \in \mathcal{O}(f)$, nor that $g \notin \mathcal{O}(f)$. |

We give some more examples.

**Example 2.2.4.** (1) If $f(n) = n^2 + 1$ and $g(n) = n^3$, then $f \in \mathcal{O}(g)$, as $\lim_{n \to \infty} (n^2 + 1)/n^3 = 0$.

(2) Let $g(n) = (2 + (-1)^n)n^2$. Clearly, we have $g \in \mathcal{O}(n^2)$ and $f = n^2 \in \mathcal{O}(g)$, while the ratio

$$\frac{(2 + (-1)^n)n^2}{n^2} = 2 + (-1)^n$$

has no limit.

(3) Consider the functions $f(n) = (n+1) + n(-1)^n$ and $g(n) = (n+1) - n(-1)^n$. Then the ratios $f/g$ and $g/f$ have no limit. Moreover, $g \notin \mathcal{O}(f)$ and $f \notin \mathcal{O}(g)$.

Indeed, the functions $f$ and $g$ are such that

$$f(n) = \begin{cases} 2n + 1 & \text{if } n \text{ is even,} \\ 1 & \text{if } n \text{ is odd,} \end{cases} \qquad g(n) = \begin{cases} 1 & \text{if } n \text{ is even,} \\ 2n + 1 & \text{if } n \text{ is odd,} \end{cases}$$

and this entails that neither function dominates the other.

The following result gives some useful properties of the symbol $\mathcal{O}$ with respect to the main algebraic operations on functions.

**Theorem 2.2.5.** *If* $f \in \mathcal{O}(g)$ *and* $c > 0$, *then* $c \cdot f \in \mathcal{O}(g)$. *Moreover, if* $f_1 \in \mathcal{O}(g_1)$ *and* $f_2 \in \mathcal{O}(g_2)$, *then*

$$f_1 + f_2 \in \mathcal{O}(g_1 + g_2), \qquad\qquad f_1 \cdot f_2 \in \mathcal{O}(g_1 \cdot g_2).$$

*In particular, if* $f_1 \in \mathcal{O}(g)$ *and* $f_2 \in \mathcal{O}(g)$, *then* $f_1 + f_2 \in \mathcal{O}(g)$.

PROOF. As an example, we just prove that if $f_1 \in \mathcal{O}(g_1)$ and $f_2 \in \mathcal{O}(g_2)$, then $f_1 + f_2 \in \mathcal{O}(g_1 + g_2)$.

Let $S$ be the domain of $f_1$ and $f_2$. As $f_1 \in \mathcal{O}(g_1)$ and $f_2 \in \mathcal{O}(g_2)$, there exist constants $k_1$, $k_2$, $c_1$, $c_2$ such that

$$f_1(x) \le k_1 \cdot g_1(x), \qquad \text{for all } x \in S, \text{ with } x > c_1,$$
$$f_2(x) \le k_2 \cdot g_2(x), \qquad \text{for all } x \in S, \text{ with } x > c_2.$$

If we define $k = \max(k_1, k_2)$ and $c = \max(c_1, c_2)$, then clearly

$$(f_1 + f_2)(x) \le k \cdot (g_1 + g_2)(x), \qquad \text{for all } x \in S, \text{ with } x > c,$$

proving the claimed formula.

In the case where $f_1 \in \mathcal{O}(g)$ and $f_2 \in \mathcal{O}(g)$, then $f_1 + f_2 \in \mathcal{O}(2 \cdot g)$ and, by the first claim of Theorem 2.2.5, $2 \cdot g \in \mathcal{O}(g)$; hence the result follows.  $\square$

**Example 2.2.6.** Let $f_1(n) = n + 3 \log n$ and $f_2(n) = 5n \log n + 28n^3$. Then $f_1 \cdot f_2 \in \mathcal{O}(n^4)$, as $n + 3 \log n \in \mathcal{O}(n)$ and $5n \log n + 28n^3 \in \mathcal{O}(n^3)$.

Notice that *estimates of $\mathcal{O}$ type* are used in everyday life too, whenever, rather than needing a precise information, we can content ourselves with knowing its *order of magnitude*. For instance, we might say that we have to walk for a dozen metres, without implying that the metres are exactly 12, but rather 10, 20 or a similar amount. If we have to go from London to Manchester, we might say that we must travel hundreds of kilometres, and so forth.

So it is natural to look for estimates of $\mathcal{O}$ type for the complexity of an algorithm too. In this case, saying that an algorithm depending on $n$ data has a complexity given by the function $f(n)$ with $f \in \mathcal{O}(g)$, means that, for $n$ big enough, the number of bit operations necessary to execute the algorithm never exceeds $k \cdot g(n)$ for some constant $k$. As the computational time is proportional to computational complexity, neither does the time exceed $k \cdot g(n)$ for some constant $k$. In general one looks for an easy way to compute the reference function $g(n)$, which clearly is not uniquely determined. For instance, if we have evaluated the complexity of an algorithm by the function $n^4 + 8n^2 + 5n + 10$, it is natural to choose as reference function the function $g = n^4$ and to say that the complexity of the algorithm is $\mathcal{O}(n^4)$.

## 2.3 Polynomial time, exponential time

In this section we start by discussing the fundamental notion, touched upon at the end of § 2, of the dependence of the complexity of an algorithm on the integers on which it operates.

As we mentioned, the larger the integers on which a given algorithm works, the larger the expected number of bit operations necessary to execute the algorithm, and so the larger its complexity. In other words, it is to be expected that the complexity of an algorithm is a function of the size of the numbers on which it operates. To attribute a mathematical meaning to this idea, it is useful to introduce a new notion, the *length* of a number, which basically is a measure of the size of a number.

**Definition 2.3.1.** *Let $n$ be an integer number and $\beta$ an integer number greater than 1. Call* length $L_\beta(n)$ *of $n$ in base $\beta$ the number $k$ such that the following holds:*

$$\beta^{k-1} \le n < \beta^k, \tag{2.2}$$

*that is to say, $n = \sum_{i=0}^{k-1} a_i \beta^i = (a_{k-1}, \ldots, a_0)_\beta$, with $a_{k-1} \ne 0$.*

*By $L(n)$ without subscript we shall mean the length of the number $n$ in base 2, that is $L(n) = L_2(n)$.*

**Example 2.3.2.** We have $L_{10}(205) = 3$, while $L_2(205) = 8$, as $205 = (11001101)_2$.

In particular, by Equation (2.2), taking logarithms, we get

$$k - 1 \le \log_\beta n < k$$

and so
$$L_\beta(k) = [\log_\beta n] + 1 \quad \text{and} \quad L_\beta(n) \in \mathcal{O}(\log_\beta n).$$

This can be rephrased by saying that the length of an integer $n$ in a fixed base $\beta$, that is the number of digits necessary to write the number in that base, is of size $\log_\beta n$.

Although the length of a number obviously depends on the base in which it is expressed, it is important to notice that it is possible to give a $\mathcal{O}$-type estimate of the length, independently of the base. Indeed, recall the well-known relation between logarithms in different bases

$$\log_\beta n = \frac{\log_\alpha n}{\log_\alpha \beta}.$$

Thus, if by $\log n$ without any subscript we denote the natural logarithm, that is the logarithm to the base $e$, we obtain

$$\log_\beta n = \frac{\log n}{\log \beta}.$$

Using this relation, we get

$$L_\beta(n) = \left[\frac{\log n}{\log \beta}\right] + 1; \tag{2.3}$$

hence, assuming the base $\beta$ in which we are expressing numbers is fixed once and for all, and so is a constant, we have

$$L_\beta(n) \in \mathcal{O}(\log n). \tag{2.4}$$

Suppose now we have an algorithm $\mathcal{A}$ operating on integers $n_1, \ldots, n_s$, and denote by $T(\mathcal{A})$ the complexity of $\mathcal{A}$, where the letter $T$ stands for *time*. Notice that $T(\mathcal{A})$ is approximated from above by a function of the *greatest length* of the integers $n_1, \ldots, n_s$ in the base in which we are operating. Thus, if $n$ is greater than all of $n_1, \ldots, n_s$, we may approximate from above $T(\mathcal{A})$ by a function of $\log n$.

We can now come back to another important question mentioned in § 2. When performing calculations, whatever instrument we use, it is necessary to assess the time the algorithms require for their execution: some of them require a *reasonable* time, while others do not possess this property. It is now possible to give a precise mathematical meaning to this claim.

Having this in mind, it is useful to recall that, asymptotically, a polynomial function *grows far more slowly* than an exponential function. In other words, if we have a function $f(n) \in \mathcal{O}(n^k)$, we may say that $f(n) \in \mathcal{O}(a^n)$, independently on the positive integer $k$ and the real number $a > 1$ (see Table 2.1 on page 111). The following two definitions are based on this simple remark; they are of fundamental importance in the analysis of algorithms.

**Definition 2.3.3.** *An algorithm $\mathcal{A}$ to perform a calculation on integer num-bers is said to be a* polynomial time *algorithm, or simply a* polynomial algo-rithm, *if there is a positive integer $d$, called the* order *of the algorithm, such that the number of bit operations required to execute the algorithm on integers of binary length smaller than or equal to $k$ is $\mathcal{O}(k^d)$.*

From what precedes we have then that $\mathcal{A}$ is polynomial if $T(\mathcal{A}) \in \mathcal{O}((\log n)^d)$, where $n$ is the greatest integer on which $\mathcal{A}$ operates.

**Definition 2.3.4.** *An algorithm $\mathcal{A}$ is said to be an* exponential time *algo-rithm, or simply an* exponential *algorithm, if the number of bit operations required to execute the algorithm on integers of binary length smaller than or equal to $k$ is the same order as $e^{ck}$, for some constant $c > 0$ (see page 90).*

*An algorithm that is not exponential is said to be* subexponential *if the number of bit operations required to execute the algorithm on integers of binary length smaller than or equal to $k$ is $\mathcal{O}(e^k)$.*

For instance, $\mathcal{A}$ is exponential if $T(\mathcal{A})$ has the same order as $n$, where $n$ is the greatest integer on which $\mathcal{A}$ operates. On the contrary, an algorithm requiring $f(k) = e^{k/\log(\log k)}$ operations is subexponential. Indeed,

$$\lim_{k \to \infty} \frac{f(k)}{e^k} = 0;$$

hence $f(k) \in \mathcal{O}(e^k)$ while, for all $c > 0$, $e^{ck} \notin \mathcal{O}(f)$, and so it is not exponen-tial.

A typical example of an exponential algorithm is the so-called *sieve of Eratosthenes* everybody knows since primary school. It is used to determine whether a number is prime, as well as to *factor* an integer. We shall take up this subject again in Chapter 4.

Basically, the difference between a polynomial time algorithm and an ex-ponential time one is the following. Suppose we have to work with very large numbers, otherwise we might perform mentally or by hand our calculations and we would not need to make too many distinctions. Then, if an algorithm is polynomial, we may expect it to be feasible, if not by hand, by using suitable computing devices, in a reasonable time. On the other hand, if the algorithm is exponential, we cannot in general obtain the answer in a reasonable time, even by using the most powerful computers: the whole life span of the uni-verse might not be enough! In other words, polynomial algorithms are *feasible*, while exponential ones are not.

Notice that there are other kinds of complexity too, such as *factorial com-plexity*, including algorithms requiring a time of the order of $k!$, where $k$ is the length of the greatest integer the algorithm operates on.

**Example 2.3.5.** Let $t$ be the time it takes to perform a single bit operation. If an algorithm is polynomial of order 2, and if we want it to work with rather large numbers, for instance in the order of $10^{100}$, the time required to perform the algorithm will be bounded by $10000 \cdot C \cdot t$, where $C$ is a suitable constant. Notice that $t$ depends on the computing power of the device we use and may be made very small. For instance, it is realistic to assume $t = 10^{-9}$ seconds, and so $10000 \cdot C \cdot t = 10^{-5} \cdot C$: this shows what is meant by the sentence "the time required to execute the algorithm is reasonable". However, notice that the larger the order of the algorithm, the smaller is its convenience. For instance, if the algorithm were of order 10, an estimate of the time required to execute it on numbers of the order of $10^{100}$ would be of $10^{11} \cdot C$ seconds, and unless $C$ is tiny this is a substantial amount of time (see Exercise B2.15 and the table 2.1 on page 111).

If on the other hand we have an exponential algorithm and we want it to work on numbers of the order of $10^{100}$, the time required will be of the order of $10^{100} \cdot t$. Even if we were able to dramatically improve our computers' performances, arriving for instance to $t = 10^{-20}$, we would have $10^{100} \cdot t = 10^{80}$, and this is an enormous number: in order to get an idea of its magnitude, recall that the number of protons in the universe is a number with 79 decimal digits. Moreover, it is clear that, no matter how we improve our computers' power, an exponential algorithm will never allow us to work with very large numbers.

However, it must be emphasised that even an exponential algorithm may be feasible, and sometimes even better than a polynomial one, if applied to small numbers. A typical example is the above mentioned sieve of Eratosthenes, which we used at school to factor small numbers.

## 2.4 Complexity of elementary operations

We shall now study the complexity of elementary operations. The complexity of any algorithm operating on integers may be obtained by applying the results from this section.

Recall the notation $T(\mathcal{A})$ to denote the complexity of an algorithm $\mathcal{A}$. For instance, we shall denote by $T(a + b)$ $[T(a - b), T(a \cdot b), T(a/b)$ and so on, respectively] the complexity of the algorithm yielding the sum [the difference, the product, the quotient and so on, respectively] of two integers $a$ and $b$.

**Proposition 2.4.1.** *Let $a$ and $b$ be two positive integers*

$$a = \sum_{i=0}^{k-1} a_i \cdot 2^i = a_{k-1} \dots a_0, \qquad b = \sum_{i=0}^{l-1} b_i \cdot 2^i = b_{l-1} \dots b_0$$

*of length $L(a) = k$ and $L(b) = l$, respectively, with $k \geq l$. Then the following relations hold:*

$$L(a + b) = \begin{cases} \max(L(a), L(b)) & \text{or} \\ \max(L(a), L(b)) + 1, \end{cases} \tag{2.5}$$

$$L(a \cdot b) = \begin{cases} L(a) + L(b) = k + l & \text{or} \\ L(a) + L(b) = k + l - 1, \end{cases} \tag{2.6}$$

$$T(a \pm b) \in \mathcal{O}(k), \tag{2.7}$$

$$T(a \cdot b) \in \mathcal{O}(L(a) \cdot L(b)) = \mathcal{O}(k \cdot l), \tag{2.8}$$

$$T\left(\frac{a}{b}\right) \in \mathcal{O}(L(a) \cdot L(b)) = \mathcal{O}(k \cdot l). \tag{2.9}$$

PROOF. We have the following inequalities:

$$2^{k-1} \le 2^{k-1} + 2^{l-1} \le a + b < 2^k + 2^l \le 2^{k+1},$$
$$2^{k+l-2} \le a \cdot b < 2^{k+l}.$$

Hence the estimates for $L(a + b)$ and for $L(a \cdot b)$ immediately follow. Further, let $q$ and $r$ be the quotient and the remainder of the division of $a$ by $b$, that is, $a = qb + r$, with $0 \le r < b$.

If $b > a$, then $L(q) = 0$, while if $b \le a$, then $L(a) = L(qb + r)$. Thus, from what precedes we immediately get

$$L(a) - L(b) - 1 \le L(q) \le L(a) - L(b) + 1.$$

Let us now compute the complexity of elementary operations.

By adding some zero digits in front of the smaller number if necessary, we may assume that $a$ and $b$ have the same number of digits even if they have different length. That is, we may write $b = \sum_{i=0}^{k-1} b_i \cdot 2^i = b_{k-1} \dots b_0$, with $b_i = 0$ if $k \ge l$. Recall that the sum is given by

$$a + b = \sum_{i=0}^{k-1} (a_i + b_i) \cdot 2^i$$

and so in order to compute the sum we have to compute $a_i + b_i$, for $0 \le i \le k - 1$. The calculation of $a_i + b_i$ corresponds to a bit operation, and so to compute $a + b$ we have to perform at least $k$ bit operations. Moreover, we have to consider the possible carries, as we know from § 1.4.3. This increases the number of bit operations required, but not to a great extent. Indeed, for each $i = 1, \dots, k$, we have to perform two bit operations: the sum $a_i + b_i$ and the sum of the number so obtained with the carry, possibly equal to zero, obtained in the previous sum $a_{i-1} + b_{i-1}$. In conclusion, to compute $a + b$ we have to perform at most $2k$ bit operations. Notice that the operation consisting of remembering the *carry* is not considered a bit operation , as it is an operation of *data storing*, which is negligible with respect to the execution of a bit operation. In conclusion, we have

$$T(a + b) \in \mathcal{O}(k) = \mathcal{O}(\max(L(a), L(b)).$$

It can be seen analogously (see Exercise A2.12) that

$$T(a - b) \in \mathcal{O}(k) = \mathcal{O}(\max(L(a), L(b)).$$

Suppose now we want to multiply $a$ and $b$. Recall from § 1.4.3 that, in order to compute this multiplication, it is necessary to perform the sum of $l$ numbers, obtained from $a$ by multiplying it by the successive digits of $b$ and by progressively moving leftwards the digits in the results, inserting to the right the same number of zeros. So we have to carry out $l - 1$ sums of numbers of length at most $k + l - 1$. However, all these numbers end in a sequence of zeros, whose addition to the number that precedes does not contribute any bit operation. Thus, the number of required bit operations can be estimated by the number of operations required in order to carry out the sum of $l - 1$ numbers of length $k$. Hence, by what has already been shown, we deduce that

$$T(a \cdot b) \in \mathcal{O}((l - 1)k) = \mathcal{O}(lk) = \mathcal{O}(L(a) \cdot L(b)).$$

Let us now consider the question of giving an estimate for the computational complexity of division, that is to say for finding the quotient $q$ and the remainder $r$ of the division of $a$ by $b$. The situation is analogous to what happens for multiplication.

Notice that when we carry out divisions in base two there is no need to proceed by trial and error: at each step the corresponding digit in the quotient is 0 or 1 depending on whether the divisor is greater than the dividend or not: the computer only has to compare the two numbers and this requires a time negligible with respect to a bit operation.

If $a < b$, then the quotient $q$ is zero and $r = a$.

So, suppose $a \geq b$. In order to find $q$ and $r$ we have to compute the first digit of $q$, which is necessarily 1, multiply $1 \cdot b$ (an operation whose complexity is $\mathcal{O}(l)$), and then subtract the result from the number consisting of the first $l$ digits of $a$ (again, an operation whose complexity is $\mathcal{O}(l)$). So the time necessary for this first step is $\mathcal{O}(2l)$. How many times is this step iterated? As many as the length of $q$ which, as already shown, is at most $k - l + 1$. In conclusion, we have

$$T\left(\frac{a}{b}\right) \in \mathcal{O}(2l(k - l + 1)) = \mathcal{O}(L(a) \cdot L(b)),$$

as $k = L(a) \geq L(b) = l$.    □


## 2.5 Algorithms and complexity

In this section we shall determine the computational complexity of some algorithms already described in Chapter 1 and we shall introduce some new algorithms, computing their complexity as well. Further algorithms shall be outlined in the exercises.

### 2.5.1 Complexity of the Euclidean algorithm

We are now going to estimate the complexity of the Euclidean algorithm $GCD_E(a, b)$ for the determination of the greatest common divisor of two integers $a$ and $b$. When we examined the Euclidean algorithm in section 1.3, we remarked that it terminates after at most $b$ steps. Using this estimate for the number $\text{div}(a, b)$ of divisions that are necessary to execute the algorithm, we would find ourselves in a mess, as we should deduce that we are dealing with an exponential time algorithm, and so an unfeasible one, its complexity being

$$\mathcal{O}(bL(a)^2) = \mathcal{O}(2^{\log_2 b} L(a)^2) = \mathcal{O}(2^{L(b)} L(a)^2),$$

where $\mathcal{O}(L(a)^2)$ is, by Proposition 2.4.1, an estimate for the cost of each division.

So we have to improve our estimate of the number $\text{div}(a, b)$. We shall need the following lemma, which determines $\text{div}(a, b)$ in the particular case in which the numbers $a$ and $b$ are two consecutive Fibonacci numbers.

**Lemma 2.5.1.** *Let $f_n$ be the $n$th Fibonacci number. The number of divisions that are necessary to compute $\text{GCD}(f_{n+1}, f_n)$ by using the Euclidean algorithm is $n - 1$.*

PROOF. We have

$$
\begin{aligned}
f_{n+1} &= f_n \cdot 1 + f_{n-1}, & 0 &< f_{n-1} < f_n, \\
f_n &= f_{n-1} \cdot 1 + f_{n-2}, & 0 &< f_{n-2} < f_{n-1}, \\
f_{n-1} &= f_{n-2} \cdot 1 + f_{n-3}, & 0 &< f_{n-3} < f_{n-2}, \\
&\;\;\vdots & & \\
f_4 &= f_3 \cdot 1 + f_2, & 0 &< f_2 < f_3, \\
f_3 &= f_2 \cdot 2 + f_0. & &
\end{aligned}
$$

As $f_0 = 0$, we have $\text{GCD}(f_{n+1}, f_n) = f_2 = 1$ and so there are exactly $n - 1$ divisions to be carried out to compute the greatest common divisor.  □

The Euclidean algorithm we have described has as its input the pair $(f_{n+1}, f_n)$ and then executes $n - 1$ steps, as the following diagram shows

$$(f_{n+1}, f_n) \to (f_n, f_{n-1}) \to \cdots \to (f_3, f_2).$$

Any other choice of a pair of integers $(a, b)$, with $a \geq b$, for which the Euclidean algorithm terminates after $n$ steps is bounded from below by the corresponding pair $(f_{n+2}, f_{n+1})$ of elements of Fibonacci sequence. This is the gist of the following theorem, due to Lamé:

**Theorem 2.5.2 (Lamé).** *Let $a$ and $b$ be two positive integers with $0 < b < a$ and let $d = \text{GCD}(a, b)$. If the Euclidean algorithm to calculate $\text{GCD}(a, b)$ terminates after $n$ steps, then $a \geq d \cdot f_{n+2}$ and $b \geq d \cdot f_{n+1}$.*

PROOF. The basis of the induction is true because if $n = 1$ then the Euclidean algorithm yields a single step: $a = qb + 0$, that is to say $b \mid a$, and so

$$\text{GCD}(a, b) = b = d = d \cdot 1 = d \cdot f_2.$$

Moreover, we have $a = qb \geq 2b = 2d = d \cdot f_3$. Assume now that the result is true when the algorithm terminates after at most $n - 1$ steps. For $n$ steps we have

$$
\begin{aligned}
a &= b\, q_1 + r_1, & 0 < r_1 < b, \\
b &= r_1\, q_2 + r_2, & 0 < r_2 < r_1, \\
r_1 &= r_2\, q_3 + r_3, & 0 < r_3 < r_2, \\
&\ \ \vdots \\
r_i &= r_{i+1}\, q_{i+2} + r_{i+2}, & 0 < r_{i+2} < r_{i+1}, \\
&\ \ \vdots \\
r_{n-3} &= r_{n-2}\, q_{n-1} + r_{n-1}, & 0 < r_{n-1} < r_{n-2}, \\
r_{n-2} &= r_{n-1}\, q_n + 0.
\end{aligned}
$$

Apply the induction hypothesis to the pair $(b, r_1)$, noticing that $\text{GCD}(b, r_1) = d = \text{GCD}(a, b)$. We obtain

$$b \geq d f_{n+1}, \qquad r_1 \geq d f_n,$$

hence

$$a = r_1 + b q_1 \geq r_1 + b \geq d(f_{n+1} + f_n) = d f_{n+2},$$

which is what was to be proved. □

The above theorem makes it possible to determine a good upper bound for $\text{div}(a, b)$.

**Proposition 2.5.3.** *Let $a$ and $b$ be two positive integers with $0 < b < a$. Then we have*

$$\text{div}(a, b) \leq 5L_{10}(b). \tag{2.10}$$

PROOF. From Theorem 2.5.2 it can be immediately deduced that for all pairs of positive integers $a$ and $b$ such that $a > b$ and $b < f_{n+1}$, the number of divisions necessary in order to get a zero remainder when applying the Euclidean algorithm is smaller than $n$.

Let then $n = \text{div}(a, b)$, and so $b \geq f_{n+1}$. Recall that if $\lambda_1 = (1 + \sqrt{5})/2$ then $f_{n+1} > \lambda_1^{n-1}$, for all $n \geq 3$ (see Exercise A1.33). Thus, from $b \geq f_{n+1}$ it follows that $b > \lambda_1^{n-1}$. As $\log_{10} \lambda_1 > 1/5$, we have that

$$\log_{10} b > (n - 1) \log_{10} \lambda_1 > \frac{n - 1}{5}$$

and so
$$n - 1 < 5 \log_{10} b.$$
By the definiton of length (2.2) we have $b < 10^{L_{10}(b)}$: so
$$n - 1 < 5L_{10}(b),$$
and (2.10) follows.    □

With the bound on the number on divisions given by Equation (2.10), the state of things is distinctly improved with respect to the previous situation which gave an exponential estimate for complexity. Indeed, now we have:

**Proposition 2.5.4.** *Let a and b be two positive integers with* $0 < b < a$. *Then*
$$T(GCD_E(a, b)) \in \mathcal{O}(L(a)^3).$$

PROOF. We have
$$T(GCD_E(a, b)) \in \mathcal{O}(\log b \log^2 a) = \mathcal{O}(\log^3 a) = \mathcal{O}(L(a)^3),$$
as $\mathcal{O}(\log b)$ gives an estimate of the complexity with regard to the *number* of divisions, while $\mathcal{O}(\log^2 a)$ estimates the complexity of *each* division.    □

This shows that the complexity of the Euclidean algorithm is *polynomial*. Let us see now an example.

**Example 2.5.5.** Give an estimate for the number of divisions to complete the Euclidean algorithm to search the GCD of the two numbers $a = 34567$ and $b = 457$.

The number $b$ has 3 digits, and so $L_{10}(b) = 3$. Then we are confident that we can bring to an end the Euclidean algorithm after at most $5 \cdot 3 = 15$ divisions. Let us see how many divisions are actually necessary:

$$34567 = 457 \cdot 75 + 292,$$
$$457 = 292 \cdot 1 + 165,$$
$$292 = 165 \cdot 1 + 127,$$
$$165 = 127 \cdot 1 + 38,$$
$$127 = 38 \cdot 3 + 13,$$
$$38 = 13 \cdot 2 + 12,$$
$$13 = 12 \cdot 1 + 1,$$
$$12 = 1 \cdot 12 + 0.$$

So 8 divisions have been sufficient, rather than 15, in order to verify that $GCD(34567, 457) = 1$.

Keeping in mind the remarks in § 1.3.3 and the estimate (2.5.4) we may conclude (see Exercise A2.13) that:

**Corollary 2.5.6.** *Let $a$ and $b$ be two positive integers with $0 < b < a$. The complexity of computing a Bézout's relation of the form (1.14) by means of the Euclidean algorithm is $\mathcal{O}(L(a)^3)$.*

Analogously, we have (see again Exercise A2.13):

**Corollary 2.5.7.** *The complexity of the algorithm to find the solution of a linear Diophantine equation of the form $ax + by = c$ by means of the algorithm given in the proof of Proposition 1.3.11 is $\mathcal{O}(k^3)$, where $k$ is the greatest length of the integers $a$, $b$, $c$.*

### 2.5.2 From binary to decimal representation: complexity

Recall that in Section 1.4 we described how the coefficients of an integer with respect to an arbitrary base can be found. In particular, if our base is 10, and $d$ is a positive integer such that $d = (d_{k-1} \ldots d_0)_{10}$, then, by denoting by $q$ the quotient and by $r$ the remainder of the divisions of $d$ by 10, we have that $d_0$ is the remainder $r$, $d_1$ is the remainder of the division of $q$ by 10 and so on.

As $10 = (1010)_2$, the division of $d$ by 10, when working in base 2, has complexity $\mathcal{O}(4L_2(d)) = \mathcal{O}(\log d)$. In this way we get the remainder $r = d_0$ in binary form, that is to say, one of the numbers in $\{0, 1, \ldots, 9\}$, in binary representation. Now repeat the procedure, dividing the quotient $q$ by 10. This operation has again complexity estimated by $\mathcal{O}(\log d)$. So we determine $d_1$, that is the remainder of the division, and the quotient becomes the starting point to compute $d_2$. This procedure is to be iterated a number of times equal to the length of $d$ in base 10, that is the number of decimal digits of $d$. This number, by Equation (2.4), is $L_{10}(d) \in \mathcal{O}(\log d)$.

In conclusion, the complexity of this algorithm to convert $d$ from its binary representation to the decimal one is

$$\mathcal{O}(L(d)^2) = \mathcal{O}(\log^2 d).$$

By following the same reasoning, the following can be verified (see Exercise A2.14):

**Proposition 2.5.8.** *Let $\alpha$ and $\beta$ two integers greater than 1. There is an algorithm to convert its representation in base $\alpha$ of a number $n$ to the representation in base $\beta$ having complexity $\mathcal{O}(\log^2 n)$.*

### 2.5.3 Complexity of operations on polynomials

We frequently work with polynomials and with operations on them. So it is useful to give an estimate of the complexity of operations on polynomials with integer coefficients. The following holds.

**Proposition 2.5.9.** *Let $f(x)$ and $g(x)$ be two polynomials with integer coefficients, of degree $n$ and $m$, with $n \geq m$; assume that their coefficients have length at most $k$. Then:*

(a) *computing the sum or the difference of $f(x)$ and $g(x)$ has complexity $\mathcal{O}(mk)$;*

(b) *computing the product of $f(x)$ and $g(x)$ has complexity $\mathcal{O}(m(n+m)(k^2 + \log m))$.*

PROOF. To carry out the sum [or the difference, respectively] of $f(x)$ and $g(x)$ it is necessary to complete $m$ additions or subtractions. Hence, from Proposition 2.4.1 claim (a) immediately follows.

If

$$f(x) = a_0 + a_1 x + a_2 x^2 + \cdots + a_n x^n, \quad g(x) = b_0 + b_1 x + b_2 x^2 + \cdots + b_m x^m,$$

then the product $f(x) \cdot g(x)$ has degree $n + m$ and the coefficient of $x^h$ is $\sum_{i+j=h} a_i b_j$. In order to compute this coefficient we need at most $m + 1$ multiplications among numbers of length at most $k$ and at most $m$ additions of numbers of length at most $2k$. As we have to perform $m$ such additions, the length of the summands may be bounded by $2k + L(m)$. As a consequence, the complexity we are determining is $\mathcal{O}((n+m+1)((m+1)k^2 + m(2k + \log m + 1)))$, hence claim (b) holds. □

It is also useful to determine the number of operations which are necessary to carry out operations on polynomials on an arbitrary field $\mathbb{K}$ in which it is meaningful to define the computational cost of the operations. More precisely, assume the following:

- computing the sum or the difference of two elements in $\mathbb{K}$ has complexity $\mathcal{O}(u)$;
- computing the product of two elements in $\mathbb{K}$ has complexity $\mathcal{O}(t)$, with $t \geq u$;
- computing the inverse of an element in $\mathbb{K}$ has complexity $\mathcal{O}(s)$, with $s \geq t$.

We shall see later that finite fields are examples of this state of things (see Chapter 5).

Thus, we have:

**Proposition 2.5.10.** *Let $\mathbb{K}$ be a field in which the above conditions hold. If $f(x) = a_0 + a_1 x + \cdots + a_n x^n$ and $g(x) = b_0 + b_1 x + \cdots + b_m x^m$ are polynomials on $\mathbb{K}$, with $n \geq m$, then:*

- *the complexity of computing $f(x) \pm g(x)$ is $\mathcal{O}(nu)$;*
- *the complexity of computing $f(x) \cdot g(x)$ is $\mathcal{O}(n^2 t)$;*
- *the complexity of computing the division of $f(x)$ by $g(x)$ is $\mathcal{O}(n^2 s)$;*
- *the complexity of computing the greatest common divisor of $f(x)$ and $g(x)$ is $\mathcal{O}(n^3 s)$;*

- *the complexity of computing a Bézout's relation for the above greatest common divisor is $\mathcal{O}(n^3 s)$.*

PROOF. The complexities of the sum, the difference, and the product of polynomials are computed as in Proposition 2.5.9 and so are omitted. Let us fix our attention on computing the remaining complexities.

From the proof of Theorem 1.3.17 it follows that, in order to carry out the division of $f(x)$ by $g(x)$, it is necessary to perform the following operations:

- compute the inverse of $b_m$;
- compute $g(x)/b_m$, which requires $m$ multiplications of the coefficients of $g(x)$ that are different from $b_m$, by $1/b_m$;
- multiply $g(x)/b_m$ by the leading coefficient of the polynomial to be divided, which requires $m+1$ multiplications of elements of $\mathbb{K}$; this is to be repeated $n - m + 1$ times;
- carry out a difference of polynomials of degree at most $n$, which requires at most $n+1$ subtractions in $\mathbb{K}$; this too is to be repeated $n - m + 1$ times.

In conclusion, the division we are studying has complexity $\mathcal{O}(s + mt + (n - m + 1)(m + 1)t + (n - m + 1)(n + 1)u)$, which, keeping in mind the hypothesis $s \geq t \geq u$, can be estimated by $\mathcal{O}(2(n + 1)^2 s)$; hence the thesis.

In order to compute the greatest common divisor, the division is to be repeated at most $m$ times; so the complexity can be estimated by $\mathcal{O}(n^3 s)$.

In order to compute a Bézout's relation, after the Euclidean algorithm its steps are again considered backwards as in the case of integers, as seen in § 1.3.3. This requires at most $m$ multiplications and additions of polynomials of degrees at most $m$. Hence the claim easily follows.     □

### 2.5.4 A more efficient multiplication algorithm

The algorithms we have described can often be improved, as they can be made more efficient by means of some clever idea. As an example, we describe in this section an algorithm which substantially reduces the number of bit operations necessary to compute the product of $a$ and $b$, where $a$, $b$ are two positive integers of length an *even* number $2n$, that is to say

$$a = (a_{2n-1} \ldots a_0)_2 \quad \text{and} \quad b = (b_{2n-1} \ldots b_0)_2.$$

Notice that this hypothesis does not actually impose any restriction. Write

$$a = A_1 \cdot 2^n + A_0 \quad \text{with } A_1 = (a_{2n-1} \ldots a_n)_2, \ A_0 = (a_{n-1} \ldots a_0)_2,$$
$$b = B_1 \cdot 2^n + B_0 \quad \text{with } B_1 = (b_{2n-1} \ldots b_n)_2, \ B_0 = (b_{n-1} \ldots b_0)_2.$$

So

$$a \cdot b = (A_1 \cdot 2^n + A_0)(B_1 \cdot 2^n + B_0) =$$
$$= A_1 B_1 \cdot 2^{2n} + (A_1 B_0 + A_0 B_1) \cdot 2^n + A_0 B_0. \tag{2.11}$$

**Example 2.5.11.** If $a = 2^3 + 2 + 1$, then $a$ is a 4-bit number; hence $a = A_1 \cdot 2^2 + A_0$, where $A_1 = (1, 0)$ and $A_0 = (1, 1)$.

So the problem reduces to carrying out the following operations on $n$-bit integers

$$A_1 B_1, \qquad A_1 B_0 + A_0 B_1, \qquad A_0 B_0.$$

Actually, Equation (2.11) yields

$$a \cdot b = A_1 B_1 (2^{2n} + 2^n) + (A_1 - A_0)(B_0 - B_1) 2^n + A_0 B_0 (2^n + 1), \qquad (2.12)$$

so the problem reduces to computing the three products on $n$-bit integers

$$A_1 B_1, \qquad (A_1 - A_0)(B_0 - B_1), \qquad A_0 B_0,$$

in addition, evidently, to some additions and digit translations.

Denote by $M(n)$ the number of bit operations that are necessary to compute the product of two positive $n$-bit integers. Then, by Equation (2.12), we have

$$M(2n) \le 3 \cdot M(n) + C_0 \cdot n, \qquad (2.13)$$

because the number of additions and translations necessary to compute $a \cdot b$ by Equation (2.11) is of class $\mathcal{O}(n)$. Now we can prove the following.

**Lemma 2.5.12.** *Set* $C = \max(M(2), C_0)$; *then* $M(2^k) \le (3^k - 2^k) \cdot C$.

PROOF. Prove the formula by induction on $k$. By the definition of $C$, we have

$$M(2) \le \max(M(2), C_0) = C = (3^1 - 2^1) \cdot C,$$

so the theorem is true for $k = 1$. Assume the theorem is true for $k$ and prove it for $k + 1$. We have

$$M(2^{k+1}) = M(2 \cdot 2^k) \le 3 \cdot M(2^k) + C_0 \cdot 2^k \le 3 \cdot (3^k - 2^k)C + C \cdot 2^k =$$
$$= (3^{k+1} - 3 \cdot 2^k + 2^k) \cdot C = (3^{k+1} - 2^{k+1}) \cdot C,$$

where the first inequality is a consequence of Equation (2.13), while the second one is true by the induction hypothesis and because $C \ge C_0$. $\qquad \square$

Thus, for $n = 2^k$ we obtain

$$M(n) = M(2^k) \le M(2^{\lceil \log_2 n \rceil + 1}) \le (3^{\lceil \log_2 n \rceil + 1} - 2^{\lceil \log_2 n \rceil + 1}) \cdot C \le$$
$$\le 3^{\lceil \log_2 n \rceil + 1} \cdot C = 3C \cdot 3^{\lceil \log_2 n \rceil} \le 3C \cdot 3^{\log_2 n} = 3C \cdot n^{\log_2 3},$$

because obviously the equality $3^{\log_2 n} = n^{\log_2 3}$ holds.

So we may conclude:

**Proposition 2.5.13.** *The complexity of the above algorithm for computing the product of two $n$-bit integers, with $n = 2^k$, is $\mathcal{O}(n^{\log_2 3})$.*

Notice that this estimate is better than the estimate $\mathcal{O}(n^2)$ obtained with the usual method, as discussed in Proposition 2.4.1, as $\log_2 3 < 2$.

## 2.5.5 The Ruffini–Horner method

Let us now turn to the study of the problem of evaluating a polynomial $p(x)$ with integer coefficient at $x = \alpha$, with $\alpha \in \mathbb{Z}$, that is of calculating $p(\alpha)$. This value is exactly the remainder of the division of $p(x)$ by $x - \alpha$. Let

$$p(x) = a_0 + a_1 x + \cdots + a_{n-1} x^{n-1} + a_n x^n \in \mathbb{Z}[x],$$

which we shall also write $p(x) = a_n x^n + a_{n-1} x^{n-1} + \cdots + a_1 x + a_0$. The most immediate way of computing the value of the polynomial in $\alpha$ is by substituting in the polynomial, wherever the indeterminate $x$ appears, the element $\alpha$, and then carrying out all the required operations. However, this method is not efficient, as it requires the following steps:

- computing $\alpha^i$, for all $i$, $2 \le i \le n$;
- computing $a_i \alpha^i$ for $1 \le i \le n$;
- computing, one after another, the sums

$$a_0 + a_1 \alpha, \quad \ldots, \quad (a_0 + \cdots + a_{n-1} \alpha^{n-1}) + a_n \alpha^n.$$

So we have to perform $n - 1$ multiplications in the first step, $n$ more multiplications in the second, and $n$ additions in the third.

A better method is the Ruffini–Horner one. It relies on the fact that we may write the polynomial in *nested* form, as follows:

$$p(x) = a_0 + (a_1 + (a_2 + \cdots + (a_{n-1} + a_n x)x)\cdots)x.$$

This expression relies on the following recursive formula:

$$p_k(x) = \begin{cases} a_n & \text{for } k = 0, \\ a_{n-k} + p_{k-1}(x) & \text{for } 0 < k \le n. \end{cases}$$

So we have

$$p_0(x) = a_n,$$
$$p_1(x) = a_{n-1} + p_0(x),$$
$$p_2(x) = a_{n-2} + p_1(x),$$

$$\vdots$$

$$p_n(x) = p(x) = a_0 + p_{n-1}(x).$$

For instance, let $p(x) = 3x^4 + 2x^3 + x^2 - x + 4$. Then

$$p_0(x) = 3,$$
$$p_1(x) = a_3 + p_0(x) = 2 + 3x,$$
$$p_2(x) = a_2 + p_1(x) = 1 + (2 + 3x)x,$$
$$p_3(x) = a_1 + p_2(x) = -1 + (1 + (2 + 3x)x)x,$$
$$p_4(x) = a_0 + p_3(x) = 4 + (-1 + (1 + (2 + 3x)x)x)x,$$

which represents the nested form of the polynomial $p(x)$.

Let us see how to actually use this algorithm. Write down two lines: in the first line put all the coefficients of $p(x)$, starting with the leading coefficient, including zero coefficients. The leftmost term of the second line is to be the leading coefficient of $p(x)$; the second term is the previous term multiplied by $\alpha$, plus the second term of the first line, that is $p_1(\alpha)$. In general, the $i$th term in the second line is equal to the previous term times $\alpha$, plus the corresponding term of the first line: this is precisely $p_{i-1}(\alpha)$. The last term so obtained is $p_n(\alpha)$, that is the value $p(\alpha)$ we were looking for.

Basically, the two lines the algorithm constructs are

$$
\begin{array}{cccccc}
a_n & a_{n-1} & \cdots & a_2 & a_1 & a_0 \\
p_0 & p_1(\alpha) & \cdots & p_{n-2}(\alpha) & p_{n-1}(\alpha) & p_n(\alpha)
\end{array}. \tag{2.14}
$$

Notice that from this table the quotient of $p(x)$ by $x - \alpha$ can be read off: it is

$$
p_0 x^{n-1} + p_1(\alpha)x^{n-2} + \cdots + p_{n-2}(\alpha)x + p_{n-1}(\alpha) \tag{2.15}
$$

(see Exercise A2.16). This method to get the quotient of the division is called *synthetic division*.

It can also be easily seen (see again Exercise A2.16) that the total number of multiplications that are necessary to write down the second line is exactly $n$. Let us see an example.

**Example 2.5.14.** Evaluate the polynomial $p(x) = x^5 - 3x^4 + x^2 + x - 1$ at $x = -3$ using this algorithm.

The two lines generated by the algorithm are

$$
\begin{array}{cccccc}
1 & -3 & 0 & 1 & 1 & -1 \\
1 & -6 & 18 & -53 & 160 & -481
\end{array}
$$

and the value we are looking for, $p(-3)$, is $-481$, that is the remainder of the division of $x^5 - 3x^4 + x^2 + x - 1$ by $x + 3$. Notice that from this table the quotient of this division can be read off too: $x^4 - 6x^3 + 18x^2 - 53x + 160$.

As for the complexity of this algorithm, the number of operations to be performed in order to evaluate at $\alpha$ the polynomial written in nested form is smaller than the number of operation which would be necessary if the polynomial were written in the usual form. Indeed, we only need $n$ multiplications and $n$ additions. To simplify things, assume $\alpha > 0$; if $\alpha$ were negative the formulas would become true by putting $|\alpha|$ in place of $\alpha$. Notice that the absolute value of

$$
p_k(\alpha) = a_{n-k} + (a_{n-k+1} + (\cdots + (a_{n-1} + a_n\alpha)\alpha)\alpha \ldots)\alpha
$$

is smaller than or equal to

$$d + (d + \cdots + (d + (d + d\alpha)\alpha)\alpha) \cdots \alpha \qquad (2.16)$$

where $d = \max\{\,|a_i|\,\}$. Suppose that $d$ has length $h$ and $\alpha$ has length $k$. Notice that the number appearing in (2.16) is the polynomial $d(1 + x + \cdots + x^k)$ evaluated at $\alpha$. So we may conclude that for all $i = 0, \ldots, n$, we have

$$|p_i(\alpha)| \le (n+1)d\alpha^n;$$

thus $L(p_i(\alpha))$ is $\mathcal{O}(nhk \log n)$. In order to pass from $p_i(\alpha)$ to $p_{i+1}(\alpha)$ we have to carry out a multiplication and an addition: so the complexity of the computation of $p_{i+1}(\alpha)$ is $\mathcal{O}(nh^2k \log n)$. Moreover, $i$ may assume values between $0$ and $n$, so the complexity is $\mathcal{O}(n^2h^2k \log n)$.

# Appendix to Chapter 2

## A2 Theoretical exercises

**A2.1.** Prove that if $f \in \mathcal{O}(g)$ and $g \in \mathcal{O}(h)$, then $f \in \mathcal{O}(h)$.

**A2.2.** Assume that $\lim_{x \to \infty}(f(n)/g(n)) = \infty$. Prove that $g \in \mathcal{O}(f)$.

**A2.3.** Assume that $\lim_{x \to \infty}(f(n)/g(n)) = a \ne 0$. Prove that, in this case, $f \in \mathcal{O}(g)$ e $g \in \mathcal{O}(f)$.

**A2.4.** Assume that $\lim_{x \to \infty}(f(n)/g(n)) = 0$. Prove that $f \in \mathcal{O}(g)$.

**A2.5.** Prove that, if $f(n) = (n + 1) + n(-1)^n$ and $g(n) = (n + 1) - n(-1)^n$, then neither of the following limits exist:

$$\lim_{n \to \infty} \frac{f(n)}{g(n)}, \qquad \lim_{n \to \infty} \frac{g(n)}{f(n)}.$$

**A2.6.** Prove that if $f \in \mathcal{O}(g)$ and $c > 0$, then $c \cdot f \in \mathcal{O}(g)$.

**A2.7.** Prove that if $f_1 \in \mathcal{O}(g_1)$ and $f_2 \in \mathcal{O}(g_2)$, then $f_1 \cdot f_2 \in \mathcal{O}(g_1 \cdot g_2)$.

**A2.8.** Suppose we want to multiply $n$ numbers of length at most $k$. Which of the following is an upper bound for the length of the result?

(a) $k^3$.
(b) $k^2$.
(c) $nk$.
(d) None of the above.

**A2.9.**\* Give an estimate of the length $L(n!)$ of $n!$ with a simple reference function, where $n$ is an integer of length $k$.

(a) $L(n!) \in \mathcal{O}(nk)$.
(b) $L(n!) \in \mathcal{O}(k^2)$.
(c) $L(n!) \in \mathcal{O}(k^3)$.
(d) None of the above.

**A2.10.**\* Give an estimate of the length $L = L\left(\binom{n}{m}\right)$, where $n$ has length $h$ and $m$ has length $k$.

(a) $L \in \mathcal{O}(m^2 k^2)$.
(b) $L \in \mathcal{O}(n^2 h^2)$.
(c) $L \in \mathcal{O}(hk)$.
(d) None of the above.

**A2.11.** Given $n$ numbers with length $k_1, \ldots, k_n$, prove that the length of their sum is smaller than or equal to the greatest length, plus $L(n)$. What about the product?

**A2.12.** Let $a$ and $b$ be integers with length $k$ and $l$ respectively, with $k \geq l$. Prove that

$$T(a - b) \in \mathcal{O}(k) = \mathcal{O}(\max\{L(a), L(b)\}).$$

**A2.13.** Prove Corollaries 2.5.6 and 2.5.7 (page 101).

**A2.14.** Prove Proposition 2.5.8.

**A2.15.** Give an estimate of the complexity (that is, the number of bit operations needed) of computing the expression of a rational number as a continued fraction.

**A2.16.** Verify that the polynomial given in Equation (2.15) is the quotient of the division of $p(x)$ by $x - \alpha$ and that the number of multiplications needed to write the second row of table (2.14) is exactly $n$.

**A2.17.** Keeping in mind Exercises A1.36, A1.40 and B1.37, prove that the algorithm to compute the greatest common divisor described in Exercise A1.40 is *faster* than the one based on the usual division algorithm. Prove that, nevertheless, the asymptotic estimate given in Proposition 2.5.4 cannot be improved in this way.

**A2.18.**\* Prove that the computational complexity of computing the row-by-column product of two $n \times n$ matrices with integer entries bounded in absolute value by a positive integer number $m$ is $\mathcal{O}(n^3(\log^2 m + \log n))$.

**A2.19.** Prove that the computational complexity of computing the product $A \cdot X$, where $A$ is a square matrix of order $n$ and $X$ is a column vector with $n$ components, when their entries are integer numbers bounded in absolute value by a positive integer number $m$, is $\mathcal{O}(n^2 \log^2 m)$.

**A2.20.**\* Prove that the number of operations that are needed to reduce a $m \times n$ matrix (with $m \leq n$) on a field in row echelon form using Gaussian elimination (see [13], Chap. 8) is $\mathcal{O}(n^3)$.

**A2.21.**\* Prove that the number of operations that are needed to compute the determinant of a square matrix of order $n$ on a field is $\mathcal{O}(n^3)$.

**A2.22.**\* Prove that the number of operations that are needed to compute the inverse of a square matrix of order $n$ on a field is $\mathcal{O}(n^3)$.

**A2.23.**\* Give an estimate for the number of bit operations necessary in order to compute $n!$, where $n$ is an integer of length $k$.

(a) $T(n!) \in \mathcal{O}(e^k)$.
(b) $T(n!) \in \mathcal{O}(n^3 k^3)$.

(c) $T(n!) \in \mathcal{O}(n^2 k^2)$.
(d) None of the above.

**A2.24.*** Give an estimate for the complexity of computing $\binom{n}{m}$.

**A2.25.*** Give an estimate for the complexity of computing $m^n$.

# B2 Computational exercises

**B2.1.** Let $f(n) = n^3 - 2n^2 - 1$. Determine the best $\mathcal{O}$-estimate for $f(n)$ among the following ones.

(a) $f(n) \in \mathcal{O}(n^4)$.
(b) $f(n) \in \mathcal{O}(n^3)$.
(c) $f(n) \in \mathcal{O}(n^2)$.
(d) $f(n) \in \mathcal{O}(n^2 \log n)$.

**B2.2.** Let $f(n) = 2 \log n + n$. Determine the best $\mathcal{O}$-estimate for $f(n)$ among the following ones.

(a) $f(n) \in \mathcal{O}(\log n)$.
(b) $f(n) \in \mathcal{O}(n^2)$.
(c) $f(n) \in \mathcal{O}(n \log n)$.
(d) $f(n) \in \mathcal{O}(n)$.

**B2.3.** Let $f(n)$ be the sum of the cubes of the first $n$ even numbers (see Exercise B1.9). Determine the best $\mathcal{O}$-estimate for $f(n)$ among the following ones.

(a) $f(n) \in \mathcal{O}(n^4)$.
(b) $f(n) \in \mathcal{O}(n^3)$.
(c) $f(n) \in \mathcal{O}(2^n)$.
(d) $f(n) \in \mathcal{O}(3^n)$.

**B2.4.** Let $f(n) = \binom{n}{4}$. Determine the best $\mathcal{O}$-estimate for $f(n)$ among the following ones.

(a) $f(n) \in \mathcal{O}(n^4)$.
(b) $f(n) \in \mathcal{O}(4^n)$.
(c) $f(n) \in \mathcal{O}(n^3)$.
(d) $f(n) \in \mathcal{O}(3^n)$.

**B2.5.** Determine the best $\mathcal{O}$-estimate for $f(n) = 3^n + n!$ among the following ones.

(a) $f(n) \in \mathcal{O}((3n)!)$.
(b) $f(n) \in \mathcal{O}(n!)$.
(c) $f(n) \in \mathcal{O}(3^n)$.
(d) $f(n) \in \mathcal{O}(n^3)$.

**B2.6.** Let $f(n) = 3 \log^5 n + \log n + 2$. Determine the best $\mathcal{O}$-estimate for $f(n)$ among the following ones.

(a) $f(n) \in \mathcal{O}(n)$.
(b) $f(n) \in \mathcal{O}(\log n)$.
(c) $f(n) \in \mathcal{O}(n \log n)$.
(d) $f(n) \in \mathcal{O}(\log^5 n)$.

**B2.7.** Which is the length of 2047 in base 2?

(a) 13.
(b) 12.
(c) 11.
(d) None of the above.

**B2.8.** Which is the length of $(110110110110)_2$ in base 10?

(a) 3.
(b) 4.
(c) 5.
(d) None of the above.

**B2.9.** Which is the length of $(110110110110)_2$ in base 16? (Remember Remark 1.4.3.)

(a) 3.
(b) 4.
(c) 5.
(d) None of the above.

**B2.10.** Which is the length of 32769 in base 8?

(a) 4.
(b) 5.
(c) 6.
(d) None of the above.

**B2.11.** Which is the length in base 10 of the hexadecimal number $(F424B)_{16}$?

(a) 7.
(b) 6.
(c) 5.
(d) None of the above.

**B2.12.** Which is the length of $(7424B)_{16}$ in base 2?

(a) 21.
(b) 20.
(c) 19.
(d) None of the above.

**B2.13.** Which is the length in base 2 of the sum of $(100100110)_2$ and $(11100101)_2$?

(a) 9.
(b) 10.
(c) 11.
(d) None of the above.

**B2.14.** Which is the length in base 2 of the product of the binary numbers 11101 and 1101?

(a) 8.
(b) 9.
(c) 10.
(d) None of the above.

**B2.15.** Assume that an algorithm is given which, receiving as input an integer of length $n$, has complexity $\log n$ (or $n$, $n \log n$, $n^2$, $2^n$, $n!$). Suppose the time needed to carry out each bit operation is $10^{-9}$ seconds.

In these hypotheses verify Table 2.1, which gives the time needed for the algorithm to come to an end.

**Table 2.1.** Time to carry out the bit operations needed

| $n$ | $\log n$ | $n$ | $n \log n$ | $n^2$ | $2^n$ | $n!$ |
|---|---|---|---|---|---|---|
| 10 | $3 \cdot 10^{-9}$ s | $10^{-8}$ s | $3 \cdot 10^{-8}$ s | $10^{-7}$ s | $10^{-6}$ s | $3 \cdot 10^{-3}$ s |
| $10^2$ | $7 \cdot 10^{-9}$ s | $10^{-7}$ s | $7 \cdot 10^{-7}$ s | $10^{-5}$ s | $4 \cdot 10^{13}$ y | * |
| $10^3$ | $1{,}0 \cdot 10^{-8}$ s | $10^{-6}$ s | $1 \cdot 10^{-5}$ s | $10^{-3}$ s | * | * |
| $10^4$ | $1{,}3 \cdot 10^{-8}$ s | $10^{-5}$ s | $1 \cdot 10^{-4}$ s | $10^{-1}$ s | * | * |
| $10^5$ | $1{,}7 \cdot 10^{-8}$ s | $10^{-4}$ s | $2 \cdot 10^{-3}$ s | 10 s | * | * |
| $10^6$ | $2 \cdot 10^{-8}$ s | $10^{-3}$ s | $2 \cdot 10^{-2}$ s | 17 m | * | * |

The first column of the table gives the size of the input, while the other columns display the complexity, and "s", "m" and "y" denote seconds, minutes and years necessary to execute the algorithm, respectively. The asterisk means that the time taken is greater than $10^{100}$ years.

**B2.16.** Order the exponential, factorial, linear, logarithmic and polynomial complexities from the most intractable to the most tractable one.

(a) Factorial, exponential, logarithmic, polynomial, and linear.
(b) Exponential, factorial, polynomial, logarithmic, and linear.
(c) Exponential, factorial, logarithmic, polynomial, and linear.
(d) Factorial, exponential, polynomial, linear, and logarithmic.

**B2.17.** Let $k$ be a positive integer and let $a$, $b$ be integers of length $[e^{2k}]$ and $[e^k]$, respectively. Which is the best estimate for the computation time of $a + b$?

(a) $\mathcal{O}(e^k)$.
(b) $\mathcal{O}(2e^k)$.
(c) $\mathcal{O}(e^{2k})$.
(d) $\mathcal{O}(e^{3k})$.

**B2.18.** Let $k$ be a positive integer and let $a$, $b$ be integers of length $k^3$ and $[\log^2 k]$, respectively. Which is the best estimate for the computation time of $a - b$?

(a) $\mathcal{O}(k)$.
(b) $\mathcal{O}(k^3)$.
(c) $\mathcal{O}(\log k)$.
(d) $\mathcal{O}(\log^2 k)$.

**B2.19.** Let $k$ be a positive integer and let $a$, $b$ be integers of length $[e^{2k}]$ and $[e^k]$, respectively. Which is the best estimate for the computation time of $a \cdot b$?

(a) $\mathcal{O}(e^{2k})$.
(b) $\mathcal{O}(e^{3k})$.
(c) $\mathcal{O}(e^{2k^2})$.
(d) $\mathcal{O}(e^{e^k})$.

**B2.20.** Let $a$, $b$ be integers of length $k^3$ and $[\log^2 k]$, respectively. Which is the best estimate for the computation time of $a \cdot b$?

(a) $\mathcal{O}(k^2 \log k)$.
(b) $\mathcal{O}(k^3 \log k)$.
(c) $\mathcal{O}(\log^3 k)$.
(d) $\mathcal{O}(k^3 \log^2 k)$.

**B2.21.** Consider the formula expressing the sum of the first $n$ squares

$$\sum_{j=1}^{n} j^2 = \frac{n(n+1)(2n+1)}{6}. \tag{2.17}$$

Which of the following is an estimate for the number of bit operations that are needed to compute the left-hand side of Equation (2.17)?

(a) $\mathcal{O}(n \log n)$.
(b) $\mathcal{O}(n^2 \log n)$.
(c) $\mathcal{O}(n \log^2 n)$.
(d) None of the above.

**B2.22.** Consider again Formula (2.17). Which of the following is an estimate for the number of bit operations that are needed to compute the right-hand side of Equation (2.17)?

(a) $\mathcal{O}(n \log n)$.
(b) $\mathcal{O}(n^2)$.
(c) $\mathcal{O}(\log^2 n)$.
(d) None of the above.

**B2.23.** In the last two exercises we saw two different methods to compute Formula (2.17). Which is the faster method?

(a) The two methods have the same complexity.
(b) The first method is the faster one.
(c) The second method is the faster one.
(d) None of the above.

**B2.24.** How many steps are necessary to compute GCD(1176, 159) using the variant to the Euclidean algorithm described in Exercise A1.40?

(a) 4.
(b) 5.
(c) 6.
(d) None of the above.

# C2 Programming exercises

**C2.1.** Write a program that computes the length of an integer number $a$ in base 2.

**C2.2.** Write a program that computes the length of an integer number $a$ in base $b$, where $b$ is an integer greater than 1.

**C2.3.** Write a program that computes the product of two integer numbers of length $2n$ using the algorithm described in §2.5.4.

**C2.4.** Write a program that evaluates a polynomial at an integer using Ruffini–Horner method.

# 3

## From infinite to finite

In this chapter we shall define an important relation on $\mathbb{Z}$: the *congruence relation modulo a positive integer $n$*. We shall give here the fundamental properties of this relation, but we shall repeatedly come back to it in later parts of the book.

## 3.1 Congruence: fundamental properties

The congruence relation modulo a positive integer $n$ identifies two integers if and only if their difference is a multiple of $n$: so it may be regarded as an "equality" up to multiples of $n$. At first sight, this might look quite strange. Actually, we use this relation in everyday life. Consider the way the time is displayed by a clock: at six p.m., that is to say 18 hours after midnight, we read the digit 6. So, with respect to keeping the time, the number 18 *equals* the number 6. This means that we are working *modulo* 12.

We are now going to give a formal treatment of the subject.

**Definition 3.1.1.** *Let $n$ be a fixed positive integer. The* congruence relation modulo $n$ *is the relation on $\mathbb{Z}$ defined as follows: $a \equiv b \pmod{n}$ if and only if there exists an integer $h$ such that $a - b = nh$, that is*

$$a \equiv b \pmod{n} \quad \text{if and only if} \quad n \mid (a - b);$$

*in this case we say that $a$ is* congruent to $b$ modulo $n$.

We shall sometimes write $a \equiv_n b$ rather than $a \equiv b \pmod{n}$, and even $\equiv$ rather than $\equiv_n$ if no misunderstanding can possibly arise.

It is easy to verify (see Exercise A3.1) that the congruence modulo any fixed positive integer $n$ is an *equivalence relation*, that is to say:

- it is *reflexive*, i.e. for every integer $a \in \mathbb{Z}$, it one has $a \equiv_n a$;
- it is *symmetric*, i.e. for every pair of integers $(a, b) \in \mathbb{Z} \times \mathbb{Z}$, one has $a \equiv_n b$ if and only if $b \equiv_n a$;

- it is *transitive*, i.e. for every triple of integers $(a, b, c) \in \mathbb{Z} \times \mathbb{Z} \times \mathbb{Z}$, if $a \equiv_n b$ and $b \equiv_n c$ then $a \equiv_n c$.

Notice that an integer is congruent to every other integer modulo 1, that is to say congruence modulo 1 is a trivial equivalence relation; for this reason in what follows we shall in general assume $n > 1$.

Given an integer $a$, we shall denote by $[a]_n$, or, if no confusion may possibly arise, by $[a]$, or by $\bar{a}$, the congruence class of $a$ modulo $n$, that is,

$$[a]_n = \{ b \in \mathbb{Z} \mid b \equiv_n a \} = \{ a + hn \mid h \in \mathbb{Z} \} =$$
$$= \{ \ldots, a - 2n, a - n, a, a + n, a + 2n, \ldots \}.$$

As we may always divide an integer $a$ by $n$ (see Proposition 1.3.1), if we denote by $r$ the remainder of the division, we immediatly obtain:

**Proposition 3.1.2.** *Every integer $a$ is congruent modulo $n$ to a unique integer $r$ such that $0 \leq r < n$.*

In particular, notice that for every integer $a$ there is a unique element $r$ of $[a]_n$ such that $0 \leq r < n$, that is to say the remainder of the division of $a$ by $n$. This element is denoted by the symbol:

$$\boxed{a \bmod n}.$$

**Remark 3.1.3.** Let us remark that, keeping in mind Proposition 2.4.1, it is possible to deduce that the complexity of computing $a \bmod n$ is $\mathcal{O}(\log a \cdot \log n)$. If we work modulo a fixed integer $n$, considering it as a constant, the complexity is $\mathcal{O}(\log a)$.

There are exactly $n$ equivalence classes, namely (see Exercise A3.4):

$$\bar{0} = \{\text{integers that, divided by } n, \text{ give a remainder equal to } 0\} =$$
$$= \{ kn \mid k \in \mathbb{Z} \} = \{\text{multiples of } n\},$$
$$\bar{1} = \{\text{integers that, divided by } n, \text{ give a remainder equal to } 1\} =$$
$$= \{ kn + 1 \mid k \in \mathbb{Z} \},$$

$$\vdots$$

$$\overline{n-1} = \{\text{integers that, divided by } n, \text{ give a remainder equal to } n - 1\} =$$
$$= \{ kn + n - 1 \mid k \in \mathbb{Z} \}.$$

This is the reason why the equivalence classes modulo $n$ are also called *residue classes* modulo $n$.

To give a graphical representation of this situation, we can envisage integer numbers as arranged at equal distances on a circumference of length $n$. So it becomes clear that all integer multiples of $n$ coincide with 0, the integers that, divided by $n$, give a remainder equal to 1 coincide with 1, those that, divided by $n$, give a remainder equal to 2 coincide with 2 and so on.

We shall denote by $\mathbb{Z}_n$ the *quotient set* of $\mathbb{Z}$ with respect to the congruence modulo $n$, that is to say, $\mathbb{Z}_n$ consists of $n$ distinct residue classes modulo $n$:

$$\mathbb{Z}_n \overset{\text{def}}{=} \{\, \bar{0}, \bar{1}, \bar{2}, \ldots, \overline{n-2}, \overline{n-1} \,\} = \mathbb{Z}/\!\equiv_n .$$

Thus, using the congruence relation modulo $n$

> we passed from the *infinite* set $\mathbb{Z}$ to the *finite* set $\mathbb{Z}_n$.

This passage from infinite to finite is often very useful, especially when calculations or checks are to be performed with a computer which, as is well known, only works with finitely many objects. We shall see examples of this fact, beginning with this chapter and even more in the following ones.

The congruence shares several properties with equality among integer numbers. The symbol $\equiv$ has been introduced by Gauss, by virtue of the analogy with the equality relation. In particular, we may *add*, *subtract* and *multiply* congruent elements preserving the congruence:

**Proposition 3.1.4.** *Let $n > 1$ be a fixed integer. If $a$, $b$, $c$, $d$ are integers with $a \equiv_n b$ and $c \equiv_n d$, then the following properties hold:*

$$a \pm c \equiv b \pm d \pmod{n}, \qquad\qquad ac \equiv bd \pmod{n}.$$

PROOF. The easy proof is left as an exercise to the reader (see Exercise A3.5). $\square$

Proposition 3.1.4 says that the congruence relation defined on $\mathbb{Z}$ is compatible with the operations of addition and multiplication defined on $\mathbb{Z}$. In other words, the two operations on $\mathbb{Z}_n$ given by

$$\bar{a} + \bar{b} \overset{\text{def}}{=} \overline{a + b}, \qquad\qquad \bar{a} \cdot \bar{b} \overset{\text{def}}{=} \overline{a \cdot b} \tag{3.1}$$

are well defined, as they do not depend on the representative elements chosen. So, exactly as $\mathbb{Z}$, $\mathbb{Z}_n$ is also an example of a commutative ring with unity, called *ring of residue classes modulo $n$* (see Exercise A3.6).

**Remark 3.1.5.** By keeping in mind Remark 3.1.3 and Proposition 2.4.1, we find that the complexity of computing the sum of two elements in $\mathbb{Z}_n$ is $\mathcal{O}(\log n)$. Indeed, we may consider those elements as numbers smaller than $n$, and so of length $\mathcal{O}(\log n)$. Their sum still has length $\mathcal{O}(\log n)$, but now we have to reduce it modulo $n$, which requires at most subtracting $n$. Hence the claim follows.

Analogously, the complexity of computing the product of two elements in $\mathbb{Z}_n$ is $\mathcal{O}(\log^2 n)$. Indeed, the product of two elements in $\mathbb{Z}_n$ has length at most $\mathcal{O}(2 \log n)$. Then it is necessary to reduce modulo $n$, and this requires a division by $n$. Hence the assertion follows.

**Remark 3.1.6.** This is a convenient moment to recall some algebraic notions, in order to put in a wider context the operation that led us from the ring $\mathbb{Z}$ to the ring $\mathbb{Z}_n$.

Let $A$ be a commutative ring, and let $I$ be an ideal of $A$ (see § 1.3.5). In $A$ it is possible to define the relation of *congruence modulo $I$*, denoted by $\equiv_I$, as follows: given $a, b \in A$, we put $a \equiv_I b$ if and only if $a - b \in I$. It is easy to verify that $\equiv_I$ is an equivalence relation (see Exercise A3.7), whose quotient set is denoted by the symbol $A/I$. If $a \in A$, its class modulo $I$ is denoted by the usual symbols $[a]_I$, $[a]$, $\bar{a}$ and also by $a + I$, it consisting of all the elements of the form $a + b$, with $b \in I$. Notice that $I$ is exactly the equivalence class of 0 modulo $I$. Also notice that $\equiv_{(0)}$ is the equality relation, and so $A/(0) = A$, while any two elements of $A$ are congruent modulo $A$, that is to say $A/A$ consists of a single element.

An analogue of Proposition 3.1.4 holds: if $a$, $b$, $c$, $d$ are elements of $A$ and $a \equiv_I b$ and $c \equiv_I d$, then we have (see Exercise A3.8)

$$a \pm c \equiv_I b \pm d, \qquad\qquad ac \equiv_I bd. \qquad (3.2)$$

This implies that the two operations on $A/I$ given by

$$\bar{a} + \bar{b} \overset{\text{def}}{=} \overline{a + b}, \qquad\qquad \bar{a} \cdot \bar{b} \overset{\text{def}}{=} \overline{a \cdot b} \qquad (3.3)$$

are well defined in $A/I$, that is they do not depend on the representative elements chosen. With these operations, $A/I$ is a commutative ring, called the *quotient ring of $A$ modulo $I$* (see Exercise A3.9).

In this setting, we can see that $Z_n$ is just the quotient of $\mathbb{Z}$ modulo the ideal $(n)$. As $\mathbb{Z}$ is an Euclidean ring (see § 1.3.5), and consequently its ideals are principal (see Theorem 1.3.14), each proper ideal of $\mathbb{Z}$ is of the form $(n)$, with $n$ positive integer and so, when $n$ varies in the set of positive integers, $\mathbb{Z}_n$ describes the set of all quotients of $\mathbb{Z}$ that are commutative ring with unity.

However, it would be wrong to assume that every property holding for the equality holds for congruences as well. For instance, in $\mathbb{Z}$ we have the following *cancellation property*:

$$ac = bc \quad \text{implies} \quad a = b,$$

as long as $c \neq 0$. In other words, $\mathbb{Z}$ is an integral domain. This property *does not hold*, in general, for congruences, as the following example shows:

$$2 \cdot 3 = 6 \equiv_6 12 = 4 \cdot 3, \qquad \text{but} \qquad 2 \not\equiv_6 4. \qquad (3.4)$$

Nevertheless, the following result holds:

**Proposition 3.1.7.** *Let $n > 1$ be an integer. If $ac \equiv_n bc$ and if $c$ and $n$ are coprime, that is if $\mathrm{GCD}(c, n) = 1$, then $a \equiv_n b$.*

PROOF. If $ac \equiv_n bc$, then $n \mid (a - b)c$. As $\mathrm{GCD}(c, n) = 1$, it follows from Corollary 1.3.9 that $n \mid a - b$, that is $a \equiv b \pmod{n}$. □

Notice that in the example (3.4) given above, the hypothesis was not satisfied, as $\mathrm{GCD}(3, 6) = 3 \neq 1$. Proposition 3.1.7 is actually a consequence of the following more general result.

**Proposition 3.1.8.** *Let $n > 1$ be an integer. If $ac \equiv bc \pmod{n}$, then $a \equiv b$ (mod $m$), where $d = \mathrm{GCD}(c, n)$ and $n = md$.*

PROOF. Notice that, by the definition of $d$, $m = n/d$ and $f = c/d$ are integer numbers. The hypothesis implies that $(a - b)c = kn$ for some integer $k$, and so, by dividing by $d$, we find that

$$(a - b)f = km.$$

Then $m$ divides the product $(a - b)f$, and $m$ and $f$ are coprime, that is $\mathrm{GCD}(m, f) = 1$ (see Corollary 1.3.6), and so $m$ divides $a - b$, as already proved in Proposition 3.1.7; thus $a \equiv_m b$.                □

Basically, Proposition 3.1.8 says that *simplifying a congruence* by cancelling out a common factor is always possible, *as long as the modulo is suitably modified.* For instance, $21 = 3 \cdot 7 \equiv_9 30 = 3 \cdot 10$ and $7 \not\equiv 10 \pmod{9}$, while $7 \equiv 10 \pmod{3}$.

**Remark 3.1.9.** As a consequence of the above, we may see, for instance, that in $\mathbb{Z}_5$ the same cancellation property holds as in $\mathbb{Z}$; in other words, $\mathbb{Z}_5$ is, like $\mathbb{Z}$, an integral domain, and consequently, since it is finite, a field (see Exercise A1.48). Indeed, if $\bar{a} \cdot \bar{b} = \bar{0}$, with $\bar{a}, \bar{b} \in \mathbb{Z}_5$ and $\bar{b} \neq \bar{0}$, then $\bar{a} = \bar{0}$ by Proposition 3.1.7, as $\mathrm{GCD}(b, 5) = 1$. So $\mathbb{Z}_5$ is an integral domain. On the other hand, the same results say nothing about $\mathbb{Z}_4$. In fact, in $\mathbb{Z}_4$ there are *zero-divisors*: we have $\bar{2} \cdot \bar{2} = \bar{0}$, con $\bar{2} \neq \bar{0}$.

**Remark 3.1.10.** Returning to Remark 3.1.6, assume $A$ is an integral domain. Keeping in mind that $[0]_I = I$, it is easy to verify (see Exercise A3.12) that $A/I$ is itself an integral domain if and only if $I$ is a *prime ideal*, that is if and only if: *for all pairs $(a, b)$ of elements of $A$ such that $ab \in I$, either $a \in I$ or $b \in I$.*

Now, in the case of $\mathbb{Z}$ we may ask for which positive integer numbers $n$ greater than 1 it is the case that the ideal $(n)$ is prime, this being a necessary and sufficient condition for $\mathbb{Z}_n$ to be an integral domain as well, and consequently a field (see Exercise A1.48). Clearly this happens if and only if: *for all pairs $(a, b)$ of integers such that $n \mid ab$, either $n \mid a$ or $n \mid b$.* In this case we say that $n$ is a *prime number.* At first sight, it is not evident that this definition of a prime number is equivalent to the definition given in § 1.3.1. Indeed, this deserves to be further investigated, and this will be done in Chapter 4.

For instance $2, 3, 5, 7$ and so forth are prime numbers, and so $\mathbb{Z}_2, \mathbb{Z}_3, \mathbb{Z}_5, \mathbb{Z}_7$ and so forth are integral domains and, being finite, they are fields.

Notice that Proposition 3.1.8 is the first result connecting two congruences modulo different numbers. We collect now further useful properties relating congruences *modulo different numbers*, leaving the easy proof to the reader (see Exercise A3.13; see also Exercise A1.47).

**Proposition 3.1.11.** *The following properties hold:*

(1) *if $a \equiv_n b$ and $d \mid n$, then $a \equiv_d b$;*
(2) *if $a \equiv_n b$ and $a \equiv_m b$, then $a \equiv b \pmod{\mathrm{lcm}(n, m)}$.*

## 3.2 Elementary applications of congruence

### 3.2.1 Casting out nines

Let us recall how the so-called *casting out nines* works: the reader is certainly familiar with this procedure since primary school. It is used to check the results of operations on integers.

Suppose we want to multiply two integers, say 123 and 456 and find the result 56088. We want to check its correctness.

We act as follows. Write down the sum of the digits of the factors, casting out nines, which are substituted by zeros, and iterate this procedure until we get a one-digit number. Multiply the one-digit numbers obtained in this way and reduce the result of this operation as well to a one-digit number $x$. Finally, the result to be checked is reduced in the same way to a one-digit number: it must be the same as $x$, or the operation was not carried out correctly.

In our case we have

$$123 \qquad \times \; 456 \qquad = 56088,$$
$$1+2+3 = 6 \quad 4+5+6 = 15 \quad 5+6+0+8+8 = 27,$$
$$6 = 6 \qquad 1+5 = 6 \qquad 2+7 = 9 = \boxed{0}.$$

Now, carry out the product $6 \cdot 6 = 36$, whose digit sum is $3+6 = 9 = 0$, which equals the sum of the digits of the result of the multiplication of the original numbers (that is, 0). If we had found a number different from 0, we would be sure to have made a mistake while multiplying. On the other hand, the agreement of the two number *does not* assure us that the multiplication was carried out correctly.

Let us now investigate the meaning of this "casting out nines" procedures, and in particular what is so special about the number 9.

As we know from section 1.4, when we write 123 in base 10 we mean the number $1 \cdot 10^2 + 2 \cdot 10 + 3 \cdot 10^0$. Now, for all $n > 0$, we have that

$$10^n - 1 = \underbrace{999 \cdots 9}_{n} = 9 \cdot \underbrace{111 \cdots 1}_{n}, \qquad \text{that is,} \qquad 10^n \equiv_9 1.$$

Then, by exploiting the properties of congruence, it follows that

$$123 = 1 \cdot 10^2 + 2 \cdot 10 + 3 \equiv_9 1 + 2 + 3,$$

that is, the number 123, as all numbers written in base 10, is congruent modulo 9 to the sum of its digits. So if $z$ is the positive integer number

$$z = (a_n a_{n-1} a_{n-2} \ldots a_1 a_0)_{10} = a_n 10^n + \cdots + a_1 \cdot 10 + a_0, \qquad (3.5)$$

then

$$z \equiv a_n + a_{n-1} + \cdots + a_0 \pmod{9}.$$

Here lies the magic of number 9 and the explanation of casting out nines. In fact, if two numbers are different modulo 9 they are plainly different, while two numbers that are congruent modulo 9 are not necessarily equal. Now, if we carry out an operation, cast out nines and get a match, we only deduce that the right result and the one we found are congruent modulo nine. So we are not certain the operation was carried out correctly. On the other hand, if the procedures fails, we are certain that the result was wrong. In other words, casting out nines is a necessary but not sufficient condition for the calculation to be correct. For instance, if we by mistake get 65817 as result of multiplying 123 and 456, we would not notice it by casting out nines, as the wrong result and the correct one are in the same class modulo 9, as both are multiples of 9.

### 3.2.2 Tests of divisibility

As another useful application, congruences yield some *tests of divisibility* without having to explicitly carry out the division. We keep writing numbers in base 10, that is to say in the form (3.5). So we have the following tests:

- *test of divisibility by 3 and by 9:* an integer number $z = (a_n a_{n-1} \ldots a_1 a_0)_{10}$ is divisible by 3 (or by 9, respectively) if and only if the sum of its digits

$$a_0 + a_1 + a_2 + \cdots + a_n$$

  is divisible by 3 (or by 9, respectively);

- *test of divisibility by 2 and by 5:* an integer number $z = (a_n a_{n-1} \ldots a_1 a_0)_{10}$ is divisible by 3 or by 5 if and only if its rightmost digit, $a_0$, is divisible by 2 or by 5;

- *test of divisibility by 4 and by 25:* an integer number $z = (a_n a_{n-1} \ldots a_1 a_0)_{10}$ is divisible by 4 (or by 25, respectively) if and only if the number $(a_1 a_0)_{10}$ formed by its last two digits is divisible by 4 (or by 25, respectively);

- *test of divisibility by $2^k$:* an integer $z = (a_n a_{n-1} a_{n-2} \ldots a_1 a_0)_{10}$ is divisible by $2^k$ if and only if $2^k$ divides the number $(a_{k-1} a_{k-2} \ldots a_1 a_0)_{10}$ formed by the last $k$ digits of $z$; so, in particular, an integer is divisible by 8 if and only if $(a_2 a_1 a_0)_{10}$ is divisible by 8;

- *test of divisibility by 11:* an integer $z = (a_n a_{n-1} a_{n-2} \ldots a_1 a_0)_{10}$ is divisible by 11 if and only if the sum of its digits taken with alternating signs

$$a_0 - a_1 + a_2 - \cdots + (-1)^n a_n$$

  is divisible by 11.

PROOF. The test of divisibility by 3 and by 9 follows from the fact that $z \equiv a_n + a_{n-1} + \cdots + a_0$ both modulo 3 and modulo 9.

Notice that $10^n \equiv 0$ both modulo 2 and modulo 5, for all $n \geq 1$. So, $z \equiv a_0$ both modulo 2 and modulo 5, and this proves the corresponding divisibility tests.

Analogously, $100 = 2^2 5^2 \equiv 0$ both modulo 4 and modulo 25, and so every integer is congruent modulo 4 and modulo 25 to the integer formed by its rightmost two digits.

More generally, $10^n = 2^n 5^n \equiv 0 \pmod{2^k}$ for all $k \geq n$, proving the divisibility test for $2^k$.

Finally, notice that $10 \equiv -1 \pmod{11}$, and so for all $p \geq 1$, we have $10^{2p} \equiv 1 \pmod{11}$ and $10^{2p+1} \equiv -1 \pmod{11}$; hence the divisibility test for 11 follows.  □

Part of the remarks discussed in this section can be extended to numbers written in an arbitrary base (see the exercises in the appendix).

## 3.3 Linear congruences

As for equalities, for congruences too the problem can be posed of solving a congruence with respect to one or more variables. The simplest equation is

$$x \equiv a \pmod{n},$$

which admits as its solutions all the integers of the form $a + hn$ with $h \in \mathbb{Z}$. We may generalise and give the following definition:

**Definition 3.3.1.** *A* linear congruence *or* linear congruence equation *in the unknown $x$ is an equation of the form*

$$ax \equiv b \pmod{n}$$

*with $a$, $b$ in $\mathbb{Z}$ and $n$ a positive integer greater than 1.*

We are going to study if and when such a congruence admits solutions, where by a solution we mean an integer $x_0$ such that $ax_0 \equiv_n b$. If a solution exists, we say that the equation is *soluble*.

The examples that follow show that both solvable linear congruences and insolvable ones exist.

**Example 3.3.2.** The equation $3x \equiv 2 \pmod 6$ does not admit solutions: if it did, we would find solutions in $\mathbb{Z}$ for the equation $3x + 6y = 2$, but Proposition 1.3.11 tells us that this equation does not admit integer solutions, as $\mathrm{GCD}(3,6) = 3$, which does not divide 2.

**Example 3.3.3.** The equation $4x \equiv 6 \pmod{10}$ admits the solution $x = 4$, which is not unique, as $x = 9$ is a solution too.

We now give the general conditions for a linear congruence to be solvable and, for solvable congruences, its solutions.

**Proposition 3.3.4.** *Consider the congruence $ax \equiv b \pmod{n}$.*

(1) *It admits solutions if and only if $d = \mathrm{GCD}(a,n)$ divides $b$.*
(2) *If the congruence is solvable, denote by $x_0$ one of its solutions; then all numbers of the form $x_0 + hm$, with $h \in \mathbb{Z}$ and $m = n/d$, are solutions, and there are no other solutions. Among these, the solutions*

$$x_0, \quad x_0 + m, \quad x_0 + 2m, \quad \ldots, \quad x_0 + (d-1)m \tag{3.6}$$

*are pairwise non-congruent and all other solutions are congruent to one of them. So the congruence admits exactly $d$ non-congruent solutions modulo $n$.*

PROOF. Solving the congruence is equivalent to finding integer solutions of the equation $ax + ny = b$, and by Proposition 1.3.11 these exist if and only if $GCD(a, n) \mid b$.

Let us prove now that $x_0 + hm$ is a solution, for all $h \in \mathbb{Z}$. Indeed,

$$a(x_0 + hm) = ax_0 + ah\frac{n}{d} = ax_0 + hn\frac{a}{d} \equiv_n ax_0 \equiv_n b.$$

We prove next that *each* solution is of this form. Let $x_0$ and $x_0'$ be two solutions. Then

$$ax_0 = b + hn, \qquad ax_0' = b + kn,$$

for suitable integers $h$ and $k$, so

$$a(x_0 - x_0') = (h - k)n.$$

By dividing both sides by $d$ we get

$$a'(x_0 - x_0') = (h - k)m,$$

where $a' = a/d$. As $a'$ and $m$ are relatively prime (see Corollary 1.3.6), by applying Proposition 1.3.9 we see that $m$ divides $x_0 - x_0'$, that is $x_0 - x_0' = zm$ for some integer $z$.

We prove now that, among the solutions $x_0 + hm$, when $h$ varies in $\mathbb{Z}$, exactly $d$ of them are non-congruent modulo $n$. To this end, we shall show that the solutions (3.6) are non-congruent modulo $n$ and that every solution is congruent to one of them. Assume by contradiction that two of the solutions (3.6) are congruent modulo $n$, that is to say,

$$x_0 + h_1 m \equiv x_0 + h_2 m \pmod{n},$$

for some $h_1, h_2 \in \mathbb{Z}$ with $0 \leq h_1 < h_2 \leq d - 1$. Then we would get

$$h_1 m \equiv h_2 m \pmod{n};$$

hence, by dividing by $m = GCD(m, n)$, it would follow that

$$h_1 \equiv h_2 \pmod{d},$$

which yields a contradiction because we assumed $0 < h_2 - h_1 < d$.

To prove that *each* solution of the form $x_0 + hm$, when $h$ varies in $\mathbb{Z}$, is congruent to one of the solutions (3.6), it suffices to divide $h$ by $d$, that is to say, defining $h = dq + r$, with $0 \leq r \leq d - 1$, we have

$$x_0 + hm = x_0 + (dq + r)m = x_0 + qn + rm \equiv_n x_0 + rm,$$

which is exactly one of the solutions (3.6).    □

**Example 3.3.5.** The congruence $4x \equiv 6 \pmod{10}$ form Example 3.3.3 admits exactly $2 = GCD(4, 10)$ solutions that are non-congruent modulo 10: $x = 4$ e $x = 9$.

**Corollary 3.3.6.** *If $a$ and $n$ are relatively prime, then the congruence $ax \equiv b$ (mod $n$) admits exactly one solution modulo $n$.*

The above results allow us, in particular, to find the invertible elements of $\mathbb{Z}_n$, that is the elements $\bar{x} \in \mathbb{Z}_n$ for which an element $\bar{y} \in \mathbb{Z}_n$ such that $\bar{x}\bar{y} = 1$ exists.

**Corollary 3.3.7.** *An element $\bar{a} \in \mathbb{Z}_n$ is invertible if and only if $a$ and $n$ are relatively prime.*

PROOF. Determining the classes $\bar{a}$ that are invertible in $\mathbb{Z}_n$ is equivalent to solving the congruence

$$ax \equiv 1 \pmod{n}.$$

Now, this congruence admits a solution, and this solution is unique, if and only if $\mathrm{GCD}(a, n) = 1$. So the invertible classes are the classes $\bar{a}$ with $0 < a < n$ and $\mathrm{GCD}(a, n) = 1$. □

**Remark 3.3.8.** To solve a linear congruence of the form $ax \equiv b$ (mod $n$) we have essentially to solve a linear Diophantine equation of the form $ax + ny = b$. Let us study the complexity of this operation. We may assume that $n$ is fixed, and so is constant. The first steps consist in computing $d = \mathrm{GCD}(a, n)$; this has complexity $\mathcal{O}(\log^3 a)$ by Proposition 2.5.4. Next, we have to divide $b$ by $\mathrm{GCD}(a, n)$, which has complexity $\mathcal{O}(\log a \log b)$ by Proposition 2.4.1. Now we are able to decide whether the equation is solvable or not, depending on $b$ being or not a multiple of $d$. Thus, the complexity of the algorithm that *decides* whether the equation is solvable has complexity $\mathcal{O}(\log^3 a + \log a \log b)$. As to actually solving the equation recall that, by the results in § 1.3.4, in order to find a solution of $ax + ny = b$, and consequently a solution $x_0$ of $ax \equiv b$ (mod $n$), it is necessary to compute a Bézout's identity for $\mathrm{GCD}(a, n)$ and to multiply it by $b/d$. By Corollary 2.5.6 and by Proposition 2.4.1 this does not increase the complexity. Neither does, for similar reasons, computing the further solutions (3.6) of the congruence, which are obtained using Proposition 3.3.4.

In particular, in order to decide if $\bar{a} \in \mathbb{Z}_n$ is invertible and, if so, to determine an inverse, the complexity is simply $\mathcal{O}(\log^3 a)$.

If, on the other hand, in this kind of problem we consider $a < n$ as fixed, that is, we take $a$ as a constant and $n$ as a variable, we get that the complexity of deciding if $\bar{a} \in \mathbb{Z}_n$ is invertible and, if so, of finding an inverse is $\mathcal{O}(\log^3 n)$.

We give now an important definition.

**Definition 3.3.9.** *For all $n \geq 1$, denote by $\varphi(n)$ the number of positive integers smaller than $n$ that are coprime to $n$. The function $\varphi : \mathbb{N} \to \mathbb{N}$ defined in this way is called* Euler function *or $\varphi$ function.*

For instance, $\varphi(24) = 8$ as the numbers that are smaller than 24 and coprime to it are: 1, 5, 7, 11, 13, 17, 19 and 23.

The Euler function will be, as we shall see, very important in what follows: we shall return to it again, in particular in Chapter 4. At present we remark that, as $\mathbb{Z}_n = \{[0], [1], \dots, [n-1]\}$, the multiplicative group $U(\mathbb{Z}_n)$ of invertible elements of $\mathbb{Z}_n$ consists of all classes $[a]$, with $0 < a < n$ such that $a$ is coprime to $n$. So $U(\mathbb{Z}_n)$ has order $\varphi(n)$.

**Example 3.3.10.** Let us see some examples of $U(\mathbb{Z}_n)$ for some $n$:

$$\mathbb{Z}_4 = \{\bar{0}, \bar{1}, \bar{2}, \bar{3}\}, \qquad\qquad U(\mathbb{Z}_4) = \{\bar{1}, \bar{3}\},$$
$$\mathbb{Z}_6 = \{\bar{0}, \bar{1}, \bar{2}, \bar{3}, \bar{4}, \bar{5}\}, \qquad\qquad U(\mathbb{Z}_6) = \{\bar{1}, \bar{5}\},$$
$$\mathbb{Z}_8 = \{\bar{0}, \bar{1}, \cdots, \bar{7}\}, \qquad\qquad U(\mathbb{Z}_8) = \{\bar{1}, \bar{3}, \bar{5}, \bar{7}\}.$$

Notice that $\varphi(4) = |U(\mathbb{Z}_4)| = 2$, $\varphi(6) = |U(\mathbb{Z}_6)| = 2$, and $\varphi(8) = |U(\mathbb{Z}_8)| = 4$.

We conclude this section with an important theorem, whose scope will be better appreciated when we shall be able to *compute* Euler function (see Proposition 4.2.3).

**Theorem 3.3.11 (Euler's Theorem).** *If $a$ and $n$ are relatively prime, then*

$$a^{\varphi(n)} \equiv 1 \pmod{n}.$$

PROOF. We have $[a] \in U(\mathbb{Z}_n)$, and the order of the group $U(\mathbb{Z}_n)$ is $\varphi(n)$. So the claim follows from general group-theoretical properties (see Exercise A3.37) which we recall in the following Remark 3.3.12 for the convenience of the reader. □

**Remark 3.3.12.** It is convenient to recall here the group-theoretical properties used in the proof of Euler's Theorem.

Let $G$ be a group and let $x$ be an element of $G$. The elements of $G$ of the form $x^m$, with $m$ an integer, that is to say the powers of $x$, form a subgroup of $G$, denoted by $\langle x \rangle$, known as the subgroup of $G$ *generated* by $x$. If there exists in $G$ an element $x$ such that $G = \langle x \rangle$, the group $G$ is said to be *cyclic*.

Some general properties of cyclic groups which will be used again later in the book are recalled in Exercises A3.23-A3.35.

Let $G$ be a group and let $x$ be an element of $G$. The element $x$ is said to have *finite order* if $\langle x \rangle$ is finite; clearly, this happens if and only if there exists a positive integer $n$ such that $x^n = 1$. The least positive integer $n$ such that $x^n = 1$ is said to be the *order* or *period* of $x$ and is denoted by $\mathrm{ord}(x)$. If $x$ has not finite order, it is said to have *infinite order*. In other words, $\mathrm{ord}(x)$ is the order of the cyclic subgroup $\langle x \rangle$ (see Exercise A3.28). If $x$ has finite order, then $x^m = 1$ for an integer $m$ if and only if $\mathrm{ord}(x) \mid m$ and so $x^i = x^j$ if and only if $i \equiv j \pmod{\mathrm{ord}(x)}$ (see Exercise A3.27).

Now let $G$ be a finite group and let $H$ be a subgroup of $G$. Recall that *Lagrange's theorem* states that the order of $H$ divides the order of $G$ (see Exercise A3.36). In particular, if $G$ is finite, the order of any of its elements divides the order of the group. So, if $G$ has order $n$ and $x \in G$ it is always true that $x^n = 1$.

The original proof by Euler relies on the Fundamental Theorem of Arithmetic and is given in the exercises.

### 3.3.1 Powers modulo $n$

Given an integer $a$ and two positive integers $m$ and $n$, we want to compute

$$a^m \pmod n.$$

It is a kind of computation we shall frequently need in the rest of the book.

Let us start with an example:

**Example 3.3.13.** Compute

$$17^{31} \bmod 58.$$

We may successively divide the exponent by 2, as follows:

$$17^{31} = (17^{15})^2 \cdot 17 = \left((17^7)^2 \cdot 17\right)^2 \cdot 17 = \left(\left((17^3)^2 \cdot 17\right)^2 \cdot 17\right)^2 \cdot 17 =$$
$$= \left(\left((17^2 \cdot 17)^2 \cdot 17\right)^2 \cdot 17\right)^2 \cdot 17. \tag{3.7}$$

Then we carry out the operations reducing each time modulo $n = 58$. The reader will easily verify (see Exercise B3.37) that

$$17^{31} \equiv 41 \pmod{58}.$$

Let us see a description of the steps we take, and from this we shall deduce a general algorithm. In base 2, the number 31 is written

$$31 = (11111)_2. \tag{3.8}$$

It is a 5-digit number. So write four ($= 5 - 1$) letters $E$

$$E\ E\ E\ E$$

and, going from left rightwards, insert among them an $M$ for each digit 1 we read in (3.8), while for each 0 simply step forward by one place. The final result will be

$$MEMEMEMEM. \tag{3.9}$$

We attribute now a meaning to the letters $M$ and $E$. $M$ denotes the multiplication by 17 (and its subsequent reduction modulo 58), while $E$ (for "exponentiation") denotes the squaring (and reduction modulo 58). Now, apply the operation (3.9) to 1 (starting from the left), that is, evaluate

$$(1)MEMEMEMEM.$$

The result is exactly the last term in equalities (3.7).

Now we describe the algorithm in general and compute its complexity. Start from the problem of reducing a power $a^m$ modulo $n$ when both the exponent $m$ and the modulo $n$ are very large. Using this algorithm, we do not need to carry out $m$ multiplications of $a$ times itself. We may assume $a < n$, and every time we carry out a multiplication, we reduce the result modulo $n$. So we shall never work with integers grater than $n^2$.

Write the exponent $m$ in base 2

$$m = n_0 + n_1 \cdot 2 + n_2 \cdot 2^2 + \cdots + n_{k-1} \cdot 2^{k-1},$$

with $n_j \in \{0,1\}$; then

$$a^m = a^{n_0 + n_1 \cdot 2 + n_2 \cdot 2^2 + \cdots + n_{k-1} \cdot 2^{k-1}} = a^{n_0}(a^2)^{n_1} \cdots \left(a^{2^j}\right)^{n_j} \cdots \left(a^{2^{k-1}}\right)^{n_{k-1}}.$$

At the beginning of the procedure we set $p_0 = 1$. Then we perform $k-1$ steps defining successively $p_1, p_2, \ldots, p_{k-1}, p_k$, finding at the end $p_k = a^m \bmod n$.

If $n_0 = 1$, then we set $p_1 = a$, that is, we set $p_1 = a \cdot p_0 \bmod n$, else we leave $p_1 = 1$.

Then we square $a$ and set $a_1 = a^2 \bmod n$. If $n_1 = 1$, then we set $p_2 = a_1 \cdot p_1 \bmod n$, else we leave $p_2 = p_1$.

Square now $a_1$ and set $a_2 = a_1^2 \bmod n$. If $n_2 = 1$, then set $p_3 = a_2 \cdot p_2 \bmod n$, else leave $p_3 = p_2$.

By continuing in this way, in the $j$th step we compute

$$a_j = a^{2^j} \bmod n.$$

If $n_j = 1$, that is, if $2^j$ appears in the binary expression of $m$, then $p_{j+1} = a_j \cdot p_j \bmod n$, else we leave $p_{j+1} = p_j$. After $k-1$ steps we get $p_k = a^m \bmod n$.

We now compute the complexity of this algorithm, called *exponentiation by squaring* or *square-and-multiply*: in each step we may have one or two multiplications of numbers smaller than $n^2$; so each step involves $\mathcal{O}(\log^2(n^2)) = \mathcal{O}(\log^2 n)$ bit operations. There are $k-1$ steps, so the estimate of the time to compute $a^m \bmod n$ is

$$\mathcal{O}((\log m)(\log^2 n)).$$

Let us see another example.

**Example 3.3.14.** Compute $3^{13} \bmod 7$ using the method just described.

We have $13 = 1 + 2^2 + 2^3$, so $n_0 = 1$, $n_1 = 0$, $n_2 = n_3 = 1$, $k-1 = 3$ and $a = 3$.

(1) Set $p_0 = 1$.
(2) As $n_0 = 1$, we set $p_1 = a = 3$.
(3) Square $a = 3$ and reduce it modulo 7, obtaining $a_1 = 2$. As $n_1 = 0$, set $p_2 = p_1 = 3$.
(4) Square $a_1$ and reduce it modulo 7, obtaining $a_2 = 4$. As $n_2 = 1$, set $p_3 = a_2 p_2 = 12 \equiv 5 \bmod 7$.
(5) Square $a_2$ and reduce it modulo 7, obtaining $a_3 = 2$. As $n_3 = 1$, set $p_4 = a_3 p_3 = 10 \equiv 3 \bmod 7$.

The result is $p_4$, that is 3. The reader may verify that the result is right using Euler's theorem (see Exercise B3.38).

**Remark 3.3.15.** As suggested in Example 3.3.13, another way of carrying out the algorithm is the following:

• write the exponent $m$ in base 2 as a sequence of 0 and 1;
• write a sequence of $L_2(m) - 1$ letters $E$

$$\underbrace{E\ E\ E\ E\ \cdots E}_{L_2(m)-1};$$

- beginning from the left, insert among the $E$s an $M$ for each digit 1 we read in the binary expression of $m$, while for each 0 simply step forward by one place; the meaning of the letters is as follows: $E$ (for "exponentiation") denotes the squaring and reduction modulo $m$, $M$ denotes the multiplication by $a$ and its subsequent reduction modulo $m$;
- the result is obtained by applying this sequence of $M$s and $E$s to 1, starting from the left.

Let us give an illustration of the above by computing $3^{13}$ (mod 7). We have $m = 13$ and $13 = (1101)_2$. As $L_2(13) = 4$, write three letters $E$

$$E \ E \ E$$

Now, by inserting as many letters $M$ from the left, as the 1s in the expression of $m$ and just stepping forward for each 0, we get

$$MEMEEM.$$

Compute now

$$(1) MEMEEM,$$

where $M$ means multiplying by 3, and $E$ squaring: we get back

$$((3^2 \cdot 3)^2)^2 \cdot 3 \equiv 3 \pmod 7.$$

We have used again the operators introduced in the previous example: the operations we need are multiplying by $a = 3$ ($M$) and squaring ($E$).

**Remark 3.3.16.** When computing $a^m \bmod n$ it may happen that $a$ and $n$ are relatively prime. In this case in order to simplify the calculations we might use Euler's theorem 3.3.11, but this implies knowing how to evaluate the Euler function $\varphi(n)$. If we know $\varphi(n)$ and if $\mathrm{GCD}(a, n) = 1$, then it may be verified that the complexity of computing $a^m \bmod n$ becomes $\mathcal{O}(\log^3 n)$ (see Exercise A3.38).

## 3.4 The Chinese remainder theorem

Suppose we have to count quickly the students in a school but we have not time for calling each of them individually. We only know that there are less than 1000 of them. What can we do?

The solution to this problem is ancient and amounts to the so-called Chinese remainder theorem, which appears in Chinese texts from the first century CE. This theorem allows us to solve a system of linear congruences, which we may translate our original problem into.

So, consider the following system:

$$\begin{cases} a_1 x \equiv b_1 \pmod{n_1}, \\ a_2 x \equiv b_2 \pmod{n_2}, \\ \vdots \\ a_s x \equiv b_s \pmod{n_s}. \end{cases} \tag{3.10}$$

Solving this system means finding an integer $x_0$ such that

$$\begin{cases} a_1 x_0 \equiv b_1 \pmod{n_1}, \\ a_2 x_0 \equiv b_2 \pmod{n_2}, \\ \vdots \\ a_s x_0 \equiv b_s \pmod{n_s}. \end{cases} \tag{3.11}$$

Such an integer $x_0$ is said to be a *solution* of system 3.10 and, if a solution exists, the system is said to be *solvable* or *compatible*. So, if even one of the equations is not solvable, the whole system does not admit solutions. The problem is easily solved when $\mathrm{GCD}(n_i, n_j) = 1$ for all pairs $(i, j)$ with $i \neq j$.

**Lemma 3.4.1.** *Consider a system of the form* (3.10) *consisting of solvable congruences and such that* $\mathrm{GCD}(n_i, n_j) = 1$ *for all pairs* $(i, j)$ *such that* $i \neq j$. *Then solving* (3.10) *is equivalent to solving a system of the form*

$$\begin{cases} x \equiv c_1 \pmod{r_1}, \\ x \equiv c_2 \pmod{r_2}, \\ \vdots \\ x \equiv c_s \pmod{r_s} \end{cases} \tag{3.12}$$

*with* $\mathrm{GCD}(r_i, r_j) = 1$ *for all pairs* $(i, j)$ *such that* $i \neq j$.

PROOF. For the system (3.10) to admit solutions we must necessarily have $d_k = \mathrm{GCD}(a_k, n_k) \mid b_k$, for all $k = 1, \ldots, s$. If these conditions are satisfied, we may divide the $k$th congruence by $d_k$, obtaining

$$\begin{cases} a_1' x \equiv b_1' \pmod{r_1}, \\ a_2' x \equiv b_2' \pmod{r_2}, \\ \vdots \\ a_s' x \equiv b_s' \pmod{r_s}, \end{cases} \tag{3.13}$$

where $a_k' = a_k/d_k$, $b_k' = b_k/d_k$ and $r_k = n_k/d_k$ and the condition $\mathrm{GCD}(r_i, r_j) = 1$ still holds for each pair $(i, j)$ such that $i \neq j$. Notice that the new system (3.13) is *equivalent* to the previous one, in the sense the they both admit the same solutions.

Notice now that $\mathrm{GCD}(a_k', r_k) = 1$, for all $k = 1, \ldots, s$; so, by Corollary 3.3.6, each of the congruences of the system admits a *unique* solution $c_k$ modulo $r_k$. Thus we may substitute (3.12) for system (3.13). □

So we study now systems of congruences of the form (3.12).

**Theorem 3.4.2 (Chinese remainder theorem).** *A systems of congruences* (3.12), *with* $\mathrm{GCD}(r_i, r_j) = 1$ *for all pairs* $(i, j)$ *with* $i \neq j$, *admits a solution. This solution is unique modulo* $R = r_1 r_2 \cdots r_s$.

PROOF. Let $R_k = R/r_k$. As $\mathrm{GCD}(r_i, r_j) = 1$ for all pairs $(i, j)$ with $i \neq j$, we have, by repeatedly applying Corollary 1.3.7, that $\mathrm{GCD}(R_k, r_k) = 1$ for all $k = 1, \ldots, s$. So, for each $k = 1, \ldots, s$, the congruence

$$R_k x_k \equiv c_k \pmod{r_k}$$

admits a unique solution modulo $r_k$ which will be denoted by $\bar{x}_k$. Thus the number

$$\bar{x} = R_1 \bar{x}_1 + R_2 \bar{x}_2 + \cdots + R_s \bar{x}_s$$

is a solution of the system (3.12). Indeed, as $R_i$ is a multiple of $r_k$ for $i \neq k$, we have

$$R_i \equiv 0 \pmod{r_k} \qquad \text{for } i \neq k,$$

and so

$$\begin{cases} \bar{x} \equiv R_1 \bar{x}_1 \equiv c_1 \pmod{r_1}, \\ \bar{x} \equiv R_2 \bar{x}_2 \equiv c_2 \pmod{r_2}, \\ \vdots \\ \bar{x} \equiv R_s \bar{x}_s \equiv c_s \pmod{r_s}. \end{cases}$$

As to the unicity modulo $R$, let $\bar{y}$ be another solution of the system; in other words, assume that

$$\bar{x} \equiv c_k \equiv \bar{y} \pmod{r_k}, \qquad \text{for } k = 1, \ldots, s.$$

Then $\bar{x} - \bar{y} \equiv 0 \pmod{r_k}$, for $k = 1, \ldots, s$; hence $\bar{x} - \bar{y} \equiv_R 0$ (see Corollary 1.3.8). □

**Remark 3.4.3.** Analogously it can be seen that, if we want to solve a system of the form

$$\begin{cases} a_1 x \equiv b_1 \pmod{r_1}, \\ a_2 x \equiv b_2 \pmod{r_2}, \\ \vdots \\ a_s x \equiv b_s \pmod{r_s}. \end{cases}$$

with $\mathrm{GCD}(a_i, r_i) = 1$ for all $i = 1, \ldots, s$ and $\mathrm{GCD}(r_i, r_j) = 1$ for all pairs $(i, j)$ such that $i \neq j$, then the solution is

$$\bar{x} = R_1 \bar{x}_1 + R_2 \bar{x}_2 + \cdots + R_s \bar{x}_s$$

where $\bar{x}_k$ satisfies $a_k R_k \bar{x}_k \equiv b_k \pmod{r_k}$, for all $k = 1, \ldots, s$.

**Example 3.4.4.** Let us study again the example about students we started with. In order to count them, it is sufficient to request the children to line up 5 abreast, then 8, then 19, and to count each time how many students are not lined up: these remainders $c_i$, for $i = 1, 2$ and 3, are smaller than 5, 8, and 19, respectively. Then the system

$$\begin{cases} x \equiv c_1 \pmod 5, \\ x \equiv c_2 \pmod 8, \\ x \equiv c_3 \pmod{19}, \end{cases}$$

is solved. By the Chinese remainder theorem, it admits a unique solution modulo $R = 5 \cdot 8 \cdot 19 = 760$. This solution is the total number of students.

Suppose, for example, that lining up the students five abreast, one of them is not lined up, when eight abreast, 2 are not lined up, and when 19 abreast, 3 are not lined up. Then $c_1 = 1$, $c_2 = 2$ and $c_3 = 3$. In this case the number of the students is found by finding the solution, unique modulo 760, of the system

$$\begin{cases} x \equiv 1 \pmod 5, \\ x \equiv 2 \pmod 8, \\ x \equiv 3 \pmod{19}. \end{cases} \tag{3.14}$$

So, how many students are there? The proof itself of the Chinese remainder theorem tells us how to find the solution. Indeed, we have $R_1 = 8 \cdot 19 = 152$, $R_2 = 5 \cdot 19 = 95$, $R_3 = 5 \cdot 8 = 40$; so, a solution is $3 \cdot 152 + 6 \cdot 95 + 11 \cdot 40 = 1466 \equiv 706 \bmod 760$. As 706 is the only solution smaller than 1000, this is the number we were looking for.

There is a useful consequence or, rather, a different interpretation of the Chinese remainder theorem. Let $\mathbb{Z}_r \times \mathbb{Z}_s$ be the Cartesian product of $\mathbb{Z}_r$ and $\mathbb{Z}_s$. As $\mathbb{Z}_r$ and $\mathbb{Z}_s$ are rings, the set $\mathbb{Z}_r \times \mathbb{Z}_s$ itself is in a natural way a *product ring*, in which the operations are defined by

$$(\bar{a}_r, \bar{b}_s) + (\bar{c}_r, \bar{d}_s) \stackrel{\text{def}}{=} (\bar{a}_r + \bar{c}_r, \bar{b}_s + \bar{d}_s),$$
$$(\bar{a}_r, \bar{b}_s) \cdot (\bar{c}_r, \bar{d}_s) \stackrel{\text{def}}{=} (\bar{a}_r \cdot \bar{c}_r, \bar{b}_s \cdot \bar{d}_s).$$

The following proposition is an immediate consequence of the Chinese remainder theorem.

**Proposition 3.4.5.** *Let $r$ and $s$ be two relatively prime integers greater than 1. The map*

$$f : \mathbb{Z}_{rs} \longrightarrow \mathbb{Z}_r \times \mathbb{Z}_s,$$

*defined by*

$$f([x]_{rs}) = ([x]_r, [x]_s),$$

*is a ring isomorphism.*

PROOF. By the Chinese remainder theorem the system of congruences

$$\begin{cases} x \equiv a \pmod r, \\ x \equiv b \pmod s \end{cases}$$

admits *exactly one* solution modulo $rs$. This implies the surjectivity and the injectivity of the map. The fact that this map preserves the operations is left to the reader (see Exercise A3.40). $\qquad\square$

**Example 3.4.6.** We write explicitly the correspondence $f$ in the case $r = 2$ and $s = 3$.

| $\mathbb{Z}_6$ | $\mathbb{Z}_2$ | $\mathbb{Z}_3$ |
|---|---|---|
| 0 | 0 | 0 |
| 1 | 1 | 1 |
| 2 | 0 | 2 |
| 3 | 1 | 0 |
| 4 | 0 | 1 |
| 5 | 1 | 2 |

The map defined in Proposition 3.4.5 may also be described as follows. Write down all the elements of $\mathbb{Z}_{rs}$ in a column. Next to it, in a second column write down $s$ times the elements of $\mathbb{Z}_r$. In a third column write down $r$ times the elements of $\mathbb{Z}_s$. The correspondence $f$ associates to each element of $\mathbb{Z}_{rs}$ the pair in $\mathbb{Z}_r \times \mathbb{Z}_s$ that can be read on its right. The reader will easily be able to check this result.

Proposition 3.4.5 has an important application in computer science, as it allows to translate calculations in $\mathbb{Z}_n$ into *independent* calculations in several $\mathbb{Z}_{n_i}$, for $i = 1, \ldots, r$, when $n = n_1 n_2 \cdots n_r$, with $\text{GCD}(n_i, n_j) = 1$ for $i \neq j$.

The following example allows us to fully appreciate the advantages of this method.

**Example 3.4.7.** Assume we have to carry out some operations in $\mathbb{Z}_{21}$: in particular we have to multiply the class $\overline{17}$ by the class $\overline{19}$. As $21 = 3 \cdot 7$, with $\text{GCD}(3, 7) = 1$, rather than performing the calculations in $\mathbb{Z}_{21}$, we may use the correspondence $f$ of Proposition 3.4.5, which we explicitly write down in table 3.1 for the convenience of the reader.

**Table 3.1.** Identifying $\mathbb{Z}_{21}$ with $\mathbb{Z}_3 \times \mathbb{Z}_7$

| $\mathbb{Z}_{21}$ | $\mathbb{Z}_3$ | $\mathbb{Z}_7$ | $\mathbb{Z}_{21}$ | $\mathbb{Z}_3$ | $\mathbb{Z}_7$ | $\mathbb{Z}_{21}$ | $\mathbb{Z}_3$ | $\mathbb{Z}_7$ |
|---|---|---|---|---|---|---|---|---|
| 0 | 0 | 0 | 7 | 1 | 0 | 14 | 2 | 0 |
| 1 | 1 | 1 | 8 | 2 | 1 | 15 | 0 | 1 |
| 2 | 2 | 2 | 9 | 0 | 2 | 16 | 1 | 2 |
| 3 | 0 | 3 | 10 | 1 | 3 | 17 | 2 | 3 |
| 4 | 1 | 4 | 11 | 2 | 4 | 18 | 0 | 4 |
| 5 | 2 | 5 | 12 | 0 | 5 | 19 | 1 | 5 |
| 6 | 0 | 6 | 13 | 1 | 6 | 20 | 2 | 6 |

It can be seen from the table that $\overline{17}$ is identified with the pair $(\overline{2}_3, \overline{3}_7)$, while $\overline{19}$ is identified with $(\overline{1}_3, \overline{5}_7)$. Then, rather than carrying out the product $\overline{17} \cdot \overline{19}$ in $\mathbb{Z}_{21}$, it suffices to carry out the product $\overline{2} \cdot \overline{1}$ in $\mathbb{Z}_3$ and the product $\overline{3} \cdot \overline{5}$

in $\mathbb{Z}_7$. So we get the pair $(\bar{2}_3, \bar{1}_7)$, which corresponds to the element $\bar{8}$ of $\mathbb{Z}_{21}$, the same class we would get by making calculations in $\mathbb{Z}_{21}$. As the operations in $\mathbb{Z}_3$ and in $\mathbb{Z}_7$ can be performed *separately*, that is, they are independent from each other, we can for example use two different computers working in parallel.

## 3.5 Examples

### 3.5.1 Perpetual calendar

As a simple application of congruences, consider the drawing up of a calendar. Congruences are well suited to this application due to the periodicity of the years' structure.

Suppose we want to know the day of the week corresponding to a given date: for instance, we might want to know which day of the week corresponded to 5 May 1851.

Clearly, the answer depends on how we measure time. Recall that a calendar is an organisation of time for civil or rerligious purposes, fixed according to astronomical phenomena. In solar calendars, like the Julian one, one year corresponds to the time it takes the Earth to complete one orbit around the Sun. This kind of calendar is used in Europe and in America.

The *Julian calendar*, introduced by Julius Caesar in 46 BCE, considered the year to be exactly 365 days long and added a day every four years, obtaining a *leap year* 366 days long, to allow for what then was deemed to be the true length of a year, that is to say 365 days and 6 hours. But this was just a rough approximation of the year's real duration, which is actually 365.2422 days, slightly shorter than the one Julius Caesar knew. Using that approximation caused, over the centuries, an ever increasing gap between the civil and the astronomical calendar, so much so that the former did not agree with the important dates of the latter: in particular, equinoxes and solstices, and so the beginning of the seasons, did not fall in the corresponding days.

So it was necessary to introduce a correction, brought about in 1582 by Pope Gregory XIII, whose astronomers observed that the civil calendar was ten days behind the astronomical one. Pope Gregory decided with a papal bull promulgated from Villa Mondragone in Frascati (near Rome), which presently is a property of "Tor Vergata" University of Rome, that these ten days were to be deleted, so that 5 October 1582 was immediately followed by 15 October 1582. Moreover, he declared the leap years to be all years divisible by 4, with the exception of those divisible by 100, that were to be leap years only if divisible by 400. So the years 1700, 1800, 1900 were not leap years, while 1600 and 2000 were. Thus, the year's measure is almost the actual one: there is an error of just 3 days per 10000 years! The resulting calendar, the one we use today, was called *Gregorian calendar*.

So we tackle our problem using the Gregorian calendar. As the extra day in leap years is added at the end of February by convention, it is convenient to give a number to each month starting from March: so January and February will be reckoned in the previous year.

For instance, we shall consider February 1952 as the twelfth month of 1951, while October 1952 will be the eighth month of 1952.

We shall do something similar for the days of the week, assigning a number to each of them beginning with zero, corresponding to Sunday. For the convenience of the reader we give explicitly the correspondence tables for days and months in Table 3.2.

**Table 3.2.** Numbering of days and months

| | | | | | |
|---|---|---|---|---|---|
| Sunday | = 0 | March | = 1 | September | = 7 |
| Monday | = 1 | April | = 2 | October | = 8 |
| Tuesday | = 2 | May | = 3 | November | = 9 |
| Wednesday | = 3 | June | = 4 | December | = 10 |
| Thursday | = 4 | July | = 5 | January | = 11 |
| Friday | = 5 | August | = 6 | February | = 12 |
| Saturday | = 6 | | | | |

As the days of the weeks repeat every seven days, in tackling our problem it will be necessary to count modulo 7. This way, the same day in each week will always be identified by the same number. Denote now by

$$d = \text{day of the month}, \qquad m = \text{month}, \qquad y = \text{year},$$
$$c = \text{century}, \qquad n = \text{years in the century}.$$

For instance, if we talk about 1 April 1673, then $d = 1$, $m = 2$ and $y = 1673 = 100c + n$, where $c = 16$ and $n = 73$. Now, denote by $x$ the day of the week corresponding the the data $(d, m, y)$. Using the first day of March 1600 as our starting point, compute successively:

- the day of the week $d_y$ of 1 March in any year $y$, with $y \geq 1600$;
- the day of the week of the first day of any month of any year;
- the day of the week of any day in any month of any year.

As there are 366 or 365 days between 1 March of year $y - 1$ and 1 March of year $y$, depending on $y$ being a leap year or not, clearly the following recurrence relation holds:
$$\begin{cases} d_y \equiv_7 d_{y-1} + 366 \equiv_7 d_{y-1} + 2 & \text{if } y \text{ is a leap year}, \\ d_y \equiv_7 d_{y-1} + 365 \equiv_7 d_{y-1} + 1 & \text{if } y \text{ is not a leap year}. \end{cases}$$
In conclusion,
$$d_y \equiv d_{1600} + (y - 1600) + l \pmod{7},$$
where $l$ is the number of leap year between 1600 exclusive (as we start from 1 March) and year $y$ inclusive, while $y - 1600$ is the number of years between 1600 and year $y$.

In order to compute $l$ we have to find all numbers of the form $1600 + p$ with $1 \leq p \leq y - 1600$ that are divisible by 4, remove from these the numbers divisible by 100 and add again those divisible by 400.

Dividing $y - 1600$ by 4 we find

$$y - 1600 = q \cdot 4 + r$$

with $0 \le r < 4$; so $\{\,4,\,8,\,\ldots,\,4q\,\}$ is the set of positive numbers smaller than or equal to $y - 1600$ that are divisible by 4. There are exactly $q$ of them, with

$$q = \left[\frac{y - 1600}{4}\right].$$

So we have

$$l = \left[\frac{y - 1600}{4}\right] - \left[\frac{y - 1600}{100}\right] + \left[\frac{y - 1600}{400}\right] =$$

$$= \left[\frac{y}{4}\right] - 400 - \left[\frac{y}{100}\right] + 16 + \left[\frac{y}{400}\right] - 4 = \left[\frac{y}{4}\right] - \left[\frac{y}{100}\right] + \left[\frac{y}{400}\right] - 388$$

so, keeping in mind that $y = 100 \cdot c + n$, in terms of $c$ and $n$ we find

$$d_y \equiv d_{1600} - 2c + n + \left[\frac{c}{4}\right] + \left[\frac{n}{4}\right] - 3 - 1600 \pmod 7 \equiv$$

$$\equiv d_{1600} - 2c + n + \left[\frac{c}{4}\right] + \left[\frac{n}{4}\right] \pmod 7.$$

Now we are able to compute $d_y$ for any year $y$, as soon as we know $d_{1600}$. But to do so, it is sufficient to have a look at a recent calendar and learn that 1 March 2006 falls on a Wednesday, that is to say, $d_{2006} = 3$. As for $y = 2006 = 100\,c + n$ we have $c = 20$, $n = 6$, from what precedes we get $3 \equiv d_{1600} - 40 + 6 + 5 + 1 \pmod 7$, and so $d_{1600} \equiv 31 \equiv 3 \pmod 7$, that is to say, 1 March 1600 was a Wednesday. So we get the general formula

$$d_y \equiv 3 - 2c + n + \left[\frac{c}{4}\right] + \left[\frac{n}{4}\right] \pmod 7. \tag{3.15}$$

Now, reasoning in the same way, we may compute the day of the week corresponding to the first day of an arbitrary month. For instance, if 1 October of year $y$ falls on a Thursday, then 1 November of the same year falls on a Sunday. Indeed, denote by $t$ the day of the week corresponding to the first of November: then we have $t \equiv 4 + 31 \pmod 7$, as October has 31 days, Thursday corresponds to the number 4, and $t = 3 + 4 = 7 \equiv 0 \pmod 7$ represents Sunday. In conclusion, the first of the month is found by adding a fixed amount of days, depending on the length of the month, to the first of previous month. In detail, we have

| | | |
|---|---|---|
| form 1 March to 1 April | we must add | 3 days, |
| from 1 April to 1 May | " | 2 days, |
| from 1 May to 1 June | " | 3 days, |
| from 1 June to 1 July | " | 2 days, |
| from 1 July to 1 August | " | 3 days, |
| from 1 August to 1 September | " | 3 days, |
| from 1 September to 1 October | " | 2 days, |
| from 1 October to 1 November | " | 3 days, |
| from 1 November to 1 December | " | 2 days, |
| from 1 December to 1 January | " | 3 days, |
| from 1 January to 1 February | " | 3 days. |

The function that represents the increment described is

$$\left[\frac{13m - 1}{5}\right] - 2.$$

This may be verified by simply substituting for $m$ the values corresponding to the months.

In conclusion, if we denote by $x$ the day of the week we intend to compute and which corresponds to the data $d$, $m$, $y$, we get the required formula by adding $d-1$ (days) to the formula that gives the day of the week corresponding to the first day of the same month and year, that is to say,

$$x \equiv d - 1 + \left[\frac{13m-1}{5}\right] - 2 + 3 - 2c + n + \left[\frac{c}{4}\right] + \left[\frac{n}{4}\right] \pmod{7} \equiv$$

$$\equiv d + \left[\frac{13m-1}{5}\right] - 2c + n + \left[\frac{c}{4}\right] + \left[\frac{n}{4}\right] \pmod{7}.$$

As an example, we show how to compute the day of the week corresponding to 9 May 1973. As $d = 9$, $m = 3$ e $1973 = 100c + n$ with $c = 19$ and $n = 73$ we get

$$x \equiv 9 + 7 - 38 + 73 + 4 + 18 \equiv 3 \pmod{7},$$

that is to say, that day was a Wednesday.

### 3.5.2 Round-robin tournaments

We want to draw up the schedule for a tournament involving $N$ football teams (or of any other team sport) in which each team is to play each other team exactly once. If $N$ is odd, as each play involves exactly two teams, we add a fictitious team and make the team playing it sit out that round. This way, no team is at an advantage and we may suppose $N$ to be an even number. Clearly, the number of rounds to be scheduled in such a way that each team may play each of the others exactly once is $N - 1$.

Assign the numbers from 1 to $N$ to the teams and suppose we want to schedule the games of the $k$th round, with $1 \le k \le N - 1$. If $1 \le i \le N - 1$, have team $i$ play team $j$ if

$$i + j \equiv k \pmod{(N-1)}. \tag{3.16}$$

This assigns to each team a single adversary team in the $k$th round, unless the team has as number the value of $i$ for which $2i \equiv k \pmod{(N-1)}$. As $N-1$ is odd, this congruence has a unique solution. This determines the single team that in the $k$th round will play team $N$.

We leave to the reader the easy task of checking that the rule just described actually has the effect of having each team play each other team exactly once in the $N - 1$ rounds (see Exercise A3.42).

## Appendix to Chapter 3

## A3 Theoretical exercises

**A3.1.** Prove that the congruence relation modulo $n$ given in Definition 3.1.1 is an equivalence relation, that is to say, it is reflexive, symmetric and transitive.

**A3.2.** Recall that a *relation* $R$ on a set $A$ is a subset of $A \times A$. It is customary to write $a\,R\,b$ to mean that $(a, b) \in R$. A relation $R$ is said to be *reflexive* if $a\,R\,a$ for all $a \in A$. We say that $R$ is *symmetric* if $a\,R\,b$ implies that $b\,R\,a$, for all $a, b \in A$. Finally, $R$ is said to be *transitive* if $a\,R\,b$ and $b\,R\,c$ imply that $a\,R\,c$, for all $a$, $b$, $c \in A$. A relation $R$ is said to be an *equivalence relation* if it is reflexive, symmetric and transitive.

Prove that the conditions defining an equivalence relation are independent, by giving examples of a:

(a) reflexive and symmetric, but not transitive, relation;
(b) reflexive and transitive, but not symmetric, relation;
(c) symmetric and transitive, but not reflexive, relation.

**A3.3.** Recall that a relation $R$ in a set $A$ is said to be *antisymmetric* if $a\,R\,b$ and $b\,R\,a$ imply that $a = b$, for all $a, b \in A$. An *order* relation is a reflexive, antisymmetric and transitive relation.

Let $R$ be the following relation defined on natural numbers: $a\,R\,b$ if and only if there exists a natural number $c$ such that $b = ac$. Is it an order relation? And what if we consider the same relation $R$ on integer numbers?

**A3.4.** Prove that the residue classes modulo the positive integer $n$ are exactly the $n$ classes described on page 116.

**A3.5.** Prove Proposition 3.1.4 on page 117.

**A3.6.** Prove that the operations defined in $\mathbb{Z}_n$, with $n \geq 1$, on page 117 are well-defined and make $\mathbb{Z}_n$ a ring with unity.

**A3.7.*** Let $A$ be a ring and $I$ an ideal of $A$. Prove that $\equiv_I$ is an equivalence relation.

**A3.8.*** Let $A$ be a ring and $I$ an ideal of $A$. Prove that Equations (3.2) hold.

**A3.9.*** Let $A$ be a ring and $I$ a proper ideal of $A$. Prove that the operations defined in $A/I$ on page 118 in Remark 3.1.6 are well-defined and make $A/I$ a ring.

**A3.10.*** Let $A$ be a commutative ring with unity and $I$ an ideal of $A$. Consider the map $\pi : A \to A/I$ that associates with each $a \in A$ its class in $A/I$. Prove that if $J$ is an ideal of $A/I$ then $\pi^{-1}(J)$ is an ideal of $A$ containing $I$ and, vice versa, that if $K$ is an ideal of $A$ containing $I$ then $\pi(K)$ is an ideal of $A/I$ such that $\pi^{-1}(\pi(K)) = K$.

**A3.11.** Prove that in $\mathbb{Z}$ the following *cancellation property* holds:

$$\text{if } ac = bc \text{ and } c \neq 0, \text{ then } a = b. \tag{3.17}$$

Prove that this cancellation property holds, more generally, in any integral domain.

**A3.12.*** Let $A$ be a ring with unity and $I$ a proper ideal of $A$. Prove that $A/I$ is an integral domain if and only if $I$ is a prime ideal.

**A3.13.** Prove Proposition 3.1.11.

**A3.14.** Is it true that $\mathbb{Z}_n$ is a field if and only if it is an integral domain?

(a) Yes, because a commutative ring with unity is a field if and only if it is an integral domain with unity.
(b) Yes.

(c) No.

(d) No, because there are integral domains that are not fields.

**A3.15.** Is it true that $\mathbb{Z}_n$ is a field if and only if $n$ is a prime number?

(a) Yes.

(b) No.

(c) No, because $\mathbb{Z}_n$ being a field does not imply $n$ being a prime.

(d) None of the above.

**A3.16.** Fix a positive integer $b$ and associate with each integer number $n$, written in base $b$ as $(a_h \ldots a_1 a_0)_b$, and consequently of length $h+1$, the polynomial $f_n(x) = a_0 + a_1 x + \cdots + a_h x^h$ of degree $h$ in $\mathbb{Z}_b[x]$. Prove that the map $f : \mathbb{N} \to \mathbb{Z}_b[x]$ so defined is a bijection but it is not true in general that

$$f_{n+m}(x) = f_n(x) + f_m(x),$$

nor

$$f_{n \cdot m}(x) = f_n(x) \cdot f_m(x).$$

**A3.17.** Prove that the complexity of adding up two polynomials of degree at most $m$ in $\mathbb{Z}_n[x]$ is $\mathcal{O}(m \log n)$.

**A3.18.** Prove that the complexity of multiplying two polynomials of degree at most $m$ in $\mathbb{Z}_n[x]$ is $\mathcal{O}(m^2 \log^2 n)$.

**A3.19.** Describe a "casting out elevens" procedure to verify computations with base 10 numbers.

**A3.20.**\* Prove that for any positive integer $n$ we have $(b-1) \mid b^n - 1$ and $(b+1) \mid (b^n + (-1)^{n+1})$. Deduce a test of divisibility for $b-1$ and one for $b+1$ based on the digits of integer numbers written in base $b$. (Hint: mimic the $b = 10$ case.)

**A3.21.** Find a procedure to carry out *casting out nines* or *casting out elevens* in an arbitrary base $b$.

**A3.22.**\* Use the congruence $40 \equiv 1 \pmod{13}$ to deduce a test of divisibility for 13.

In the exercises that follow some useful properties of cyclic groups are recalled. The group operation will usually be written multiplicatively.

**A3.23.** Prove that $(\mathbb{Z}, +)$ is an infinite cyclic group. Prove that $(\mathbb{Z}_n, +)$ is a cyclic group of order $n$.

**A3.24.**\* Prove that an infinite cyclic group is isomorphic to the additive group of $\mathbb{Z}$, while a cyclic group of order $n$ is isomorphic to the additive group of $\mathbb{Z}_n$.

**A3.25.**\* Prove that a subgroup of a cyclic group is itself cyclic.

**A3.26.**\* If $G$ is a cyclic group generated by an element $x$ and has finite order $n$, prove that the elements of $G$ are $1, x, x^2, \ldots, x^{n-1}$.

**A3.27.**\* Let $x$ be an element of period $m$ of a group $G$. Prove that $x^n = 1$ if and only if $m \mid n$.

**A3.28.*** Let $x$ be an element of finite order $m$ of a group $G$. Prove that the powers of $x$ form a cyclic subgroup of $G$ of order $m$.

**A3.29.*** If $G$ is a cyclic group generated by an element $x$ and has finite order $n$, prove that $x^h = x^k$ if and only if $h \equiv k \pmod{n}$. In particular, $x^m = 1$ if and only if $n \mid m$.

**A3.30.*** If $G$ is a cyclic group generated by an element $x$ and has finite order $n$, the element $x^m$ is a generator of $G$ if and only if $\mathrm{GCD}(m, n) = 1$. Deduce that in $G$ there are $\varphi(n)$ distinct generators.

**A3.31.*** If $G$ is an infinite cyclic group generated by an element $x$, then the only generators are $x$ and $x^{-1}$.

**A3.32.*** If $G$ is a cyclic group generated by an element $x$ and has finite order $n$, and if $d$ is a divisor of $n$, then there exists exactly one subgroup of $G$ of order $d$.

**A3.33.*** Let $G$ be a cyclic group generated by an element $x$ and having finite order $n$. Let $n = dm$. The element $x^h$ has order $d$ if and only if $h \equiv mi \pmod{n}$ with $\mathrm{GCD}(i, d) = 1$. Deduce that in $G$ there are $\varphi(d)$ distinct elements of order $d$.

**A3.34.*** Let $G$ be a cyclic group generated by an element $x$ and having finite order $n$, and let $y_1 = x^{n_1}$ and $y_2 = x^{n_2}$ be two elements of $G$. Find the integers $m$, if any, such that $y_1^m = y_2$.

**A3.35.*** If $G$ is an infinite cyclic group, every subgroup of $G$ other than $(1)$ is infinite cyclic.

**A3.36.*** Prove Lagrange's theorem: if $G$ is a finite group and $H$ is a subgroup of $G$, then the order of $H$ divides the order of $G$.

**A3.37.*** Prove Euler's Theorem as a consequence of the fact that the period of an element divides the order of a finite group.

**A3.38.** Prove that if we know $\varphi(n)$ and if $\mathrm{GCD}(a, n) = 1$, then the complexity of computing $a^m \bmod n$ is $\mathcal{O}(\log^3 n)$.

**A3.39.** Consider the system of congruences

$$\begin{cases} x \equiv a \pmod{n}, \\ x \equiv b \pmod{m}, \end{cases}$$

where we are not assuming $\mathrm{GCD}(n, m) = 1$. Find the conditions for the system to be solvable. (Hint: the solvability is equivalent to the existence of an integer $t$ such that $a - b + tn \equiv 0 \pmod{m}$. Then apply Proposition 3.3.4).

**A3.40.** Prove that the map $f$ defined in Proposition 3.4.5 preserves the sum and the product, that is to say $f(y + z) = f(y) + f(z)$ and $f(yz) = f(y)f(z)$ for all $y, z \in \mathbb{Z}_{rs}$.

**A3.41.** Extend Proposition 3.4.5 to more than two factors, that is to say, prove that if $r_1, \ldots, r_s$ are positive integers relatively prime to each other, if we define $r = r_1 \cdots r_s$, then the map

$$[x]_r \in \mathbb{Z}_r \longrightarrow ([x]_{r_1}, \ldots, [x]_{r_s}) \in \mathbb{Z}_{r_1} \times \cdots \times \mathbb{Z}_{r_s}$$

is a ring isomorphism.

**A3.42.** Verify that, following the rule described on page 136, each of the $N$ teams play all the other teams once each over the $N-1$ rounds of a round-robin tournament.

# B3 Computational exercises

**B3.1.** How many elements are there in $\mathbb{Z}_{1472}$?

(a) No element.
(b) 1471.
(c) 1472.
(d) None of the above.

**B3.2.** Which of the following equals 36728 mod 11?

(a) 10.
(b) 5.
(c) 1.
(d) None of the above.

**B3.3.** Are there zero-divisors in $\mathbb{Z}_6$?

(a) No.
(b) Yes, there are exactly two zero-divisors.
(c) Yes, there are exactly three zero-divisors.
(d) None of the above.

**B3.4.** Are there zero-divisors in $\mathbb{Z}_{19}$?

(a) No.
(b) Yes, there are exactly three zero-divisors.
(c) Yes, there are exactly nine zero-divisors.
(d) None of the above.

**B3.5.** Is it true that $\mathbb{Z}_{27}$ is a field?

(a) Yes.
(b) No.
(c) No, it is not, even if it is zero-divisor free.
(d) None of the above.

**B3.6.** Determine the last digit of the number $725843^{594}$.

**B3.7.** Find the residue class modulo 9 of $74^{6^h}$, when $h$ varies in $\mathbb{N}$.

**B3.8.** Find the remainder of the division by 10 of the number $43816^{20321}$.

**B3.9.** Find the remainder of the division by 6 of the number $29345^{362971}$.

**B3.10.** Find the remainder of the division by 6 of the number $362971^{29345}$.

**B3.11.** In which residue class modulo 9 is $725843^{594}$?

**B3.12.** Which is the remainder of the division of $239487^{192387}$ by 6?

(a) 3.
(b) 2.
(c) 1.
(d) None of the above.

**B3.13.** Which is the remainder of the division of $7574632^{2845301}$ by 7?

(a) 2.
(b) 3.
(c) 4.
(d) None of the above.

**B3.14.** Which is the remainder of the division of $7574632^{2845301}$ by 11?

(a) 0.
(b) 1.
(c) 10.
(d) None of the above.

**B3.15.** Find the inverse, if it exists, of the following classes in $\mathbb{Z}_8$: $\bar{6}$, $\bar{3}$.

**B3.16.** Which is the inverse of 4 modulo 9?

(a) 4 is not invertible.
(b) 5.
(c) 7.
(d) None of the above.

**B3.17.** Which is the inverse of 4 modulo 18?

(a) 4 is not invertible.
(b) 11.
(c) 15.
(d) None of the above.

**B3.18.** Find the last two digits of all the numbers $x \geq 0$ such that

$$\begin{cases} 3x \equiv 1 \pmod 4, \\ 2x \equiv 3 \pmod{25}. \end{cases}$$

**B3.19.** Without carrying out the division, determine whether the number

$$129873498712738753786$$

is divisible by 11.

(a) No, it is not divisible by 11.
(b) It is impossible to say without carrying out the division.
(c) Yes, it is divisible by 11.
(d) None of the above.

**B3.20.** Is the operation $273498 \cdot 587234 = 160607234532$ correct?

(a) Yes, it is correct, as it satisfies the 'casting out nines' procedure.
(b) Yes, it is correct.
(c) No, it is not correct, as it does not satisfy the 'casting out nines' procedure.
(d) No, it is not correct, even though it satisfies the 'casting out nines' procedure.

**B3.21.** Determine whether the base 2 number 100100110 is divisible by 2, 4, 3.

**B3.22.** Determine whether the base 3 number 12002122 is divisible by 2, 3, 9, 4.

**B3.23.** Determine a test of divisibility by 3 that uses the digits of numbers written in base 9.

**B3.24.** Determine all the solutions of the congruence

$$3x \equiv 5 \pmod 4.$$

(a) $x = 3 + 4k$, for all $k \in \mathbb{Z}$.
(b) $x = 1 + 2k$, for all $k \in \mathbb{Z}$.
(c) It does not admit any solution.
(d) None of the above.

**B3.25.** Determine all the solutions of the congruence

$$3x \equiv 9 \pmod 6.$$

(a) $x = 3 + 6k$, for all $k \in \mathbb{Z}$.
(b) $x = 1 + 2k$, for all $k \in \mathbb{Z}$.
(c) It does not admit any solution.
(d) None of the above.

**B3.26.** Determine all the solutions of the congruence

$$4x \equiv 7 \pmod 9.$$

(a) $x = -5 + 9k$, for all $k \in \mathbb{Z}$.
(b) $x = 1 + 3k$, for all $k \in \mathbb{Z}$.
(c) $x = 5 + 9k$, for all $k \in \mathbb{Z}$.
(d) None of the above.

**B3.27.** Determine all the solutions of the congruence

$$6x \equiv 8 \pmod 9.$$

(a) $x = 3 + 9k$, for all $k \in \mathbb{Z}$.
(b) $x = -1 + 3k$, for all $k \in \mathbb{Z}$.
(c) It does not admit any solution.
(d) None of the above.

**B3.28.** Determine all the solutions of the congruence

$$4x \equiv 3 \pmod{385}.$$

(a) $x = 74 + 385k$, for all $k \in \mathbb{Z}$.
(b) $x = 75 + 385k$, for all $k \in \mathbb{Z}$.
(c) $x = -1 + 385k$, for all $k \in \mathbb{Z}$.
(d) None of the above.

**B3.29.** Construct, if possible, a linear congruence of the form

$$ax \equiv b \pmod{319}$$

admitting exactly 11 solutions that are not congruent to each other modulo 319. Then determine all the solutions of such a congruence.

**B3.30.** Determine two integers $a$ and $b$ (if it is possible) in such a way that the congruence

$$ax \equiv b \pmod{299}$$

has exactly 13 non-congruent solutions modulo 299. If such $a$ and $b$ are found, solve the congruence and determine all the solutions.

**B3.31.** Determine all the solutions of the linear congruence

$$128x \equiv 10 \pmod{17}.$$

(a) It does not admit any solution.
(b) $3 + 17k$, per ogni $k \in \mathbb{Z}$.
(c) $5 + 17k$, per ogni $k \in \mathbb{Z}$.
(d) None of the above.

**B3.32.** Determine all the solutions of the linear congruence

$$3128x \equiv 3 \pmod{1024}.$$

(a) It does not admit any solution.
(b) $513 + 1024k$, for all $k \in \mathbb{Z}$.
(c) $1007 + 1024k$, for all $k \in \mathbb{Z}$.
(d) None of the above.

**B3.33.** Determine all the solutions of the linear congruence

$$2047x \equiv 3 \pmod{1024}.$$

(a) $3 + 1024k$, for all $k \in \mathbb{Z}$.
(b) $319 + 1024k$, for all $k \in \mathbb{Z}$.
(c) $1021 + 1024k$, for all $k \in \mathbb{Z}$.
(d) None of the above.

**B3.34. B3.35.** Determine all the solutions of the linear congruence

$$2047x \equiv 1022 \pmod{1024}.$$

(a) $2 + 1024k$, for all $k \in \mathbb{Z}$.
(b) $3 + 1024k$, for all $k \in \mathbb{Z}$.
(c) $-1 + 1024k$, for all $k \in \mathbb{Z}$.
(d) None of the above.

**B3.36.** Find the remainder of the division by 17 of

$$190^{597}.$$

**B3.37.** Verify that $17^{31} \equiv 41 \pmod{58}$.

**B3.38.** Prove that $3^{13} \equiv 3 \pmod 7$ by using Euler's Theorem.

**B3.39.** Which of the following equals $40^{65}$ mod 199?

(a) 5.
(b) 71.
(c) 193.
(d) None of the above.

**B3.40.** Which of the following equals $27^{33}$ mod 157?

(a) 2.
(b) 64.
(c) 82.
(d) None of the above.

**B3.41.** Which of the following equals $31^{96}$ mod 359?

(a) 1.
(b) 79.
(c) 283.
(d) None of the above.

**B3.42.** Consider the system of congruences

$$\begin{cases} x \equiv 3 \pmod 5, \\ x \equiv 7 \pmod 9. \end{cases}$$

Does this system admit solutions? If it does, which are the solutions?

(a) Yes, the solutions are $x = -34 + 45k$, for all $k \in \mathbb{Z}$.
(b) Yes, the solutions are $x = 34 + 45k$, for all $k \in \mathbb{Z}$.
(c) Yes, the solutions are $x = 43 + 45k$, for all $k \in \mathbb{Z}$.
(d) None of the above.

**B3.43.** Consider the system of congruences

$$\begin{cases} 1025x \equiv 5312065 \pmod 8, \\ 36x \equiv 322 \pmod 5, \\ 4x \equiv 7 \pmod 3. \end{cases}$$

Does this system admit solutions? If it does, which are the solutions?

(a) Yes, the solutions are $x = 77 + 120k$, for all $k \in \mathbb{Z}$.
(b) Yes, the solutions are $x = -21 + 120k$, for all $k \in \mathbb{Z}$.
(c) Yes, the solutions are $x = 82 + 120k$, for all $k \in \mathbb{Z}$.
(d) None of the above.

**B3.44.** Determine all the solutions of the system of linear congruences

$$\begin{cases} 3x \equiv 65 & (\text{mod } 7), \\ 11x \equiv 4 & (\text{mod } 17). \end{cases}$$

(a) $111 + 119k$, for all $k \in \mathbb{Z}$.
(b) $11 + 119k$, for all $k \in \mathbb{Z}$.
(c) $-12 + 119k$, for all $k \in \mathbb{Z}$.
(d) None of the above.

**B3.45.** Determine all the solutions of the system of linear congruences

$$\begin{cases} 2x \equiv 1 & (\text{mod } 3), \\ 8x \equiv 7 & (\text{mod } 11), \\ 5x \equiv 3 & (\text{mod } 13). \end{cases}$$

(a) There are no solutions.
(b) $368 + 429k$, for all $k \in \mathbb{Z}$.
(c) $5 + 429k$, for all $k \in \mathbb{Z}$.
(d) None of the above.

**B3.46.** Determine all the solutions of the system of linear congruences

$$\begin{cases} 2x \equiv 4 & (\text{mod } 6), \\ 8x \equiv 3 & (\text{mod } 13), \\ 12x \equiv 1 & (\text{mod } 18). \end{cases}$$

(a) There are no solutions.
(b) $107 + 351k$, for all $k \in \mathbb{Z}$.
(c) $-107 + 351k$, for all $k \in \mathbb{Z}$.
(d) None of the above.

**B3.47.** Determine all the solutions of the system of linear congruences

$$\begin{cases} 4x + 2 \equiv 3x - 1 & (\text{mod } 5), \\ 6x - 3 \equiv 2(x - 1) & (\text{mod } 7), \\ 2x \equiv 1 & (\text{mod } 3). \end{cases}$$

(a) $57 + 105k$, for all $k \in \mathbb{Z}$.
(b) $101 + 105k$, for all $k \in \mathbb{Z}$.
(c) $-103 + 105k$, for all $k \in \mathbb{Z}$.
(d) None of the above.

**B3.48.** Each Saturday morning, Mrs Smith does the shopping for her family, while every five days she baths her puppy. This week she bathed her puppy on a Sunday (11 March). When will she first (if ever) have to do the shopping and bathing the puppy on the same day?

**B3.49.** We are rearranging the books of a library. By stacking them 11 at a time, four books are left out; stacking them 13 at a time, five books are left out; and, by stacking them 17 at a time, a single book is left out. If we know that in the library there are less than 2500 books, is it possible to find the exact number of books?

**B3.50.** We have a tray full of sweets and we know that there are less then 200 of them. Moreover, we know that by dividing them in 11 shares two sweets are left out, dividing them in 5 shares three are left out, while dividing them in three shares no one is left out. Find the number of sweets on the tray.

(a) 123.
(b) 138.
(c) 153.
(d) None of the above.

**B3.51.** If from a box full of buttons they are removed two at a time, three at a time, four at a time, five at a time, or six at a time, in the box is always left a single button. If they are removed seven at a time, no button remains. Which is the least number of buttons the box can possibly contain?

(a) 241.
(b) 601.
(c) 301.
(d) None of the above.

**B3.52.** Solve the system of congruences

$$\begin{cases} -11 \cdot x \equiv 117 & (\text{mod } 4), \\ 395 \cdot x \equiv -7 & (\text{mod } 3), \\ 18 \cdot x \equiv -25 & (\text{mod } 7). \end{cases}$$

**B3.53.** What day of the week did 31 December 2000 fall on?

(a) Thursday.
(b) Friday.
(c) Saturday.
(d) Sunday.

**B3.54.** What day of the week did 28 February 2004 fall on?

(a) Thursday.
(b) Friday.
(c) Saturday.
(d) Sunday.

**B3.55.** What day of the week did 14 July 1789 fall on?

(a) Saturday.
(b) Sunday.
(c) Monday.
(d) Tuesday.

**B3.56.** What day of the week will 1 January 3001 fall on?

(a) Sunday.
(b) Monday.
(c) Tuesday.
(d) Thursday.

**B3.57.** Draw up explicitly the calendar of a round-robin tournament with 6 teams, following the method described in this chapter.

# C3 Programming exercises

**C3.1.** Write a program that computes the sum of two integer numbers $a$ and $b$ modulo a positive integer $n$.

**C3.2.** Write a program that computes the product of two integer numbers $a$ and $b$ modulo a positive integer $n$.

**C3.3.** Write a program that computes the inverse of a number $a$ modulo $n$, that is the number $x$ such that $ax \equiv 1 \pmod{n}$.

**C3.4.** Write a program that computes $a^m$ mod $n$ using the method described in the text.

**C3.5.** Write a program that verifies if a linear congruence $ax \equiv b \pmod{n}$ admits solutions and that, if it does, finds them.

**C3.6.** Write a program that verifies if a system of $m$ linear congruences admits solutions and that, if it does, finds them.

**C3.7.** Write a program that, given $r$, $s$ and $x$, computes the image of $x$ under the function $f : \mathbb{Z}_{rs} \to \mathbb{Z}_r \times \mathbb{Z}_s$ defined in Proposition 3.4.5.

**C3.8.** Write a program that, given two relatively prime integers $r$ and $s$ and a pair $(y, z)$, computes the preimage of $(y, z)$ under the function $f : \mathbb{Z}_{rs} \to \mathbb{Z}_r \times \mathbb{Z}_s$ defined in Proposition 3.4.5.

**C3.9.** Write a program that computes the day of the week of a date according the Gregorian calendar.

**C3.10.** Write a program that draws up the calendar of a round-robin tournament with $n$ teams.

# 4

## Finite is not enough: factoring integers

In the previous chapters, we met more than once the notion of a *prime number* (see § 1.3.1, Remark 3.1.10 and so forth). In this chapter we are going to investigate further this notion, first by proving the Fundamental Theorem of Arithmetic 4.1.2, well known since primary school, which basically states that prime numbers are the building blocks from which, by multiplication, all integer numbers can be obtained. This theorem will yield several important consequences: we shall for instance deduce a formula to compute Euler function $\varphi(n)$ (see § 3.3), and much more. Moreover, we shall frame the notion of prime number in a more general context, studying similar notions in an arbitrary ring, among them the ring of polynomials over a field.

As we shall see, prime numbers and the Fundamental Theorem of Arithmetic play a basic role in the applications to cryptography and to coding theory. But how can we tell if a number is prime? And how can we write an arbitrary number as a product of prime numbers? In this chapter we shall discuss some classic, and therefore elementary, algorithms for these tasks. We shall see that, unfortunately, these algorithms are in general exponential and are so of little practical use. It will then be necessary to find more efficient algorithms requiring more advanced mathematical tools: we shall discuss such algorithms later (see Chapter 6).

## 4.1 Prime numbers

Recall (see Section 1.3) that a prime number $p$ is an integer greater than 1 having no divisors other than $\pm p$ and $\pm 1$.

A prime number is sometimes also called *irreducible*, in agreement with the more general definition we shall discuss later (see Definition 4.5.1). Clearly, a number that is not prime is said to be *reducible* or *composite*.

### 4.1.1 The Fundamental Theorem of Arithmetic

As is well known since primary school, by multiplying prime numbers we can get in an essentially unique way each positive integer number. This is the gist of the Fundamental Theorem of Arithmetic, which Euclid already knew and expounded in Book VII of his Elements. Before discussing the Fundamental Theorem of Arithmetic, it is necessary to state the following simple but crucial characterisation of prime numbers, which we shall return to for some important remarks.

**Proposition 4.1.1.** *A positive integer number $p > 1$ is irreducible if and only if the following property holds:*

$(P)$ *whenever $p$ divides a product $ab$, $p$ divides $a$ or $b$.*

PROOF. We assume that $p$ is irreducible and prove that $(P)$ holds. So, assume that $p$ divides $ab$ and that $p$ does not divide $a$. As $p$ does not divide $a$, and as $p$ has no factors other than $\pm p$ and $\pm 1$, we have that $p$ and $a$ have no non-trivial common factors, that is to say, $\mathrm{GCD}(a, p) = 1$. So there exist integers $s$ and $t$ such that $1 = sa + tp$. By multiplying both sides by $b$ we obtain $b = sab + tpb$. As $p \mid ab$ and $p \mid p$, we may conclude that $p \mid b$.

Vice versa, assume that $(P)$ holds. If $p$ were not prime, we would have $p = hk$ with $h$, $k$ positive integers smaller than $p$. But we have $p \mid hk = p$, so either $p \mid h$ or $p \mid k$, both of which are impossible because $h$ and $k$ are smaller than $p$.                                                                                    □

We are now in a position to prove the:

**Theorem 4.1.2 (Fundamental Theorem of Arithmetic).** *Let $n$ be an integer greater than 1. Then*

$$n = p_1^{h_1} p_2^{h_2} p_3^{h_3} \cdots p_s^{h_s}, \qquad (4.1)$$

*where $p_1, p_2, \ldots, p_s$ are distinct prime numbers and the exponents $h_j$ are positive, for all $j = 1, \ldots, s$. Moreover, the representation (4.1) for $n$, called* prime decomposition *or* factorisation *of $n$, is unique up to the order of the factors.*

PROOF. We shall prove separately the existence and the uniqueness of the factorisation.

• *Existence of a factorisation.*

The proof is by induction on the integer $n$ to be factored. If $n = 2$, there is nothing to prove. So, we may assume that the existence of a factorisation has been proved for each positive integer $k$ with $2 \leq k < n$, and prove the same for $n$. If $n$ is prime, again there is nothing to prove. Thus, let $n$ be reducible, so it may be written as $n = ab$, with positive $a$ and $b$, both greater than 1 and therefore smaller than $n$. Then, by the induction hypothesis, $a$ and $b$ are factorisable as products of primes

$$a = p_1 p_2 \cdots p_r, \qquad b = \bar{p}_1 \bar{p}_2 \cdots \bar{p}_s.$$

Then,

$$n = p_1 p_2 \cdots p_r \bar{p}_1 \bar{p}_2 \cdots \bar{p}_s.$$

Clearly, it suffices to group together, in the right-hand side, equal prime numbers in order to get the result in the form (4.1).

- *Uniqueness of the factorisation.*

In order to prove the uniqueness of the factorisation for any integer $n$, we proceed by induction, this time on the number $m$ of irreducible factors in any factorisation of $n$. Notice that the number of factors appearing in the factorisation (4.1) is $m = h_1 + h_2 + \cdots + h_s$. If $m = 1$, then the number $n$ having that factorisation is a prime number $p$. Assume that $n = p$ has another factorisation

$$p = q_1^{k_1} q_2^{k_2} \cdots q_t^{k_t}.$$

As $p$ is a prime that divides the right-hand side, it divides one of the factors of the right-hand side, for instance $p \mid q_1$ (see Proposition 4.1.1). But $q_1$ is prime as well, so it has no non-trivial factors: hence $p = q_1$. By the cancellation property, which holds in $\mathbb{Z}$, we get

$$1 = q_1^{k_1-1} q_2^{k_2} \cdots q_t^{k_t}.$$

This relation implies that all the right-hand side exponents are zero, otherwise we would have a product equal to 1 of integers greater than 1. Then the original right-hand side equals $q_1$, so $p = q_1$ is the only factorisation of $n$. So we have proved the *basis* of the induction.

Assume now that the uniqueness of the factorisation has been proved for all integers admitting a factorisation into $m - 1$ irreducible factors. Let $n$ be an integer admitting a factorisation into $m$ irreducible factors. Let then

$$n = p_1^{h_1} p_2^{h_2} \cdots p_s^{h_s} = q_1^{k_1} q_2^{k_2} \cdots q_t^{k_t}$$

be two factorisations of $n$ into irreducible factors, the first one consisting of $m$ irreducible factors, that is to say, $h_1 + h_2 + \cdots + h_s = m$. Now, $p_1$ is a prime dividing the right-hand side, so it divides for instance $q_1$ (see again Proposition 4.1.1). As before, we have $p_1 = q_1$ and then, by the cancellation property, we get

$$p_1^{h_1-1} p_2^{h_2} \cdots p_s^{h_s} = q_1^{k_1-1} q_2^{k_2} \cdots q_t^{k_t}$$

where the number of irreducible factors in the left-hand side is $m - 1$. By the induction hypothesis, in this case the uniqueness of the factorisation holds, so the primes $q_j$ and the primes $p_i$ are the same, up to the order. Then the factorisation of $n$ is also unique. □

## 4.1.2 The distribution of prime numbers

How many prime numbers are there? Euclid already knew the following theorem which can be proved in several ways. We give here the completely elementary proof dating back to Euclid himself.

**Theorem 4.1.3.** *There are infinitely many prime numbers.*

PROOF. Assume the set of prime numbers to be finite, consisting for instance of the numbers $p_1 < p_2 < \cdots < p_n$. Consider the number $N = p_1 \cdots p_n + 1$. This number is not prime, as it is greater than $p_n$ which, by hypothesis is the greatest prime number. So $N$ has a prime decomposition which can be written as

$$N = p_1^{h_1} \cdots p_n^{h_n}$$

with at least one of the numbers $h_1, \ldots, h_n$ positive. Assume $h_i > 0$. Then $p_i \mid N$. Moreover, $p_i \mid (N - 1) = p_1 \cdots p_n$. Thus, $p_i \mid 1 = N - (N - 1)$, which is not possible, as $p_i > 1$. □

As already mentioned, there are several proofs of this theorem. We give now another proof, which relies on a famous theorem by Euler (for still another proof, again by Euler, see Exercise A4.6).

**Theorem 4.1.4 (Euler).** *The sum of the reciprocals of the primes diverges. In other words, if $\{p_1, \ldots, p_n, \ldots\}$ is the sequence, possibly finite, of all prime numbers in increasing order, we have*

$$\sum_{n=1}^{\infty} \frac{1}{p_n} = \infty.$$

It is not hard to give a sketch of the proof of this theorem relying on the properties of the famous *zeta function* defined by

$$\zeta(s) = \sum_{n=1}^{\infty} \frac{1}{n^s}.$$

First of all, notice that:

(1) $\zeta$ is well defined and converges if $s \in \mathbb{R} \cap (1, +\infty)$;
(2) $\lim_{s \to 1}(s - 1)\zeta(s) = 1$, and then, clearly,
(3) $\lim_{s \to 1} \zeta(s) = +\infty$.

To prove (1) it is sufficient to remark that

$$\int_1^{\infty} \frac{dt}{t^s} = \lim_{x \to +\infty} \int_1^x \frac{dt}{t^s} = \lim_{x \to \infty} \frac{x^{1-s} - 1}{1 - s} = \frac{1}{s - 1},$$

so it converges.

To prove (2) notice that

$$\left| \sum_{t=1}^{n} \frac{1}{t^s} - \int_1^{n+1} \frac{dt}{t^s} \right| < 1. \tag{4.2}$$

Indeed, it is sufficient to consider the graph of the function $y = 1/x^s$ and to interpret geometrically the difference in the left-hand side: we leave the details to the reader (see Exercise A4.5).

Computing the integral in the left-hand side we find

$$\left| \sum_{t=1}^{n} \frac{1}{t^s} - \frac{1}{s-1}\left(1 - \frac{1}{(n+1)^{s-1}}\right) \right| < 1;$$

taking the limit as $n$ approaches to $\infty$ we get

$$\left| \zeta(s) - \frac{1}{s-1} \right| < 1,$$

which concludes the proof of (2).

Before going further, notice that the zeta function has a fundamental importance in number theory. For further information, see [59].

Notice now that:

**Lemma 4.1.5.** *If* $\{p_n\}_{n \in \mathbb{N}}$ *is the sequence in increasing order of the prime numbers, we have*

$$\zeta(s) = \prod_n \frac{1}{1 - \dfrac{1}{p_n^s}}.$$

PROOF. One has

$$\frac{1}{1 - \dfrac{1}{p_n^s}} = \left(1 + \frac{1}{p_n^s} + \frac{1}{p_n^{2s}} + \cdots\right)$$

and so

$$\prod_n \frac{1}{1 - \dfrac{1}{p_n^s}} = \prod_n \left(1 + \frac{1}{p_n^s} + \frac{1}{p_n^{2s}} + \cdots\right).$$

By carrying out the product and keeping in mind the Fundamental Theorem of Arithmetic we get the claim.  □

Now we may give the:

PROOF OF EULER'S THEOREM 4.1.4.   Keeping in mind that $\log(1 + t) = \sum_n (-1)^{n+1} t^n / n$, compute:

$$\log \zeta(s) = \sum_n \log \frac{1}{1 - \dfrac{1}{p_n^s}} = -\sum_n \log\left(1 + \left(-\frac{1}{p_n^s}\right)\right) =$$

$$= \sum_n \left(\frac{1}{p_n^s} + \frac{1}{2p_n^{2s}} + \cdots\right) = \sum_n \frac{1}{p_n^s} + E.$$

If we prove that the *error* $E$ is bounded, the theorem follows from property (3) of $\zeta(s)$. Indeed, we have

$$E = \sum_n \left( \frac{1}{2p_n^{2s}} + \frac{1}{3p_n^{3s}} + \cdots \right) \le \sum_n \frac{1}{2p_n^{2s}} \left( 1 + \frac{1}{p^s} + \frac{1}{p^{2s}} + \cdots \right) =$$

$$= \sum_n \frac{1}{2p_n^{2s}} \frac{1}{1 - \frac{1}{p_n^s}} = \sum_n \frac{1}{2p_n^s(p_n^s - 1)} \le \sum_n \frac{1}{p_n^{2s}} < \sum_n \frac{1}{p_n^2} < \zeta(2). \qquad \square$$

The similarity of this theorem and of the well-known fact that the *harmonic series* $\sum_{n=1}^{\infty} 1/n$ diverges will not escape the reader. The other proof, again by Euler, of Theorem 4.1.3 relies on this similarity (see Exercise A4.6).

Here follows another famous result which also can be used to prove that prime numbers are infinitely many.

**Theorem 4.1.6 (Dirichlet).** *Let $n$ and $k$ be relatively prime positive integers. Then there are infinitely many prime numbers $p$ such that $p \equiv k \pmod{n}$.*

The proof can be found in [26].

Unfortunately, no known proof of Theorem 4.1.3 is constructive. In other words, none of them gives even a hint on how to find a *formula* for the $n$th prime number $p_n$, nor a recursive formula to get infinitely many prime numbers. Notice, however, that such a formula would certainly be, in any case, very elaborate, as the following proposition suggests.

**Proposition 4.1.7.** *No non-constant polynomial $f(x)$ in one indeterminate $x$, with coefficients in $\mathbb{Z}$, may assume prime values for all sufficiently large $n \in \mathbb{N}$.*

PROOF. Assume $f(x)$ is not constant and that $f(y) = p$ is a prime number. By using Taylor's formula (1.26), it is clear that $p$ divides $f(y + kp)$ for all $k \in \mathbb{N}$. If $f(n)$ were a prime number for all sufficiently large $n \in \mathbb{N}$, we would have $f(y + kp) = p$ for all sufficiently large $k$. On the other hand, $g(t) = f(y + tp)$ is a polynomial in $t$ having the same degree as $f(x)$. As $g(k) = p$ for all sufficiently large $k$, it follows that $g(t)$ is a constant, that is to say, it has degree 0. Then $f(x)$ would have degree 0 too, so it would be a constant, a contradiction. $\qquad \square$

Notice however that there exist integer-valued polynomials assuming prime values for *many* consecutive values of $n$. For instance, it may be verified that the polynomial

$$x^2 - x + 41$$

takes prime values for $x = n$, for all $n \in \{0, \ldots, 40\}$, but of course not for $x = 41$ (see Exercise C4.3).

A further, important remark is that no proof of Theorem 4.1.3 allows us to tell *how many* prime numbers are smaller than a fixed number. In particular, if for all real numbers $x > 0$ we denote by $\pi(x)$ the number of primes $p$ such that $p \le x$, there is no formula to compute this function in terms of *elementary functions*. However, Gauss, in 1792, when he was 15 years old, and Legendre before that, conjectured that $\pi(x)$ is *asymptotically* equal to a very simple function, that is, $x/(\log x)$; in other words they conjectured that the following holds:

$$\lim_{x \to \infty} \frac{\pi(x)}{\dfrac{x}{\log x}} = 1.$$

An important step toward proving this conjecture was made in 1851 by Chebyshev, who proved the following result:

**Theorem 4.1.8 (Chebyshev).** *There exist two numbers $A$ and $B$, with $0 < A \leq 1$ and $1 \leq B \leq 2$, such that for all sufficiently large $n \in \mathbb{N}$ the following relation holds*

$$A \frac{n}{\log n} \leq \pi(n) \leq B \frac{n}{\log n}.$$

The conjecture by Legendre and Gauss was proved only at the end of the 19th century by Hadamard and de la Vallée Poussin using analytic methods: the theorem is known as Prime Number Theorem (see [59]).

The Prime Number Theorem may be given a *probabilistic* interpretation: given a positive integer $n$, the probability that a positive integer $p < n$ randomly chosen between 2 and $n$ is prime is $\pi(n)/n$ which, for large $n$, is of the same magnitude as $1/\log n$.

Let us look at a simple example: suppose we are interested in finding a 100-digit prime number, that is to say, a number of the order of $n = 10^{100}$. The probability that a randomly chosen number among those with at most 100 digits is prime is, according to what precedes, $1/\log(10^{100}) \sim 1/230$. This means that, in 230 tries, we may hope to find a prime number of that magnitude. We may improve our chances by excluding even numbers (one half of all numbers) and multiples of 3 ($1/3$ of the numbers): in conclusion, the fraction of numbers we may exclude is $2/3 = 1/2 + 1/6$ (the multiples of two, plus the odd multiples of three). So the probability of finding a prime numbers becomes $1/77$.

To demonstrate how things work, we give the values of $\pi(x)$, of $[x/(\log x)]$ and of their ratio for $x = 10^3$, $x = 10^6$ and $10^9$:

| $x$ | $10^3$ | $10^6$ | $10^9$ |
|---|---|---|---|
| $\pi(x)$ | 168 | 78,498 | 50,847,478 |
| $[x/(\log x)]$ | 145 | 72,382 | 48,254,942 |
| $[\pi(x)(\log x)/x]$ | $1.159\ldots$ | $1.084\ldots$ | $1.053\ldots$ |

A further simple remark complements what precedes: *given a positive integer $n$, there exist $n$ consecutive integers that are not prime*. In fact, it is sufficient to consider the consecutive integers

$$(n+1)! + 2, \ (n+1)! + 3, \ (n+1)! + 4, \ \ldots, \ (n+1)! + (n+1).$$

Each of these numbers is reducible: the first one is divisible by 2, the second one by 3, up to the last one, which is divisible by $n+1$.

This would seem to mean that the larger the prime numbers become, the more apart they are. But... Is this really the case?

To illustrate how little is known about this simple question, we need yet another definition:

**Definition 4.1.9.** *Given two positive integers $p$ and $p+2$, if both of them are prime they are said to be* twin prime numbers.

Examples of pairs of twin primes are:

$$(3,5), \ (5,7), \ (11,13), \ (17,19), \ (41,43), \ \ldots, \ (2111,2113).$$

The number of twin prime numbers smaller than $10^{11}$ is 224,376,048 (see for instance [48]). There are isolated examples of twin primes with more than one thousand digits. Nonetheless, we do not yet know whether there exist infinitely many pairs of twin primes or not, but it is conjectured that this is the case. Contrast this with the following theorem, to be compared with Euler's theorem 4.1.4 above (again, see [59]):

**Theorem 4.1.10 (Brun, 1919).** *The sum of the reciprocals of the twin primes converges. In other words, if*

$$\{\, p_1, \ p_1 + 2, \ \ldots, \ p_n, \ p_n + 2, \ \ldots \}$$

$\{p_1, \ldots, p_n, \ldots\}$ *is the possibly finite sequence of all twin prime numbers in increasing order, we have*

$$\sum_{n=1}^{\infty} \left( \frac{1}{p_n} + \frac{1}{p_n + 2} \right) < \infty.$$

Notice that this does not imply at all that the sequence of twin primes is finite.

The investigation about the existence of arbitrarily large twin primes is not simply a matter of curiosity: it is relevant in computer science, as the possibility of some efficient systems for computer data storing is connected with the existence of suitable twin primes (see [49], § 4.4).

Let us close this section by stating the most famous conjecture still unsettled about prime numbers, the *Goldbach conjecture* (1742):

> every even positive number greater than 2 is a sum of two primes.

The conjecture has been verified for all even numbers smaller than $3 \cdot 10^{17}$. Let us see some examples for small numbers:

$$
\begin{array}{ll}
4 = 2 + 2, & 14 = 7 + 7, \\
6 = 3 + 3, & 16 = 3 + 13, \\
8 = 3 + 5, & 18 = 5 + 13, \\
10 = 5 + 5, & 20 = 3 + 17, \\
12 = 5 + 7, & 22 = 11 + 11.
\end{array}
$$

The reader might find it entertaining to read the fine novel by A. Doxiadis [21] about this famous conjecture.

### 4.1.3 The sieve of Eratosthenes

It is important to remark that the Fundamental Theorem of Arithmetic assures us that every positive integer can be written in a unique way as a product of prime numbers, but does not give any algorithm to find the factorisation of a given number. It is a typical example of purely *theoretical*, rather than *constructive*, result. Actually, finding the prime decomposition of a number is far from easy; in fact, by the present-day mathematical and technological knowledge it is practically unfeasible when we are dealing with very large numbers. This is a very important point to keep in mind, now and in what follows!

As we shall see, no polynomial time *factoring algorithm* that could be implemented on an existing computer is known. It is precisely on the difficulty of factoring a number into prime factors that the *current* security of some cryptographic techniques relies, that is to say, of being able to transmit secret data between two subjects without allowing third parties to eavesdrop. We shall return on these topics in the next chapters.

Another important question, partially connected to the previous one is: how do we tell if a number is prime? The algorithms to recognise whether a given number is prime or not are called *primality tests*. The most ancient method is the so-called *sieve of Eratosthenes* which we are dealing with in this section. As we shall see, this is an exponential algorithm, so it is quite inefficient when applied to large numbers, as it is basically a factoring algorithm more than a simple primality test. However, there are different, far more efficient primality tests: among them a polynomial one recently found by Agrawal, Kayal, and Saxena (see [2]). We shall return on this subject in Chapter 6.

Let us see now the sieve of Eratosthenes (276-194 BCE), which allows, in principle, to determine all prime numbers smaller than or equal to a fixed positive integer number $n$.

The most natural, but by far not the most efficient way to determine whether a number $n$ is prime consists in verifying that it is not divisible by any number preceding it, that is to say, by 2, 3, 4, ..., $n - 1$. The following result, based on the Fundamental Theorem of Arithmetic, reduces the number of necessary divisions.

**Proposition 4.1.11.** *If a positive integer $n$ is not divisible by any prime number smaller than or equal to $\sqrt{n}$, then $n$ is prime.*

PROOF. Assume $n$ to be reducible, that is to say that $n = ab$ with $a$ and $b$ integers such that $1 < a < n$ and $1 < b < n$. One of the factors, $a$ or $b$, is necessarily smaller than or equal to $\sqrt{n}$: otherwise we would have $n = ab > \sqrt{n} \cdot \sqrt{n} = n$, which is impossible. So $n$ has a factor, say $a$, smaller than or equal to $\sqrt{n}$. If it is prime, the proposition is proved. Otherwise, $a$ has a prime factor $p$, and $p < a \leq \sqrt{n}$. $\qquad\square$

**Example 4.1.12.** To prove that the number 397 is prime, it is sufficient to prove that it is not divisible by any prime smaller than or equal to $\sqrt{397}$, that is to say that it is not divisible by 2, 3, 5, 7, 11, 13, 17 and 19. The reader may easily check that this is the case.

The remark contained in Proposition 4.1.11 might look unimportant when dealing with small numbers, but when the numbers we are working with are large the time saved in the computations necessary to check the primality becomes evident.

In order to use the last result, however, we need to *know* the prime numbers smaller than or equal to $\sqrt{n}$, and this may result troublesome. So we may use that result in a weaker form, and check if $n$ is divisible by 2, 3, 4, 5, ..., $[\sqrt{n}]$, that is, by *all* integer numbers smaller than or equal to $\sqrt{n}$. In conclusion:

(1) if $n$ is not divisible by any of these numbers, then $n$ is a prime number;
(2) else, denoting by $n_1$ a factor of $n$, apply the same procedure to $n_1$ and $n/n_1$ (which is an integer), arriving finally to the complete factorisation of $n$ in prime factors.

Unfortunately, this is not an efficient method, as its complexity is exponential. Let us give an estimate of the number of bit operations necessary to verify whether a number $n$ is prime or not using this method. According to the prime number theorem, there are approximately

$$\frac{\sqrt{n}}{\log \sqrt{n}} = \frac{2\sqrt{n}}{\log n}$$

prime numbers smaller than or equal to $\sqrt{n}$. Keeping in mind formula (2.9), we see that the number of bit operations necessary to verify if $n$ is prime by dividing it by all prime numbers smaller than or equal to $\sqrt{n}$ is at least

$$\mathcal{O}\left(\frac{2\sqrt{n}}{\log n} \log^2 n\right) = \mathcal{O}(\sqrt{n}\log n) = \mathcal{O}(ke^{ck}),$$

where $k$ is the length of $n$ and $c$ is a constant.

Thus, the algorithm is exponential.

However, it is necessary to stress the fact that this procedure is more then just a primality test, as it does not only determine if a given number is prime, but also provides a factorisation if the number turns out to be composite.

To determine just the prime numbers smaller than or equal to a given number $n$, we can proceed in a slightly different way. Write down all the numbers smaller than or equal to $n$, starting with 2; underline it, as it is prime. Then delete all the multiples of 2 (as they are not prime). Underline the first number that has not been deleted, that is 3, and then delete all the multiples of 3, because again they are not prime, and so forth *until there are no non-deleted nor non-underlined numbers smaller than or equal to $\sqrt{n}$.*

Now, all the underlined numbers, together with all the numbers that have not been deleted, provide the complete list of all prime numbers smaller than or equal to $n$ The non-deleted numbers are prime because, not being divisible by any prime smaller than or equal to $\sqrt{n}$, they have no non-trivial divisor, by Proposition 4.1.11.

The term *sieve* describes precisely this procedure of successive deletion of non-prime numbers.

Let us see step by step how the procedure works when we want to find all the primes smaller than or equal to 100.

- *Step 1.* Write down all the numbers from 2 to 100:

$$
\begin{array}{cccccccccc}
 & 2 & 3 & 4 & 5 & 6 & 7 & 8 & 9 & 10 \\
11 & 12 & 13 & 14 & 15 & 16 & 17 & 18 & 19 & 20 \\
21 & 22 & 23 & 24 & 25 & 26 & 27 & 28 & 29 & 30 \\
31 & 32 & 33 & 34 & 35 & 36 & 37 & 38 & 39 & 40 \\
41 & 42 & 43 & 44 & 45 & 46 & 47 & 48 & 49 & 50 \\
51 & 52 & 53 & 54 & 55 & 56 & 57 & 58 & 59 & 60 \\
61 & 62 & 63 & 64 & 65 & 66 & 67 & 68 & 69 & 70 \\
71 & 72 & 73 & 74 & 75 & 76 & 77 & 78 & 79 & 80 \\
81 & 82 & 83 & 84 & 85 & 86 & 87 & 88 & 89 & 90 \\
91 & 92 & 93 & 94 & 95 & 96 & 97 & 98 & 99 & 100
\end{array}
$$

- *Step 2.* Underline 2 and remove all the multiples of 2:

$$
\begin{array}{ccccc}
\underline{2} & 3 & 5 & 7 & 9 \\
11 & 13 & 15 & 17 & 19 \\
21 & 23 & 25 & 27 & 29 \\
31 & 33 & 35 & 37 & 39 \\
41 & 43 & 45 & 47 & 49 \\
51 & 53 & 55 & 57 & 59 \\
61 & 63 & 65 & 67 & 69 \\
71 & 73 & 75 & 77 & 79 \\
81 & 83 & 85 & 87 & 89 \\
91 & 93 & 95 & 97 & 99
\end{array}
$$

- *Step 3.* Underline the next prime number, 3, and remove all the multiples of 3:

$$
\begin{array}{ccccc}
\underline{2} & \underline{3} & 5 & 7 & \\
11 & 13 & & 17 & 19 \\
 & 23 & 25 & & 29 \\
31 & & 35 & 37 & \\
41 & 43 & & 47 & 49 \\
 & 53 & 55 & & 59 \\
61 & & 65 & 67 & \\
71 & 73 & & 77 & 79 \\
 & 83 & 85 & & 89 \\
91 & & 95 & 97 & \\
91 & & 95 & 97 &
\end{array}
$$

- *Step 4.* Underline the next prime number, 5, and remove all its multiples:

$$
\begin{array}{cccc}
\underline{2}\ \underline{3} & \underline{5} & 7 & \\
11\quad 13 & & 17 & 19 \\
23 & & & 29 \\
31 & & 37 & \\
41\quad 43 & & 47 & 49 \\
53 & & & 59 \\
61 & & 67 & \\
71\quad 73 & & 77 & 79 \\
83 & & & 89 \\
91 & & 97 &
\end{array}
$$

- *Step 5.* Underline the next prime number, 7, and remove all its multiples:

$$
\begin{array}{cccc}
\underline{2}\ \underline{3} & \underline{5} & \underline{7} & \\
11\quad 13 & & 17 & 19 \\
23 & & & 29 \\
31 & & 37 & \\
41\quad 43 & & 47 & \\
53 & & & 59 \\
61 & & 67 & \\
71\quad 73 & & & 79 \\
83 & & & 89 \\
& & 97 &
\end{array}
$$

Now we are done, as the next prime is 11, which is greater than $10 = \sqrt{100}$. The underlined numbers, together with those that survived the deletions are the prime number smaller than 100.

Unfortunately, this is another inefficient method to find the prime numbers smaller than a given number: a computation analogous to that given above shows that it is exponential (see Exercise A4.7). For instance, in 1979 it has been proved with completely different methods that the 13395-digit number $2^{44497} - 1$ is prime: had we used the sieve of Eratosthenes, a computer carrying out one million multiplications per second would have taken $10^{6684}$ years to obtain this result!

## 4.2 Prime numbers and congruences

We begin this section by going back to an important question left open in Section § 3.3: how to compute Euler function $\varphi$.

### 4.2.1 How to compute Euler function

We need a definition first.

**Definition 4.2.1.** *A function* $f : \mathbb{N} \setminus \{0\} \to \mathbb{N} \setminus \{0\}$ *is said to be* multiplicative *if*

$$f(r \cdot s) = f(r) \cdot f(s) \qquad \forall \, r, s : \text{GCD}(r, s) = 1.$$

*The function* $f$ *is said to be* completely multiplicative *if*

$$f(r \cdot s) = f(r) \cdot f(s) \qquad \forall \, r, s \in \mathbb{N} \setminus \{0\}.$$

The definition itself of multiplicative function and the Fundamental Theorem of Arithmetic imply that:

**Proposition 4.2.2.** *Let* $f : \mathbb{N} \setminus \{0\} \to \mathbb{N} \setminus \{0\}$ *be a multiplicative function. If* $n = p_1^{h_1} p_2^{h_2} \cdots p_s^{h_s}$, *with* $p_i$ *distinct primes for* $i = 1, \ldots, s$, *then*

$$f(n) = f(p_1^{h_1}) f(p_2^{h_2}) \cdots f(p_s^{h_s}). \tag{4.3}$$

In other words, we may compute the value $f(n)$ of a multiplicative function $f$ for an integer $n$ if:

- we know the factorisation of $n$;
- we know the value of $f$ on all the prime powers.

The fundamental property of Euler function is that it is multiplicative, and this is the key ingredient to compute Euler function for all integers we know the prime decomposition of:

**Proposition 4.2.3.** *The Euler function* $\varphi$ *is multiplicative. Thus, if* $n = p_1^{h_1} p_2^{h_2} \cdots p_s^{h_s}$, *with* $p_i$ *distinct primes for* $i = 1, \ldots, s$, *then*

$$\varphi(n) = \varphi(p_1^{h_1}) \varphi(p_2^{h_2}) \cdots \varphi(p_s^{h_s}). \tag{4.4}$$

*Moreover, if* $p$ *is a prime number, then*

$$\varphi(p^h) = p^h - p^{h-1}. \tag{4.5}$$

*In particular,* $\varphi(p) = p - 1$. *So we have*

$$\varphi(n) = n \cdot \left(1 - \frac{1}{p_1}\right) \cdots \left(1 - \frac{1}{p_s}\right). \tag{4.6}$$

PROOF. Formula (4.4) follows from Proposition 4.2.2, once the multiplicativity of the function $\varphi$ is known. Formula (4.6) immediately follows from Equations (4.4) and (4.5).

So we have to prove the multiplicativity of the function $\varphi$. Let $n = rs$, with $\text{GCD}(r, s) = 1$. The numbers $m$ such that $0 \le m < n$ can be represented (see Proposition 3.4.5) as pairs of the form

$$(m \bmod r, m \bmod s).$$

The reader may also verify (Exercise A4.2) that an integer $m$ and the product $rs$ of $r$ and $s$ are relatively prime if and only if

$$\mathrm{GCD}(m \bmod r, r) = 1 \quad \text{and} \quad \mathrm{GCD}(m \bmod s, s) = 1.$$

So the total number $\varphi(rs)$ of elements $m$ modulo $rs$ that are coprime with $rs$ is $\varphi(r) \cdot \varphi(s)$, as there are $\varphi(r)$ elements modulo $r$ that are coprime with $r$ in the first element of the pair and $\varphi(s)$ elements modulo $s$ that are coprime with $s$ in the second element of the pair.

In order to complete the proof, we have to prove Formula (4.5). To this end, it is sufficient to remark that the only numbers that are *not* coprime with $p^h$ are the multiples of $p$, which are of the form

$$p \cdot i, \quad \text{con } 1 \leq i \leq p^{h-1},$$

so there are $p^{h-1}$ of them.    □

This proposition allows us to compute Euler function *for every integer $n$ for which the prime decomposition is known*. For instance,

$$\varphi(108) = \varphi(2^2 \cdot 3^3) = \varphi(2^2)\varphi(3^3) = (2^2 - 2)(3^3 - 3^2) = 36.$$

An immediate consequence of Proposition 4.2.3 and of Euler's theorem 3.3.11 is the following result (see also the Exercises A3.14, A3.15):

**Proposition 4.2.4.** *Let $n > 1$ be a positive integer. Then $\mathbb{Z}_n$ is a field if and only if $n$ is prime.*

PROOF. Clearly, $\mathbb{Z}_n$ is a field if and only if $U(\mathbb{Z}_n) = \mathbb{Z}_n \setminus \{0\}$. By Corollary 3.3.7, $U(\mathbb{Z}_n)$ has order $\varphi(n)$ and so $\mathbb{Z}_n$ is a field if and only if $\varphi(n) = n - 1$; by Proposition 4.2.3, this happens if and only if $n$ is prime.    □

### 4.2.2 Fermat's little theorem

Keeping in mind Euler's theorem 3.3.11, we may immediately deduce from it the following result:

**Theorem 4.2.5 (Fermat's little theorem).**  *Let $a$ be an integer and $p$ a prime number. Then*

$$a^p \equiv a \pmod{p} \tag{4.7}$$

*and, if $a$ is not divisible by $p$:*

$$a^{p-1} \equiv 1 \pmod{p}. \tag{4.8}$$

PROOF. If $a$ is divisible by $p$, then (4.7) is trivially true, as $a^p \equiv a \equiv 0 \pmod{p}$. If $a$ is not divisible by $p$, then (4.8) follows from Euler's theorem, and (4.7) follows by multiplying both sides by $a$.    □

Notice that it is possible to give elementary, direct proofs of Fermat's little theorem and of Euler's theorem without using group-theoretic properties, only relying on the Fundamental Theorem of Arithmetic (see Exercises A4.11, A4.12).

Fermat's little theorem may be used to create a primality test or, more precisely, to verify if a number *is not* prime, as

> if there exists a number $a$ such that $a^n \not\equiv a \pmod{n}$, then $n$ is not prime.

For instance, $n = 10$ is not prime because $2^{10} \not\equiv 2 \pmod{10}$.

Generally, we look for a small $a$ (for instance, $a = 2$) to keep computations as simple as possible.

Notice that this *non-primality* test, if it works, guarantees that the number *is not prime*, but does not give a factorisation of $n$.

The ancient Chinese mathematicians believed a number to be prime *if and only if*

$$2^{n-1} \equiv 1 \pmod{n}. \tag{4.9}$$

However, this is not true, as the following example shows.

**Example 4.2.6.** The composite number $341 = 11 \cdot 31$ satisfies congruence $2^{340} \equiv 1 \pmod{341}$. Indeed, by Fermat's little theorem, we have $2^{10} \equiv 1 \pmod{11}$, and so

$$2^{340} = (2^{10})^{34} \equiv 1 \pmod{11};$$

moreover,

$$2^{340} = (2^5)^{68} = 32^{68} \equiv 1 \pmod{31}.$$

The two congruences imply the result, by Proposition 3.1.11.

This means that (4.9) is a necessary, but not sufficient, condition for a number to be prime.

The following result shows that, by adding an extra hypothesis, Formula (4.9) becomes a sufficient condition for $n$ to be prime.

We need a definition first.

**Definition 4.2.7.** *Let $n$ be a positive integer and let $a$ be an integer coprime with $n$. The order of $[a]$ as an element of the group $U(\mathbb{Z}_n)$, that is to say the least positive integer $m$ such that $a^m \equiv 1 \pmod{n}$, will be called* Gaussian of $n$ with respect to $a$ *and will be denoted by* $\mathrm{Gss}(n, a)$.

For instance, it is clear that $\mathrm{Gss}(7, 2) = 3$. Computing the Gaussian of two numbers is not at all trivial. Keeping in mind what was mentioned in Remark 3.3.12, we immediately obtain:

**Corollary 4.2.8.** *Let $a$ and $n$ be relatively prime numbers. Then:*

- $\mathrm{Gss}(n, a)$ *divides* $\varphi(n)$;

- *if $m$ is an integer, then $a^m \equiv 1 \pmod{n}$ if and only if $\mathrm{Gss}(n,a) \mid m$;*
- *if $i, j \in \mathbb{Z}$, then $a^i \equiv a^j \pmod{n}$ if and only if $i \equiv j \pmod{\mathrm{Gss}(n,a)}$.*

For some further elementary properties of the Gaussian, see Exercise A4.30.

We may now state the following result.

**Proposition 4.2.9 (Lucas, 1876).** *Let $n > 1$. If there exists an integer $a$, with $1 < a < n$, such that $\mathrm{Gss}(n,a) = n - 1$, that is to say such that*

(1) $a^{n-1} \equiv 1 \pmod{n}$, *and*
(2) $a^m \not\equiv 1 \pmod{n}$, *for all $m = 1, 2, \ldots, n-2$,*

*then $n$ is prime.*

PROOF. Firstly, notice that $\mathrm{GCD}(a, n) = 1$, as this is the necessary and sufficient condition for the congruence $ax \equiv 1 \pmod{n}$ to admit solutions (see (3.3.4)).

So, suppose by way of contradiction that $n$ is not prime. Then $\varphi(n)$ is strictly smaller than $n - 1$ and, as $\mathrm{GCD}(a, n) = 1$, by Euler's theorem the order of $a$ would divide $\varphi(n) < n - 1$, contradicting the hypothesis. So $n$ is prime.    □

In conclusion, we have

> $n$ is prime if and only if there is an element $a$ such that $\mathrm{Gss}(n, a) = n - 1$.

Indeed, we may say that if there exists an integer $a$ such that $a^{n-1} \not\equiv 1 \pmod{n}$, then $n$ is *not* prime by Fermat's little theorem. On the other hand, if there exists an integer $a$ of order $n - 1$ modulo $n$, then $n$ is prime.

This test is not efficient, as it is necessary to carry out $n - 2$ multiplications to get the powers of $a$, and to find their remainders modulo $n$, which entails an exponential complexity. However, Lucas himself improved this criterion, by remarking that:

**Proposition 4.2.10.** *Let $n > 1$. If there exists an integer $a$, with $1 < a < n$, such that*

(1) $a^{n-1} \equiv 1 \pmod{n}$, *and*
(2) $a^m \not\equiv 1 \pmod{n}$, *for all $m < n$ dividing $n - 1$,*

*then $n$ is prime.*

PROOF. Analogous to the proof of Proposition 4.2.9.    □

It is clear that by using this proposition it is possible to reduce the number of multiplications and of reductions modulo $n$. However, in order to apply this test it is necessary *to know all the factors* of $n - 1$, and this is feasible only when $n - 1$ is of some special form. So, unless $n - 1$ is easily factorisable, this test is exponential too.

Although these tests have exponential complexity, we shall see later that they can nevertheless be used to construct new primality tests of a different type: the *probabilistic* ones (see § 6.1).

### 4.2.3 Wilson's theorem

**Theorem 4.2.11 (Wilson).** *If $p$ is a prime number, then*

$$(p - 1)! \equiv -1 \pmod{p}.$$

PROOF. For $p = 2$ and $p = 3$ the theorem is straightforward. So, assume $p > 3$. Consider the congruence $ax \equiv 1 \pmod{p}$, where $a$ is one of the integers $1, 2, \ldots, p-1$. As $\mathrm{GCD}(a, p) = 1$, this congruence admits exactly one solution $a'$ modulo $p$, with $1 \le a' < p$. For which values of $a$ does $a = a'$ hold? When $a = a'$ the congruence becomes $a^2 \equiv 1 \pmod{p}$, which amounts to saying that $p$ divides $(a + 1)(a - 1)$. So, either $p$ divides $a + 1$ or $p$ divides $a - 1$. It follows that either $a - 1 \equiv 0 \pmod{p}$, that is, $a = 1$, or $a + 1 \equiv 0$, that is, $a = p - 1$. Leaving out these two extremal values, the other elements $2, 3, \ldots, p - 2$ can be paired up into $(p - 3)/2$ pairs $\{a, a'\}$, with $a \neq a'$ such that $aa' \equiv 1 \pmod{p}$. By multiplying together all the congruences, we obtain

$$2 \cdot 3 \cdots (p - 2) = (p - 2)! \equiv 1 \pmod{p}.$$

Now, by multiplying both sides by $p - 1 \equiv -1 \pmod{p}$, we find that $(p - 1)! \equiv p - 1 \pmod{p}$, which is our claim. $\qquad\square$

The inverse of this theorem is also true:

**Proposition 4.2.12.** *If $(n - 1)! \equiv -1 \pmod{n}$, then $n$ is prime.*

PROOF. If $n$ were not a prime, it would have a divisor $c$, with $1 < c < n$, which, being a divisor of $n$, would divide $(n - 1)! + 1$ as well. But $c$ should appear among the factors of $(n - 1)!$, as $1 < c < n$, so $c \mid (n - 1)!$. These two relations imply that $c$ divides 1, which is a contradiction. $\qquad\square$

Wilson's theorem 4.2.11, together with its inverse 4.2.12, give us another *characterisation of prime numbers*:

> $n$ is prime if and only if $(n - 1)! + 1$ is divisible by $n$.

For instance, $n = 5$ is prime, as $(5 - 1)! = 24 \equiv -1 \pmod{5}$, that is to say, $24 + 1$ is divisible by 5.

Just like Fermat's little theorem, Wilson's theorem provides a primality test. In principle, the test works as follows: given $n$, compute $(n - 1)! + 1$ and verify whether it is divisible by $n$ or not. If it is not, and only in this case, $n$ is prime.

However, such a test is unfeasible in practice, because it relies on an exponential algorithm. Indeed, given $n$ we have to compute $(n - 1)!$ first, and, if $n$ is an integer of length $k$, we have $T(n!) \in \mathcal{O}(n^2 k^2)$ (see Chapter 2, in particular Exercises A2.9 and A2.23). In other words, computing $(n - 1)!$ is exponential. It is sufficient to consider that, even if $n$ is a number with, say, 4 decimal digits, the number $(n - 1)!$ may have up to 35652 digits: an impressive and intractable number.

## 4.3 Representation of rational numbers in an arbitrary base

In this section we prove Proposition 1.4.7, which was left unproved.

PROOF OF PROPOSITION 1.4.7. Given a rational number $a$ such that $0 < a < 1$, we shall show that $a$ is $\beta$–recurring, that is to say that it has an expression of the form

$$a = (0, c_1 \ldots c_m \overline{c_{m+1} \ldots c_{m+h}})_\beta \tag{4.10}$$

in base $\beta$. Moreover, we shall show that

$$\frac{(c_1 \ldots c_m c_{m+1} \ldots c_{m+h})_\beta - (c_1 \ldots c_m)_\beta}{(\underbrace{\beta - 1, \ldots, \beta - 1}_{h}, \underbrace{0, \ldots, 0}_{m})_\beta} =$$
$$= (0, c_1 \ldots c_m \overline{c_{m+1} \ldots c_{m+h}})_\beta. \tag{4.11}$$

Let us prove the first claim. Given $a = c/d$, where the fraction is reduced, we shall distinguish three cases:

(1) every prime divisor of $d$ is a divisor of $\beta$ too;
(2) $d$ and $\beta$ are relatively prime;
(3) $d$ and $\beta$ are not relatively prime, but there is some prime divisor of $d$ that does not divide $\beta$.

*Case* (1). In this case there are powers of $\beta$ that are divisible by $d$. Let $\beta^m$ be the least such power, so we have $\beta^m = dk$. Then

$$a = \frac{c}{d} = \frac{ck}{dk} = \frac{ck}{\beta^m}.$$

On the other hand,

$$ck = (a_{m-1} \ldots a_0)_\beta = a_0 + a_1\beta + \cdots + a_{m-1}\beta^{m-1},$$

as $a < 1$ and so $ck < \beta^m$. In conclusion, we have

$$a = \frac{ck}{\beta^m} = \frac{a_0 + a_1\beta + \cdots + a_{m-1}\beta^{m-1}}{\beta^m} = (0, a_{m-1}a_{m-2} \ldots a_0)_\beta,$$

that is Equation (4.10) holds, with $h = 0$ and $c_i = a_{m-i}, i = 1, \ldots, m$.

*Case* (2). In this case $a$ cannot be expressed as a fraction with a power of $\beta$ as its denominator. However, given a positive integer $h$, there exists exactly one fraction of the form $a_h/\beta^k$ that approximates $a$ from below within $1/\beta^h$, that is, such that

$$\frac{a_h}{\beta^h} < a < \frac{a_h + 1}{\beta^h} :$$

clearly, this is the fraction having $\beta^h$ as its denominator and $\lfloor \beta^h a \rfloor$ as it numerator, where by $\lfloor x \rfloor$ we denote the greatest integer less than or equal to $x$.

Consider a particular value of $h$, which certainly exists, such that $\beta^h \equiv 1$ (mod $d$): for instance we may take $h = \mathrm{Gss}(d, \beta)$ (see § 4.2.2) and determine $b := a_h$. Clearly, we get

$$b = \left\lfloor \beta^h \frac{c}{d} \right\rfloor = (\beta^h - 1)\frac{c}{d},$$

that is to say,

$$a = \frac{c}{d} = \frac{b}{\beta^h - 1}. \tag{4.12}$$

Keeping in mind the identity

$$\beta^{hk} - 1 = (\beta^h - 1)(1 + \beta^h + \cdots + \beta^{hk-h}),$$

which holds for every pair of non-negative integers $h$, $k$ (see polynomial equality (4.13) below), we immediately find

$$a = \frac{b}{\beta^h} + \frac{b}{\beta^{2h}} + \cdots + \frac{b}{\beta^{kh}} + \frac{1}{\beta^{kh}} a$$

for all positive integers $k$. Define

$$A_k = \frac{b}{\beta^h} + \frac{b}{\beta^{2h}} + \cdots + \frac{b}{\beta^{kh}}.$$

This is a $\beta$–defined number, and precisely, if $b = (a_{m-1} \ldots a_0)_\beta$, then

$$A_k = (0, \underbrace{a_0 \ldots a_{m-1} a_0 \ldots a_{m-1} \ldots a_0 \ldots a_{m-1}}_{k})_\beta.$$

Moreover, we have

$$0 < a - A_k = \frac{a}{\beta^{kh}} < \frac{1}{\beta^{kh}}$$

and so $\lim_{k \to \infty} A_k = a$. Keeping in mind Theorem 1.4.4, we may conclude that $a = 0, \overline{a_0 \ldots a_{m-1}}$ is a simple recurring number. Equation (4.12) is just Equation (4.11) in this case.

*Case* (3). In this case let $d = p_1^{h_1} \cdots p_s^{h_s}$ be a factorisation of $d$ with $p_1, \ldots, p_s$ distinct prime numbers, and let $p_1, \ldots, p_k$ be the only primes among those dividing $\beta$ as well. Define $b = p_1^{h_1} \cdots p_k^{h_k}$, so $d' = d/b$ and $\beta$ are relatively prime, and we may consider the least positive integer $m$ such that $b \mid \beta^m$. We have

$$\beta^m a = \frac{\beta^m}{b} \frac{c}{d'}.$$

It follows from this relation that the denominator of the reduced fraction of $\beta^m a - [\beta^m a]$ is divisible by $d'$ and so is relatively prime with $\beta$, while $[\beta^m a] < \beta^m$. Thus, keeping in mind case (2), we get an expression of the form

$$\beta^m a = (c_1 \ldots c_m, \overline{c_{m+1} \ldots c_{m+h}})_\beta$$

and from it (4.10) follows.

Finally, we prove Equation (4.11):

$$\frac{(c_1 \ldots c_m c_{m+1} \ldots c_{m+h})_\beta - (c_1 \ldots c_m)_\beta}{(\underbrace{\beta - 1, \ldots, \beta - 1}_{h}, \underbrace{0, \ldots, 0}_{m})_\beta} =$$

$$= \frac{(c_1 \ldots c_m)_\beta \beta^h + (c_{m+1} \ldots c_{m+h})_\beta - (c_1 \ldots c_m)_\beta}{(\beta - 1)\beta^m(1 + \beta + \cdots + \beta^h)} =$$

$$= \frac{(c_1 \ldots c_m)_\beta}{\beta^m} + \frac{(c_{m+1} \ldots c_{m+h})_\beta}{(\underbrace{\beta - 1, \ldots, \beta - 1}_{h})_\beta \beta^m} ;$$

keeping in mind case (2), this number coincides with $0, c_1 \ldots c_m \overline{c_{m+1} \ldots c_{m+h}}$.

□

**Remark 4.3.1.** This proof shows that cases (1), (2) and (3) correspond to the three cases in which the rational number is $\beta$–defined, simple recurring or mixed recurring, respectively. Moreover, the proof itself shows how to find the mantissa and the number of recurring digits in the three cases.

## 4.4 Fermat primes, Mersenne primes and perfect numbers

### 4.4.1 Factorisation of integers of the form $b^n \pm 1$

Although the problem of factoring a number is in general very hard, it may be far easier for numbers of particular forms. We show next an example.

Consider the following two identities among polynomials on $\mathbb{Z}$:

$$x^n - 1 = (x - 1)(x^{n-1} + x^{n-2} + \cdots + x^2 + x + 1), \qquad (4.13)$$

$$x^n + 1 = (x + 1)(x^{n-1} - x^{n-2} + \cdots + x^2 - x + 1). \qquad (4.14)$$

The first one holds for every positive integer $n$ and the second one holds for every odd positive integer $n$ (see Exercises A4.32 and A4.33). It follows from them that:

**Proposition 4.4.1.** *(i) If $m$ is an integer of the form $a^n - 1$, it may be decomposed as*

$$a^n - 1 = (a - 1)(a^{n-1} + a^{n-2} + \cdots + a^2 + a + 1).$$

*Moreover, if $n = rs$, then $m$ may be further factored as follows:*

$$a^{rs} - 1 = (a^r - 1)(a^{r(s-1)} + a^{r(s-2)} + \cdots + a^{2r} + a^r + 1).$$

*In particular, if $m = a^n - 1$ is prime, then $a = 2$ and $n$ is prime.*
*(ii) If $m$ is an integer of the form $a^n + 1$, with odd $n$, it may be decomposed as*

$$a^n + 1 = (a + 1)(a^{n-1} - a^{n-2} + \cdots + a^2 - a + 1).$$

*Moreover, if $n = rs$, with odd $s$, then $m$ may be further factored as follows:*

$$a^{rs} + 1 = (a^r + 1)(a^{r(s-1)} - a^{r(s-2)} + \cdots + a^{2r} - a^r + 1).$$

*In particular, if $m = a^n + 1$ is prime, then $n$ is a power of two.*

**Example 4.4.2.** We want to factor the number 16383. It is useful to keep at hand the values of the powers of 2 up to, say, $2^{14}$. Inspecting the list of these powers, we find that $16383 = 2^{14} - 1$. Then

$$16383 = 2^{14} - 1 = (2^7 - 1)(2^7 + 1) = 127 \cdot 129 = 127 \cdot 3 \cdot 43,$$

which is a prime decomposition.

Clearly, we could have immediately checked that 16383 is divisible by 3:

$$16383 = 3 \cdot 5461.$$

However, we would have come across the problem of factoring 5461, which would probably have taken more time. When a number is recognised as having the form $a^n - 1$, it is always convenient to use this information. The problem, of course, lies exactly in *recognising* that the number has this form.

Here follows another useful result.

**Proposition 4.4.3.** *Let $p$ be a prime dividing $a^n - 1$. Then, either*

*(i) $p$ divides $a^d - 1$ for some proper divisor $d$ of $n$, or*
*(ii) $p \equiv 1 \pmod{n}$.*

*If both $p$ and $n$ are odd, then $p \equiv 1 \pmod{2n}$.*

PROOF. By hypothesis, $a^n \equiv 1 \pmod{p}$, and by Fermat's little theorem we also have $a^{p-1} \equiv 1 \pmod{p}$. Then (see Exercise A4.8) $a^d \equiv 1 \pmod{p}$, with $d = \text{GCD}(n, p-1)$, that is $p \mid (a^d - 1)$. If $d$ is strictly smaller than $n$, then $(i)$ holds, otherwise, if $d = n$, it follows from $d \mid (p-1)$ that $p \equiv 1 \pmod{d}$, that is, $(ii)$. Further, if both $p$ and $n$ are odd, $n \mid (p-1)$ implies that $2n \mid (p-1)$. $\square$

Let us see how one can apply this proposition to factor integers of the form $a^n - 1$.

**Example 4.4.4.** We want to factor the number $2^{21} - 1 = 2097151$. By Proposition 4.4.3, we know that if a prime $p$ is a factor of $2^{21} - 1$, then either $p$ divides $2^d - 1$, with $d$ a proper divisor of 21, that is to say, it divides $2^3 - 1$ or $2^7 - 1$, or $p \equiv 1 \pmod{21}$. In the first case $p$ may be either 7 or 127. Let us check if 2097151 is divisible by 7 or by 127.

It may be seen that it is divisible by both numbers, and so we get the factorisation

$$2097151 = 7 \cdot 127 \cdot 2359.$$

Now, if 2359 were a prime number, as it does not divide $2^7 - 1$ nor $2^3 - 1$, we would have $2359 \equiv 1 \pmod{42}$, as in this case both $p$ and $n$ are odd. But it may be checked that this is not the case, and so 2359 is not a prime. It is now easily seen that $2359 = 7 \cdot 337$. The prime decomposition is

$$2097151 = 7^2 \cdot 127 \cdot 337.$$

We shall shortly see further applications of Proposition 4.4.1.

### 4.4.2 Fermat primes

How are prime numbers found? As we have already said, there are no known recurrence formulas to compute prime numbers, and so there is no efficient method to *generate* prime numbers. Still, in the past, several mathematicians have dreamed of the possible existence of such formulas, perhaps even quite simple ones. Among them there were Fermat and Mersenne, both of whom lived in the 17th century.

Fermat remarked what follows, as an immediate consequence of our Proposition 4.4.1, (ii):

**Corollary 4.4.5.** *If the number $n = 2^k + 1$ is prime, then $k$ has no odd factor and so it is a power of 2.*

We have the following definition.

**Definition 4.4.6.** *A Fermat number is an integer of the form*

$$F_n = 2^{2^n} + 1.$$

The Fermat numbers corresponding to the first few values of $n$ are

$$F_0 = 3, \quad F_1 = 5, \quad F_2 = 17, \quad F_3 = 257, \quad F_4 = 65537.$$

It is not difficult to verify that each of these five numbers is a prime number. Fermat believed that *all* the numbers of the form $2^{2^n} + 1$ are prime. His conjecture was disproved by Euler, who proved that the next Fermat number, $F_5$, is not prime, by showing the factorisation

$$F_5 = 4294967297 = 641 \cdot 6700417,$$

which is quite easy to find. In fact,

$$641 = 5 \cdot 2^7 + 1 = 2^4 + 5^4$$

and so

$$F_5 = 2^{32} + 1 = 2^4 \cdot 2^{28} + 1 = (641 - 5^4) \cdot 2^{28} + 1 =$$
$$= 641 \cdot 2^{28} - (5 \cdot 2^7)^4 + 1 = 641 \cdot 2^{28} - (641 - 1)^4 + 1 =$$
$$= 641(2^{28} - 641^3 + 4 \cdot 641^2 - 6 \cdot 641 + 4).$$

The problem of finding Fermat numbers that are *prime* is still open. The largest known prime Fermat number is $F_4$. The largest known non-prime Fermat number is $F_{23471}$. The factorisations of $F_5$, $F_6$, $F_7$, $F_8$, $F_9$ and $F_{11}$ are known. $F_{10}$ and $F_{14}$ are known not to be prime, but their complete factorisation is not known. It is still an open problem whether there are infinitely many Fermat primes or not, and whether there are infinitely many composite Fermat numbers or not. Surprisingly, Fermat primes appear in the solution to the geometric problem of constructing regular polygons with ruler and compasses. In fact, the following result from Galois theory holds (see [4], [32], [45]).

**Theorem 4.4.7.** *A regular polygon with $n$ sides can be constructed with ruler and compasses if and only if $n = 2^a p_1 \cdots p_t$, where $p_1, \ldots, p_t$ are distinct Fermat primes and $a$ is a non-negative integer.*

So, among the ruler-and-compasses constructible regular polygons there are: the triangle, the hexagon, the pentagon, the decagon, the 17-gon, *but not* a regular polygon with nine sides (as 9 is the product of two *equal* Fermat numbers), nor a 7-gon or a 25-gon.

There are some primality tests for Fermat numbers, like the following *Pépin's test*.

**Theorem 4.4.8.** *The Fermat number $F_n$ is prime if and only if there exists an integer $a$ such that*

$$a^{2^{(2^n-1)}} = a^{(F_n-1)/2} \equiv -1 \pmod{F_n}. \tag{4.15}$$

PROOF. We prove now only a part of theorem, that is, that if (4.15) holds then $F_n$ is prime. We shall complete the proof of the theorem in Section 5.2.4 (see Proposition 5.2.33), as an application of the law of quadratic reciprocity. From Equation (4.15) we deduce $a^{F_n-1} \equiv 1 \pmod{F_n}$. Thus, if $p$ is a prime dividing $F_n$ we have $a^{F_n-1} \equiv 1 \pmod{p}$ and so $\mathrm{Gss}(p, a) \mid F_n - 1 = 2^{2^n}$. On the other hand, as Equation (4.15) holds, we have $a^{2(2^n-1)} \equiv -1 \pmod{p}$ as well, and so $\mathrm{Gss}(p, a) \nmid F_n - 1 = 2^{2^n-1}$: then $\mathrm{Gss}(p, a) = F_n - 1 = 2^{2^n}$. Moreover, $\mathrm{Gss}(p, a) < p$, so $F_n \leq p$, implying $F_n = p$.          $\square$

It is interesting to notice that if $n \neq m$ then $\mathrm{GCD}(F_n, F_m) = 1$ (see Exercise A4.39), which yields a new proof that there are infinitely many prime numbers (see Exercise A4.40).

### 4.4.3 Mersenne primes

Mersenne observed something which, again, is a consequence of Proposition 4.4.1, (i): if a number of the form $2^p - 1$ is prime, then $p$ is prime. So we may define:

**Definition 4.4.9.** *A* Mersenne number *is a number of the form*

$$M_p = 2^p - 1, \qquad \text{with } p \text{ prime.}$$

If $M_p$ is itself a prime number, then it is called a *Mersenne prime*. However, not for all primes $p$ the number $M_p$ is prime. For instance, if $p = 11$, we have

$$M_{11} = 2^{11} - 1 = 2047 = 23 \cdot 89.$$

Mersenne proved in 1644 that $M_p$ is prime for the following values of $p$:

$$2, 3, 5, 7, 13, 17, 19, 31, 67, 127, 257,$$

claiming further that $M_p$ is reducible for all other primes smaller than 257. More than 200 years later, this claim was proved false, as it has been shown that there are other values of $p$ smaller than 257, such as $p = 61, 89, 107$, for which the number $M_p$ is prime. For Mersenne numbers we have the following *ad hoc* primality test, devised by Lucas at the end of the 19th century (its proof can be found in [34]):

**Theorem 4.4.10 (Lucas test).** *The number* $M_p = 2^p - 1$, *with* $p$ *a prime greater than 2, is a prime number if and only if* $M_p$ *divides* $S_p$, *where* $S_k$ *is defined recursively as follows:*

$$S_2 = 4, \qquad S_k = S_{k-1}^2 - 2.$$

For instance,

$$S_2 = 4, \ S_3 = 14, \ S_4 = 194, \ S_5 = 37634.$$

Lucas test 4.4.10 is quite efficient. Indeed, we have:

**Corollary 4.4.11.** *Lucas test determines if* $M_p = 2^p - 1$, *with* $p$ *a prime greater than 2, is a prime number using* $\mathcal{O}(p^3)$ *bit operations.*

PROOF. It is necessary to compute $p - 1$ squares modulo $M_p$, and each of them requires $\mathcal{O}(\log^2 M_p) = \mathcal{O}(p^2)$ bit operations.    □

Only 43 Mersenne primes are known. The last Mersenne prime was found in December 2005: it is $M_{30402457}$ and has 9152052 digits. It is also the largest known prime.

Exactly as for Fermat numbers, for Mersenne numbers too there are several open problems: are there infinitely many Mersenne primes? Are there infinitely many composite Mersenne numbers? It is conjectured that the answer to both questions is positive. We do not dwell further on Mersenne primes, just mentioning the large quantity of such numbers that has been found, which suggests that many among Mersenne numbers are prime (see [48]).

### 4.4.4 Perfect numbers

Given a positive integer $n$ we define

$$\sigma(n) = \sum_{m>0, m|n} m;$$

notice that for all $n \in \mathbb{N}$ we have $\sigma(n) \geq n + 1$. Further, $n$ is prime if and only if $\sigma(n) = n + 1$. It is useful to give a definition:

**Definition 4.4.12.** *An integer $n > 0$ is said to be* perfect *if $\sigma(n) = 2n$.*

In other words, $n$ is perfect if and only if the sum of its proper divisors is equal to $n$.

For instance, 6 and 28 are perfect numbers, as $\sigma(6) = 1 + 2 + 3 + 6 = 12$ e $\sigma(28) = 1 + 2 + 4 + 7 + 14 + 28 = 56$.

Apparently, the ancient Greeks already knew the following theorem; it provides an unexpected relation between perfect numbers and Mersenne primes.

**Theorem 4.4.13.** *A positive, even integer $n$ is a perfect number if and only if there exists a Mersenne prime $M_p$ such that $n = 2^{p-1} \cdot M_p$.*

The proof is not hard, but we omit it and give a short sketch of it in Exercise A4.41.

So this theorem reduces the determination of even perfect numbers to that of Mersenne primes. As for the odd ones, things are far more difficult: it is not even known whether odd perfect numbers exist or not. It is known, for instance, that there are none smaller than $10^{200}$.

## 4.5 Factorisation in an integral domain

In this section we shall recast the results of Section 4.1 in a more general frame, which will turn out to be quite useful later. Basically, we shall define a class of rings, the so-called *factorial rings* or *unique factorisation domains*, which have the property that for them a suitable form of the Fundamental Theorem of Arithmetic holds. This section is a natural continuation of Section 1.3.5.

### 4.5.1 Prime and irreducible elements in a ring

Let $A$ be a ring with unity. We shall need the following definitions.

**Definition 4.5.1.** *A non-zero, non-invertible element $a \in A$ is said to be* irreducible *if whenever $a$ is written as a product $a = bc$ with $b$ and $c$ in $A$, either the element $b$ or the element $c$ is invertible.*

**Definition 4.5.2.** *A non-zero, non-invertible element $a \in A$ is said to be* prime *if whenever $a$ divides a product $bc$ with $b$ and $c$ in $A$, it divides one of the factors. In other words,*

$$a \mid bc \text{ implies either } a \mid b \text{ or } a \mid c.$$

Notice that associate elements are either both irreducible or non-irreducible, as they share the same divisors. Analogously, associate elements are either both prime or non-prime.

We have already remarked that in $\mathbb{Z}$ these two notions coincide: this is the content of Proposition 4.1.1. It is useful to remark here two facts the reader will easily be persuaded of by going again over the results in § 4.1:

- the Fundamental Theorem of Arithmetic is an immediate consequence of Proposition 4.1.1;
- Proposition 4.1.1 is, in its turn, a consequence of the fact that in $\mathbb{Z}$ Bézout's identity for the greatest common divisor holds; this, in turn, follows from $\mathbb{Z}$ being a principal ideal domain, that is, from the existence of the Euclidean algorithm.

As to the relationship between the notions of prime and irreducible element, we have immediately the following lemma.

**Lemma 4.5.3.** *In an integral domain every prime element is also irreducible.*

The proof is left as an exercise for the reader, who will only have to repeat part of the proof of Proposition 4.1.1 (see Exercise A4.43).

However, it is not true that in an arbitrary integral domain every irreducible element is prime. Here follows an example.

**Example 4.5.4.** Consider the following set of complex numbers:

$$\mathbb{Z}[\sqrt{-3}] = \{a + b\sqrt{-3} \mid a, b \in \mathbb{Z}\}.$$

It is easy to see that $\mathbb{Z}[\sqrt{-3}]$ is *stable* under the operations of addition and multiplication as defined in $\mathbb{C}$, that is to say, by adding and multiplying elements of $\mathbb{Z}[\sqrt{-3}]$, we obtain elements of $\mathbb{Z}[\sqrt{-3}]$. Clearly, with these operations, $\mathbb{Z}[\sqrt{-3}]$ is an integral domain with unity (see Exercise A4.45 and A4.46).

Let us determine the invertible elements of $\mathbb{Z}[\sqrt{-3}]$. For each element $\alpha = a + b\sqrt{-3}$ of $\mathbb{Z}[\sqrt{-3}]$ its complex norm is defined: $||\alpha|| = a^2 + 3b^2$. If $\alpha = a + b\sqrt{-3}$ is invertible, its norm must be 1. Indeed, if $\alpha$ is invertible, there exists a $\beta = c + d\sqrt{-3}$ such that $\alpha \cdot \beta = 1$. But then $||\alpha|| \cdot ||\beta|| = ||\alpha \cdot \beta|| = ||1|| = 1$. As $||\alpha||$ is a non-negative integer, it must be $||\alpha|| = 1$. Now, the relation $a^2 + 3b^2 = 1$ with $a, b \in \mathbb{Z}$ implies $b = 0$ and $a = \pm 1$. Vice versa, if $||\alpha|| = 1$ then $\alpha = \pm 1$ is invertible. Thus,

> the invertible elements in $\mathbb{Z}[\sqrt{-3}]$ are $\pm 1$

or

> the invertible elements in $\mathbb{Z}[\sqrt{-3}]$ are those with norm equal to 1.

Having established this, we want to prove that in $\mathbb{Z}[\sqrt{-3}]$ there exist irreducible elements that are not prime; for instance, the number $1 + \sqrt{-3} \in \mathbb{Z}[\sqrt{-3}]$ has this property.

We prove first that $1+\sqrt{-3}$ is *irreducible*. Suppose that $1+\sqrt{-3} = (a+b\sqrt{-3})(c+d\sqrt{-3})$ in $\mathbb{Z}[\sqrt{-3}]$. Taking the norms, we have

$$4 = ||a + b\sqrt{-3}|| \cdot ||c + d\sqrt{-3}|| = (a^2 + 3b^2)(c^2 + 3d^2).$$

The only possible case is either $a^2 + 3b^2 = 1$ or $a^2 + 3b^2 = 4$, as it is not possible that $a^2 + 3b^2 = 2$. The first relation implies, as we have seen, that $a = \pm 1$ and $b = 0$, that is $a + b\sqrt{-3}$ invertible, while in the second case $c + d\sqrt{-3}$ is invertible. As $1 + \sqrt{-3}$ is non-zero and non-invertible, we have proved that $1 + \sqrt{-3}$ is irreducible.

We now prove that $1 + \sqrt{-3}$ *is not prime*. From the obvious relation

$$\left(1 + \sqrt{-3}\right)\left(1 - \sqrt{-3}\right) = 2 \cdot 2, \tag{4.16}$$

it follows that $1 + \sqrt{-3}$ divides $2 \cdot 2$, but does not divide 2. Indeed, if $1 + \sqrt{-3}$ divided 2, then $2 = (1 + \sqrt{-3})(a + b\sqrt{-3})$, hence, taking the norms, we would find the following equality in $\mathbb{N}$:

$$4 = 4 \cdot (a^2 + 3b^2),$$

so $a^2 + 3b^2 = 1$, that is $a = \pm 1$ and $b = 0$, which means that $a + b\sqrt{-3} = \pm 1$, implying

$$2 = \pm\left(1 + \sqrt{-3}\right),$$

a contradiction.

### 4.5.2 Factorial domains

So we may ask: which are the integral domains where each irreducible element is also prime? In order to answer this question, we give a definition.

**Definition 4.5.5.** *An integral domain is said to be a* factorial domain *or a* unique factorisation domain *or a* factorial ring *if every non-zero, non-invertible element a admits a factorisation into irreducible elements, that is to say, can be written as a product of irreducible elements*

$$a = p_1 p_2 \cdots p_n$$

*in such a way that, if*

$$a = p_1' p_2' \cdots p_m'$$

*is another factorisation into irreducibles, then $m = n$ and each of the $p_i'$ is associate to a $p_j$.*

Basically, this definition identifies the class of domains in which an analogue of the Fundamental Theorem of Arithmetic holds. We have the following result.

**Theorem 4.5.6.** *An integral domain $A$ is factorial if and only if:*

(1) *every non-zero, non-invertible element of $A$ can be written as a product of irreducible elements;*

(2) *the prime elements of $A$ coincide with the irreducible elements.*

PROOF. If (1) and (2) hold, then $A$ is factorial: this can be proved exactly as we did for the Fundamental Theorem of Arithmetic.

Vice versa, assume $A$ is factorial. Property (1) is clearly true. Assume now that $a$ is irreducible and that it divides $bc$, that is $bc = ap$. If $p$ is invertible, then $ap$ is irreducible as $a$, and so either $b$ or $c$ is invertible. If $c$ [$b$, respectively] is invertible then $a$ is associate to $b$ [to $c$, resp.], and so divides it. So we may assume that $p$ is not invertible. Moreover, as $a$, which is not invertible, divides $bc$, $bc$ is not invertible. So, either $b$ or $c$ is non-invertible. Assume both elements are non-invertible: the other case is analogous and is left to the reader. Let

$$b = p_1 p_2 \cdots p_n, \quad c = q_1 q_2 \cdots q_m, \quad p = r_1 r_2 \cdots r_s$$

be the factorisations of $b, c, p$. Then we have

$$p_1 p_2 \cdots p_n q_1 q_2 \cdots q_m = bc = ap = a r_1 r_2 \cdots r_s.$$

By the uniqueness of the factorisation, $a$ has to be associate to one of the irreducibles in the left-hand side, and so it must divide either $b$ or $c$.    □

**Example 4.5.7.** Keeping in mind Example 4.5.4 and Theorem 4.5.6, we see that $\mathbb{Z}[\sqrt{-3}]$ is not a unique factorisation domain. For instance, formula (4.16) shows two factorisations of the same element of $\mathbb{Z}[\sqrt{-3}]$ into irreducibles. The two factorisations are *different*, in the sense that 2 is not associate either to $1 + \sqrt{-3}$ or to $1 - \sqrt{-3}$.

We further remark that in $\mathbb{Z}[\sqrt{-3}]$ part (1) of Theorem 4.5.6 holds. Indeed, if $\alpha \in \mathbb{Z}[\sqrt{-3}]$ is non-zero and non-invertible, either it is irreducible, or $\alpha = \alpha_1 \beta_1$, with $\alpha_1, \beta_1$ non-invertible and so with norm greater than 1. Thus, $0 < ||\alpha_1|| < ||\alpha||, 0 < ||\beta_1|| < ||\alpha||$. By iterating this argument on $\alpha_1$ and $\beta_1$, we arrive to a factorisation.

**Remark 4.5.8.** Let $A$ be a factorial ring and let $a \in A$ be a non-zero element. Consider the factorisation of $a$ into irreducible elements

$$a = p_1^{h_1} \cdots p_s^{h_s},$$

where $p_1, \ldots, p_s$ are pairwise non-associate irreducibles and $h_1, \ldots, h_s$ are positive integers. It is clear that all divisors of $a$, and no other number, are associate to elements of the form $p_1^{k_1} \cdots p_s^{k_s}$, with $0 \le k_i \le h_i, i = 1, \ldots, s$.

It follows from this that in $A$ the greatest common divisor of two non-zero elements $a, b$ always exists. Indeed, we always may write the factorisations of $a$ and $b$ as

$$a = p_1^{h_1} \cdots p_s^{h_s}, \qquad b = p_1^{k_1} \cdots p_s^{k_s},$$

where $p_1, \ldots, p_s$ are pairwise non-associate irreducibles and $h_1, \ldots, h_s, k_1, \ldots, k_s$ are non-negative integers. Then

$$\mathrm{GCD}(a, b) = p_1^{n_1} \cdots p_s^{n_s},$$

where $n_i = \min\{h_i, k_i\}, i = 1, \ldots, s$.

### 4.5.3 Noetherian rings

Let us momentarily go back to condition (1) in Theorem 4.5.6, in order to define an interesting class of rings where it is always satisfied. Recall the following two definitions:

**Definition 4.5.9.** *A commutative ring $A$ is said to be* Noetherian *if every ideal of $A$ is finitely generated.*

**Definition 4.5.10.** *A commutative ring $A$ is said to satisfy the* ascending chain condition *if every ascending chain*

$$I_1 \subseteq I_2 \subseteq \cdots \subseteq I_n \subseteq \cdots \qquad (4.17)$$

*of ideals of $A$ is* stationary, *that is there is an element $m \in \mathbb{N}$ such that $I_n = I_m$ for all $n \in \mathbb{N}$ such that $n \geq m$.*

The relation among these two definitions is given by the following proposition.

**Proposition 4.5.11.** *A commutative ring $A$ is Noetherian if and only if it satisfies the ascending chain condition.*

PROOF. Assume $A$ to be Noetherian. Let (4.17) be an ascending chain of ideals of $A$. Set $I = \cup_{n \in \mathbb{N}} I_n$. It is clear that $I$ is an ideal of $A$ (see Exercise A4.49). As $A$ is Noetherian, we have $I = (x_1, \ldots, x_n)$. So there exists an integer $m$ such that $x_i \in I_m$, for all $i = 1, \ldots, n$. Thus, $I = I_m$ and so, $I_n = I_m$ for all $n \in \mathbb{N}$ such that $n \geq m$.

Assume that $A$ satisfies the ascending chain condition and let $I$ be an ideal of $A$. If $I$ were not finitely generated, we could find a sequence $\{x_n\}_{n \in \mathbb{N}}$ of elements of $I$ such that, if $I_n = (x_1, \ldots, x_n)$, we would have $I_n \subset I_{n+1}$ and $I_n \neq I_{n+1}$, for all $n \in \mathbb{N}$ (see Exercise A4.50). Clearly, this contradicts the fact that $A$ satisfies the ascending chain condition. □

We have some lemmas.

**Lemma 4.5.12.** *Let $A$ be a Noetherian integral domain. Let $\{x_n\}_{n \in \mathbb{N}}$ be a sequence of non-zero elements in $A$ such that $x_{n+1} \mid x_n$ for all $n \in \mathbb{N}$. Then there exists an integer $m \in \mathbb{N}$ such that, for all $n \geq m$, $x_n$ is associate to $x_m$.*

PROOF. Define $I_n = (x_n)$. Clearly, (4.17) holds, and so, by Proposition 4.5.11, there exists an element $m \in \mathbb{N}$ such that $I_n = I_m$ for all $n \in \mathbb{N}$ such that $n \geq m$. Hence the claim follows (see Exercise A1.46). □

**Lemma 4.5.13.** *Let $A$ be a Noetherian integral domain. Then every non-zero, non-invertible element of $A$ is divisible by some irreducible element.*

PROOF. Let $a$ be a non-zero, non-invertible element of $A$. If $a$ is irreducible, we are done. Otherwise, we can find an element $a_1$ of $A$, not associate to $a$, such that $a_1 \mid a$. If $a_1$ is irreducible, then we are done. If $a_1$ is reducible, we iterate the previous argument. Keeping in mind Lemma 4.5.12, we see that this procedure ends, proving the claim. □

We may now prove that condition (1) of Theorem 4.5.6 is always verified in Noetherian rings:

**Proposition 4.5.14.** *Let $A$ be a Noetherian integral domain. Then every non-zero, non-invertible element of $A$ can be written as a product of irreducible elements.*

PROOF. Let $a$ be a non-zero, non-invertible element of $A$. We want to prove that $a$ can be written as a product of irreducible elements. In order to do so, we use the following procedure. If $a$ is irreducible, there is nothing to prove. Otherwise, by Lemma 4.5.13, we have $a = a_1 b_1$ with $a_1$ irreducible. If $b_1$ is invertible, there is nothing more to prove and our procedure terminates. Otherwise, we repeat the procedure. The proposition is true if the procedure terminates. If it did not, we would find in this way two sequences $\{a_n\}_{n \in \mathbb{N}}$ and $\{b_n\}_{n \in \mathbb{N}}$ of elements of $A$, such that $a_n$ is irreducible and $b_n$ is non-zero and non-invertible for all $n \in \mathbb{N}$, and

$$b_n = a_{n+1} b_{n+1}. \tag{4.18}$$

By Lemma 4.5.12, there exists a positive integer $m$ such that $b_n$ is associate to $b_m$ for all $n \geq m$. In particular, $b_n$ and $b_{n+1}$ would be associate for all $n \geq m$, and so, from (4.18) it follows that $a_{n+1}$ is invertible for all $n \geq m$, which is a contradiction.    □

As a consequence we have that several remarkable rings are factorial. For instance:

**Theorem 4.5.15.** *Every principal ideal ring is factorial.*

PROOF. As already remarked in Section 4.5.1, part (2) of Theorem 4.5.6 can be proved by reasoning as in the proof of Proposition 4.1.1 and keeping in mind that in $A$ a Bézout's identity holds for the greatest common divisor (see Proposition 1.3.13).    □

Moreover:

**Corollary 4.5.16.** *Every Euclidean ring is factorial.*

PROOF. Recall Theorem 1.3.14.    □

Let us put on record the following important theorem concerning Noetherian rings.

**Theorem 4.5.17 (Hilbert's basis theorem).** *If $A$ is a Noetherian commutative ring, then $A[x_1, \ldots, x_n]$ is Noetherian too.*

PROOF. It is sufficient to prove that if $A$ is Noetherian, then $A[x]$ is too, and then to proceed by induction. Then, let $I$ be an ideal of $A[x]$. For every integer $n \in N$ let $J_n$ be the set of elements of $A$ that are leading coefficients of some polynomial of degree $n$ in $I$. It is clear that $J_n$ is an ideal of $A$ and that $J_n \subseteq J_{n+1}$ for all $n \in \mathbb{N}$. Then there exists an $m \in \mathbb{N}$ such that $J_n = J_m$ for all integers $n \geq m$.

Let $a_{i,1}, \ldots, a_{i,h_i}$ be a system of generators of $J_i$ and let $f_{i,j}(x)$ be a polynomial of $I$ of degree $i$ having $a_{i,j}$ as its leading coefficient, for all $i = 1, \ldots, m$ and $j = 1 \ldots, h_i$. We want to show that $I$ coincides with the ideal $I'$ generated by all polynomials $f_{i,j}(x)$ for $i = 1, \ldots, m$ and $j = 1, \ldots, h_i$. Reasoning by contradiction, assume that this is not the case, and let $f(x) \in I$ be a polynomial of least degree not in $I'$ and let $a$ be its leading coefficient.

Suppose $f(x)$ has degree $r \leq m$. Then $a \in J_r$ and so there is a relation of the form $a = \alpha_1 a_{r,1} + \cdots + \alpha_{h_r} a_{r,h_r}$ with $\alpha_j \in A$, $j = 1, \ldots, h_r$. So the polynomial

$$f(x) - (\alpha_1 f_{r,1}(x) + \cdots + \alpha_{h_r} f_{r,h_r}(x))$$

has degree smaller than $r$ and is in $I$. By the minimality of $r$, it is in $I'$ and hence it immediately follows that $f(x)$ is in $I'$ too, which is a contradiction.

Assume $f(x)$ has degree $r > m$. Then $a \in J_r = J_m$ and so there is a relation of the form $a = \alpha_1 a_{m,1} + \cdots + \alpha_{h_m} a_{m,h_m}$ with $\alpha_j \in A$, $j = 1, \ldots, h_m$. Again, the following polynomial has degree smaller than $r$ and is in $I$:

$$f(x) - x^{r-m}(\alpha_1 f_{m,1}(x) + \cdots + \alpha_{h_m} f_{m,h_m}(x)).$$

Reasoning as above we get a contradiction.                                     $\square$

### 4.5.4 Factorisation of polynomials over a field

Theorem 1.3.17 tells us the the ring of polynomials over a field is Euclidean and so, by Corollary 4.5.16, we have:

**Corollary 4.5.18.** *The ring $\mathbb{K}[x]$ of polynomials over a field $\mathbb{K}$ is factorial.*

Then, given a field $\mathbb{K}$, the problem arises of finding a factorisation of a given polynomial as a product of irreducible polynomials. Recall that a polynomial in $\mathbb{K}[x]$ is invertible if and only if it is a non-zero constant. But which polynomials are irreducible in $\mathbb{K}[x]$ or, as is commonly said, *irreducible over* $\mathbb{K}$? The answer to this question depends not so much on the polynomials as on the properties of the field $\mathbb{K}$.

We begin by giving some remarks which can be proved by the reader as an exercise (see Exercise A4.51):

**Lemma 4.5.19.** *Let $\mathbb{K}$ be an arbitrary field. Then:*
(1) *every polynomial is associate to a monic polynomial;*
(2) *all polynomials of degree one in $\mathbb{K}[x]$ are irreducible;*
(3) *a non-zero polynomial $f(x) \in \mathbb{K}[x]$ has a factor of degree one if and only if it has a root in $\mathbb{K}$.*

Recall that a field $\mathbb{K}$ is said to be *algebraically closed* if every polynomial of positive degree over $\mathbb{K}$ has some root in $\mathbb{K}$.

**Example 4.5.20.** The polynomial $x^2 - 2$ is *reducible over* $\mathbb{R}$ as

$$x^2 - 2 = (x + \sqrt{2})(x - \sqrt{2}),$$

and $x + \sqrt{2}$ and $x - \sqrt{2}$ are irreducible polynomials with coefficients in $\mathbb{R}$, while it is *irreducible over* $\mathbb{Q}$, as the polynomial $x^2 - 2$ has no roots in $\mathbb{Q}$, that is,

$$\sqrt{2} \notin \mathbb{Q} \quad \text{and so } x \pm \sqrt{2} \notin \mathbb{Q}[x].$$

The polynomial $x^2 + 1$ is reducible over $\mathbb{C}$ and

$$x^2 + 1 = (x - i)(x + i)$$

is a factorisation into irreducibles, but it is irreducible over $\mathbb{R}$ and over $\mathbb{Q}$ as $x^2 + 1$ has no roots in $\mathbb{R}$, and therefore it cannot factor (into linear factors).

The following well known theorem says that $\mathbb{C}$ is an algebraically closed field:

**Theorem 4.5.21 (Fundamental theorem of algebra).** *A polynomial $f(x)$ in $\mathbb{C}[x]$ of degree $n \geq 1$ admits at least one root in $\mathbb{C}$.*

PROOF. We prove first that, *given a polynomial $f(x) \in \mathbb{C}[x]$, the function $|f(x)|$ has a minimum in $\mathbb{C}$.*

Indeed, we have $\lim_{|x| \to +\infty} |f(x)| = +\infty$ (see Exercise A4.52). Let $c \in \mathbb{R}$ be the infimum of $|f(x)|$. There exists a sequence $\{x_n\}_{n \in \mathbb{N}}$ of complex numbers such that $\lim_{n \to \infty} |f(x_n)| = c$. On the other hand, the sequence $\{|x_n|\}_{n \in \mathbb{N}}$ is not unbounded. So we can extract from $\{x_n\}_{n \in \mathbb{N}}$ a sequence converging to a point $x \in \mathbb{C}$ and so $|f(x)| = c$.

The theorem is now a consequence of the following claim: *given a non-constant polynomial $f(x) \in \mathbb{C}[x]$, let $z$ be a point such that $f(z) \neq 0$. Then $|f(z)|$ is not the minimum of $|f(x)|$.*

We may perform the change of variables $x \to x + z$ and assume that $z = 0$. Moreover, we may multiply $f(x)$ by a suitable constant and assume that $f(0) = 1$. Then we have

$$f(x) = 1 + ax^k + \text{terms of degree greater than } k.$$

If we perform the change of variables $x \to \alpha x$ with $\alpha^k = -1/a$ (see Exercise A4.53), we have just to consider the case

$$f(x) = 1 - x^k + x^{k+1} g(x)$$

with $g(x)$ a suitable polynomial. Let now $x$ be a positive real number very close to 0. Then

$$|f(x)| \leq |1 - x^k| + x^{k+1}|g(x)| = 1 - x^k + x^{k+1}|g(x)| = 1 - x^k(1 - x|g(x)|).$$

As $\lim_{x \to 0} x|g(x)| = 0$, we may suppose that $x^k(1 - x|g(x)|) > 0$ and so for $x$ a positive real number very close to 0 we have $|f(x)| < 1 = f(0)$.    □

If $\mathbb{K}$ is an algebraically closed field, every polynomial $f(x) \in \mathbb{K}[x]$ of degree $n$ admits in $\mathbb{K}$ exactly $n$ roots, counting each with its multiplicity. That is to say, the decomposition of $f(x)$ into irreducible factors is

$$f(x) = a(x - a_1)^{n_1} \cdots (x - a_h)^{n_h},$$

where $a$ is the leading coefficient of $f(x)$, $a_1, \ldots, a_h$ are the distinct roots of $f(x)$, $n_1, \ldots, n_h$ are their multiplicities, and $n = n_1 + \cdots + n_h$ (see Exercise A1.56).

We may conclude that $\mathbb{K}$ is an algebraically closed field if and only if

> *the irreducible polynomials in $\mathbb{K}[x]$ are exactly the polynomials of degree one*

that is, if and only if

> *every polynomial with coefficients in $\mathbb{K}$ decomposes into linear factors.*

The above holds, as already said, for polynomials over $\mathbb{C}$. As regards a polynomial over $\mathbb{R}$, notice that, if $\alpha \in \mathbb{C}$ is one of its roots and is not real, then its conjugate $\overline{\alpha}$ is a root as well (see Exercise A1.57). Then, by the factor theorem, $f(x)$ is divisible by the real polynomial $(x - \alpha)(x - \overline{\alpha}) = x^2 - 2Re(\alpha)x + |\alpha|^2$, having discriminant $\Delta < 0$ as this polynomial has no real roots. Whence we can deduce (see Exercise A4.60) that

> the irreducible polynomials in $\mathbb{R}[x]$ are exactly the polynomials of degree one and the ones of degree two with $\Delta < 0$.

**Example 4.5.22.** We have that

$$x^3 + x^2 + 5x + 5 = (x + 1)(x^2 + 5)$$

is a factorisation into irreducibles over $\mathbb{R}$, while over $\mathbb{C}$ it factors into irreducibles as

$$x^3 + x^2 + 5x + 5 = (x + 1)(x - i\sqrt{5})(x + i\sqrt{5}).$$

**Remark 4.5.23.** Notice that a polynomial with coefficients in $\mathbb{R}$ with no real roots may nevertheless be reducible. For instance, the polynomial

$$x^4 + 8x^2 + 15$$

has no real roots, but it factors as follows

$$x^4 + 8x^2 + 15 = (x^2 + 3)(x^2 + 5).$$

However notice the following:

**Lemma 4.5.24.** Let $\mathbb{K}$ be a field and let $f(x) \in \mathbb{K}[x]$ be a polynomial of degree 2 or 3. Then $f(x)$ is irreducible in $\mathbb{K}[x]$ if and only if it has no roots in $\mathbb{K}$.

PROOF. Indeed, if $f(x)$ had a factorisation, at least one of its factors would have degree 1, so it would have a root in $\mathbb{K}$. □

To sum up:

> if a polynomial $f(x)$ of degree $> 1$ with coefficients in a field $\mathbb{K}$ has a root in $\mathbb{K}$, then $f(x)$ has a non trivial factorisation (and a factor is linear).

On the other hand,

> a polynomial may well factor even if it has no roots in the field (apart from degree 2 and 3 polynomials, for which the existence of a root is equivalent to reducibility).

### 4.5.5 Factorisation of polynomials over a factorial ring

Corollary 4.5.18 is a special case of the following general result:

**Theorem 4.5.25.** *Consider a factorial integral domain $A$. Then $A[x_1, \ldots, x_n]$ is factorial too.*

This section is devoted to the proof of this result. Clearly, it is sufficient to prove that if $A$ is factorial then $A[x]$ is factorial, and then proceed by induction.

We begin with the following remark. Given an integral domain $A$, we may consider its *field of fractions* $\mathbb{Q}(A)$, that is, the smallest field containing $A$. It consists of all *fractions* $a/b$ with $a, b \in A$ and $b \neq 0$. Clearly, $a/b = c/d$ if and only if $ad = bc$. If $A$ is factorial, we always may assume the fractions $a/b$ to be *reduced*, that is to say, $\mathrm{GCD}(a, b)$ to be invertible.

For instance, $\mathbb{Q}$ is the field of fractions of $\mathbb{Z}$. If $\mathbb{K}$ is an arbitrary field, the field of fractions of $\mathbb{K}[x]$ is denoted by $\mathbb{K}(x)$, is called *field of rational functions* over $\mathbb{K}$, and consists of all the fractions $f(x)/g(x)$ with $f(x), g(x)$ relatively prime polynomials with $g(x)$ different from zero.

Of course, $A[x]$ is a subring of $\mathbb{Q}(A)[x]$, which is a factorial ring. However, we cannot immediately deduce from this that $A[x]$ is factorial, because, as we shall see shortly, reducibility or irreducibility in $\mathbb{Q}(A)[x]$ and in $A[x]$ are distinct concepts.

Before going on, we need a definition:

**Definition 4.5.26.** *Let $A$ be a factorial ring and let $f(x) \in A[x]$ be a non-zero polynomial. The* divisor, *or* content, *of $f(x)$ is a greatest common divisor $c(f(x))$ of its coefficients. A polynomial is said to be* primitive *if its content is invertible.*

Notice that, given a polynomial $f(x) \in A[x]$, we have

$$f(x) = c(f(x))f^*(x)$$

with $f^*(x)$ a primitive polynomial.

We may prove the following:

**Proposition 4.5.27.** *Let $A$ be a factorial ring and let $f(x) \in \mathbb{Q}(A)[x]$. Then*

$$f(x) = \frac{d}{m}f^*(x), \tag{4.19}$$

*where $d, m \in A$ and $\mathrm{GCD}(d, m) = 1$, and $f^*(x) \in A[x]$ is associate to $f(x)$ in $\mathbb{Q}[x]$, primitive and uniquely determined up to multiplication by an invertible element of $A$.*

PROOF. Let $f(x) = q_0 + q_1 x + q_2 x^2 + \cdots + q_n x^n$, where $q_i = b_i/c_i \in \mathbb{Q}(A)$ and $b_i, c_i$ in $A$, for all $i = 0, \ldots, n$. Thus,

$$f(x) = \frac{b_0}{c_0} + \frac{b_1}{c_1}x + \cdots + \frac{b_n}{c_n}x^n.$$

Denoting by $m'$ a common multiple of $c_0, c_1, \ldots, c_n$, the polynomial

$$\phi(x) = m'f(x) = b'_0 + b'_1 x + \cdots + b'_n x^n$$

is a polynomial with coefficients in $A$. We have

$$\phi(x) = c(\phi(x))f^*(x)$$

with $f^*(x) = \phi^*(x) \in A[x]$ primitive. So,

$$f(x) = \frac{c(\phi(x))}{m'}f^*(x)$$

and by reducing the fraction $c(\phi(x))/m'$ we have (4.19).

Suppose now that

$$f(x) = \frac{d}{m}f^*(x) = \frac{d'}{m'}g^*(x),$$

with $d', m' \in A$ and $\mathrm{GCD}(d', m') = 1$ and $g^*(x) \in A[x]$ primitive. So we have

$$m'df^*(x) = md'g^*(x).$$

As $f^*(x)$ and $g^*(x)$ are primitive, it follows that $m'd$ and $md'$ are associate, so $d/m = \epsilon d'/m'$, with $\epsilon \in U(A)$. Hence the claim follows.                    □

**Example 4.5.28.** Consider the following polynomial with coefficients in $\mathbb{Q}$

$$f(x) = \frac{5}{7} + \frac{5}{8}x - \frac{10}{3}x^2.$$

If $m = 7 \cdot 8 \cdot 3 = 168$, we have

$$168f(x) = 24 \cdot 5 + 21 \cdot 5x - 56 \cdot 10x^2 = 5(24 + 121x - 112x^2).$$

So

$$f(x) = \frac{5}{7} + \frac{5}{8}x - \frac{10}{3}x^2 = \frac{5}{168}(24 + 121x - 112x^2),$$

or

$$f(x) = \frac{5}{168}f^*(x),$$

where $f^*(x) = 24 + 121x - 112x^2$ is primitive, as $\mathrm{GCD}(24, 121, 112) = 1$.

Let us now deal with the following general question: how are factorisation in $A[x]$ and those in $\mathbb{Q}(A)[x]$ related? First, we give some examples.

**Example 4.5.29.** Consider the *primitive* polynomial in $\mathbb{Z}[x]$

$$f(x) = x^4 + 7x^2 + 10.$$

Suppose we know a factorisation over $\mathbb{Q}$, for instance the following one:

$$f(x) = \left(\frac{5}{3}x^2 + \frac{10}{3}\right)\left(\frac{3}{5}x^2 + 3\right).$$

Then

$$f(x) = \frac{5}{3}(x^2 + 2)\frac{3}{5}(x^2 + 5) = (x^2 + 2)(x^2 + 5),$$

which is a factorisation over $\mathbb{Z}$. Notice that the polynomial $x^2 + 2$, which has coefficients in $\mathbb{Z}$, is the primitive polynomial associate to the polynomial $(5/3)x^2 + 10/3$, and the polynomial $x^2 + 5$ is the primitive associate polynomial to the polynomial $(3/5)x^2 + 3$.

Consider now with a *non primitive* polynomial

$$f(x) = 3x^4 + 30x^2 + 72$$

and let

$$f(x) = \left(2x^2 + 8\right)\left(\frac{3}{2}x^2 + 9\right)$$

be a factorisation of $f(x)$ over $\mathbb{Q}$. In order to find a factorisation over $\mathbb{Z}$ we may proceed as follows. Write first

$$f(x) = df^*(x) = 3(x^4 + 10x^2 + 24), \qquad d = \mathrm{GCD}(3, 30, 72).$$

Then

$$f^*(x) = \frac{1}{3}f(x) = \frac{1}{3}\left(2x^2 + 8\right)\left(\frac{3}{2}x^2 + 9\right).$$

As $f^*(x)$ is primitive, we may repeat what we did above, that is,

$$f^*(x) = \frac{1}{3}\,2\left(x^2 + 4\right)\frac{3}{2}\left(x^2 + 6\right) = (x^2 + 4)(x^2 + 6),$$

which is a factorisation of $f^*(x)$ over $\mathbb{Z}$. Then we have

$$f(x) = 3f^*(x) = 3(x^2 + 4)(x^2 + 6).$$

In these examples we were able to find a factorisation over $\mathbb{Z}$ of polynomial with integer coefficients when we had a factorisation over $\mathbb{Q}$. This is possible in general, and this is the content of Gauss Theorem 4.5.32 we shall prove shortly. To this end, we need some preliminary results.

**Proposition 4.5.30 (Gauss lemma).** *Let $A$ be a factorial ring. The product of two primitive polynomials in $A[x]$ is again a primitive polynomial.*

PROOF. Let

$$f(x) = a_0 + a_1 x + \cdots + a_n x^n, \quad g(x) = b_0 + b_1 x + \cdots + b_m x^m$$

be the two primitive polynomials. Assume by contradiction that $f(x)g(x)$ is not primitive. Then there exists an irreducible element $p \in A$ dividing *all* the coefficients of $f(x)g(x)$. This element $p$ cannot divide all the coefficients of $f(x)$ and of $g(x)$ by the hypothesis of primitivity. Let $a_h$ and $b_k$ be the smallest-indexed coefficients of $f(x)$ and $g(x)$, respectively, that are not divided by $p$. Look at the coefficient with index $h + k$ in $f(x)g(x)$. It is

$$c_{h+k} = a_h b_k + (a_{h-1}b_{k+1} + \cdots + a_0 b_{h+k}) + (a_{h+1}b_{k-1} + \cdots + a_{h+k}b_0).$$

Now, $p$ divides $c_{h+k}$ and both summands in the parentheses, and so it divides $a_h b_k$. Thus, it divides one of the two factors, contradicting the hypothesis. $\qquad\square$

**Corollary 4.5.31.** *Let $A$ be a factorial ring. The content of the product of two polynomials over $A$ equals the product of the contents of the two polynomials.*

PROOF. Let $f(x) = g(x)h(x)$, where $f(x)$, $g(x)$ and $h(x)$ are polynomials in $A[x]$. Extract the content of the polynomials

$$f(x) = c(f(x))f^*(x) = c(g(x)) \cdot g^*(x) \cdot c(h(x)) \cdot h^*(x) = c(g(x))c(h(x)) \cdot g^*(x)h^*(x).$$

As $g^*(x)h^*(x)$ is primitive, we have $c(f(x)) = c(g(x))c(h(x))$, up to product by invertibles, as was to be proved.    □

The following result is the key step to prove the Theorem 4.5.25:

**Theorem 4.5.32 (Gauss theorem).** *Let $A$ be a factorial ring. If a polynomial $f(x) \in A[x]$ factors as a product of two polynomials $g(x), h(x)$ with coefficients in $\mathbb{Q}(A)$, then it also factors as the product of two polynomials with coefficients in $A$, associate in $\mathbb{Q}(A)[x]$ to $g(x)$ and to $h(x)$.*

PROOF. Assume first that $f(x)$ is primitive. So let $f(x) = g(x)h(x)$, $g(x), h(x) \in \mathbb{Q}(A)[x]$. By Proposition 4.5.27 we have $g(x) = (d_1/m_1)g^*(x)$ and $h(x) = (d_2/m_2)h^*(x)$, with $d_i, m_i \in A$ relatively prime, and $g^*(x), h^*(x)$ primitive polynomials in $A[x]$. As a consequence, $f(x) = (d/m)g^*(x)h^*(x)$, $d = d_1 d_2$, $m = m_1 m_2$. By Gauss lemma, the polynomial $f^*(x) = g^*(x)h^*(x)$ is primitive and we have

$$mf(x) = df^*(x),$$

with $f(x)$, $f^*(x)$ primitive. The left-hand-side polynomial and the right-hand-side one have the same content. But as $f(x)$ and $f^*(x)$ are primitive, we have $m = d$ and so

$$f(x) = g^*(x)h^*(x),$$

which is a factorisation over $A$ with associate factors in $\mathbb{Q}(A)[x]$ to $g(x), h(x)$.

When $f(x) \in A[x]$ is not primitive, we write $f(x) = c(f(x))f^*(x)$ with $f^*(x)$ primitive. If $f(x) = c(f(x))f^*(x) = g(x)h(x)$ is a factorisation of $f(x)$ over $\mathbb{Q}(A)$, then

$$f^*(x) = \frac{1}{c(f(x))}g(x)h(x)$$

is a factorisation over $\mathbb{Q}(A)$ and $f^*(x)$ is primitive. Thus, by what has already been proved, $f^*(x)$ is factorisable over $\mathbb{Z}$ as well, that is,

$$f^*(x) = \bar{g}(x)\bar{h}(x)$$

with $\bar{g}(x), \bar{h}(x) \in A[x]$ associate in $\mathbb{Q}(A)[x]$ to $g(x), h(x)$. But this implies that

$$f(x) = c(f(x))f^*(x) = c(f(x))\bar{g}(x)\bar{h}(x)$$

is a factorisation of $f(x)$ over $A$ as required.    □

**Corollary 4.5.33.** *Let $A$ be a factorial ring and let $f(x)$, $g(x)$ be polynomials with coefficients in $A$. If $f(x)$ is primitive and divides $g(x)$ in $\mathbb{Q}(A)[x]$ then this happens in $A[x]$ too.*

PROOF. If $f(x)$ divides $g(x)$ in $\mathbb{Q}(A)[x]$, by Gauss theorem, there is a polynomial in $A[x]$ associate to $f(x)$ in $\mathbb{Q}(A)[x]$, dividing $g(x)$. Such a polynomial is of the form $(a/b)f(x)$, with $\mathrm{GCD}(a,b) = 1$. As $(a/b)f(x) \in A[x]$, $b$ has to divide every coefficient of $f(x)$, and so $b$ is invertible because $f(x)$ is primitive. So $a/b \in A$, and from this the claim easily follows.    □

Now somebody might want to venture the hypothesis that a polynomial with coefficients in a factorial domain $A$ is irreducible *if and only if* it is irreducible over $\mathbb{Q}(A)$. Beware! This is not true. The reason is that $A$ is not a field, so it contains non-invertible elements. Let us demonstrate this point with an example:

**Example 4.5.34.** Consider the polynomial

$$f(x) = 3x^2 + 6,$$

which is *irreducible over* $\mathbb{Q}$, as it is associate to the polynomial $x^2 + 2$ which is irreducible over $\mathbb{Q}$. However, $f(x)$ is *reducible* over $\mathbb{Z}$, as the factorisation

$$3x^2 + 6 = 3(x^2 + 2)$$

*is a non-trivial factorisation over* $\mathbb{Z}$ because 3 *is not* invertible over $\mathbb{Z}$!

In other words, the two polynomials $3x^2 + 6$ and $x^2 + 2$ are not associate in $\mathbb{Z}[x]$, as the invertible elements in $\mathbb{Z}[x]$ *are not* all non-zero constants, but only $\pm 1$. It should now be clear why *reducibility of polynomials over* $\mathbb{Z}$ *does not imply their reducibility over* $\mathbb{Q}$. On the other hand, if a polynomial is reducible over $\mathbb{Q}$, then it is reducible over $\mathbb{R}$. Indeed, both $\mathbb{Q}$ and $\mathbb{R}$ are fields, one contained in the other, and an element of $\mathbb{Q}$ is invertible in $\mathbb{Q}$ if and only if it is invertible in $\mathbb{R}$. Thus, if a polynomial in $\mathbb{Q}[x]$ is irreducible over $\mathbb{R}$ it is obviously also irreducible over $\mathbb{Q}$.

It is also clear that:

**Lemma 4.5.35.** *Let $A$ be a factorial ring. A primitive polynomial $f(x) \in A[x]$ is irreducible over $A$ if and only if it is irreducible over $\mathbb{Q}(A)$.*

The easy proof of this fact is left to the reader (see Exercise A4.54).

We are now in a position to give the:

PROOF OF THEOREM 4.5.25. First of all, we prove that every $f(x) \in A[x]$ can be written as a product of irreducible elements. Clearly, it suffices to consider the case in which $f(x)$ is primitive. If $f(x)$ is irreducible, there is nothing to prove. Otherwise, by Lemma 4.5.35, $f(x)$ is reducible in $\mathbb{Q}(A)$, and so, by Gauss theorem, we have $f(x) = g(x)h(x)$ with $g(x), h(x) \in A[x]$, both of positive degree. By iterating this reasoning, we arrive to a decomposition into irreducibles.

We have yet to prove that every element $f(x)$ irreducible in $A[x]$ is prime. If $f(x)$ is a constant, the claim follows from the fact that $A$ is factorial. If $f(x)$ has positive degree, the hypothesis of it being irreducible implies that $f(x)$ is primitive and that it is irreducible in in $\mathbb{Q}(A)$ too (see Lemma 4.5.35). Suppose now that $f(x)$ divides $g(x)h(x)$. Then it divides one of the two factors in $\mathbb{Q}[x]$, as $\mathbb{Q}(A)[x]$ is factorial. By Corollary 4.5.33 we may conclude that $f(x)$ divides the same factor in $A[x]$.     □

**Remark 4.5.36.** Theorem 4.5.25 implies in particular that, if $\mathbb{K}$ is a field, the ring of polynomials $\mathbb{K}[x_1, \ldots, x_n]$ in $n \geq 2$ variables over $\mathbb{K}$ is factorial. However, it is not at all easy, not even when $\mathbb{K}$ is algebraically closed, to determine the irreducible polynomials, and even less so to factor into irreducible polynomials an arbitrary polynomial of $\mathbb{K}[x_1, \ldots, x_n]$. On the other hand, it is clear that all degree one polynomials are irreducible (see Exercise A4.56). As regards degree two polynomials, see [13], Chapter 22.

We close this section with a useful remark. Let $A$ be a factorial ring. In order to decide about the reducibility of a polynomial $f(x) \in A[x]$ over $\mathbb{Q}(A)$ it is useful in the first place to have a criterion to establish whether the polynomial has roots in $\mathbb{Q}(A)$ or not. If it has roots, it is sure to be reducible. The following proposition teaches us something useful in this regard.

**Proposition 4.5.37 (Newton).** *Let $A$ be a factorial ring. Let $f(x) = a_0 + a_1 x + a_2 x^2 + \cdots + a_n x^n \in A[x]$. Let $\alpha = r/s \in \mathbb{Q}(A)$ be a root of $f(x)$, with $r$ and $s$ relatively prime. Then $r \mid a_0$ and $s \mid a_n$.*

PROOF. If $\alpha = r/s$ is a root of $f(x)$, we find: $0 = f(r/s) = a_0 + a_1 r/s + \cdots + a_n r^n/s^n$. Multiplying by $s^n$ we get

$$0 = s^n a_0 + s^{n-1} a_1 r + \cdots + a_n r^n =$$
$$= s(s^{n-1} a_0 + s^{n-2} a_1 r + \cdots + a_{n-1} r^{n-1}) + a_n r^n =$$
$$= s^n a_0 + r(s^{n-1} a_1 + \cdots + a_n r^{n-1}).$$

From the relation $0 = s(s^{n-1} a_0 + \cdots + a_{n-1} r^{n-1}) + a_n r^n$ it follows that $s \mid a_n r^n$, and from $0 = s^n a_0 + r(s^{n-1} a_1 + \cdots + a_n r^{n-1})$ it follows that $r \mid a_0 s^n$. As $\mathrm{GCD}(r,s) = 1$, we may conclude that $s \mid a_n$ and $r \mid a_0$.    □

**Corollary 4.5.38.** *Let $A$ be a factorial ring. If a polynomial with coefficients in $A$ has a root in $A$, this root divides the constant term.*

**Corollary 4.5.39.** *Let $A$ be a factorial ring. If a monic polynomial with coefficients in $A$ has a root in $\mathbb{Q}(A)$, this root is in $A$.*

Proposition 4.5.37 gives a set of elements of $\mathbb{Q}(A)$ which are *candidates* to be roots of the polynomial $f(x) \in A[x]$. This set can be used to construct an algorithm, which unfortunately is not efficient in the case $A = \mathbb{Z}$ (see Exercise A4.59), to determine the rational roots of $f(x)$: find these candidates, which are finitely many, first, and then compute the value of $f(x)$ in these candidates, to verify if they are actually roots or not.

**Example 4.5.40.** Determine whether the polynomial

$$f(x) = 3x^3 - 4x^2 + 2$$

has rational roots. By applying the algorithm described above, the possible rational roots of $f(x)$ *must* lie in the set

$$\left\{ \pm 1, \pm 2, \pm \frac{1}{3}, \pm \frac{2}{3} \right\}.$$

As it may be easily verified, none of these rational numbers is a root of $f(x)$, so $f(x)$ admits no rational roots. Being of degree 3, the polynomial is irreducible over $\mathbb{Q}$.

### 4.5.6 Polynomials with rational or integer coefficients

The results given in the previous section imply a strong connection between the factorisation of polynomials over $\mathbb{Q}$ and over $\mathbb{Z}$. However, in general it is not easy to solve either problem. For instance, we have characterised the irreducible polynomials over $\mathbb{C}$ and over $\mathbb{R}$, but there is no similar result for polynomials over $\mathbb{Q}$ or $\mathbb{Z}$. There are only a few criteria to determine whether a given polynomial is irreducible over $\mathbb{Q}$ or $\mathbb{Z}$ and some indications on how to attack the general problem.

As for criteria to determine whether a polynomial of arbitrary degree is irreducible over $\mathbb{Q}$, here is one of them.

**Proposition 4.5.41 (Eisenstein's irreducibility criterion).** *Consider a polynomial* $f(x) = a_0 + a_1 x + \cdots + a_n x^n \in \mathbb{Z}[x]$. *Let* $p$ *be a prime number such that:*

*(1)* $p \nmid a_n$;
*(2)* $p \mid a_i$ *for all* $i = 0, \ldots, n-1$;
*(3)* $p^2 \nmid a_0$.

*Then* $f(x)$ *is irreducible over* $\mathbb{Q}$.

PROOF. By Gauss theorem 4.5.32 it is sufficient to prove that $f(x)$ is irreducible over $\mathbb{Z}$. Assume by contradiction that $f(x) = g(x)h(x)$, with

$$g(x) = b_0 + b_1 x + \cdots + b_r x^r, \quad h(x) = c_0 + c_1 x + \cdots + c_s x^s,$$

polynomials with integer coefficients of degree $r < n$ and $s < n$ respectively. Then $r + s = n$ and $b_0 c_0 = a_0$. As $p \mid a_0$, we have $p \mid b_0$ or $p \mid c_0$. Notice that $p \mid b_0$ and $p \mid c_0$ cannot hold simultaneously, or else we would have $p^2 \mid a_0$. So assume that $p$ divides, say, $b_0$ but not $c_0$. Notice that $p$ cannot divide all the $b_i$s, or else $p$ would divide all the $a_i$s, contradicting the hypothesis. Let $b_i$ be the lowest-indexed coefficient not divided by $p$. Then, for $i \leq r < n$,

$$a_i = b_i c_0 + b_{i-1} c_1 + \cdots + b_0 c_i.$$

Notice that $p$ divides $a_i$ ($i < n$) and also divides all the $b_k$s with $k = 0, \ldots, i-1$. Hence $p$ divides $b_i c_0$ and, as it may not divide $b_i$, it has to divide $c_0$, yielding a contradiction, derived from assuming $f(x)$ to be reducible. □

Notice that Eisenstein's criterion gives a *sufficient* condition for irreducibility, but not a *necessary* one.

Eisenstein's criterion extends to polynomials over factorial domains.

If Eisenstein's criterion does not directly apply to a polynomial, it might nevertheless be possible to apply it after having suitably modified the polynomial. Given a constant $\alpha$, it is for instance possible to consider, rather than the polynomial $f(x)$, a new polynomial $f(x - \alpha)$ or, for $\alpha \neq 0$, the polynomial $f(x/\alpha)$, by observing that $f(x) \in \mathbb{K}[x]$ is irreducible over the field $\mathbb{K}$ if and only if $f(x - \alpha)$ and $f(x/\alpha)$ with $\alpha \in \mathbb{K}$ are as well (see Exercise A4.57). As an application of this idea, we prove the following result:

**Proposition 4.5.42.** *If* $p$ *is a prime number, the polynomial*

$$x^{p-1} + x^{p-2} + \cdots + x^2 + x + 1$$

*is irreducible over* $\mathbb{Q}$.

PROOF. Notice that $x^{p-1}+x^{p-2}+\cdots+x^2+x+1 = (x^p - 1)/(x - 1)$. So, substituting $x + 1$ for $x$, we get

$$(x + 1)^{p-1} + (x + 1)^{p-2} + \cdots + (x + 1)^2 + (x + 1) + 1 = \frac{(x + 1)^p - 1}{(x + 1) - 1} =$$

$$= \frac{1}{x}\left(\sum_{k=0}^{p}\binom{p}{k}x^{p-k} - 1\right) = x^{p-1} + \binom{p}{1}x^{p-2} + \binom{p}{2}x^{p-3} + \cdots + p.$$

We may now apply Eisenstein's criterion with respect to the prime $p$, and so prove that the originary polynomial is irreducible.                                               $\square$

Another method, which is practical only when the degree of the polynomial under consideration is not too large, amounts to looking directly for a factorisation. For instance, if $f(x)$ is primitive (as we may suppose without loss of generality) of degree 5 and if we have previously checked that the polynomial has no rational roots, then, if $f(x)$ is reducible, it may only decompose as the product of a degree two polynomial and a degree three one. So, writing down the factors with indeterminate coefficients and equating the coefficients of the polynomial and those of the product, we obtain a system and we are interested in its *integer* solutions: indeed, as we know, we may always reduce to a factorisation in $\mathbb{Z}[x]$. If the system is not compatible, then the originary polynomial is irreducible.

**Example 4.5.43.** Prove, using the method just described, that $x^4 + 1$ is irreducible over $\mathbb{Q}$. Indeed, the polynomial has no rational roots, so it may only factor into the product of two degree two polynomials. Keeping in mind that the leading coefficient of the product is the product of the leading coefficients, and the constant term of the product is the product of the constant terms of the factors, we may write

$$x^4 + 1 = (x^2 + \alpha x \pm 1)(x^2 + \beta x \pm 1),$$

where either both plus signs or both minus signs will be chosen. The system obtained by equating the coefficients is

$$\begin{cases} \alpha + \beta = 0, \\ \alpha\beta = \mp 2, \end{cases}$$

which does not admit integer solutions. Notice that, over $\mathbb{R}$, $x^4 + 1$ factors as $x^4 + 1 = (x^2 + \sqrt{2}x + 1)(x^2 - \sqrt{2}x + 1)$.

This method is not at all computationally efficient (see Exercise A4.61): in particular, it becomes more and more infeasible as the degree of the polynomial increases.

Another useful method is the following. Let $f(x) = \sum_{i=0}^{n} a_i x^i$ be a primitive polynomial with coefficients in $\mathbb{Z}$. Reduce the coefficients modulo a prime number $p$, that is to say, consider the polynomial $f(x)$ as having coefficients in $\mathbb{Z}_p$. Denote by $\bar{f}(x) \in \mathbb{Z}_p[x]$ the new polynomial, which we shall call the *reduction* of $f(x)$ modulo $p$. Notice that if we choose a value of $p$ not dividing $a_n$, then $f(x)$ and $\bar{f}(x)$ have the same degree. If $f(x) = g(x)h(x)$, with $g(x)$ and $h(x)$ of positive degree in $\mathbb{Z}[x]$, then we also have $\bar{f}(x) = \bar{g}(x)\bar{h}(x)$. If $p \nmid a_n$, then $p$ does not divide the leading coefficients of $g(x)$ and $h(x)$, so $\bar{g}(x)$ and $\bar{h}(x)$ have positive degree too. In conclusion, if $f(x)$ is reducible in $\mathbb{Q}$, then $\bar{f}(x)$ it is reducible in $\mathbb{Z}_p[x]$. So we may deduce that:

> *if $\bar{f}(x)$ is irreducible over $\mathbb{Z}_p$ for some $p \nmid a_n$,*
> *then $f(x)$ is irreducible over $\mathbb{Q}$ as well.*

Notice that the converse is not true, that is to say, *it is not the case* that if, for some $p$, $\bar{f}(x)$ is *reducible* over $\mathbb{Z}_p$, then $f(x)$ is reducible over $\mathbb{Q}$.

**Example 4.5.44.** We have seen that $x^4 + 1$ is irreducible over $\mathbb{Q}$. Still, $x^4 + 1$ is reducible over $\mathbb{Z}_2$; indeed, $x^4 + 1 = (x^2 + 1)(x^2 + 1)$ in $\mathbb{Z}_2[x]$.

Now, as the irreducibility test in $\mathbb{Z}_p[x]$ is a *finite test*, working modulo $p$ is convenient and not computationally expensive (see Exercise A4.62). Let us see an example to show how this method works.

**Example 4.5.45.** We shall verify that the polynomial $5x^4 - 2x^3 + 9x - 1$ is irreducible over $\mathbb{Q}$. If we consider it as a polynomial with coefficients in $\mathbb{Z}_2$, the polynomial is $x^4 + x + 1$. This polynomial has no roots in $\mathbb{Z}_2$, so, if it factors, it does so into the product of two second degree factors: that is, we would get

$$x^4 + x + 1 = (x^2 + ax + 1)(x^2 + bx + 1),$$

or

$$x^4 + x + 1 = x^4 + (a + b)x^3 + abx^2 + (a + b)x + 1,$$

which is clearly impossible. So $x^4 + x + 1$ is irreducible over $\mathbb{Z}_2$, which implies that the originary polynomial is irreducible over $\mathbb{Q}$.

For the convenience of the reader, we give now a summary of the methods we have seen to study the reducibility or irreducibility of a polynomial with rational coefficients.

---
**Methods to study the irreducibility of a polynomial over $\mathbb{Q}$:**
---

- consider without loss of generality only primitive polynomials $f(x) \in \mathbb{Z}[x]$;
- use the test for the existence of rational roots based upon Proposition 4.5.37 to conclude that $f(x)$ is *reducible*: the existence of roots implies the reducibility of the polynomial; in particular, if $f(x)$ is of degree 2 or 3, it is irreducible over $\mathbb{Q}$ if and only if it has no roots in $\mathbb{Q}$;
- if possible, especially for low degrees, look for a factorisation into polynomials with integer coefficients;
- apply Eisenstein's criterion, when there is a prime $p$ verifying the hypotheses, to conclude that $f(x)$ is irreducible over $\mathbb{Q}$;
- use transformations of the form $x \to x + \alpha$ to be able to apply Eisenstein's criterion;
- consider $\bar{f}(x) \in \mathbb{Z}_p[x]$ rather than $f(x) \in \mathbb{Z}[x]$: if there exists a $p$ not dividing the leading coefficient of $f(x)$ and such that $\bar{f}(x)$ is irreducible over $\mathbb{Z}_p$, then $f(x)$ is irreducible over $\mathbb{Q}$.

## 4.6 Lagrange interpolation and its applications

In this section we are going to describe a classical idea, that is, the so-called *Lagrange interpolation*. In order to do so, we shall start with a concrete cryptographic problem, the *secret sharing* problem, which can be solved using Lagrange interpolation. In the next section we shall see another application of Lagrange interpolation to the factorisation of polynomials.

So, we begin by describing a concrete problem which apparently has nothing to do with polynomials.

Suppose that the chairman of a pharmaceutical company has the formula for a new chemical compound of the utmost importance. To prevent industrial espionage, he hides the formula into a safe, the access code for which nobody knows but him. However, it is necessary that the people working on the molecule will be able to open the safe even when he is away. His problem is a security one: he does not want all the information to be in the hands of just one or two persons, but neither it is advisable that, in order to open the safe, all the heads of the laboratories have to be present, their number $n$ being quite high. So the chairman would like to divide the information about the access code into $n$ blocks of information, sharing them among $n$ people, but in such a way that $k$ people, with $k \leq n$, are sufficient to reconstruct the secret. The number $k$ is, for instance, the least number of people necessary to perform the experiment to produce the important molecule.

A solution to this problem is given by *Lagrange interpolation* (for a different kind of solution to the same problem, using integers rather than polynomials, see Exercise A4.66). Before seeing how it works, we give some useful results.

We have seen that many properties of the integers also hold for polynomials. For instance, for polynomials too a Chinese remainder theorem holds, analogous to the one for integers (see Theorem 3.4.2 on page 129). It will just be stated here, the proof being left as an exercise to the reader, who can mimic the argument given in the case of integers (see Exercise A4.67).

**Proposition 4.6.1 (Chinese remainder theorem for polynomials).** *Let $\mathbb{K}$ be a field. Let $a_1(x), a_2(x), \ldots, a_t(x)$ be arbitrary polynomials in $\mathbb{K}[x]$ and let $m_1(x), m_2(x), \ldots, m_t(x)$ be pairwise relatively prime polynomials. Then there exists a unique polynomial $f(x) \in \mathbb{K}[x]$, of degree smaller than the degree of $m_1(x)m_2(x) \cdots m_t(x)$ such that*

$$\begin{cases} f(x) & \equiv a_1(x) \pmod{m_1(x)}, \\ f(x) & \equiv a_2(x) \pmod{m_2(x)}, \\ \vdots \\ f(x) & \equiv a_t(x) \pmod{m_t(x)}. \end{cases}$$

Clearly, $f(x) \equiv g(x) \pmod{h(x)}$ means that $f(x) - g(x)$ is divisible by $h(x)$. The following result is a consequence of the Chinese remainder theorem (for another proof, see Exercise A4.63).

**Corollary 4.6.2.** *If $a_0, a_1, \ldots, a_M$ are distinct elements of a field $\mathbb{K}$ and $s_0, s_1, \ldots, s_M$ are elements of $\mathbb{K}$, there exists a unique polynomial $p(x) \in \mathbb{K}[x]$ of degree $n \leq M$ such that*

$$p(a_i) = s_i, \qquad i = 0, 1, \ldots, M. \tag{4.20}$$

PROOF. It suffices to remark that $p(x) \equiv b \pmod{(x - a)}$ if and only if $p(a) = b$ and to apply the Chinese remainder theorem with $m_i(x) = x - a_i$.

□

The unique polynomial satisfying the conditions of the corollary is called the *Lagrange interpolation polynomial* with respect to the data $a_0, a_1, \ldots, a_M$ and $s_0, s_1, \ldots, s_M$. How is this polynomial found? Here follows an algorithm to solve this problem. It is called the *Lagrange interpolation algorithm*.

Consider the polynomial of degree $M + 1$

$$h(x) = (x - a_0)(x - a_1) \cdots (x - a_M)$$

with its derivative $h'(x)$ (see § 1.3.6).

The polynomial

$$(x - a_0)(x - a_1) \cdots \widehat{(x - a_i)} \cdots (x - a_M)$$

(where by $\widehat{(x - a_i)}$ we mean that the factor $(x - a_i)$ is omitted) has degree $M$, its value in $a_i$ is different from zero and coincides with $h'(a_i)$. The polynomial

$$L_i(x) = \frac{h(x)}{(x - a_i)h'(a_i)} = \frac{\prod_{i \neq j}(x - a_j)}{\prod_{i \neq j}(a_i - a_j)}$$

has degree $M$ and is such that

$$\begin{cases} L_i(a_j) = 1 & \text{if } i = j, \\ L_i(a_j) = 0 & \text{if } i \neq j. \end{cases}$$

The polynomials $L_i(x)$ are called the *Lagrange polynomials* with respect to $a_0, a_1, \ldots, a_M$.

The polynomial

$$L(x) = \sum_{i=0}^{M} s_i L_i(x)$$

has degree $M$ and is such that $L(a_i) = s_i$ for all $i = 0, \ldots, M$. So $L(x) = p(x)$ is the Lagrange interpolating polynomial we were looking for: so we have proved that

$$\boxed{p(x) = \sum_{i=0}^{M} s_i L_i(x).}$$

Here follows an example:

**Example 4.6.3.** Determine a polynomial $p(x) \in \mathbb{Q}[x]$ of degree at most 3 satisfying the following conditions:

$$p(0) = 1, \qquad p(1) = -1, \qquad p(-1) = 4, \qquad p(2) = 3.$$

We compute the polynomial $h(x)$ first:

$$h(x) = x(x-1)(x+1)(x-2)$$

and the values $h'(a_i)$:

$$h'(0) = (0-1)(0+1)(0-2) = 2,$$
$$h'(1) = (1-0)(1+1)(1-2) = -2,$$
$$h'(-1) = (-1-0)(-1-1)(-1-2) = -6,$$
$$h'(2) = (2-0)(2-1)(2+1) = 6.$$

Next, we compute the Lagrange polynomials $L_i(x)$:

$$L_0 = \frac{(x-1)(x+1)(x-2)}{2}, \qquad L_1 = \frac{x(x+1)(x-2)}{-2},$$
$$L_2 = \frac{x(x-1)(x-2)}{-6}, \qquad L_3 = \frac{x(x-1)(x+1)}{6}.$$

The polynomial we are looking for is

$$p(x) = 1 \cdot L_0 + (-1) \cdot L_1 + 4 \cdot L_2 + 3 \cdot L_3 = \frac{5}{6}x^3 + \frac{1}{2}x^2 - \frac{10}{3}x + 1.$$

Let us come back to the problem of the chairman of the pharmaceutical company.

**Example 4.6.4.** Suppose the secret, that is the access code of the safe, is represented by a number, and to simplify computations assume this number to be $-1$.

Recall that the chairman of the pharmaceutical company wants to divide the information, that is, the access code, into $n$ blocks of information $P_0, P_1, \ldots, P_{n-1}$ and distribute them among $n$ of his officers in such a way that it is necessary for at least $k$ people to be present ($k \le n$) to reconstruct the secret. Which is the most convenient way to choose $k$?

The chairman knows that the experiment to produce the important chemical compound requires at least four people to be present, so he decides to choose $k = 4$. He decides next to *hide* the secret key $-1$ among the coefficients of a polynomial of degree $k - 1 = 3$, for instance

$$p(x) = x^3 - x^2 + 1.$$

In this way he has chosen a polynomial of degree 3 with arbitrary integer coefficients, *except for the degree two coefficient, which equals* $-1$, that is, the key to access the secret formula. Now the sharing of the $n$ blocks of information $P_i$ among the $n$ officers may be done as follows: the chairman randomly chooses $n$ distinct numbers $a_0, a_1, \ldots, a_{n-1}$ and gives the $i$th officer $(i = 0, 1, \ldots, n-1)$ the information

$$P_i = (a_i, p(a_i)).$$

So each officer receives a pair of numbers and knows that the second element of the pair $P_i$ he has received is the value the unknown degree three polynomial assumes in the first element of the pair. Moreover, each officer knows that the secret key is the coefficient of the degree two term.

No single officer, nor any two or three of them, has enough information to solve the problem of finding the degree three polynomial. But if four people collaborate, they can discover the secret, that is, reconstruct the polynomial.

Indeed, let $C_0$, $C_1$, $C_2$ and $C_3$ be the four people and suppose they have received the following pieces of information:

$$P_0 = (-1, -1), \quad P_1 = (0, 1), \quad P_2 = (2, 5), \quad P_3 = (3, 19).$$

Let us see how they can reconstruct the polynomial chosen by the chairman: they know that the unknown polynomial $p(x)$ is such that

$$
\begin{aligned}
p(a_0) &= -1, &\qquad \text{with } a_0 &= -1, \\
p(a_1) &= 1, &\qquad \text{with } a_1 &= 0, \\
p(a_2) &= 5, &\qquad \text{with } a_2 &= 2, \\
p(a_3) &= 19, &\qquad \text{with } a_3 &= 3.
\end{aligned}
$$

So they have to construct a degree three polynomial for which the values it assumes at 4 distinct points are known. This may be done by applying Lagrange interpolation algorithm. Indeed

$$h(x) = (x + 1)x(x - 2)(x - 3)$$

can be determined, and hence

$$
\begin{aligned}
h'(-1) &= (-1 - 0)(-1 - 2)(-1 - 3) = -12, \\
h'(0) &= (0 + 1)(0 - 2)(0 - 3) = 6, \\
h'(2) &= (2 + 1)(2 - 0)(2 - 3) = -6, \\
h'(3) &= (3 + 1)(3 - 0)(3 - 2) = 12.
\end{aligned}
$$

Then

$$L_0(x) = \frac{x(x-2)(x-3)}{-12} = -\frac{1}{12}(x^3 - 5x^2 + 6x),$$

$$L_1(x) = \frac{(x+1)(x-2)(x-3)}{6} = \frac{1}{6}(x^3 - 4x^2 + x + 6),$$

$$L_2(x) = \frac{(x+1)x(x-3)}{-6} = -\frac{1}{6}(x^3 - 2x^2 - 3x),$$

$$L_3(x) = \frac{(x+1)x(x-2)}{12} = \frac{1}{12}(x^3 - x^2 - 2x),$$

$$\sum_{i=0}^{3} s_i L_i(x) = (-1)L_0 + 1L_1(x) + 5L_2(x) + 19L_3(x) =$$

$$= \frac{1}{12}(12x^3 - 12x^2 + 0 \cdot x + 12) = x^3 - x^2 + 1.$$

So the four officers discover that the secret key is $-1$, may open the safe containing the formula, and perform the experiment.

We emphasise the fact that five or more people could have opened the safe as well, but that would have been impossible for three or less.

And if the four people, rather than $C_0, C_1, C_2, C_3$, were $C_0', C_1', C_2', C_3'$, each with the following pieces of information

$$P_0' = (-2, -11), \ P_1' = (-3, -35), \ P_2' = (1, 1), \ P_3' = (0, 1),$$

how could they proceed to find the key? The reader may now answer on his own to this question (see Exercise A4.68).

## 4.7 Kronecker's factorisation method

In this section we shall describe a factorisation method found by Kronecker, which is a consequence of Lagrange interpolation. It gives the factorisation on $\mathbb{Z}$ of a polynomial with integer coefficients.

Let $f(x) \in \mathbb{Z}[x]$ be the polynomial to be factorised. Notice that, if $f(x)$ has degree $n$ and is not irreducible, then it certainly has a factor of degree $\leq n/2$. Define $M = \lfloor n/2 \rfloor$.

Pick $M+1$ distinct integer 'samples' $a_0, a_1, \ldots, a_M$. Set:

$$r_0 = f(a_0), \quad r_1 = f(a_1), \quad \ldots, \quad r_M = f(a_M).$$

Clearly, the $r_i$s are themselves integer, as the polynomial $f(x)$ has coefficients in $\mathbb{Z}$. Construct the following $(M+1)$–tuples of numbers:

$$s = (s_0, s_1, \ldots, s_M) \quad \text{with } s_i \mid r_i, \text{ for any } i = 0, \ldots, M.$$

For each of these $(M+1)$–tuple $s$, there exists exactly one polynomial $f_s(x) \in \mathbb{Q}[x]$ of degree $\leq M$ such that

$$f_s(a_i) = s_i, \qquad \text{for all } i = 0, \dots, M.$$

Now, each $r_i$ has a finite number of divisors, so there are finitely many $(M+1)$–tuples $s = (s_0, \dots, s_M)$, where $s_i$ are integers dividing $r_i$. Corresponding to these $(M+1)$–tuples $s$ there are finitely many polynomials $f_s(x)$. The following result, basically, says that the divisors of $f(x)$ are to be searched for among these polynomials $f_s(x)$.

**Proposition 4.7.1.** *Let $f(x)$ be a polynomial of degree $n$ with coefficients in $\mathbb{Z}$ and let $M = \lfloor n/2 \rfloor$. Chosen $M + 1$ distinct integers $a_0, a_1, \dots, a_M$, let $S$ be the set of all $(M + 1)$–tuples $s = (s_0, s_1, \dots, s_M)$ of integers $s_i$ such that $s_i \mid r_i = f(a_i)$ for all $i = 0, \dots, M$. Then, if $f_s(x)$ is the unique polynomial with coefficients in $\mathbb{Q}$ such that $f_s(a_i) = s_i$ for all $i = 0, \dots, M$, each divisor of $f(x)$ of degree $\leq M$ is one of the polynomials $f_s(x)$, with $s \in S$.*

PROOF. Let $p(x) \in \mathbb{Z}[x]$ be a factor of $f(x)$ of degree at most $M$. So, $f(x) = p(x)q(x)$ for some $q(x) \in \mathbb{Z}[x]$.

For all $i = 0, 1, \dots, M$, if $p(a_i)q(a_i) = f(a_i)$, then $p(a_i) \mid f(a_i) = r_i$. Thus, the $(M+1)$–tuple $s = (p(a_0), p(a_1), \dots, p(a_M))$ belongs to $S$, as each $p(a_i) \mid r_i$. By the Chinese remainder theorem, there exists exactly one polynomial $f_s(x)$ such that

$$f_s(a_0) = p(a_0), \ f_s(a_1) = p(a_1), \ \dots, \ f_s(a_M) = p(a_M).$$

The two polynomials $f_s(x)$ and $p(x)$ have degree at most $M$; so they are equal. $\qquad\square$

This proposition enables us to find the factors of a polynomial $f(x)$ of $\mathbb{Z}[x]$. Indeed, it suffices to divide $f(x)$ by each of the $f_s(x)$. If no $f_s(x)$ divides $f(x)$, it follows that $f(x)$ is irreducible. Otherwise, proceeding by induction on the degree of $f(x)$, it si possible to completely factorise $f(x)$ in a finite number of steps.

Notice that the size of the set $S$ of all strings may be very large, and the computing time required by this algorithm is exponential: it may not be efficient for polynomials of degrees as low as 5. Nevertheless, it is a useful method for low degrees.

**Example 4.7.2.** Factorise using Lagrange interpolation the polynomial $f(x) = x^4 + 3x + 1$, or prove that it is irreducible.

Here $M = 2$. Choose as sample integers the following ones:

$$a_0 = 0, \qquad a_1 = 1, \qquad a_2 = -1.$$

Then

$$r_0 = f(0) = 1, \qquad r_1 = f(1) = 5, \qquad r_2 = f(-1) = -1.$$

The set $S$ consists then of the following triples $s = (s_0, s_1, s_2)$:

$$(\pm1, \pm5, \pm1), \qquad (\pm1, \pm1, \pm1).$$

Compute next

$$h(x) = x(x-1)(x+1), \quad h'(a_0) = -1, \quad h'(a_1) = 2, \quad h'(a_2) = 2.$$

The Lagrange interpolation polynomial $f_s(x)$ is

$$f_s(x) = s_0 L_0(x) + s_1 L_1(x) + s_2 L_2(x),$$

where

$$L_i(x) = \frac{h(x)}{(x-a_i)h'(a_i)} = \frac{\prod_{i \neq j}(x-a_j)}{\prod_{i \neq j}(a_i - a_j)}.$$

So

$$f_s(x) = s_0 \frac{(x-1)(x+1)}{-1} + s_1 \frac{x(x+1)}{2} + s_2 \frac{x(x-1)}{2}.$$

Corresponding to each of the 16 triples we find the following candidate divisors of $f(x)$.

| | | | |
|---|---|---|---|
| $(1,5,1)$ | $2x^2 + 2x + 1$ | $(1,1,1)$ | $1$ |
| $(1,5,-1)$ | $x^2 + 3x + 1$ | $(1,1,-1)$ | $-x^2 + x + 1$ |
| $(-1,5,1)$ | $4x^2 + 2x - 1$ | $(-1,1,1)$ | $2x^2 - 1$ |
| $(-1,5,-1)$ | $3x^2 + 3x - 1$ | $(-1,1,-1)$ | $x^2 + x - 1$ |
| $(1,-5,1)$ | $-3x^2 - 3x + 1$ | $(1,-1,1)$ | $-x^2 - x + 1$ |
| $(1,-5,-1)$ | $-4x^2 - 2x + 1$ | $(1,-1,-1)$ | $-2x^2 + 1$ |
| $(-1,-5,1)$ | $-x^2 - 3x - 1$ | $(-1,-1,1)$ | $x^2 - x - 1$ |
| $(-1,-5,-1)$ | $-2x^2 - 2x - 1$ | $(-1,-1,-1)$ | $-1$ |

As we began with a primitive polynomial $f(x)$ we may assume that all factors of $f(x)$ are primitive. Moreover, as $f(x)$ is monic, its factors have leading coefficient $\pm1$. So we may exclude from the list all the polynomials that are not primitive (in this case, all the polynomials are primitive) and those with leading coefficient different from $\pm1$. Excluding $\pm1$ as well, the following polynomials remain:

$$x^2 + 3x + 1, -x^2 - 3x - 1, -x^2 + x + 1, x^2 + x - 1, -x^2 - x + 1, x^2 - x - 1.$$

Finally, notice that three of these polynomials coincide with the remaining three up to a factor $-1$, so it suffices to consider the following polynomials:

$$x^2 + 3x + 1, \quad x^2 + x - 1, \quad x^2 - x - 1.$$

None of them divides $f(x)$, so $f(x)$ is irreducible.

In the previous example we were lucky, as the number of triples $s$, and so of possible divisors, was quite small, due to the fact that each triple includes two elements equal to 1, which has as only divisors $\pm1$. Let us see how the situation changes with a different polynomial, even of the same degree 4.

**Example 4.7.3.** Consider the polynomial

$$f(x) = x^4 + 3x^3 + 4x^2 + 9x + 3.$$

Then $M = 2$. Chose $M + 1 = 3$ sample integers, for instance

$$a_0 = 0, \qquad a_1 = 1, \qquad a_2 = -1.$$

We have

$$f(0) = 3, \qquad f(1) = 20, \qquad f(-1) = -4.$$

The number 3 has 4 divisors, namely $\pm 1$ and $\pm 3$; 20 has 12 divisors: $\pm 1$, $\pm 2$, $\pm 4$, $\pm 5$, $\pm 10$ and $\pm 20$; $-4$ has 6 divisors: $\pm 1$, $\pm 2$ and $\pm 4$. There are 288 triples $s$ on the whole. We might change our sample integers trying to improve the situation, but it will still be difficult. It is often convenient to compute $f(a_i)$ for more than $M + 1$ sample values $a_i$ so to be able to choose those values of $f(a_i)$ having as few divisors as possible.

Finally, this method can be used to look just for divisors having a given degree $k \leq M$. In this case, the sample integers have to be $k + 1$ rather than $M + 1$: if we are looking for the linear factors, we are taking $k = 1$, and consequently 2 sample integers; for the second degree divisors, 3 sample integers, and so on.

# Appendix to Chapter 4

## A4 Theoretical exercises

**A4.1.**[*] Given two positive integers $a, b$, we may consider their *least common multiple* $\mathrm{lcm}(a, b)$, that is to say, the least element in the set $M(a, b)$ of the positive numbers $n$ that are multiples of both $a$ and $b$. Prove that if

$$a = p_1^{h_1} p_2^{h_2} \cdots p_r^{h_r}, \qquad b = p_1^{k_1} p_2^{h_2} \cdots p_r^{k_r}$$

are the factorisations of $a$ and $b$ with $p_1, \ldots, p_r$ distinct prime numbers and $h_1, \ldots, h_r, k_1, \ldots, k_r$ non-negative numbers, we have

$$\mathrm{lcm}(a, b) = p_1^{s_1} p_2^{s_2} \cdots p_r^{s_r},$$

where $s_i = \max\{h_i, k_i\}, i = 1, \ldots, r$. Deduce that

$$\mathrm{lcm}(a, b) = \frac{ab}{\mathrm{GCD}(a, b)}.$$

**A4.2.** Prove that the product $rs$ of two integers $r$ and $s$ is relatively prime with another integer $m$ if and only if $r$ is relatively prime with $(m \bmod r)$ and $s$ is relatively prime $(m \bmod s)$.

**A4.3.** Let $f(x)$ be a non-constant polynomial with integer coefficients. Assume that, for some $y$, $f(y)$ is a prime number $p$. Prove that in this case $p$ divides $f(y + kp)$ for all $k \in \mathbb{N}$.

**A4.4.** Using the Fundamental Theorem of Arithmetic, prove that if $p$ is a prime integer number, then $\sqrt{p}$ is an irrational number.

**A4.5.** * Prove formula (4.2).

**A4.6.** * In this exercise we outline a different proof, found by Euler, of Theorem 4.1.3. Recall that the *geometric series* $\sum_{n=0}^{\infty} x^n$ converges to $1/(1 - x)$ for all positive $x$ smaller than 1. Deduce that, if $p_1, \ldots, p_r$ are integer numbers greater than 1, we have

$$\prod_{i=1}^{r} \sum_{n=0}^{\infty} \frac{1}{p_i^n} = \prod_{i=1}^{r} \frac{p_i}{p_i - 1}.$$

Notice that, if there were no prime numbers other than $p_1, \ldots, p_r$, then, by the Fundamental Theorem of Arithmetic, the left-hand side would be exactly $\sum_{m=0}^{\infty} 1/m$, that is the harmonic series which, as is well known, diverges. Deduce that there are infinitely many prime numbers.

**A4.7.** * Verify that the complexity of the sieve of Eratosthenes to find all prime numbers smaller than a given positive integer number $n$ is $\mathcal{O}(\sqrt{n} \log n)$, so it is exponential.

**A4.8.** Prove that if $a, n, m, p$ are positive integers and if $a^n \equiv 1 \pmod p$ and $a^m \equiv 1 \pmod p$, then $a^d \equiv 1 \pmod p$, where $d = \mathrm{GCD}(n, m)$.

**A4.9.** Prove that, if $p$ is prime, the binomial coefficient $\binom{p}{k}$ is multiple of $p$ for all $p > k > 0$.

**A4.10.** Verify that for every prime number $p$ and for all pair of integers $x$, $y$ the following congruence holds:

$$(x + y)^p \equiv x^p + y^p \pmod p.$$

This formula is called the *Freshman's dream*.

**A4.11.** * In this exercise we give an elementary proof of Fermat's little theorem 4.2.5 based on the Fundamental Theorem of Arithmetic.

Fix $p$ and prove first the theorem for $a \geq 0$. Proceed by induction on $a$, the theorem being true for $a = 0$. Apply the previous exercise to prove that

$$(a + 1)^p \equiv a^p + 1^p \pmod p.$$

Apply next the induction hypothesis and conclude. In the case $a < 0$, notice that

$$0 \equiv 0^p = (a + (-a))^p \equiv a^p + (-a)^p \pmod p$$

And conclude by applying what has already been proved in the case $a \geq 0$.

**A4.12.** * In this exercise we give an elementary proof of Euler's Theorem 3.3.11 based on the Fundamental Theorem of Arithmetic.

Let $n = p_1^{k_1} p_2^{k_2} \cdots p_r^{k_r}$ be an integer such that $\mathrm{GCD}(a, n) = 1$. Assume we have proved that

$$a^{\varphi(p_i^{k_i})} \equiv 1 \pmod{p_i^{k_i}}, \qquad \forall\, i = 1, 2, \ldots, r.$$

As $\varphi(n)/\varphi(p_i^{k_i})$ is an integer (for all $i = 1, \ldots, r$), by the multiplicativity of Euler function (see Proposition 4.2.3) it follows that

$$\left(a^{\varphi(p_i^{k_i})}\right)^{\varphi(n)/\varphi(p_i^{k_i})} = a^{\varphi(n)} \equiv 1 \pmod{p_i^{k_i}}$$

and so

$$a^{\varphi(n)} \equiv 1 \pmod{n}.$$

Now it remains to be proved that, if $p$ is a prime not dividing $a$, then

$$a^{\varphi(p^k)} \equiv 1 \pmod{p^k}. \tag{4.21}$$

Proceed by induction on $k$. For $k = 1$, (4.21) follows from Fermat's little theorem. Assume that (4.21) is true for $k$ and prove it for $k + 1$. Equation (4.21) can be written as

$$a^{\varphi(p^k)} = 1 + hp^k$$

for some $h \in \mathbb{Z}$. Also notice that

$$\varphi(p^{k+1}) = p^{k+1} - p^k = p(p^k - p^{k-1}) = p \cdot \varphi(p^k).$$

Conclude the proof in this case by remarking that

$$a^{\varphi(p^{k+1})} = a^{p \cdot \varphi(p^k)} = (1 + hp^k)^p =$$

$$= 1 + php^k + \binom{p}{2} h^2 p^{2k} + \cdots + ph^{p-1} p^{k(p-1)} + h^p p^{kp} \equiv 1 \pmod{p^{k+1}}.$$

**A4.13.** Prove that if $n$ is a product of two distinct prime numbers, knowing $\varphi(n)$, the two prime factors of $n$ can be obtained and vice versa.

In the following exercises, from A4.14 to A4.29, we shall give some important properties of the Euler function and, more in general, of multiplicative functions:

**A4.14.** Let $n$ and $d$ be positive integers, with $d \leq n$. Let $C_d$ be the subset of $\{1, \ldots, n\}$ consisting of all the integers $m \in \{1, \ldots, n\}$ such that $\mathrm{GCD}(m, n) = d$, that is to say, such that $\mathrm{GCD}(m/d, n/d) = 1$. Prove that the order of $C_d$ is $\varphi(n/d)$. Deduce that:

$$n = \sum_{d|n} \varphi\left(\frac{n}{d}\right) = \sum_{d|n} \varphi(d).$$

**A4.15.*** An integer $n > 1$ is said to be *square-free* if it is not divisible by any square of an integer $m \neq \pm 1$. If $n$ is square-free, then its factorisation has the form $n = p_1, \ldots, p_r$ with $p_1, \ldots, p_r$ distinct prime numbers. Define $\mu(n) = (-1)^r$. Define next $\mu(1) = 1$ and $\mu(n) = 0$ if $n$ is not square-free. So we have defined the *Möbius function* $\mu : \mathbb{N} \setminus \{0\} \to \mathbb{N} \setminus \{0\}$. Verify that $\mu$ is multiplicative.

**A4.16.*** Prove that if $n > 1$ the following holds:

$$\sum_{d|n} \mu(d) = 0.$$

**A4.17.**\* Given two functions $f : \mathbb{N} \setminus \{0\} \to \mathbb{N} \setminus \{0\}$, $g : \mathbb{N} \setminus \{0\} \to \mathbb{N} \setminus \{0\}$, define the *Dirichlet product* of $f$ and $g$ as

$$f * g = \sum_{d|n} f(d) g\left(\frac{n}{d}\right).$$

Prove that this product is associative and commutative.

**A4.18.**\* Consider the function $\Pi : \mathbb{N} \setminus \{0\} \to \mathbb{N} \setminus \{0\}$ defined by $\Pi(1) = 1$ and $\Pi(n) = 0$ if $n > 1$. Prove that for every function $f : \mathbb{N} \setminus \{0\} \to \mathbb{N} \setminus \{0\}$, one has $\Pi * f = f * \Pi = f$.

**A4.19.**\* Consider the function $I : \mathbb{N} \setminus \{0\} \to \mathbb{N} \setminus \{0\}$ defined by $I(n) = 1$ for all $n \geq 1$. For every function $f : \mathbb{N} \setminus \{0\} \to \mathbb{N} \setminus \{0\}$, consider the function $E_f := I * f(n) = f * I(n) = \sum_{d|n} f(d)$, for all $n \geq 1$. Prove that $E_\varphi = \iota$, where $\iota$ is the identity function on $\mathbb{N} \setminus \{0\}$.

**A4.20.**\* Prove that if $f$ and $g$ are multiplicative, so is $f * g$. Prove that if the function $f : \mathbb{N} \setminus \{0\} \to \mathbb{N} \setminus \{0\}$ is multiplicative, so is $E_f$.

**A4.21.**\* Prove that $I * \mu = \mu * I = \Pi$, where $I$ and $\Pi$ are the functions defined in Exercises A4.19 and A4.18, respectively.

**A4.22.**\* Prove *Möbius inversion theorem*, which states that for every function $f : \mathbb{N} \setminus \{0\} \to \mathbb{N} \setminus \{0\}$, one has $f = \mu * E_f$, that is to say, for all positive integers $n$:

$$f(n) = \sum_{d|n} \mu(d) E_f\left(\frac{n}{d}\right).$$

**A4.23.**\* Prove that $f$ is multiplicative if and only if $E_f$ is multiplicative. Deduce that $\varphi$ is multiplicative.

**A4.24.**\* Apply Möbius inversion theorem to the function $\varphi$ to deduce another proof of Equation (4.4).

**A4.25.**\* Define $\sigma = E_\iota$, that is to say, for all positive integers $n$ we put $\sigma(n) = \sum_{d|n} d$. Deduce that $\sigma$ is multiplicative. Prove that for all positive integers $n$ we have

$$n = \sum_{d|n} \sigma(d) \mu\left(\frac{n}{d}\right).$$

**A4.26.**\* Define $\nu = E_I$, that is to say, for all positive integers $n$ we put $\nu(n) = \sum_{d|n} 1$, that is, $\nu(n)$ is the number of divisors of $n$. Deduce that $\nu$ is multiplicative. Prove that for all positive integers $n$ we have

$$\sum_{d|n} \nu(d) \mu\left(\frac{n}{d}\right) = 1.$$

**A4.27.**\* Let $n = p_1^{h_1} \cdots p_r^{h_r}$ be the prime decomposition of $n$. Prove that

$$\nu(n) = (h_1 + 1) \cdots (h_r + 1), \qquad \sigma(n) = \frac{p_1^{h_1+1} - 1}{p_1 - 1} \cdots \frac{p_r^{h_r+1} - 1}{p_r - 1}.$$

**A4.28.**\* Let $n = p_1 \cdots p_r$ be the prime decomposition of a square-free number $n$. Prove that

$$\nu(n) = 2^r, \qquad \sigma(n) = (p_1 + 1) \cdots (p_r + 1).$$

**A4.29.**\* Prove the following identities:

$$\frac{1}{\zeta(s)} = \sum_{n=1}^{\infty} \frac{\mu(n)}{n^s}, \qquad \zeta(s)^2 = \sum_{n=1}^{\infty} \frac{\nu(n)}{n^s}, \qquad \zeta(s)\zeta(s-1) = \sum_{n=1}^{\infty} \frac{\sigma(n)}{n^s}.$$

**A4.30.**\* Let $n, m, a$ be positive numbers pairwise relatively prime. Prove that

$$\mathrm{Gss}(nm, a) = \mathrm{lcm}(\mathrm{Gss}(n, a), \mathrm{Gss}(m, a)).$$

**A4.31.**\* Let $p$ be a prime number and $a$ a positive integer not divisible by $p$. Prove that for all positive integers $r$ the following holds:

$$\mathrm{Gss}(p, a^r) = \frac{\mathrm{Gss}(p, a)}{\mathrm{GCD}(r, \mathrm{Gss}(p, a))}.$$

If moreover $b$ is a positive integer not divisible by $p$ and of $\mathrm{GCD}(\mathrm{Gss}(p, a), \mathrm{Gss}(p, b)) = 1$, then prove that

$$\mathrm{Gss}(p, ab) = \mathrm{Gss}(p, a) \cdot \mathrm{Gss}(p, b).$$

**A4.32.** Verify identity (4.13).

**A4.33.** Verify identity (4.14).

**A4.34.** Prove that if $m \mid n$ then the polynomial $x^m - 1$ divides the polynomial $x^n - 1$ in $\mathbb{Z}[x]$.

**A4.35.** Prove that if $m \mid n$ and $n$ is odd, then the polynomial $x^m + 1$ divides the polynomial $x^n + 1$ in $\mathbb{Z}[x]$.

**A4.36.** Prove that if a polynomial $h(x) \in \mathbb{Z}[x]$ divides $x^n - 1$ and if $m \equiv m' \pmod{n}$ then $h(x)$ also divides $x^m - x^{m'}$ in $\mathbb{Z}[x]$.

**A4.37.** Give another proof of Proposition 4.4.1 by writing numbers in base $a$.

**A4.38.** Prove that for all integers $n \geq 2$ one has $F_0 F_1 \cdots F_{n-1} = F_n - 2$.

**A4.39.** Prove that if $n \neq m$, then $\mathrm{GCD}(F_n, F_m) = 1$.

**A4.40.** Deduce from Exercise A4.39 a new proof of the fact that there are infinitely many prime numbers.

**A4.41.**\* Prove Theorem 4.4.13.

**A4.42.**\* A number $n$ is said to be *multiplicatively perfect* if the product of its positive divisors equals $n^2$. Prove that a number is multiplicatively perfect if and only if either it is the product of two distinct prime numbers, or it is the cube of a prime number.

**A4.43.** Prove that in an integral domain a prime element is irreducible.

**A4.44.** Prove that in a factorial ring $A$ the least common multiple of two non-zero elements $a, b$ always exists.

**A4.45.** Prove that $\mathbb{Z}[\sqrt{-3}]$ is closed under addition and multiplication as defined on $\mathbb{C}$.

**A4.46.** Prove that $\mathbb{Z}[\sqrt{-3}]$ is an integral domain with unity.

**A4.47.**\* Determine the invertible elements in the ring $\mathbb{Z}[\sqrt{-6}]$.

**A4.48.**\* Let $A$ be a Noetherian integral domain and let $a, b$ be two non-zero elements of $A$; assume $a$ not to be invertible. Prove that the set of positive integers $n$ such that $a^n \mid b$ is bounded.

**A4.49.**\* Let $A$ be a Noetherian commutative ring. Let (4.17) be an ascending chain of ideals of $A$. Prove that $I = \cup_{n \in \mathbb{N}} I_n$ is an ideal of $A$.

**A4.50.**\* Let $A$ be a commutative ring and let $I$ be an ideal of $A$. Prove that, if $I$ is not finitely generated, then there exists a sequence $\{x_n\}_{n \in \mathbb{N}}$ of elements of $I$ such that, if $I_n = (x_1, \ldots, x_n)$ then $I_n \subset I_{n+1}$ e $I_n \neq I_{n+1}$, for all $n \in \mathbb{N}$.

**A4.51.** Prove Lemma 4.5.19.

**A4.52.** Prove that, given a polynomial $f(x) \in \mathbb{C}[x]$, we have $\lim_{|x| \to +\infty} |f(x)| = +\infty$.

**A4.53.** Let $f(x) = 1 + ax^k + $ (terms of degree greater than $k$) be a polynomial with complex coefficients. Verify that, after the change of variables $x = \alpha y$ with $\alpha^k = -1/a$ one has $f(x) = f(\alpha y) = \phi(y)$ with $\phi(y) = 1 - y^k + y^{k+1}g(y)$, where $g(y)$ is a suitable polynomial.

**A4.54.** Prove Lemma 4.5.35.

**A4.55.** Consider a field $\mathbb{K}$. Denote by $\mathbb{K}(x_1, \ldots, x_n)$ the field of fractions of $\mathbb{K}[x_1, \ldots, x_n]$, called *field of rational functions* in $n$ variables on $\mathbb{K}$. Let $f(x_1, \ldots, x_n)$ be a polynomial on $\mathbb{K}$. Prove that, if $f = gh$ with $g, h \in \mathbb{K}(x_1, \ldots, x_{n-1})[x_n]$, both of positive degree, then $f$ is reducible in $\mathbb{K}[x_1, \ldots, x_n]$.

**A4.56.** Let $\mathbb{K}$ be a field. Prove that every linear polynomial in $\mathbb{K}[x_1, \ldots, x_n]$ is irreducible.

**A4.57.** Consider a polynomial $f(x)$ with coefficients in a field $\mathbb{K}$ and an element $\alpha \in \mathbb{K}$ different from zero. Prove that $f(x)$ is irreducible (on the field $\mathbb{K}$) if and only if $f(x - \alpha)$ and $f(x/\alpha)$ are irreducible.

**A4.58.** Let $p(x)$ be a fixed polynomial in $\mathbb{K}[x]$. Prove that the map $\mathbf{T}_p : \mathbb{K}[x] \to \mathbb{K}[x]$ defined by $\mathbf{T}_p(f(x)) = f(p(x))$ for all $f(x) \in \mathbb{K}[x]$ is a ring homomorphism.

**A4.59.**\* Compute the complexity of the algorithm based on Proposition 4.5.37 to find the rational roots of a polynomial in $\mathbb{Z}[x]$. Verify that this complexity is exponential in the length of the coefficients of the polynomial.

**A4.60.** Prove that all the polynomials in $\mathbb{R}[x]$ that are irreducible over $\mathbb{R}$ are the linear ones and the second degree ones with negative discriminant.

**A4.61.** Let $f(x)$ be a polynomial of degree $n$ in $\mathbb{Z}[x]$ with coefficients bounded by $N$. Prove that the algorithm described on page 189 to verify if $f(x)$ is irreducible has complexity $\mathcal{O}(N^n)$.

**A4.62.**\* Let $f(x)$ be a polynomial of degree $n$ in $\mathbb{Z}[x]$ with coefficients of length at most $N$. Let $\bar{f}(x) \in \mathbb{Z}_p[x]$ be its reduction modulo $p$. Prove that the algorithm described on page 190 to verify if $f(x)$ is irreducible has complexity $\mathcal{O}(N)$. However, notice that the algorithm is exponential in $p$ and more than exponential in $n$.

**A4.63.**\* Find a proof of Corollary 4.6.2 that does not use the Chinese remainder theorem.

**A4.64.**\* If $x_1, \ldots, x_m$ are distinct elements of a field $\mathbb{K}$, consider the determinant $V(x_1, \ldots, x_m)$ of the matrix

$$
\begin{pmatrix}
1 & 1 & \cdots & 1 \\
x_1 & x_2 & \cdots & x_m \\
x_1^2 & x_2^2 & \cdots & x_m^2 \\
\vdots & \vdots & \ddots & \vdots \\
x_1^{m-1} & x_2^{m-1} & \cdots & x_m^{m-1}
\end{pmatrix},
$$

called the *Vandermonde determinant* of $x_1, \ldots, x_m$. Prove that it is different from zero and that

$$
V(x_1, \ldots, x_m) = \prod_{i>j}(x_i - x_j).
$$

**A4.65.**\* Let $n$ be a positive integer and let $a_1, \ldots, a_h$ be elements of a field $\mathbb{K}$. Fix $h$ non-negative integers $m_1, \ldots, m_h$ such that $m_1 + \cdots + m_h + h = n+1$ and assign $n+1$ elements $f_{i,j} \in \mathbb{K}$, with $i = 1, \ldots, h$ and $j = 0, \ldots, m_h$. Prove that there exists exactly one polynomial $f(x) \in \mathbb{K}[x]$ of degree at most $n$ such that $D^{(j)}(a_i) = f_{i,j}$ for all $i = 1, \ldots, h$ and $j = 0, \ldots, m_h$.

**A4.66.** Consider the security problem regarding the formula owned by the chairman of the pharmaceutical company, as described on page 191. Suppose that the code of the safe containing the formula is a very large integer $N$. The chairman wants to share the knowledge of $N$ among $n$ people in such a way that $k$ of them, together, are able to know $N$, but this is impossible to any $k-1$ of them.

In order to do so, the chairman chooses $n$ distinct prime numbers $p_1, \ldots, p_n$, each of them greater than $\sqrt[k]{N}$ but much smaller than $\sqrt[k-1]{N}$. Next, he lets the $i$th person know $p_i$ and $N \bmod p_i$. Prove how and why $k$ of these people, pooling their information, learn $N$, while $k-1$ of them do not. (Hint: consider the case $k=3$ first; use the Chinese remainder theorem.)

**A4.67.** Prove the the Chinese remainder theorem for polynomials 4.6.1 stated on page 191. (Hint: follow the proof of the Chinese remainder theorem for integer numbers.)

**A4.68.** Explain in detail how the four officers of the pharmaceutical company may proceed in order to discover the coefficient of the degree two term of the polynomial interpolating the following data (see page 195):

$$
P_0' = (-2, -11), \ P_1' = (-3, -35), \ P_2' = (1, 1), \ P_3' = (0, 1).
$$

# B4 Computational exercises

**B4.1.** Verify that 91 is not prime.

**B4.2.** Decompose 151 into prime factors. How many distinct prime factors are found?

(a) Exactly one, namely 151 itself, a prime.
(b) Two distinct primes.
(c) Three distinct primes.
(d) None of the above.

**B4.3.** Verify that 397 is prime, by computing all remainders of the divisions of 397 by 2, 3, 5, 7, 11, 13, 17 and 19.

**B4.4.** Decompose 1411 into prime factors. How many distinct prime factors are found?

(a) Exactly one, namely 1411 itself, a prime.
(b) Two distinct primes.
(c) Three distinct primes.
(d) None of the above.

**B4.5.** Decompose 1369 into prime factors. How many distinct prime factors are found?

(a) Exactly one, namely 1369 itself, a prime.
(b) Two distinct primes.
(c) Three distinct primes.
(d) None of the above.

**B4.6.** Decompose 65535 into prime factors. How many distinct prime factors are found?

(a) 2.
(b) 3.
(c) 4.
(d) None of the above.

**B4.7.** Which is the least common multiple of 150 and 345?

(a) 3450.
(b) 51750.
(c) 10350.
(d) None of the above.

**B4.8.** Compute the greatest common divisor of 34891 and 3977.

(a) 1.
(b) 41.
(c) 53.
(d) None of the above.

**B4.9.** Compute the greatest common divisor of 4096 and 2854673912301.

(a) 1.
(b) 7.

(c) 19.
(d) None of the above.

**B4.10.** Compute the greatest common divisor of 1296 and 122678322.

(a) 1.
(b) 6.
(c) 8.
(d) None of the above.

**B4.11.** Does the pair $(59, 61)$ consist of twin primes?

(a) Yes.
(b) No, because 59 is not prime.
(c) No, because 61 is not prime.
(d) None of the above.

**B4.12.** Which of the following is a pair of twin primes?

(a) $(131, 133)$.
(b) $(107, 109)$.
(c) $(103, 105)$.
(d) None of the above.

**B4.13.** Which of the following is a pair of twin primes?

(a) $(497, 499)$.
(b) $(509, 511)$.
(c) $(521, 523)$.
(d) None of the above.

**B4.14.** Find all the primes between 200 and 250 using the sieve of Eratosthenes. How many are there?

(a) 7.
(b) 8.
(c) 10.
(d) None of the above.

**B4.15.** Find all the primes between 300 and 400 using the sieve of Eratosthenes. How many are there?

(a) 13.
(b) 14.
(c) 15.
(d) None of the above.

**B4.16.** Compute $\varphi(151)$, $\varphi(1411)$, $\varphi(1369)$.

**B4.17.** Find $\bar{a}^{1000}$ for all $\bar{a} \in \mathbb{Z}_7$.

**B4.18.** Compute $4^{50}$ modulo 7.

**B4.19.** Compute $\mathrm{Gss}(8, a)$ for every odd $a$ (with $a \neq 1$).

**B4.20.** Compute $\mathrm{Gss}(16, a)$ for every odd $a$.

**B4.21.** Compute 22! modulo 23.

**B4.22.** Let $n$ be a positive integer greater than 2. Prove that $n$ is prime if and only if $(n - 2)! \equiv 1 \pmod{n}$.

**B4.23.** Consider the fraction $161/621$ in base 10. Determine the bases $\beta \leq 10$ in which this number is $\beta$–defined.

**B4.24.** Consider the fraction $162/621$ in base 10. Determine the bases $\beta \leq 10$ in which this number is a simple recurring number.

**B4.25.** Consider the fraction $163/621$ in base 10. Determine the bases $\beta$, with $20 \leq \beta \leq 100$ and $\beta$ not divisible by 3, in which this number is a mixed recurring number.

**B4.26.** Consider the reduced fraction $a = c/d$, and let $p$ be a prime number and $n$ a positive integer. Prove that $a$ is $p^n$-defined if and only if $d$ is a power of $p$; that it is simple recurring in base $p^n$ if and only if it is not divisible by $p$; that it is mixed recurring in base $p^n$ if and only if $d$ is divisible by $p$ but is not a power of $p$.

**B4.27.** Let $p, q$ be distinct prime numbers. Which are the $pq$-defined reduced fractions?

**B4.28.** Prove that the number $2^{35} - 1$ is divisible by 35 and 127.

**B4.29.*** Factor $2^{11} - 1$.

**B4.30.** Is it possible to construct with ruler and compasses a regular 20-sided polygon?

(a) Yes.
(b) No.
(c) No, as 20 is not a product of distinct Fermat numbers.
(d) None of the above.

**B4.31.** What is known about perfect numbers?

(a) There are as many of them as Mersenne primes.
(b) There are infinitely many of them.
(c) There are no odd perfect numbers.
(d) None of the above.

**B4.32.*** Let $\mathbb{Z}[i]$ be the ring of Gaussian integers (see Exercise A1.49). Prove that it is a factorial ring.

**B4.33.** Are the elements 2 and $2 + 3i$ prime in $\mathbb{Z}[i]$?

**B4.34.** Consider the ring $\mathbb{Z}[\sqrt{-6}]$. Which one of the following properties does it verify?

(a) It is a field.
(b) It is an integral domain.
(c) It is a unique factorisation domain.
(d) None of the above.

**B4.35.** Consider $\mathbb{Z}[\sqrt{-6}]$. List some irreducible element. Verify which of those elements are prime as well.

**B4.36.** Consider $10 \in \mathbb{Z}[\sqrt{-6}]$. Is it irreducible?

(a) Yes.
(b) No, because it is not prime.
(c) No.
(d) None of the above.

**B4.37.** Find a primitive polynomial with integer coefficients associate to the polynomial

$$f(x) = \frac{5}{3}x^3 + \frac{3}{5}x^2 - \frac{1}{4}x?$$

(a) $60x^3 + 36x - 15x$.
(b) $20x^3 + 12x - 5x$.
(c) $20x^3 + 36x - 15x$.
(d) None of the above.

**B4.38.** Find a primitive polynomial with integer coefficients associate to the polynomial

$$f(x) = \frac{15}{7} + 3x - \frac{21}{13}x^3?$$

(a) $195 + 293x - 147x^3$.
(b) $195 + 271x - 147x^3$.
(c) $49x^3 - 91x - 65$.
(d) None of the above.

**B4.39.** Is the polynomial $x^3 - 1$ reducible?

(a) Yes.
(b) No.
(c) No, because it is irreducible in the field $\mathbb{Z}_{37}$.
(d) It is impossible to answer if the field in which the coefficients are to be considered is not specified.

**B4.40.** Is the polynomial $x^3 - 3x + 1$ irreducible over $\mathbb{Q}$?

(a) Yes, because it is irreducible over $\mathbb{Z}_2$.
(b) Yes, because it is irreducible over $\mathbb{Z}_3$.
(c) Yes, because it is irreducible over $\mathbb{Z}_{19}$.
(d) No.

**B4.41.** Is the polynomial $x^3 + x^2 + x + 1$ irreducible over $\mathbb{Q}$?

(a) Yes, by Eisenstein's criterion by substituting $x + 1$ for $x$.
(b) Yes, because it has no roots.
(c) No, as it does not verify Eisenstein's criterion.
(d) No, as it has a root.

**B4.42.** Is the polynomial $x^4 + x^3 + x^2 + x + 1$ irreducible over $\mathbb{Q}$?
(a) Yes, by Eisenstein's criterion by substituting $x + 1$ for $x$.
(b) Yes, because it has no roots.
(c) No, as it does not verify Eisenstein's criterion.
(d) No, as it has a root.

**B4.43.** Is the polynomial $x^5 - x + 1$ irreducible over $\mathbb{Q}$?
(a) Yes, by Eisenstein's criterion.
(b) Yes, because it is irreducible over $\mathbb{Z}_3$.
(c) No, because it is reducible over $\mathbb{Z}_2$.
(d) No, as it has a root.

**B4.44.** Is the polynomial $x^4 - x^2 - 1$ irreducible over $\mathbb{Q}$?
(a) Yes, by Eisenstein's criterion.
(b) Yes, because it is irreducible over $\mathbb{Z}_2$.
(c) No, because it is reducible over $\mathbb{Z}_3$.
(d) None of the above.

**B4.45.** Is the polynomial $x^6 + x^5 + x^4 + x^3 + x^2 + x + 1$ irreducible over $\mathbb{Q}$?
(a) Yes, by Eisenstein's criterion by substituting $x + 1$ for $x$.
(b) Yes, because it has no roots.
(c) No, as it does not verify Eisenstein's criterion.
(d) No, as it has a root.

**B4.46.** How many irreducible factors has the polynomial $x^5 - x + 1$ over $\mathbb{Z}_2$?
(a) 1.
(b) 2.
(c) 3.
(d) 5.

**B4.47.** How many irreducible factors has the polynomial $x^3 - 3x + 1$ over $\mathbb{Z}_3$?
(a) Three, one of them with multiplicity two and the other with multiplicity one.
(b) Three, all equal to a same factor, which has multiplicity three.
(c) Two.
(d) One.

**B4.48.** How many irreducible factors has the polynomial $x^4 - x^2 - 1$ over $\mathbb{Z}_3$?
(a) 1.
(b) 2.
(c) 3.
(d) 4.

**B4.49.** How many distinct irreducible factors has the polynomial $x^6 + x^5 + x^4 + x^3 + x^2 + x + 1$ over $\mathbb{Z}_2$?
(a) 1.
(b) 2.
(c) 3.
(d) 6.

**B4.50.** How many distinct irreducible factors has the polynomial $x^6 + x^5 + x^4 + x^3 + x^2 + x + 1$ over $\mathbb{Z}_7$?
(a) 1.
(b) 2.
(c) 3.
(d) 6.

**B4.51.** Exactly one of the following claims is false. Which one?

(a) A polynomial that is irreducible over $\mathbb{R}$ is so over $\mathbb{Q}$ too.
(b) A polynomial that is irreducible over $\mathbb{Z}_p$ is so over $\mathbb{Q}$ too.
(c) A third degree polynomial over $\mathbb{C}$ is always reducible.
(d) A third degree polynomial over $\mathbb{R}$ is always reducible.

**B4.52.** Which is a polynomial $p(x) \in \mathbb{Q}[x]$ of degree at most three such that $p(0) = 0$, $p(1) = 3$ e $p(2) = 8$?

(a) Such a polynomial does not exist, as a third degree polynomial is determined by at least four values.
(b) Such a polynomial exists but is not uniquely determined, so we cannot use Lagrange interpolation polynomial to compute it.
(c) $5x^2 - 2x$.
(d) $x^2 + 2x$.

**B4.53.** Which ones are the Lagrange interpolation polynomials $L_i$ needed to interpolate a degree two polynomial for which the values in $a_0 = 2$, $a_1 = 3$ and $a_2 = 4$ are known?

(a) $L_0 = (x - 3)(x - 4)/2$, $L_1 = (x - 2)(4 - x)$, $L_2 = (x - 2)(x - 3)/2$.
(b) $L_0 = (x + 3)(x + 4)/2$, $L_1 = (x + 2)(x + 4)$, $L_2 = (x + 2)(x + 3)/2$.
(c) It is impossible to answer, as a degree two polynomial is not uniquely determined if only three values are known.
(d) It is impossible to answer unless we know the values the polynomial takes in $a_0$, $a_1$ and $a_2$.

**B4.54.** Determine the degree two polynomial over $\mathbb{R}$ that vanishes at 0 with derivative equal to 1 there, and takes the value 3 at 2.

**B4.55.** Which is the polynomial $p(x) \in \mathbb{Q}[x]$ of degree at most two such that $p(2) = 0$, $p(3) = 3$ and $p(4) = 8$?

(a) Such a polynomial exists, but it is not unique, as a second degree polynomial is not uniquely determined if only three of its values are known.
(b) $3(x - 2)(4 - x) + 8(x - 2)(x - 3)$.
(c) $3(x - 2)(x - 4) - 4(x - 2)(x - 3)$.
(d) $x^2 - 2x$.

**B4.56.** Which ones are the Lagrange interpolation polynomials $L_i$ needed to interpolate a degree four polynomial for which the values in $a_0 = 0$, $a_1 = -1$, $a_2 = 1$ and $a_3 = 2$ are known?

(a) $L_0 = (x - 1)(x + 1)(x - 2)/2$, $L_1 = x(x + 1)(2 - x)/2$, $L_2 = x(x - 1)(2 - x)/6$ and $L_3 = x(x - 1)(x + 1)/6$.
(b) $L_0 = (x - 1)(x + 1)(x - 2)/2$, $L_1 = x(x + 1)(x - 2)/2$, $L_2 = x(x - 1)(x - 2)/6$ and $L_3 = x(x - 1)(x + 1)/6$.
(c) It is impossible to answer, as a fourth degree polynomial is not uniquely determined if only four values are known.
(d) It is impossible to answer unless we know the values the polynomial takes in $a_0$, $a_1$, $a_2$ and $a_3$.

**B4.57.** Which is a polynomial $p(x) \in \mathbb{Q}[x]$ of degree at most four such that $p(0) = 1$, $p(1) = -1$, $p(-1) = 4$ and $p(2) = 3$?

(a) Such a polynomial does not exist, as a fourth degree polynomial is not determined by just four values.

(b) Such a polynomial exists but is not uniquely determined, so we cannot use Lagrange interpolation polynomial to compute it.

(c) $(5/6)x^3 + (1/2)x^2 - (10/3)x + 1$.

(d) $(5/6)x^3 + (1/2)x^2 - (11/3)x + 1$.

**B4.58.** Which is the derivative of the polynomial $x^5 + 2x^4 - 5x^3 - 10x^2 + 6x + 12$ in $\mathbb{Z}_2[x]$?

(a) $-x^2 - x^4$.

(b) $x - x^2 + x^4$.

(c) $x + x^2 - x^5$.

(d) None of the above.

**B4.59.** Which is the derivative of the polynomial $2x^5 - x^2 + 3$ in $\mathbb{Z}_3[x]$?

(a) $x^4 - x$.

(b) $x - x^4$.

(c) $x + x^4$.

(d) None of the above.

# C4 Programming exercises

**C4.1.** Write a program that factors an integer $n$ into primes, by trying to divide it by all integers smaller than $\sqrt{n}$.

**C4.2.** Write a program that computes all primes smaller than a fixed integer $n$, using the sieve of Eratosthenes.

**C4.3.** Write a program that verifies that the polynomial $x^2 - x + 41$ assumes prime values for $x = n$, for all $n \in \{0, \ldots, 40\}$.

**C4.4.** Write a program that verifies if a number is prime by applying Fermat's little theorem.

**C4.5.** Write a program that verifies if a number is prime by applying Wilson's theorem.

**C4.6.** Write a program that computes Euler function.

**C4.7.** Write a program that determines whether a given fraction written in base 10 is a defined, simple recurring or mixed recurring number in a given base $\beta$.

**C4.8.** Write a program that computes all perfect numbers smaller than 1000.

**C4.9.** Write a program that determines all irreducible polynomials of degree smaller than 100 over $\mathbb{Z}_p$, for all primes $p$ smaller than 20.

**C4.10.** Write a program that associates to a polynomial $f(x)$ with rational coefficients a primitive polynomial $f^*(x)$ with integer coefficients such that $f(x) = qf^*(x)$, where $q \in \mathbb{Q}$.

**C4.11.** Write a program that finds the rational roots of a polynomial with integer coefficients using Newton's method.

**C4.12.** Write a program that implements the factorisation method for polynomials, based on reduction modulo $p$.

**C4.13.** Write a program that determines the unique complex polynomial of degree 10 that assumes given values at 11 given points of $\mathbb{C}$.

**C4.14.** Write a program that implements Kronecker's factorisation method.

# 5

## Finite fields and polynomial congruences

The subjects covered in this chapter, that is to say, finite fields and the law of quadratic reciprocity, are both a natural development of notions seen in the previous chapters and important in their own right. However they are mostly significant in connection with some of their applications, for instance to cryptography and codes, to the problem of factoring integers and to primality tests, as we shall see in the next chapters.

### 5.1 Some field theory

In this section we shall mostly deal with finite fields, that is to say, with fields containing finitely many elements. We have already seen that every ring $\mathbb{Z}_p$, with $p$ a prime, is a finite field. Our goal is to describe all finite fields. In order to do so, we need to recall some notions from field theory.

#### 5.1.1 Field extensions

Consider two fields $A, B$, with $A$ a subfield of $B$. Then $B$ is said to be an *extension* of $A$.

For instance, if $b_1, \ldots, b_n$ are elements of $B$, we denote by $A(b_1, \ldots, b_n)$ the smallest subfield of $B$ containing $A$ and $b_1, \ldots, b_n$. Clearly, $A(b_1, \ldots, b_n)$ is an extension of $A$, consisting of all elements of $B$ of the form

$$\frac{f(b_1, \ldots, b_n)}{g(b_1, \ldots, b_n)}, \tag{5.1}$$

where $f(x_1, \ldots, x_n), g(x_1, \ldots, x_n) \in A[x_1, \ldots, x_n]$ and $g(b_1, \ldots, b_n) \neq 0$ (see Exercise A5.3). These elements are called *rational expressions* in $b_1, \ldots, b_n$ with coefficients in $A$. Rational expressions are said to be *integral* if $g(x_1, \ldots, x_n) = 1$.

As $A$ is included in $B$, the latter can be seen as a vector space over $A$, so it makes sense to consider the dimension of $B$ as a vector space over $A$. This dimension is denoted by $[B : A]$ and is called *degree* of $B$ over $A$. Clearly, $[B : A] = 1$ if and only if $A = B$.

However, $[B : A]$ is not necessarily finite. In any case it is obvious that if $A, B, C$ are fields and $A \subseteq B \subseteq C$, then, if $[C : A]$ is finite, $[B : A]$ and $[C : B]$ are finite too (see Exercise A5.5). Moreover, in this case an interesting result holds.

**Theorem 5.1.1 (Multiplicativity of degrees).** *Let $A$, $B$, $C$ fields such that $A \subseteq B \subseteq C$ and suppose that $[C : A]$ is finite. Then:*

$$[C : A] = [C : B] \cdot [B : A].$$

PROOF. Let $n = [B : A]$ and $m = [C : B]$; let $\{a_1, \ldots, a_n\}$ be a basis of $B$ as a vector space over $A$ and $\{b_1, \ldots, b_m\}$ a basis of $C$ as a vector space over $B$. It is sufficient to prove that the elements $a_i b_j, i = 1, \ldots, n, j = 1, \ldots, m$ form a basis of $C$ as a vector space over $A$. The proof is not hard and is left as an exercise for the reader (see Exercise A5.6). □

### 5.1.2 Algebraic extensions

We begin with two important definitions.

**Definition 5.1.2.** *Let $A \subset B$ be a field extension. An element $b \in B$ is said to be* algebraic *over $A$ if there exists a non-zero polynomial $f(x) \in A[x]$ such that $f(b) = 0$. Else, $b$ is said to be* transcendental *over $A$.*

The elements of $\mathbb{C}$ that are algebraic over $\mathbb{Q}$ are called *algebraic numbers*. The set of algebraic numbers is countable (see Exercise A5.16), so in $\mathbb{C}$ and also in $\mathbb{R}$ there are certainly transcendental numbers. However, it is not easy to verify whether a given number is transcendental or not. For instance, it has been proved only quite recently that $e$ (Hermite, 1873) and $\pi$ (Lindemann, 1882) are transcendental (see [26], pages 170–177).

If $b \in B$ is algebraic over $A$, we may consider the non-zero ideal $I_b$ of $A[x]$ consisting of all polynomials $f(x) \in A[x]$ such that $f(b) = 0$. It is exactly the kernel of the homomorphism

$$v_b : f(x) \in A[x] \rightarrow f(b) \in A(b),$$

called *valuation* of polynomials in $b$, which is clearly a ring homomorphism (see Exercise A5.7).

Notice that, as $A[x]$ is a Euclidean ring, and so its ideals are principal (see § 1.3.6), $I_b$ is principal, and so is generated by a polynomial $f_b(x)$, which is unique if it is taken to be monic. This polynomial is called the *minimal polynomial* of $b$ over $A$. It is straightforward to verify that $f_b(x)$ is irreducible over $A$ or, equivalently, that $I_b$ is a prime ideal (see Exercise A5.8). The degree $n$ of $f_b(x)$ is called *degree* of $b$ over $A$. Given $b, b' \in B$ algebraic elements over $A$, they are said to be *conjugate* with respect to $A$ if $f_b(x) = f_{b'}(x)$, that is to say, if they have the same minimal polynomial over $A$.

Consider now the quotient ring $A[x]/I_b$ (see Remark 3.1.6). The existence of the map $v_b$ implies the existence of an injective map

$$w_b : A[x]/I_b \to A(b).$$

Moreover:

**Lemma 5.1.3.** *The map $w_b$ is an isomorphism.*

PROOF. As $w_b$ is injective, it is sufficient to prove that it is surjective too. The image of $w_b$ includes $A$ and $b$, so it is enough to show that this image is a field, that is, that $A[x]/I_b$ is a field. To obtain this, we show that each non-zero element $\xi$ of $A[x]/I_b$ has an inverse. Clearly, this is true if $(\xi) = A[x]/I_b$. Let then $\xi \in A[x]/I_b$ be a non-zero element and suppose $(\xi) \neq A[x]/I_b$. Let $\pi : A[x] \to A[x]/I_b$ be the canonical projection map. Then $I = \pi^{-1}((\xi))$ is a proper ideal of $A[x]$ properly including $I_b$. Let $g(x)$ be a generator of this ideal. Then $g(x)$ divides $f_b(x)$, which is irreducible. Then either $g(x)$ is a constant or $g(x)$ and $f_b(x)$ are associates. In the first case $I = A[x]$, in the second one $I = I_b$, obtaining in each case a contradiction. □

As a consequence, we get the following corollary.

**Corollary 5.1.4.** *Let $A \subseteq B$ be a field extension and let $b \in B$ be algebraic over $A$ of degree $n$. Then $[A(b) : A] = n$ and a basis of $A(b)$ over $A$ is given by $\{1, b, \ldots, b^{n-1}\}$.*

PROOF. From Lemma 5.1.3 it follows that a system of generators of $A(b)$ is given by a sequence $\{b^m\}_{m \in \mathbb{N}}$. On the other hand, as in $A(b)$ we have $f_b(b) = 0$, it follows that $b^n$ depends linearly on $\{1, b, \ldots, b^{n-1}\}$.

We next prove by induction that $b^m$ depends linearly on $\{1, b, \ldots, b^{n-1}\}$ for every $m \geq n$. Indeed, this is true for $m = n$. Assume it is true for a given integer $m > n$. Then we have a relation of the form

$$b^m = a_0 + a_1 b + \cdots + a_{n-1} b^{n-1}.$$

Multiplying both sides by $b$, we have

$$b^{m+1} = a_0 b + a_1 b^2 + \cdots + a_{n-1} b^n.$$

But $b^n$ depends linearly on $\{1, b, \ldots, b^{n-1}\}$ too, hence $b^{m+1}$ depends linearly on $\{1, b, \ldots, b^{n-1}\}$.

Finally, it is clear from the definition of minimal polynomial that $\{1, b, \ldots, b^{n-1}\}$ are linearly independent over $A$. □

The above remarks suggest how to proceed the other way around. Suppose we have a field $A$ and an irreducible monic polynomial $f(x) \in A[x]$. We ask if it is possible to find a field $B$ such that:

- $A$ is a subfield of $B$;
- there exists an element $c \in B$, algebraic over $A$, such that $f_c(x) = f(x)$ and $B = A(c)$.

Such a field $B$ is said to be obtained by *adjoining the root $c$ of $f(x)$ to $A$*.

The answer to the question is clearly yes: it suffices to take $B = A[x]/(f(x))$, and as $c$ the image of $x$ in $B$. We leave to the reader the details (see Exercise A5.11).

Consider now the following situation. We have two fields $A_1, A_2$ and an isomorphism $f : A_1 \rightarrow A_2$. The map $f$ induces in a natural way an isomorphism $\phi_f : A_1[x] \rightarrow A_2[x]$, acting on the constants as $f$ and mapping $x$ in $x$. In this situation the following holds:

**Theorem 5.1.5.** *Let $A_1, A_2$ be two fields and $f : A_1 \rightarrow A_2$ an isomorphism. Let $f_i(x) \in A_i[x]$, $i = 1, 2$, be two irreducible polynomials such that $f_2(x) = \phi_f(f_1(x))$. Let $A_i(b_i)$ be obtained by adjoining to $A_i$ a root $b_i$ of $f_i(x)$, $i = 1, 2$. Then there exists a unique isomorphism $\varphi_f : A_1(b_1) \rightarrow A_2(b_2)$ that restricted to $A_1$ coincides with $f$ and such that $\varphi_f(b_1) = b_2$.*

The easy proof is left as an exercise for the reader (see Exercise A5.12).

**Example 5.1.6.** The field obtained from $\mathbb{R}$ by adjoining a root of $x^2 + 1$ is the complex field $\mathbb{C}$. If we denote by $i$ and $-i$ the roots of $x^2 + 1$, we may consider the extensions $\mathbb{R}(i)$ and $\mathbb{R}(-i)$. The isomorphism existing between these two fields according to Theorem 5.1.5 is the conjugation in $\mathbb{C}$.

**Definition 5.1.7.** *A field extension $A \subseteq B$ is said to be* algebraic *if every element of $B$ is algebraic over $A$; otherwise it is said to be* transcendental.

For instance $\mathbb{R} \subset \mathbb{C}$ is an algebraic extension, while $\mathbb{Q} \subset \mathbb{R}$ is transcendental.

**Lemma 5.1.8.** *Let $A \subseteq B$ be an extension. Given $c \in B$, $c$ is algebraic over $A$ if and only if $[A(c) : A]$ is finite, and in this case $[A(c) : A]$ is the degree of $c$ over $A$. Finally, if $c \in B$ is algebraic over $A$, then $A(c)$ is algebraic over $A$.*

PROOF. If $[A(c) : A]$ is finite, $c$ is algebraic, or else the sequence $\{c^n\}_{n \in \mathbb{N}}$ would consist of elements that are independent on $A$. The second claim follows from Corollary 5.1.4.

Finally, if $c$ is algebraic over $A$ and if $d \in A(c)$, multiplicativity of degrees implies that $[A(d) : A] < [A(c) : A]$ and so $d$ is algebraic over $A$. □

As a consequence we have:

**Proposition 5.1.9.** *Let $A \subseteq B$ be an extension of finite degree. Then every element $c \in B$ is algebraic over $A$ and its degree over $A$ divides $[B : A]$.*

PROOF. Considering the extensions $A \subseteq A(c) \subseteq B$, we find that $[A(c) : A]$ is finite. The claim follows from the previous Lemma and the multiplicativity of degrees Theorem. □

### 5.1.3 Splitting field of a polynomial

Here is another definition.

**Definition 5.1.10.** *Let $A$ be a field and $f(x)$ a polynomial of degree $n$ over $A$. An extension $A \subseteq B$ is said to be a* splitting field *of $f(x)$ over $A$ if there exist elements $b_1, \ldots, b_n \in B$ and $a \in A$ such that*

- $f(x) = a(x - b_1) \cdots (x - b_n)$;
- $B = A(b_1, \ldots, b_n)$.

  Clearly, $[B : A]$ is finite. Moreover, if $n \leq 1$ of course $A = B$.
  Splitting fields always exist:

**Theorem 5.1.11.** *Let $A$ be a field and $f(x)$ a polynomial of degree $n$ over $A$. Then there exists a splitting field $B$ of $f(x)$ over $A$.*

PROOF. The claim is true if $n \leq 1$, because in this case it suffices to take $B = A$. Proceed by induction on $n$. Let $A' = A(c)$ be the field obtained adjoining to $A$ a root $c$ of $f(x)$. Then in $A'[x]$ the polynomial $f(x)$ is divisible by $x - c$, that is, we have $f(x) = (x - c)g(x)$ with $g(x) \in A'[x]$, of degree $n - 1$. By induction a splitting field $B$ of $g(x)$ over $A'$ exists, and clearly it also is a splitting field of $f(x)$ over $A$. □

**Example 5.1.12.** Let $\mathbb{K}$ be a field and $a$ an element of $\mathbb{K}$ that *is not a square*, that is to say, there is no $b \in \mathbb{K}$ such that $b^2 = a$. If we denote by $\sqrt{a}$ a root of the polynomial $x^2 - a$, the splitting field of $x^2 - a$ over $\mathbb{K}$ is $\mathbb{K}(\sqrt{a})$.

For instance, the splitting field of $x^2 - 5$ over $\mathbb{Q}$ is the field $\mathbb{Q}(\sqrt{5})$, consisting of all real numbers $a + b\sqrt{5}$, with $a, b \in \mathbb{Q}$.

The splitting field of a polynomial is *essentially unique*. This is the gist of the following theorem.

**Theorem 5.1.13.** *Let $A_1, A_2$ be two fields and $f : A_1 \rightarrow A_2$ an isomorphism. Let $f_i(x) \in A_i[x]$, $i = 1, 2$, be irreducible polynomials such that $f_2(x) = \phi_f(f_1(x))$. Let $B_i$ be a splitting field of $f_i(x)$ over $A_i$, $i = 1, 2$. Then there exists an isomorphism $\varphi_f : B_1 \rightarrow B_2$ that, restricted to $A_1$, coincides with $f$.*

PROOF. Let $n = [B_1 : A_1]$. If $n = 1$, then $A_1 = B_1$ and $f_1(x)$ is a product of linear factors in $A_1[x]$. As $f$ is an isomorphism, so is $f_2(x)$, implying $B_2 = A_2$. Thus, is is sufficient to take $\varphi_f = f$.

If $n > 1$ we proceed by induction. In this case there exists in $B_1$ a root $b_1$ of $f_1(x)$ that is not in $A_1$. Let $g_1(x)$ be its minimal polynomial, which is an irreducible polynomial dividing $f_1(x)$. Set $g_2(x) = \phi_f(g_1(x))$. This is an irreducible polynomial dividing $f_2(x)$. So in $B_2$ there is a root $b_2$ of $g_2(x)$ not belonging to $A_2$. By Theorem 5.1.5 we may extend $f$ to an isomorphism between $A_1(b_1)$ and $A_2(b_2)$. Notice now that $B_i$ is also a splitting field of $f_i(x)$ over $A_i(b_i)$, $i = 1, 2$. By induction we may easily conclude. The details are left to the reader (see Exercise A5.17). □

**Remark 5.1.14.** Notice that the isomorphism $\varphi_f$ described in the statement of Theorem 5.1.13 maps the roots of $f_1(x)$ to those of $f_2(x)$.

In particular, we have:

**Corollary 5.1.15.** *Let $A$ be a field and let $f(x)$ be a polynomial of degree $n$ over $A$. The splitting field of $f(x)$ over $A$ is unique up to isomorphisms.*

### 5.1.4 Roots of unity

Let $\mathbb{K}$ be a field. If $\mathbb{K}$ has characteristic zero (see Remark 1.3.26), then the fundamental subring of $\mathbb{K}$ is $\mathbb{Z}$, and so $\mathbb{K}$ includes $\mathbb{Q}$.

If the characteristic $\mathbb{K}$ is not zero, then its characteristic is a prime number $p$. This means that its fundamental subring is $\mathbb{Z}_p$, which is, as we know, itself a field. So every field of characteristic $p$ is an extension of $\mathbb{Z}_p$. We shall say that $\mathbb{K}$ is a *prime field*, or *fundamental*, if $\mathbb{K} = \mathbb{Q}$ or $\mathbb{K} = \mathbb{Z}_p$ with $p$ a prime number.

Consider now the polynomial $f_n(x) = x^n - 1$ over a prime field $\mathbb{K}$ and let $F_n$ be a splitting field of $f_n(x)$. The roots of $f_n(x)$ are called *$n$th roots of unity*. It is straightforward to verify (see Exercise A5.20) that the set $R_n$ of $n$th roots of unity is a subgroup of the multiplicative group of $F_n$. Notice that if $m \mid n$ then $R_m \subseteq R_n$ (see Exercise A5.23). We want to study the structure of the group $R_n$. The following holds:

**Proposition 5.1.16.** *If $\mathbb{K}$ has characteristic $0$ or a number $p$ coprime with $n$, then $R_n$ is a cyclic group of order $n$.*

PROOF. We have $D(f_n(x)) = nx^{n-1}$. In our hypotheses, $n \neq 0$ in $\mathbb{K}$, so $D(f_n(x))$ has the unique root $0$, with multiplicity $n-1$. By the factor theorem, the polynomial $f_n(x)$ has exactly $n$ distinct roots, that is to say, $R_n$ has order $n$.

We now prove that $R_n$ is cyclic. First of all, consider the case in which $n = q^m$ is a power of a prime number $q$. If in $R_n$ there were no element of order $n$, then for all $\xi \in R_n$ an integer $m_\xi < m$ would exist such that $\xi$ has order $q^{m_\xi}$. Let $m' = \max\{m_\xi \mid \xi \in R_n\}$. Clearly $m' < m$, so all elements of $R_n$ would have order $n' = q^{m'} < n$. Consequently we would have $R_n \subset R_{n'}$, which yields a contradiction, as $R_n$ has order $n$ while $R_{n'}$ has order $n' < n$.

Suppose now that the factorisation of $n$ is $n = q_1^{m_1} \cdots q_h^{m_h}$. Call $n_i = q_i^{m_i}$, $i = 1, \ldots, h$. We have $R_{n_i} \subset R_n$ and $R_{n_i}$ is cyclic of order $n_i$; let $\xi_i$ be a generator of $R_{n_i}$ for all $i = 1, \ldots, h$. If $\xi = \xi_1 \cdots \xi_h$, one can easily verify that the order of $\xi$ is exactly $n$ (see Exercise A5.24). □

The main general properties of cyclic groups are recalled in Exercises A3.23-A3.35.

In the hypotheses of Proposition 5.1.16, a generator of the group $R_n$ is said to be a *primitive $n$th root* of unity.

**Example 5.1.17.** If $\mathbb{K} = \mathbb{C}$, the $n$th roots of unity are the numbers

$$\xi_{n,j} = \cos\frac{2j\pi}{n} + i\sin\frac{2j\pi}{n}, \qquad j = 0, \ldots, n-1$$

and $\xi_{n,j}$ is a primitive root if and only if $\mathrm{GCD}(j, n) = 1$.

**Proposition 5.1.18.** *If $\mathbb{K}$ has characteristic $p$ and $p$ divides $n$, setting $n = mp^h$, with $p$ coprime with $m$, we have $R_n = R_m$, that is, every $n$th root is $m$th as well, and has multiplicity $p^h$ as a root of $f_n(x)$.*

PROOF. In this case the freshman's dream (see Exercise A4.10) tells us that $f_n(x) = (x^m - 1)^{p^h}$. Hence the claim immediately follows. □

### 5.1.5 Algebraic closure

Recall that a field $A$ is said to be *algebraically closed* if every polynomial of positive degree over $A$ has some root in $A$.

We leave as an exercise to the reader (see Exercise A5.25) the following proposition.

**Proposition 5.1.19.** *Let $A$ be a field. The following are equivalent:*

- *$A$ is algebraically closed;*
- *the irreducible polynomials over $A$ are exactly those of degree one;*
- *for each polynomial $f(x) \in A[x]$ of degree $n$ we have*

$$f(x) = a(x - a_1)^{n_1}(x - a_2)^{n_2} \cdots (x - a_h)^{n_h},$$

*where $a$ is the leading coefficient of $f(x)$, $a_1, \ldots, a_h$ are its distinct roots, and $n_1, \ldots, n_h$ their multiplicities, so $n_1 + \cdots + n_h = n$.*

**Example 5.1.20.** As the Fundamental theorem of algebra 4.5.21 showed us, $\mathbb{C}$ is algebraically closed, but neither $\mathbb{Q}$ nor $\mathbb{R}$ are.

If $p$ is a prime, $\mathbb{Z}_p$ is not algebraically closed. Indeed, consider the polynomial $f(x) = x^n - 1$, with $n$ not divisible by $p$. Then $D(f(x)) = nx^{n-1}$ is different from zero and has as its only root 0, which is not a root of $f(x)$. So $f(x)$ has no multiple roots, and so it has $n$ distinct roots. If $n > p$, it is clear that $\mathbb{Z}_p$ cannot contain all the roots of $f(x)$ (see Exercise A5.26 for the obvious extension to general finite fields).

**Definition 5.1.21.** *An extension $A \subseteq B$ is said to be an algebraic closure of $A$ if it is an algebraic extension and if $B$ is algebraically closed. We also say that $B$ is an algebraic closure of $A$.*

**Example 5.1.22.** The field $\mathbb{C}$ is an algebraic closure of $\mathbb{R}$ but *is not* an algebraic closure of $\mathbb{Q}$, as it includes elements that are transcendental on $\mathbb{Q}$ (see Exercise A5.27).

The following two theorems are very important. Their proofs are omitted as they will not be used in what follows.

**Theorem 5.1.23.** *Every field has an algebraic closure.*

**Theorem 5.1.24.** *Let $A_1, A_2$ be two fields and $f : A_1 \to A_2$ an isomorphism. Let $A_i \subseteq B_i$ be an algebraic closure of $A_i$, $i = 1, 2$. Then there exists an isomorphism $\varphi_f : B_1 \to B_2$, whose restriction to $A_1$ coincides with $f$. In particular, all algebraic closures of a given field are isomorphic.*

### 5.1.6 Finite fields and their subfields

We are now going to study finite fields. Let $\mathbb{F}$ be such a field. By Remark 1.3.26, $\mathbb{F}$ has characteristic $p$, where $p$ is a prime number. This means that its fundamental subring is $\mathbb{Z}_p$, which, as we know, is also a field.

**Remark 5.1.25.** Notice that if a field is finite its characteristic is positive, but the converse is not true, that is, it is not true that if a field's characteristic is positive, then the field is necessarily finite. It suffices to consider the example of the field $\mathbb{Z}_p(x)$ of rational functions on $\mathbb{Z}_p$ (see § 4.5.5). This field is obviously infinite and, as it contains $\mathbb{Z}_p$, it has characteristic $p$ (see Exercise A5.2).

So every finite field is an extension of $\mathbb{Z}_p$ for some prime $p$. As $\mathbb{F}$ is finite, clearly $[\mathbb{F} : \mathbb{Z}_p]$ is finite too.

**Proposition 5.1.26.** *Let $\mathbb{F}$ be a finite field of characteristic $p$ and let $f = [\mathbb{F} : \mathbb{Z}_p]$. Then the order of $\mathbb{F}$ is $p^f$.*

PROOF. By definition, $f = [\mathbb{F} : \mathbb{Z}_p]$ is the dimension of $\mathbb{F}$ as a vector space over $\mathbb{Z}_p$. So $\mathbb{F}$ is isomorphic, as a vector space, to $\mathbb{Z}_p^f$. Hence the thesis follows. $\qquad\square$

**Proposition 5.1.27.** *Let* $\mathbb{F}$ *be a finite field of characteristic $p$ and let $f = [\mathbb{F} : \mathbb{Z}_p]$. Then $\mathbb{F}$ consists exactly of the roots of the polynomial $x^{p^f} - x$ over $\mathbb{Z}_p$: so it is a splitting field of this polynomial over $\mathbb{Z}_p$.*

PROOF. Consider the polynomial $g_n(x) = x^n - x$ over $\mathbb{Z}_p$. If $n = p^f$, then $D(g_n(x)) = -1$, so $g_n(x)$ has $n = p^f$ distinct roots $a_1, \ldots, a_n$. It is easy to verify that every element of $\mathbb{F}$ is a root of $g_n(x)$ and $\mathbb{F} = \{a_1, \ldots, a_n\}$ (see Exercise A5.29). This proves the claim.    □

**Corollary 5.1.28.** *Finite fields of the same order are isomorphic.*

PROOF. This immediately follows from the previous Proposition and from Theorem 5.1.13.    □

The following theorem completes and, in a sense, inverts Proposition 5.1.26:

**Theorem 5.1.29.** *Let $p$ be a prime number. For every positive integer $f$ there exists a field $\mathbb{F}$ of order $p^f$.*

PROOF. Keeping in mind Proposition 5.1.27 and its proof, it suffices to remark that a splitting field of the polynomial $x^{p^f} - x$ over $\mathbb{Z}_p$ has order $p^f$.    □

A remarkable piece of information regarding finite fields is given by the following theorem.

**Theorem 5.1.30.** *The multiplicative group $\mathbb{F}^*$ of a finite field $\mathbb{F}$ is cyclic.*

PROOF. Consider a finite field $\mathbb{F}$ of order $p^f$; it is a splitting field of the polynomial $x^{p^f} - x$ over $\mathbb{Z}_p$. The multiplicative group $\mathbb{F}^*$ consists of the non-zero roots of $x^{p^f} - x$, so of the roots of $x^{p^f - 1} - 1$; thus, $\mathbb{F}^* = R_{p^f - 1}$. The claim follows from Proposition 5.1.16.    □

**Corollary 5.1.31.** *Let $\mathbb{F}$ be a finite field of order $p^f$. For all divisors $m$ of $p^f - 1$, $\mathbb{F}^*$ contains $R_m$ and so contains a primitive $m$th root of unity.*

PROOF. From the proof of Theorem 5.1.30 follows the fact that $\mathbb{F}^* = R_{p^f - 1}$. The proof is concluded keeping in mind the result of Exercise A5.23.    □

A generator of the cyclic *group* $\mathbb{F}^*$ is also called a *generator* of the field $\mathbb{F}$.

Notice now an immediate consequence of the theorem about the multiplicativity of degrees:

**Proposition 5.1.32.** *Let $\mathbb{F} \subset \mathbb{F}'$ be an extension of finite fields of characteristic $p$, and let $f = [\mathbb{F} : \mathbb{Z}_p]$, $f' = [\mathbb{F}' : \mathbb{Z}_p]$. Then $f$ divides $f'$, and precisely:*

$$f' = f \cdot [\mathbb{F}' : \mathbb{F}].$$

This result too may be completed and, in a sense, inverted:

**Theorem 5.1.33.** *Let $\mathbb{F}$ be a finite field of order $p^f$ with $p$ a prime number. For all $f' > 0$ divisors of $f$ there exists a unique subfield $\mathbb{F}'$ of $\mathbb{F}$ of order $p^{f'}$.*

PROOF. As we have seen, $\mathbb{F}$ is the splitting field of the polynomial $x^{p^f} - x$ over $\mathbb{Z}_p$. Notice now that $x^{p^{f'}} - x = x(x^{p^{f'}-1} - 1)$ divides $x^{p^f} - x = x(x^{p^f-1} - 1)$ (see Exercise A4.34); so, every root of $x^{p^{f'}} - x$ is also a root of $x^{p^f} - x$. Thus $\mathbb{F}$ also contains a splitting field $\mathbb{F}'$ of $x^{p^{f'}} - x$ over $\mathbb{Z}_p$, having order $p^{f'}$. This field is unique because it consists of 0 and $R_{p^{f'}-1}$.    □

**Remark 5.1.34.** As finite fields of the same order $n$ are isomorphic, it is customary to denote each such field by a unique symbol. Usually they are denoted by the symbol $\mathbb{F}_n$. Clearly, $n$ has to be of the form $n = p^f$ with $p$ a prime number.

### 5.1.7 Automorphisms of finite fields

Let $A$ be a field of characteristic $p > 0$. Consider the map

$$\phi_A : x \in A \to x^p \in A,$$

called *Frobenius map*. The freshman's dream (see Exercise A4.10) implies that $\phi_A$ is an homomorphism. As $\phi_A$ is clearly injective, if $A$ is finite then $\phi_A$ is an automorphism.

**Theorem 5.1.35.** *Let $\mathbb{F}$ be a finite field of order $p^f$. Its automorphism group $\mathrm{Aut}(\mathbb{F})$ is cyclic of order $f$, generated by $\phi_\mathbb{F}$.*

PROOF. As a first thing, notice that each automorphism of $\mathbb{F}$ fixes each element of $\mathbb{Z}_p$, as it fixes 1.

Further, $\mathbb{F} = \mathbb{Z}_p(a)$ with $a$ a generator of $\mathbb{F}$. As $[\mathbb{F} : \mathbb{Z}_p] = f$, the minimal polynomial $f_a(x)$ of $a$ has degree $f$. Now, an automorphism $\phi$ of $\mathbb{F}$ is uniquely determined by its value $\phi(a)$. Clearly, $\phi(a)$ is, just like $a$, a root of $f_a(x)$, that is to say, $\phi(a)$ is a conjugate of $a$. Let $\mathcal{C}_a$ be the set of conjugates of $a$ in $\mathbb{F}$, a set having size at most $f$, the degree of $f_a(x)$. So there is an injective map $\mathrm{Aut}(\mathbb{F}) \to \mathcal{C}_a$, implying that the order of $\mathrm{Aut}(\mathbb{F})$ is at most $f$. On the other hand it is clear that $\phi_\mathbb{F}$ has exactly order $f$, proving the claim.    □

The following result is proved in an analogous way.

**Corollary 5.1.36.** *Let $\mathbb{F}$ be a finite field of order $p^f$. For all $a \in \mathbb{F}$ the conjugates of $a$ are exactly the elements of the form $a, a^p, a^{p^2}, \ldots, a^{p^{f-1}}$.*

### 5.1.8 Irreducible polynomials over $\mathbb{Z}_p$

Let $\mathbb{F}$ be a finite field of order $p^f$ with $p$ a prime number. As we have seen in Section 5.1.6, $\mathbb{F}$ is a splitting field of the polynomial $x^{p^f} - x$ over $\mathbb{Z}_p$. However, this polynomial is not irreducible over $\mathbb{Z}_p$; for instance it is divisible by $x$ and by $x - 1$. More precisely, we have:

**Proposition 5.1.37.** *For every prime number $p$ and every positive integer $f$, the polynomial $x^{p^f} - x$ is divisible by all polynomials of degree $d \mid f$ that are irreducible over $\mathbb{Z}_p$, but not by their squares.*

PROOF. Recall that all the roots of $x^{p^f} - x$ are simple; so for every prime factor of $x^{p^f} - x$, its square does not divide $x^{p^f} - x$.

If $g(x)$ is an irreducible factor of degree $d$ of $x^{p^f} - x$ over $\mathbb{Z}_p$, a splitting field of $g(x)$ is included in $\mathbb{F}$ and has degree $d$ over $\mathbb{Z}_p$ and consequently has order $p^d$. Thus, $d \mid f$.

Vice versa, let $g(x)$ be an irreducible polynomial over $\mathbb{Z}_p$ of degree $d \mid f$. Then, by adjoining to $\mathbb{Z}_p$ a root $b$ of $g(x)$ we obtain a field of order $p^d$ which, by Theorem 5.1.33 and Theorem 5.1.5, may be assumed to be a subfield of $\mathbb{F}$. So every root of $g(x)$ is also a root of $x^{p^f} - x$; hence, $g(x)$ divides $x^{p^f} - x$.  □

This enables us to *count* the irreducible monic polynomials of degree $d$ over $\mathbb{Z}_p$. Let their number be $n_{d,p}$.

**Corollary 5.1.38.** *For every prime number $p$ and every positive integer $f$ the following holds:*

$$p^f = \sum_{d \mid f} d n_{d,p}.$$

*In particular, if $f$ is prime, we have*

$$n_{f,p} = \frac{p^f - p}{f}.$$

PROOF. It suffices to remark that, for all $d \mid f$, the polynomial $x^{p^f} - x$ has $n_{d,p}$ simple monic factors of degree $d$, and these are all the monic factors of $x^{p^f} - x$.  □

**Remark 5.1.39.** If $f = 2$ we have $n_{2,p} = p(p-1)/2$. It is immediate to compute $n_{2,p}$. The monic polynomials of degree two over $\mathbb{Z}_p$ are of the form $x^2 + ax + b$, with $a, b \in \mathbb{Z}_p$, so they are $p^2$. The reducible ones are all polynomials of the form $(x - \alpha)(x - \beta)$, with $\alpha, \beta \in \mathbb{Z}_p$. The number of these may be computed like this: there are $p$ of them with $\alpha = \beta$, all distinct, and $p(p-1)/2$ distinct with $\alpha \neq \beta$. Hence the formula for $n_{2,p}$ immediately follows.

Notice that Theorem 5.1.37 and Corollary 5.1.38 may be easily extended to compute the number $n_{d,p^f}$ of irreducible monic polynomials of degree $d$ over $\mathbb{F}_{p^f}$ (see Exercises A5.31 and A5.32).

**Example 5.1.40.** We know that $n_{2,2} = 1$. The monic polynomials of degree two that are reducible over $\mathbb{Z}_2$ are $x^2$, $(x - 1)^2 = x^2 + 1$, $x(x - 1) = x^2 - x = x^2 + x$ and so the only irreducible monic polynomial of degree two is $x^2 + x + 1$.

Analogously, we know that $n_{2,3} = 3$. The monic polynomials of degree two that are reducible over $\mathbb{Z}_3$ are $x^2$, $(x - 1)^2 = x^2 + x + 1$, $(x + 1)^2 = x^2 - x + 1$, $x(x - 1) = x^2 - x$, $x(x + 1) = x^2 + x$, $(x + 1)(x - 1) = x^2 - 1$; so, the irreducible monic polynomials of degree two are $x^2 + 1$, $x^2 - x - 1$, $x^2 + x - 1$.

### 5.1.9 The field $\mathbb{F}_4$ of order four

In this section and in the next ones we shall give explicit examples of finite fields. We begin by describing the field $\mathbb{F}_4$. It is an extension of degree two of $\mathbb{Z}_2$. Its elements are the roots of the polynomial

$$x^4 - x.$$

Notice that

$$x^4 - x = x(x - 1)(x^2 + x + 1).$$

So the roots are 0, 1, and the two roots of the polynomial $x^2 + x + 1$. Denoting by $\alpha$ a root of $x^2 + x + 1$, we find that, in agreement with Corollary 5.1.36, the other root is $\alpha^2 = \alpha + 1$, as in a field of characteristic 2 we have $1 = -1$: indeed, $(\alpha + 1)^2 + (\alpha + 1) + 1 = \alpha^2 + 1 + \alpha + 1 + 1 = 0$. Thus,

$$\mathbb{F}_4 = \{0, 1, \alpha, \beta\},$$

where $\beta = \alpha + 1 = \alpha^2$. The addition and multiplication tables of the field are as follows:

| + | 0 | 1 | $\alpha$ | $\beta$ |
|---|---|---|---|---|
| 0 | 0 | 1 | $\alpha$ | $\beta$ |
| 1 | 1 | 0 | $\beta$ | $\alpha$ |
| $\alpha$ | $\alpha$ | $\beta$ | 0 | 1 |
| $\beta$ | $\beta$ | $\alpha$ | 1 | 0 |

| $\cdot$ | 0 | 1 | $\alpha$ | $\beta$ |
|---|---|---|---|---|
| 0 | 0 | 0 | 0 | 0 |
| 1 | 0 | 1 | $\alpha$ | $\beta$ |
| $\alpha$ | 0 | $\alpha$ | $\beta$ | 1 |
| $\beta$ | 0 | $\beta$ | 1 | $\alpha$ |

We may determine the field of order 4 in another way as well, without resorting to finding the roots of the polynomial $x^4 - x$. It is sufficient to find an irreducible polynomial $f(x)$ of degree 2 over $\mathbb{Z}_2$ and to adjoin to $\mathbb{Z}_2$ a root of $f(x)$. By Corollary 5.1.38, there is a unique irreducible polynomial of degree 2 over $\mathbb{Z}_2$, and precisely $x^2 + x + 1$ (see Example 5.1.40). Thus, $\mathbb{F}_4$ is isomorphic to $\mathbb{Z}_2[x]/(x^2 + x + 1)$. In other words, $\mathbb{F}_4$ is simply $\mathbb{Z}_2[x]$ with the condition that $x^2 + x + 1 = 0$, or $x^2 = x + 1$. So we are identifying two polynomials $f(x)$ and $g(x)$ when their difference is a multiple of $x^2 + x + 1$:

$$f(x) \equiv g(x) \pmod{x^2 + x + 1}. \tag{5.2}$$

As already remarked for $x = \alpha$, we have

$$x^2 \equiv x + 1 \pmod{x^2 + x + 1}. \tag{5.3}$$

Moreover:

$$x^3 \equiv 1 \pmod{x^2 + x + 1}$$

because $x^3 - 1 = (x - 1)(x^2 + x + 1)$. Further,

$$x^4 = x \cdot x^3 \equiv x \cdot 1 = x \pmod{x^2 + x + 1}$$

and
$$x^5 + x^4 + 1 \equiv 0 \pmod{x^2 + x + 1}$$

as $x^5 + x^4 + 1 = (x^3 + x + 1)(x^2 + x + 1)$.

So, every time we find $x^2$ in a polynomial, we may simply substitute $x + 1$ for it; when we find $x^3$ we may substitute 1 for it; when we find $x^4$ we may substitute $x$ for it and so on. For instance, the polynomial

$$x^4 + x + 1$$

is identified with the polynomial 1. The reasoning is analogous when working modulo any other polynomial.

Relation (5.2) is an equivalence relation and, keeping in mind formula (5.3), it may be seen that in every equivalence class there is a linear polynomial (see Exercise A5.36). The quotient ring is denoted by

$$\mathbb{Z}_2[x]/(x^2 + x + 1)$$

and it is a field whose elements are the four classes

$$\bar{0},\ \bar{1},\ \bar{x},\ \overline{1 + x}.$$

## 5.1.10 The field $\mathbb{F}_8$ of order eight

It is an extension of degree 3 of $\mathbb{Z}_2$. Its elements are all the roots of the polynomial $x^8 - x$.

The factorisation of $x^8 - x$ into irreducible factors over $\mathbb{Z}$ is

$$x^8 - x = x(x - 1)(x^6 + x^5 + x^4 + x^3 + x^2 + x + 1) \tag{5.4}$$

(see Exercise B5.15). But the last factor splits into two irreducible factors over $\mathbb{Z}_2$, that is, the factorisation of $x^8 - x$ over $\mathbb{Z}_2$ is (see Exercise B5.16):

$$x^8 - x = x(x - 1)(x^3 + x + 1)(x^3 + x^2 + 1). \tag{5.5}$$

Rather than finding all the roots of this polynomial, it is more convenient to choose one of the two irreducible polynomials of degree three over $\mathbb{Z}_2$ in the factorisation of $x^8 - x$, for instance $x^3 + x + 1$, and notice that $\mathbb{F}_8$ is isomorphic to $\mathbb{Z}_2[x]/(x^3 + x + 1)$.

Analogously to what happened in the previous case, every class modulo $x^3 + x + 1$ is represented by a polynomial of degree at most two, as we may identify $x^3$ with $x + 1$, $x^4$ with $x^2 + x$, $x^5$ with $x^2 + x + 1$, $x^6$ with $x^2 + 1$, $x^7$ with 1 and so on. So, if we denote by $\alpha$ a root of $x^3 + x + 1$, $\alpha$ satisfies

$$\alpha^3 = \alpha + 1,$$

and the field $\mathbb{F}_8$ will consist of the following elements:

$$\mathbb{F}_8 = \{a_0 + a_1\alpha + a_2\alpha^2 \mid a_i \in \mathbb{Z}_2\} =$$
$$= \{0, 1, \alpha, \alpha^2, 1 + \alpha, 1 + \alpha^2, \alpha + \alpha^2, 1 + \alpha + \alpha^2\}.$$

We leave to the reader the task of writing down the addition and multiplication tables of these elements (see Exercise B5.5). Moreover, notice that $\mathbb{F}_4$ is not a subfield of $\mathbb{F}_8$: indeed this would contradict the multiplicativity of degrees, as $[\mathbb{F}_8 : \mathbb{F}_2] = 3$ and $[\mathbb{F}_4 : \mathbb{F}_2] = 2$, and 2 does not divide 3. The reader may directly verify from the addition and multiplication tables he has written that the only subfield of $\mathbb{F}_8$ is $\mathbb{F}_2$ and that the same holds for $\mathbb{F}_4$.

### 5.1.11 The field $\mathbb{F}_{16}$ of order sixteen

It is an extension of degree 4 of $\mathbb{Z}_2$. Its elements are the roots of the polynomial

$$x^{16} - x.$$

The factorisation of $x^{16} - x$ into irreducible factors in $\mathbb{Z}[x]$ is as follows (see Exercise B5.17):

$$x^{16} - x = x(x - 1)(x^2 + x + 1)(x^4 + x^3 + x^2 + x + 1) \cdot$$
$$\cdot (x^8 - x^7 + x^5 - x^4 + x^3 - x + 1), \tag{5.6}$$

but over $\mathbb{Z}_2$ we have the following factorisation into irreducible polynomials (see Exercise B5.18):

$$x^{16} - x = x(x - 1)(x^2 + x + 1)(x^4 + x + 1) \cdot$$
$$\cdot (x^4 + x^3 + 1)(x^4 + x^3 + x^2 + x + 1). \tag{5.7}$$

So $x^4 + x^3 + 1$ is irreducible of degree 4 over $\mathbb{Z}_2$ and $\mathbb{F}_{16}$ is just $\mathbb{Z}_2[x]/(x^4 + x^3 + 1)$. Proceeding as above, we may see that each class in $\mathbb{Z}_2[x]/(x^4 + x^3 + 1)$ has a representative of degree at most three. We leave to the reader the task of writing down the elements of $\mathbb{F}_{16}$, as well as its addition and multiplication tables (see Exercise B5.6).

The reader will easily verify that $\mathbb{F}_8$ is not a subfield of $\mathbb{F}_{16}$, while $\mathbb{F}_4$ is (see Exercise B5.14).

### 5.1.12 The field $\mathbb{F}_9$ of order nine

Start with $\mathbb{Z}_3$ and consider the polynomial

$$x^9 - x = x(x^4 - 1)(x^4 + 1) =$$
$$= x(x - 1)(x + 1)(x^2 + 1)(x^4 + 1) = \tag{5.8}$$
$$= x(x - 1)(x + 1)(x^2 + 1)(x^2 + x - 1)(x^2 - x - 1)$$

(see Proposition 5.1.37 and Example 5.1.40). To obtain $\mathbb{F}_9$ it suffices to consider, analogously to what precedes, $\mathbb{Z}_3[x]/(x^2 + 1)$. Once more, each equivalence class modulo $x^2 + 1$ contains a polynomial of degree at most 1. Denoting by $i$, as in $\mathbb{C}$, a root of $x^2 + 1$, we get

$$\mathbb{F}_9 = \{a_0 + a_1 i|\ a_i \in \mathbb{Z}_3\} =$$
$$= \{0, 1, i, -1, -i, 1 + i, 1 - i, -1 + i, -1 - i\}.$$

The reader will find no difficulty in writing down the addition and multiplication tables of the elements of $\mathbb{F}_9$ written in this way (see Exercise B5.7). Alternatively, we may consider $\mathbb{F}_9$ as $\mathbb{Z}_3[x]/(x^2 - x - 1)$. In this case too, each equivalence class in $\mathbb{Z}_3[x]/(x^2 - x - 1)$ contains a polynomial of degree at most 1. So, if we denote by $\alpha$ a root of the polynomial $x^2 - x - 1$, which is irreducible over $\mathbb{Z}_3$, we have an element such that $\alpha^2 = \alpha + 1$, and it is easily seen that $\mathbb{F}_9$ consists of the following 9 elements:

$$0,\ 1,\ 2,\ \alpha,\ 2\alpha,\ 1 + \alpha,\ 1 + 2\alpha,\ 2 + \alpha,\ 2 + 2\alpha.$$

So we have $\alpha = -1 \pm i$. It is interesting to observe that all non-zero elements of $\mathbb{F}_9$ are powers of $\alpha$ (see Exercise A5.37), that is, $\alpha$ is a generator of $\mathbb{F}_9$, while the same is not true of $i$, as the distinct powers of $i$ are just 4: $i^0 = 1$, $i^1 = i$, $i^2 = -1$, $i^3 = -i$, while $i^4 = 1$.

### 5.1.13 About the generators of a finite field

It is very useful to know a generator of a finite field, as in this case every element of the field may be written in a simple way as a power of this generator. On the other hand, what has been said about $\mathbb{F}_9$ shows that the finite field constructions that might seem the *most obvious* ones do not yield in a natural way a generator of the multiplicative group of the field. In general, to determine such a generator is a quite delicate computational problem, even when the field is $\mathbb{Z}_p$ with $p$ prime. Here we shall restrict ourselves to a couple of remarks. The first one is given by the following proposition.

**Proposition 5.1.41.** *Let $\mathbb{F}$ be a finite field of order $q$. If $a$ is a generator, and if $i$ is a positive integer, then $a^i$ is a generator if and only if $\mathrm{GCD}(i, q-1) = 1$. In particular, the number of generators of $\mathbb{F}_q$ is $\varphi(q - 1)$.*

This is a consequence of a well-known result about cyclic groups (see Exercise A3.30).

Keeping in mind the previous proposition we get:

**Corollary 5.1.42.** *Let $p$ be a prime number and $a < p$ any positive number. The probability for the class of $a$ in $\mathbb{Z}_p$ to be a generator of $\mathbb{Z}_p$ is*

$$\pi_p = \frac{\varphi(p - 1)}{p - 1} \prod_{q \text{ prime, } q|p-1} \left(1 - \frac{1}{q}\right).$$

So the behaviour of the probability $\pi_p$ depends on the factorisation of $p-1$. However, not too much can be hoped from a random choice. Indeed, it may be proved that:

**Proposition 5.1.43.** *There exists a sequence $\{p_n\}_{n\in\mathbb{N}}$ of prime numbers such that the sequence of probabilities $\{\pi_{p_n}\}_{n\in\mathbb{N}}$ converges to zero.*

PROOF. By Dirichlet's Theorem 4.1.6, for every positive integer $n$ there exists a prime $p_n$ such that $p_n \equiv 1 \pmod{n!}$. Then the primes dividing $p_n-1$ include all the primes dividing $n$, and so

$$\pi_{p_n} \leq \prod_{q \text{ prime}, \, q|n} \left(1 - \frac{1}{q}\right).$$

To conclude it suffices to remark that, for $n \to \infty$, the right-hand side converges to 0 (see Lemma 4.1.5).    $\square$

### 5.1.14 Complexity of operations in a finite field

In Chapter 3 we have discussed the computational cost of operations in $\mathbb{Z}_n$ and consequently of operations in the fields $\mathbb{Z}_p$, with $p$ a prime.

Recall that (see Remark 3.1.5 and Section 3.3):

- computing the sum or the difference of two elements in $\mathbb{Z}_p$ has complexity $\mathcal{O}(\log p)$;
- computing the product of two elements in $\mathbb{Z}_p$ has complexity $\mathcal{O}(\log^2 p)$;
- computing the inverse of an element in $\mathbb{Z}_p$ has complexity $\mathcal{O}(\log^3 p)$.

Here we intend to extend these results to operations in finite fields.

Let $\mathbb{F}$ be a finite field of order $q = p^f$ with $p$ a prime. First of all, we have to keep in mind how we are representing the elements of $\mathbb{F}$. Recall that $\mathbb{F} = \mathbb{Z}_p[x]/(f(x))$, with $f(x)$ an irreducible polynomial of degree $f$ over $\mathbb{Z}_p$. So we may represent the elements of $\mathbb{F}$ as polynomials in $\mathbb{Z}_p[x]$ of degree at most $f - 1$.

Keeping this in mind, we prove the following proposition, which extends the results of Chapter 3 about $\mathbb{Z}_p$:

**Proposition 5.1.44.** *Let $\mathbb{F}$ be a finite field of order $q = p^f$ with $p$ a prime. Then:*

*(a) the addition of two elements of $\mathbb{F}$ has complexity $\mathcal{O}(\log^2 q)$;*
*(b) the multiplication of two elements of $\mathbb{F}$ has complexity $\mathcal{O}(\log^3 q)$;*
*(c) computing the inverse of a non-zero element of $\mathbb{F}$ has complexity $\mathcal{O}(\log^3 q)$;*
*(d) if $h$ is a positive integer and $\xi$ is an element of $\mathbb{F}$, computing $\xi^h$ has complexity $\mathcal{O}(\log h \log^3 q)$.*

PROOF. We shall only prove the claim about multiplication, as that about addition is proved in an analogous way.

We may consider two elements of $\mathbb{F}$ as polynomials of degree at most $f - 1$ over $\mathbb{Z}_p$. Keeping in mind Proposition 2.5.10, we know that computing the product of these polynomials has complexity $\mathcal{O}(f^2 \log^2 p)$. We next have to divide the result by $f(x)$ and to take the remainder of the division. Again by Proposition 2.5.10, the latter operation has complexity $\mathcal{O}(f^2 \log^3 p)$. In conclusion, the complexity is $\mathcal{O}(f^3 \log^3 p) = \mathcal{O}(\log^3 q)$.

In order to prove the claim about the computation of the inverse of an element, we may argue as follows. Suppose we are given a non-zero element $\xi$ of $\mathbb{F}$, represented by a polynomial $g(x)$ of degree at most $f - 1$ over $\mathbb{Z}_p$. As $f(x)$ is irreducible over $\mathbb{Z}_p$, we have $\mathrm{GCD}(f(x), g(x)) = 1$, so there exists a Bézout relation $A(x)f(x) + B(x)g(x) = 1$, which may be found using the Euclidean algorithm applied to $g(x)$ and $f(x)$. The class of $B(x)$ in $\mathbb{F}$ is the inverse of $\xi$. Again by Proposition 2.5.10, the computational cost is once more $\mathcal{O}(f^3 \log^3 p)$, so we may conclude as above.

As regards computing powers, the reasoning is analogous to that of § 3.3.1. So $\mathcal{O}(\log h)$ successive multiplications of elements of $\mathbb{F}$ are needed, taking $\mathcal{O}(\log h \log^3 q)$ bit operations. □

## 5.2 Non-linear polynomial congruences

Consider a congruence of the form

$$f(x) \equiv 0 \pmod{m}, \tag{5.9}$$

where $f(x) = \sum_{i=0}^{n} a_i x^i$ is a polynomial in $x$ with integer coefficients: we are looking for solutions of the congruence modulo an arbitrary positive integer $m$. Such a congruence is said to be *polynomial*. Clearly, we may assume $f(x)$ to be *reduced modulo* $m$, that is, the coefficients may be reduced modulo $m$; hence $f(x)$ may be supposed to be a polynomial in $\mathbb{Z}_m[x]$. So it has a degree which is said to be the *degree* of the polynomial congruence (5.9). In other words, the degree of the polynomial congruence (5.9) is the greatest exponent of $x$ in $f(x)$ such that its coefficient is not divisible by $m$.

The following is an example of polynomial congruence:

$$x^2 + 2x - 3 \equiv 0 \pmod{121};$$

its degree is 2.

In Chapter 3 we have studied linear congruences, that is to say, polynomial congruences of degree 1. Now we shall examine *non-linear polynomial congruences*. Their solution is one of the greatest problems in number theory and involves many still unsolved questions. We shall mainly dwell on congruences of degree 2.

We start with the case in which the modulus is a prime number $p$. As an immediate consequence of the factor theorem, a polynomial with coefficients in a field cannot have more roots than its degree (see Exercise A1.55). So the congruence

$$f(x) \equiv 0 \pmod{p},$$

where $p$ is prime, cannot have more than $n$ solutions, if $n$ is the degree of $f(x) \in \mathbb{Z}_p[x]$.

This is not true modulo a non-prime number. For instance, the congruence

$$x^2 - 1 \equiv 0 \pmod{8}$$

admits 4 solutions: $x = 1,\ 3,\ 5$ and $7$.

Let us see some examples.

**Example 5.2.1.** Solve the congruence $x^3 + 2x^2 - 1 \equiv 0 \pmod{5}$.

In order to find all the solutions, it suffices to substitute in $f(x) = x^3 + 2x^2 - 1$ the values $x = 0, 1, 2, 3, 4$ and check whether the integer so obtained is congruent to 0 modulo 5 or not. So we have

| $x$ | 0 | 1 | 2 | 3 | 4 |
|---|---|---|---|---|---|
| $f(x)$ | $-1$ | 2 | 15 | 44 | 95 |

and the values of $x$ that are solutions of $f(x) \equiv 0 \pmod{5}$ are $x = 2$ and $x = 4$. Notice that, as regards the computations involved, it would have been simpler to choose $x = 1, 2, -1, -2$, obtaining

| $x$ | 0 | 1 | 2 | $-2$ | $-1$ |
|---|---|---|---|---|---|
| $f(x)$ | $-1$ | 2 | 15 | $-1$ | 0 |

Thus, we find the solutions $x = 2$ e $x = -1 \equiv 4 \pmod{5}$.

**Example 5.2.2.** Solve the congruence $x^3 + 2x^2 - 1 \equiv 0 \pmod{7}$.

Compute $f(x) = x^3 + 2x^2 - 1$ for the values $x = 0, 1, 2, 3, 4,\ 5, 6$, or $x = 0, 1, 2, 3, -3, -2, -1$, modulo 7:

| $x$ | 0 | 1 | 2 | 3 | $-3$ | $-2$ | $-1$ |
|---|---|---|---|---|---|---|---|
| $f(x)$ | $-1$ | 2 | $15 \equiv 1$ | $44 \equiv 2$ | $-10 \equiv 4$ | $-1$ | 0 |

The congruence admits the unique solution $x = -1$, that is, $x = 6$.

**Example 5.2.3.** Solve the congruence $x^3 + 2x^2 - 1 \equiv 0 \pmod{10}$.

We could proceed as above, that is to say, evaluating $f(x) = x^3 + 2x^2 - 1$ for $x = 0, 1, 2, \ldots, 9$ modulo 10, and checking for which values of $x$ $f(x)$ is found to be congruent to 0. However, it is more convenient to proceed as follows. Recall that if $m$ and $n$ are relatively prime, then $a \equiv b \pmod{mn}$ if and only if $a \equiv b \pmod{m}$ and $a \equiv b \pmod{n}$. This means that the congruence

$$x^3 + 2x^2 - 1 \equiv 0 \pmod{10}$$

is equivalent to the *system* of two congruences

$$\begin{cases} x^3 + 2x^2 - 1 \equiv 0 \pmod{2}, \\ x^3 + 2x^2 - 1 \equiv 0 \pmod{5}. \end{cases}$$

Now, it is easy to verify that the first congruence admits the unique solution $x \equiv 1 \pmod{2}$ while the second one admits, as seen in the first example, the solutions $x \equiv 2 \pmod{5}$ and $x \equiv 4 \pmod{5}$. So $x$ is a solution of $x^3 + 2x^2 - 1 \equiv 0 \pmod{10}$ if and only if

$$x \equiv 1 \pmod{2} \quad \text{and} \quad x \equiv 2 \text{ or } 4 \pmod{5}.$$

Now we apply the Chinese remainder theorem 3.4.2 (see page 129). We know that for each pair of solutions modulo 2 and modulo 5 there is a unique solution modulo 10, as $\text{GCD}(2, 5) = 1$. So we only have to solve the two systems of linear congruences

$$\begin{cases} x \equiv 1 \pmod{2}, \\ x \equiv 2 \pmod{5}, \end{cases} \quad \text{and} \quad \begin{cases} x \equiv 1 \pmod{2}, \\ x \equiv 4 \pmod{5}. \end{cases}$$

So we find the two solutions $x = 7$ and $x = 9$ modulo 10. The reader may find again the same result by evaluating $f(x)$ for $x = 0, 1, 2, \ldots, 9$ (see Exercise B5.38).

**Example 5.2.4.** Solve the congruence $x^3 + 6x^2 + 1 \equiv 0 \pmod{12}$.
    As $12 = 3 \cdot 4$ and $\text{GCD}(3, 4) = 1$, it follows that $x$ is a solution of $x^3 + 6x^2 + 1 \pmod{12}$ if and only if

$$\begin{cases} x^3 + 6x^2 + 1 \equiv 0 \pmod{3}, \\ x^3 + 6x^2 + 1 \equiv 0 \pmod{4}; \end{cases}$$

then we proceed as in the previous example (see Exercise B5.39).

    In these examples we were able to reduce the original congruences to congruences modulo primes or *prime powers*. In the following lemma we shall prove that this is always the case. Indeed, we shall prove that, in general, given a congruence of the form

$$f(x) \equiv 0 \pmod{m},$$

it is always possible solve it through the solution of polynomial congruences of the form

$$f(x) \equiv 0 \pmod{p^\alpha}, \qquad \text{with } p \text{ a prime}, \tag{5.10}$$

and an application, as in the previous examples, of the Chinese remainder theorem.

**Lemma 5.2.5.** *Let* $m = p_1^{\alpha_1} p_2^{\alpha_2} \cdots p_k^{\alpha_k}$ *be the factorisation of the positive integer* $m$ *into distinct primes. Then the polynomial congruence*

$$f(x) \equiv 0 \pmod{m} \tag{5.11}$$

*is soluble if and only if each of the congruences* $f(x) \equiv 0 \pmod{p_i^{\alpha_i}}$, *is soluble, for* $i = 1, 2, \ldots, k$.

PROOF. Let $x_0$ be a solution of the congruence (5.11), that is, $f(x_0) \equiv 0$ (mod $m$). As $p_i^{\alpha_i} \mid m$ for all $i = 1, \ldots, k$, it follows that $f(x_0) \equiv 0 \pmod{p_i^{\alpha_i}}$ for all $i = 1, \ldots, k$.

   Vice versa, if there exists $x_i$ such that $f(x_i) \equiv 0 \pmod{p_i^{\alpha_i}}$, for $i = 1, \ldots, k$, by the Chinese remainder theorem there exists $x$ such that $x \equiv x_i$ (mod $p_i^{\alpha_i}$), for $i = 1, \ldots, k$, and so $x$ is a solution (5.11).    □

**Remark 5.2.6.** By applying the previous lemma and the Chinese remainder theorem, it is clear that if every congruence

$$f(x) \equiv 0 \pmod{p_i^{\alpha_i}}$$

admits $t_i$ solutions, then congruence (5.11) has $\prod_{i=1}^{k} t_i$ solutions.

   We shall now show that, if the solutions of the congruence

$$f(x) \equiv 0 \pmod{p}$$

are known, it is possible to find the solutions of $f(x) \equiv 0 \pmod{p^\alpha}$, enabling us to reduce every polynomial congruence to a congruence modulo a prime number. We shall prove this fact by showing that, for every positive integer $\alpha$, the solutions of $f(x) \equiv 0 \pmod{p^{\alpha+1}}$ can be obtained from the solutions of $f(x) \equiv 0 \pmod{p^\alpha}$.

   We start by discussing a simple example.

**Example 5.2.7.** Solve the congruence

$$f(x) = x^3 + 3x + 2 \equiv 0 \pmod{49}. \tag{5.12}$$

   We need not compute $f(\bar{x})$ for all $\bar{x} = 0, 1, \ldots, 48$, by noticing that a solution of (5.12) is clearly also a solution of

$$f(x) = x^3 + 3x + 2 \equiv 0 \pmod{7}. \tag{5.13}$$

But (5.13) has no solutions (see Exercise B5.40), so (5.12) has no solutions either.

**Example 5.2.8.** Solve the congruence $x^4 + x + 3 \equiv 0 \pmod{25}$.
   The solutions of $x^4 + x + 3 \equiv 0 \pmod{25}$ must also be solutions of $x^4 + x + 3 \equiv 0 \pmod{5}$. This congruence $x^4 + x + 3 \equiv 0 \pmod{5}$ has the unique solution $x = 1$, which is not a solution of $x^4 + x + 3 \equiv 0 \pmod{25}$.

We now prove the following general result:

**Proposition 5.2.9.** *Let $f(x)$ be a polynomial with integer coefficients, $p$ a prime and $\alpha$ a positive integer. If $x_\alpha$, with $0 \le x_\alpha < p^\alpha$, is a solution of*

$$f(x) \equiv 0 \pmod{p^\alpha} \tag{5.14}$$

*and $t$, with $0 \le t < p$, is a solution of*

$$\frac{f(x_\alpha)}{p^\alpha} + t f'(x_\alpha) \equiv 0 \pmod{p} \tag{5.15}$$

*(with $f'$ the derivative polynomial of $f$), then $x_{\alpha+1} = x_\alpha + tp^\alpha$ is a solution of*

$$f(x) \equiv 0 \pmod{p^{\alpha+1}}. \tag{5.16}$$

*Vice versa, every solution $x_{\alpha+1}$, with $0 \le x_{\alpha+1} < p^{\alpha+1}$, of the congruence (5.16) can be obtained in this way.*

PROOF. As regards the first part of the proposition, let $n$ be the degree of $f(x)$. By Taylor's formula 1.3.25 we can write

$$f(x_{\alpha+1}) = f(x_\alpha + tp^\alpha) =$$

$$= f(x_\alpha) + tp^\alpha f'(x_\alpha) + (tp^\alpha)^2 \frac{f^{(2)}(x_\alpha)}{2} + \cdots + (tp^\alpha)^n \frac{f^{(n)}(x_\alpha)}{n!} \equiv$$

$$\equiv f(x_\alpha) + tp^\alpha f'(x_\alpha) \pmod{p^{\alpha+1}},$$

as $p^{\alpha+1}$ divides $p^{h\alpha}$ for all $h \ge 2$. Moreover, notice that $f^{(2)}(x_\alpha)/2, \ldots$, $f^{(n)}(x_\alpha)/n!$ are integers (see Exercise A1.63). On the other hand, by Equation (5.15) we have $f(x_\alpha)/p^\alpha + t f'(x_\alpha) = Np$, where $N$ is an integer. Thus, $f(x_\alpha) + tp^\alpha f'(x_\alpha) = (Np)p^\alpha = Np^{\alpha+1}$, and so

$$f(x_{\alpha+1}) \equiv f(x_\alpha) + tp^\alpha f'(x_\alpha) \equiv 0 \pmod{p^{\alpha+1}}.$$

Vice versa, as clearly every solution of (5.16) is a solution of (5.14) as well, if $x_{\alpha+1}$, with $0 \le x_{\alpha+1} < p^{\alpha+1}$, satisfies $f(x_{\alpha+1}) \equiv 0 \pmod{p^{\alpha+1}}$, then there exists a $x_\alpha$, with $0 \le x_\alpha < p^\alpha$, such that $f(x_\alpha) \equiv 0 \pmod{p^\alpha}$, with $x_{\alpha+1} \equiv x_\alpha$ $\pmod{p^\alpha}$, that is to say, $x_{\alpha+1} = x_\alpha + tp^\alpha$, with $0 \le t < p$. Using again Taylor's formula as above, we obtain

$$f(x_\alpha) + tp^\alpha f'(x_\alpha) \equiv 0 \pmod{p^{\alpha+1}}.$$

As $f(x_\alpha) \equiv 0 \pmod{p^\alpha}$, $f(x_\alpha)/p^\alpha$ is an integer $N$, that is, $f(x_\alpha) = Np^\alpha$; hence

$$Np^\alpha + tp^\alpha f'(x_\alpha) \equiv 0 \pmod{p^{\alpha+1}}.$$

Dividing by $p^\alpha$ we get $N + t f'(x_\alpha) \equiv 0 \pmod{p}$. $\qquad\qquad \square$

The previous proposition says that if the congruence of degree $n$

$$f(x) \equiv 0 \pmod{p} \tag{5.17}$$

has $\nu$ ($\leq n$) distinct solutions modulo $p$, then *in general* the equation

$$f(x) \equiv 0 \pmod{p^\alpha}$$

also has $\nu$ distinct solutions modulo $p^\alpha$. For instance, if $\alpha = 2$, for every solution $x_\alpha$ of (5.17), a unique solution $f(x) \equiv 0 \pmod{p^2}$ is found, as long as $f'(x_\alpha)$ is not divisible by $p$.

**Example 5.2.10.** Consider the equation

$$x^2 \equiv n^2 \pmod{m} \tag{5.18}$$

with $m = p_1^{\alpha_1} p_2^{\alpha_2} \cdots p_k^{\alpha_k}$ the factorisation of $m$. Suppose $m$ to be odd and $n \not\equiv 0 \pmod{m}$. Notice that the equation $x^2 \equiv n \pmod{p_i}$ has exactly two solutions $\pm x_i$ for all $i = 1, \dots, k$. Moreover, clearly $\pm 2x_i \not\equiv 0 \pmod{p_i}, i = 1, \dots, k$. Proposition 5.2.9 and Remark 5.2.6 imply that (5.18) has exactly $2^k$ distinct solutions modulo $m$.

**Example 5.2.11.** Solve the congruence $x^2 + 8 \equiv 0 \pmod{121}$.
    We reduce to solving the equation

$$x^2 + 8 \equiv 0 \pmod{11},$$

whose solutions are $x \equiv 5, 6 \pmod{11}$. They lead to the following linear congruences in $t$:

$$3 - t \equiv 0 \pmod{11}, \qquad 4 + t \equiv 0 \pmod{11},$$

whose solutions are $t \equiv 3 \pmod{11}$ and $t \equiv 7 \pmod{11}$, respectively. Correspondingly we find the solutions $5 + 3 \cdot 11 = 38 \pmod{121}$ and $6 + 7 \cdot 11 = 83$ modulo 121 of the original congruence.

**Remark 5.2.12.** Notice that Equation (5.15) has a unique solution if and only if $f'(x_\alpha) \not\equiv 0 \pmod{p}$. If on the other hand $f'(x_\alpha) \equiv 0 \pmod{p}$, then Equation (5.15) has no solutions, unless one has $f(x_\alpha)/p^\alpha \equiv 0 \pmod{p}$, that is, $f(x_\alpha) \equiv 0 \pmod{p^{\alpha+1}}$. In this case Equation (5.15) is satisfied by all $t = 0, \dots, p - 1$, so there are $p$ solutions of (5.16).

## 5.2.1 Degree two congruences

We are now going to study in detail polynomial congruences of *degree two* or, as they are called, *quadratic* congruences, that is to say, of the form

$$ax^2 + bx + c \equiv 0 \pmod{m}.$$

Using the results of the previous section we may assume the modulus $m$ to be a prime number $p$. Further, as the case $p = 2$ is trivial (see Exercise B5.37), we may directly assume $p$ to be an *odd prime*. So, consider the congruence

$$ax^2 + bx + c \equiv 0 \pmod{p}, \qquad p \text{ odd prime, } p \nmid a. \tag{5.19}$$

It may be solved by the classic method of *completing the square* we are now going to recall. Let $a' \in \mathbb{Z}$ be such that $2aa' \equiv 1 \pmod{p}$: such an $a'$ always exists, as $2a \neq 0$ in $\mathbb{Z}_p$. Multiplying Equation (5.19) by $2a'$ we get an equivalent congruence of the form

$$x^2 + 2b'x + c' \equiv 0 \pmod{p}.$$

It can be written in the equivalent form

$$(x + b')^2 \equiv b'^2 - c' \pmod{p}.$$

By setting $y = x + b'$ and $k = b'^2 - c'$, Equation (5.19) reduces to the form

$$\boxed{y^2 \equiv k \pmod{p}.} \tag{5.20}$$

In order to solve Equation (5.19) it suffices then to solve $x + b' \equiv y \pmod{p}$, where $y$ is a solution of $y^2 \equiv k \pmod{p}$. Notice that the congruence $x + b' \equiv y$ $\pmod{p}$ always admits a unique solution. On the other hand, equation $y^2 \equiv k$ $\pmod{p}$ may either admit no solution or admit two solutions, opposite each other, which are the two *square roots* of $k$ in $\mathbb{Z}_p$. These two solutions may *coincide*, and yield a solution with multiplicity two, if $k \equiv 0 \pmod{p}$. In the first case the original equation (5.19) has no solutions in $\mathbb{Z}_p$, while in the second one it has two solutions (see Exercise A5.42).

Gauss was the first to observe that to solve quadratic congruences it suffices to solve congruence of the particular form (5.20).

**Example 5.2.13.** Find the solutions of the quadratic congruence

$$2x^2 + x + 3 \equiv 0 \pmod{5}. \tag{5.21}$$

Using the same notation as above, we find $a' \in \mathbb{Z}$ such that $2 \cdot 2a' \equiv 1$ $\pmod{5}$, and precisely $a' = -1$. Multiplying the congruence by $2a' = -2$ we get

$$x^2 - 2x - 1 \equiv 0 \pmod{5},$$

or

$$(x - 1)^2 \equiv 2 \pmod{5}.$$

Setting $y = x - 1$, the congruence to be solved becomes

$$y^2 \equiv 2 \pmod{5}.$$

This congruence has no solutions, so neither has the original congruence (5.21).

**Example 5.2.14.** Find the solutions of the quadratic congruence

$$3x^2 + x + 1 \equiv 0 \pmod 5. \tag{5.22}$$

The congruence is equivalent to

$$(x + 1)^2 - 4 \equiv 0 \pmod 5,$$

or

$$y^2 \equiv 4 \pmod 5,$$

which admits two solutions, $y_1 = 2$ and $y_2 = 3$. So the original congruence (5.22) has the solutions $x_1 = y_1 - 1 = 1$ and $x_2 = y_2 - 1 = 2$.

In the above examples we found a congruence of the form (5.20) that did not admit solutions and one that did. We would like to find general criteria enabling us to decide when a quadratic congruence of the form (5.20) admits solutions and when it does not admit any. This is the purpose of the next section.

### 5.2.2 Quadratic residues

**Definition 5.2.15.** *Let $p$ be an odd prime and $a$ an integer such that $p \nmid a$. If the congruence*

$$x^2 \equiv a \pmod p$$

*is soluble, then $a$ is said to be a* quadratic residue *of $p$, else $a$ is said to be a* quadratic non-residue *of $p$.*

In other words, a quadratic residue of $p$ is an element of the group $\mathbb{Z}_p^* = \mathbb{Z}_p \setminus \{0\}$ that is a square. So we shall indifferently speak about quadratic residues of $p$, or of squares modulo $p$, or of squares in $\mathbb{Z}_p^*$. Looking at the above examples, we see that 2 is not a square modulo 5, while 4 is.

The squares in $\mathbb{Z}_5^*$ are

$$\boxed{1} = 1^2 = 4^2, \quad \boxed{4} = 2^2 = 3^2.$$

The squares in $\mathbb{Z}_7^*$ are

$$\boxed{1} = 1^2 = 6^2, \quad \boxed{2} = 3^2 = 4^2, \quad \boxed{4} = 2^2 = 5^2.$$

The squares in $\mathbb{Z}_{11}^*$ are

$$\boxed{1} = 1^2 = 10^2, \qquad \boxed{3} = 5^2 = 6^2, \qquad \boxed{4} = 2^2 = 9^2,$$
$$\boxed{5} = 4^2 = 7^2, \qquad \boxed{9} = 3^2 = 8^2.$$

One may guess from these examples that the squares are exactly one half of all elements of $\mathbb{Z}_p^*$. Indeed, the following proposition holds.

**Proposition 5.2.16.** *Let $p$ an odd prime. Then there are exactly $(p-1)/2$ quadratic residues of $p$:*

$$1^2, 2^2, \ldots, \left(\frac{p-3}{2}\right)^2, \left(\frac{p-1}{2}\right)^2.$$

PROOF. We can write the $p$ elements of $\mathbb{Z}_p$ as follows:

$$-\frac{p-1}{2}, \ -\frac{p-3}{2}, \ldots, -1, 0, 1, \ldots, \frac{p-3}{2}, \frac{p-1}{2}.$$

Now, in a field the relation $a^2 - b^2 = (a-b)(a+b) = 0$ implies $a = \pm b$. So, in a field, if two squares $a^2$ and $b^2$ coincide, then $a = \pm b$. This tells us that the squares of the elements $1, \ldots, (p-3)/2, (p-1)/2 \in \mathbb{Z}_p^*$ are all distinct and there are $(p-1)/2$ of them. □

**Example 5.2.17.** Determine all the quadratic residues of 23.

By Proposition 5.2.16, the squares modulo $p = 23$ are the squares of the integers from 1 to $(p-1)/2 = 11$, that is

$$1^2, \ 2^2, \ 3^2, \ 4^2, \ 5^2, \ 6^2, \ 7^2, \ 8^2, \ 9^2, \ 10^2, \ 11^2,$$

or

$$1, \ 4, \ 9, \ 16, \ 2, \ 13, \ 3, \ 18, \ 12, \ 8, \ 6.$$

For not too large values of $p$ we may easily find the squares (and so the non-squares) using Proposition 5.2.16. But what about large moduli? Are there criteria to recognise whether an integer is a square modulo a prime $p$ or not? Here follows a quite simple remark.

**Lemma 5.2.18.** *Let $g \in \mathbb{Z}_p^*$ be a generator of the multiplicative group $\mathbb{Z}_p^*$. Then $a = g^j$ (for $0 \le j < p-1$) is a quadratic residue if and only if $j$ is even.*

The proof is left to the reader (see Exercise A5.43).

The following result, due to Euler, is simple but important.

**Proposition 5.2.19 (Euler's criterion).** *Let $p$ be an odd prime not dividing $a$. If $a$ is a quadratic residue of $p$, then*

$$a^{(p-1)/2} \equiv 1 \pmod{p}.$$

*If, on the other hand, $a$ is not a quadratic residue of $p$, then*

$$a^{(p-1)/2} \equiv -1 \pmod{p}.$$

PROOF. We have

$$(a^{(p-1)/2})^2 = a^{p-1} \equiv 1 \pmod{p},$$

so
$$a^{(p-1)/2} \equiv \pm 1 \pmod{p}.$$

If $g \in \mathbb{Z}_p^*$ is a generator of the multiplicative group and $a = g^j \pmod{p}$, then

$$a^{(p-1)/2} \equiv g^{j(p-1)/2} \pmod{p}.$$

On the other hand,
$$g^{j(p-1)/2} \equiv 1 \pmod{p}$$

if and only if $j(p-1)/2$ is divisible by $p-1$, that is, if and only if $j$ is even, that is, if and only if $a$ is a quadratic residue.    □

**Remark 5.2.20.** Euler's criterion is quite expensive computationally when $p$ is very large. Indeed, the computational complexity of computing $a^{(p-1)/2}$ modulo $p$ is $\mathcal{O}(\log^3 p)$ (see § 3.3.1 or Proposition 5.1.44). However, we shall see in the next chapter how it may nevertheless be used as the theoretical basis for some primality tests.

### 5.2.3 Legendre symbol and its properties

A criterion simpler than Euler's is obtained by exploiting the *law of quadratic reciprocity*, which will shortly be stated. To this purpose, we define a symbol, the so-called *Legendre symbol*, to denote whether an integer $a$ is a square modulo an odd prime $p$ or not.

**Definition 5.2.21.** *Let $p$ be an odd prime and $a$ an integer. Define the Legendre symbol as follows:*

$$\left(\frac{a}{p}\right) = \begin{cases} 1 & \text{if } a \text{ is a quadratic residue of } p, \\ 0 & \text{if } a \text{ is divisible by } p, \\ -1 & \text{if } a \text{ is not a quadratic residue of } p. \end{cases}$$

For instance,
$$\left(\frac{2}{11}\right) = -1$$

as 2 is not a quadratic residue of 11, while

$$\left(\frac{3}{11}\right) = 1$$

as $3 \equiv 5^2 \pmod{11}$. Analogously,

$$\left(\frac{4}{13}\right) = 1$$

as $4 = 2^2$. Finally,

$$\left(\frac{6}{3}\right) = 0$$

as 6 is divisible by 3.

Our goal is to compute Legendre symbol $\left(\frac{a}{p}\right)$ to decide whether the congruence

$$x^2 \equiv a \pmod{p}$$

has solutions or not. To this purpose, we shall need some basic properties of the Legendre symbol, collected in the following proposition.

**Proposition 5.2.22.** *Let $p$ be an odd prime and $a, b$ two integers coprime with $p$. Then the Legendre symbol enjoys the following properties:*

*(1) if $a \equiv b \pmod{p}$, then $\left(\frac{a}{p}\right) = \left(\frac{b}{p}\right)$;*

*(2) $\left(\frac{a^2}{p}\right) = 1$;*

*(3) $\left(\frac{a}{p}\right) \equiv a^{(p-1)/2} \pmod{p}$;*

*(4) $\left(\frac{ab}{p}\right) = \left(\frac{a}{p}\right)\left(\frac{b}{p}\right)$;*

*(5) $\left(\frac{1}{p}\right) = 1$ and $\left(\frac{-1}{p}\right) = (-1)^{(p-1)/2}$.*

We leave the easy proof of these properties to the reader (see Exercise A5.44). An immediate consequence is the following corollary, also left to the reader (see Exercise A5.45):

**Corollary 5.2.23.** *If $p$ is an odd prime, then*

$$\left(\frac{-1}{p}\right) = \begin{cases} 1 & \text{if } p \equiv 1 \pmod{4}, \\ -1 & \text{if } p \equiv 3 \pmod{4}. \end{cases}$$

**Example 5.2.24.** Solve the congruence $x^2 \equiv -46 \pmod{17}$.

Use the properties of the Legendre symbol to check whether $-46$ is a square in $\mathbb{Z}_{17}^*$ or not. We have to compute the Legendre symbol

$$\left(\frac{-46}{17}\right).$$

We have

$$\left(\frac{-46}{17}\right) = \left(\frac{-1}{17}\right)\left(\frac{46}{17}\right) = \left(\frac{46}{17}\right)$$

because $17 \equiv 1 \pmod{4}$. As $46 \equiv 12 \pmod{17}$,

$$\left(\frac{-46}{17}\right) = \left(\frac{12}{17}\right) = \left(\frac{3 \cdot 2^2}{17}\right) = \left(\frac{3}{17}\right).$$

Finally,

$$\left(\frac{3}{17}\right) \equiv 3^{16/2} = 3^8 \equiv -1 \pmod{17}.$$

In conclusion, $-46$ is *not* a square modulo 17.

Another useful information about the Legendre symbol is given by the following proposition.

**Proposition 5.2.25 (Gauss Lemma).** *Let $p$ be an odd prime and let $a \in \mathbb{Z}$ be such that $\mathrm{GCD}(a,p) = 1$. Then*

$$\left(\frac{a}{p}\right) = (-1)^{n(a,p)},$$

*where $n(a,p)$ denotes the number of integers in the set*

$$A := \left\{ a, 2a, \ldots, \frac{(p-1)a}{2} \right\}$$

*such that the remainder of the division by $p$ is greater than $p/2$.*

PROOF. First of all, notice that the size of $A$ modulo $p$ is $(p-1)/2$. Indeed, as $\mathrm{GCD}(a,p) = 1$, no element in $A$ is congruent to zero modulo $p$ and all of them are distinct. Then divide the elements of $A$ into two disjoint subsets $A_1$ and $A_2$. Put in subset $A_1$ the elements of $A$ that are congruent, modulo $p$, to positive numbers smaller than $(p-1)/2$, and in subset $A_2$ the remaining elements of $A$, that is to say, those congruent to numbers between $p/2$ and $p$. Denote by $m$ the size of $A_1$, while the size of $A_2$ is exactly $n = n(a,p)$. Thus, $n + m = (p-1)/2$.

Denote now by $r_1, r_2, \ldots, r_m$ the remainders modulo $p$ of the elements of $A_1$ and by $s_1, s_2, \ldots, s_n$ the remainders modulo $p$ of the elements of $A_2$. Then the integers $r_1, \ldots, r_m, p - s_1, \ldots, p - s_n$ satisfy the following:

$$0 < r_1, \ldots, r_m, p - s_1, \ldots, p - s_n < \frac{p}{2}. \tag{5.23}$$

We shall prove that they are all distinct, and so coincide, up to the ordering, with the integers $1, 2, 3, \ldots, (p-1)/2$. Indeed, suppose by contradiction that for some choice of $h$ and $k$

$$p - s_k = r_h.$$

By definition there exist two integers $\alpha, \beta$, with $1 \le \alpha, \beta \le (p-1)/2$ and such that

$$r_h \equiv \alpha a \pmod{p}, \qquad e \qquad s_k \equiv \beta a \pmod{p},$$

hence

$$(\alpha + \beta)a \equiv r_h + s_k = p \equiv 0 \pmod{p},$$

which implies $\alpha + \beta \equiv 0 \pmod{p}$. But this relation yields a contradiction, as $0 < \alpha + \beta \le p - 1$.

As all of $r_1, \ldots r_m, p - s_1, \ldots, p - s_n$ are distinct, it follows from the relations (5.23) that they coincide, up to the ordering, with the integers $1, 2, 3, \ldots, (p-1)/2$. So we have

$$\left(\frac{p-1}{2}\right)! = r_1 \cdot r_m (p - s_1) \dots (p - s_n) \equiv (-1)^n r_1 \dots r_m s_1 \dots s_n \pmod{p}.$$

As the numbers $r_h$ and $s_k$, for $h = 1, \dots, m$ and $k = 1, \dots, n$, are the remainders modulo $p$ of the elements of $A$, we have

$$\left(\frac{p-1}{2}\right)! \equiv (-1)^n a \cdot 2a \cdot 3a \cdots \left(\frac{p-1}{2}\right) a \equiv (-1)^n a^{\frac{p-1}{2}} \left(\frac{p-1}{2}\right)! \pmod{p}.$$

As $\mathrm{GCD}(((p-1)/2)!, p) = 1$, we get

$$a^{(p-1)/2} \equiv (-1)^n \pmod{p}.$$

It is now possible to conclude immediately by using part (3) of Proposition 5.2.22. □

The following proposition can be proved in a similar way as Gauss Lemma:

**Proposition 5.2.26.** *Let $p$ be an odd prime number and let $a$ be an odd integer such that $a$ and $p$ are relatively prime. Define:*

$$\tau(a,p) = \sum_{i=1}^{(p-1)/2} \left[\frac{ia}{p}\right].$$

*Then:*

$$\left(\frac{a}{p}\right) = (-1)^{\tau(a,p)}.$$

PROOF. By Gauss Lemma, it is sufficient to prove that

$$\tau(a,p) \equiv n(a,p) \pmod{2}. \tag{5.24}$$

Dividing the integer $ia$, $i = 1, \dots, (p-1)/2$, by $p$, we get

$$ia = p \left[\frac{ia}{p}\right] + \rho_i, \tag{5.25}$$

where $\rho_i$ is the remainder of the division. Using the same notation as in the proof of Gauss Lemma, $\rho_i = r_h$ or $\rho_i = s_k$ depending on $ia$ being in $A_1$ or in $A_2$. Summing Equation (5.25) for $i = 1, \dots, (p-1)/2$, we obtain

$$\sum_{i=1}^{(p-1)/2} ia = p \sum_{i=1}^{(p-1)/2} \left[\frac{ia}{p}\right] + \sum_{h=1}^{m} r_h + \sum_{k=1}^{n} s_k, \tag{5.26}$$

where $n = n(a,p)$. As we saw in the proof of Gauss Lemma, the integers $r_1, \dots r_m, p - s_1, \dots p - s_n$ coincide, up to the ordering, with the integers $1, 2, 3, \dots, (p-1)/2$. So we have

$$\sum_{i=1}^{(p-1)/2} i = \sum_{h=1}^{m} r_h + \sum_{k=1}^{n}(p - s_k) = p \cdot n(a, p) + \sum_{h=1}^{m} r_h - \sum_{k=1}^{n} s_k. \qquad (5.27)$$

By subtracting Equation (5.27) from Equation (5.26), we get

$$(a - 1)\sum_{i=1}^{(p-1)/2} i = p \cdot \tau(a, p) - p \cdot n(a, p) + 2\sum_{k=1}^{n} s_k.$$

Considering it modulo 2, as $a$ and $p$ are odd, (5.24) is found. □

As a consequence of Gauss Lemma we have the following corollary.

**Corollary 5.2.27.** *If $p$ is an odd prime, then*

$$\left(\frac{2}{p}\right) = (-1)^{(p^2 - 1)/8} = \begin{cases} 1 & \text{if } p \equiv \pm 1 \pmod 8, \\ -1 & \text{if } p \equiv \pm 3 \pmod 8. \end{cases}$$

PROOF. Apply Gauss Lemma to the case $a = 2$, where $A = \{2, 4, \ldots, p - 1\}$ and so we have to find the number $n$ of elements of $A$ that are greater than $p/2$. An integer $2i$, with $1 \le i \le (p-1)/2$ is smaller than $p/2$ if and only if $1 \le i \le [p/4]$. Then $n = (p-1)/2 - [p/4]$. We prove next that

$$\frac{p-1}{2} - \left[\frac{p}{4}\right] \equiv \frac{p^2 - 1}{8} \pmod 2. \qquad (5.28)$$

We must simply check that both sides of (5.28) have the same parity. During this check, we shall also prove that $(p^2 - 1)/8$ is even or odd depending on $p \equiv \pm 1 \pmod 8$ or $p \equiv \pm 3 \pmod 8$. Indeed:

- if $p \equiv \pm 1 \pmod 8$, that is, if $p = 8h \pm 1$, with $h$ a positive integer, we have

$$\frac{p^2 - 1}{8} = \frac{(8h \pm 1)^2 - 1}{8} = \frac{64h^2 \pm 16h}{8} = 8h^2 \pm 2h \equiv 0 \pmod 2;$$

- if $p \equiv \pm 3 \pmod 8$, that is, if $p = 8h \pm 3$, with $h$ a positive integer, we have

$$\frac{p^2 - 1}{8} = \frac{(8h \pm 3)^2 - 1}{8} = \frac{64h^2 \pm 48h + 8}{8} = 8h^2 \pm 6h + 1 \equiv 1 \pmod 2;$$

On the other hand:

- if $p \equiv 1 \pmod 8$, that is, if $p = 8h + 1$, with $h$ a positive integer, we have

$$\frac{p-1}{2} - \left[\frac{p}{4}\right] = 4h - \left[\frac{8h+1}{4}\right] = 4h - 2h \equiv 0 \pmod 2;$$

- if $p \equiv -1 \equiv 7 \pmod 8$, that is, if $p = 8h + 7$, with $h$ a positive integer, we have

$$\frac{p-1}{2} - \left[\frac{p}{4}\right] = 4h + 3 - \left[\frac{8h+7}{4}\right] = 4h - 2h + 2 \equiv 0 \pmod 2;$$

- if $p \equiv 3 \pmod 8$, that is, if $p = 8h + 3$, with $h$ a positive integer, we have

$$\frac{p-1}{2} - \left[\frac{p}{4}\right] = 4h + 1 - \left[\frac{8h+3}{4}\right] = 4h - 2h + 1 \equiv 1 \pmod 2;$$

- if $p \equiv -3 \equiv 5 \pmod 8$, that is, if $p = 8h + 5$, with $h$ a positive integer, we have

$$\frac{p-1}{2} - \left[\frac{p}{4}\right] = 4h + 2 - \left[\frac{8h+5}{4}\right] = 4h - 2h + 1 \equiv 1 \pmod 2.$$

So the proof is complete.  $\square$

A different proof of Corollary 5.2.27, which exploits some properties of finite fields, will be outlined in the exercises (see Exercise A5.50).

### 5.2.4 The law of quadratic reciprocity

The next theorem is a milestone of number theory, and is, together with Proposition 5.2.22 and Corollary 5.2.27, the fundamental tool to compute efficiently the Legendre symbols. This theorem, conjectured in the 18th century by Euler and Legendre, who did not succeed in giving a complete proof, was proved by Gauss when he was 19 years old. Since then, several more proofs have been found. We show here a very simple proof, referring to the exercises for a more technical one, based upon the properties of finite fields (see Exercises A5.47-A5.54).

**Theorem 5.2.28 (Law of quadratic reciprocity).** *Let $p$ and $q$ be distinct odd primes. Then*

$$\left(\frac{p}{q}\right) = \left(\frac{q}{p}\right)$$

*unless $p \equiv q \equiv 3 \pmod 4$, in which case*

$$\left(\frac{p}{q}\right) = -\left(\frac{q}{p}\right).$$

*Equivalently,*

$$\left(\frac{p}{q}\right)\left(\frac{q}{p}\right) = (-1)^{(p-1)(q-1)/4}.$$

PROOF. Consider the *open* rectangle $R$ in the plane $xy$, having vertices $(0,0), (p/2,0), (0,q/2), (p/2,q/2)$. Clearly, it includes $(p-1)(q-1)/4$ points having integer coordinates, that is, all the points of the form $(i,j)$, with $1 \le i \le (p-1)/2, 1 \le j \le (q-1)/2$.

Let $D$ be the diagonal from $(0,0)$ to $(p/2,q/2)$ of the rectangle, having equation $py = qx$. As $\mathrm{GCD}(p,q) = 1$, no point with integer coordinates in $R$ lies on $D$. Denote by $R_1$ the subset of $R$ lying below $D$, that is, where $py < qx$, and by $R_2$ the subset of $R$ lying above $D$, that is, where $py > qx$.

Count the points of $R_1$ having integer coordinates. They are all the points of the form $(i,j)$, with $1 \le i \le (p-1)/2, 1 \le j \le [qi/p]$. So the number of points of $R_1$ having integer coordinates is exactly $\tau(q,p) = \sum_{i=1}^{(p-1)/2}[qi/p]$. Analogously, the number of points of $R_2$ having integer coordinates is $\tau(p,q) = \sum_{i=1}^{(q-1)/2}[pi/q]$.

So we have

$$\frac{(p-1)(q-1)}{4} = \tau(p,q) + \tau(q,p).$$

The theorem immediately follows from Proposition 5.2.26.    □

**Remark 5.2.29.** The Legendre symbol $\left(\frac{p}{q}\right)$ is connected with the solution of the congruence $x^2 \equiv p \pmod q$, while $\left(\frac{q}{p}\right)$ is connected with the solution of the congruence $x^2 \equiv q \pmod p$. The law of quadratic reciprocity connects the two congruences, modulo different integers, $x^2 \equiv q \pmod p$ and $x^2 \equiv p \pmod q$, which a priori have not much in common. As can be guessed, and as we shall shortly see, the law of quadratic reciprocity helps greatly in computing the Legendre symbols, and so enables us to find easily the quadratic residues.

**Example 5.2.30.** Compute $\left(\frac{59}{131}\right)$.

Using by turns the properties of the Legendre symbol and the law of quadratic reciprocity, we obtain:

$$\left(\frac{59}{131}\right) = -\left(\frac{131}{59}\right) = -\left(\frac{13}{59}\right) = -\left(\frac{59}{13}\right) = -\left(\frac{7}{13}\right) = -\left(\frac{13}{7}\right) =$$

$$= -\left(\frac{-1}{7}\right) = 1.$$

This theorem has many consequences in several fields of mathematics. In the next chapter we shall see an application to the problem of recognising whether a number is a prime or not. Let us now see two more elementary examples.

**Example 5.2.31.** Characterise the odd primes $p$ such that the congruence

$$x^2 \equiv 5 \pmod p$$

admits a solution.

By the law of quadratic reciprocity, the equation $x^2 \equiv 5 \pmod{p}$ admits a solution if and only if the congruence

$$x^2 \equiv p \pmod{5}.$$

admits a solution. This congruence has solutions only if $p = \pm 1 \pmod{5}$ (see Exercise B5.41). So the odd primes for which $x^2 \equiv 5 \pmod{p}$ admits solutions are those of the form $p = \pm 1 + 5h$; for instance 11, 19, 29. Recall that, by Dirichlet's Theorem 4.1.6, there are infinitely many such primes.

**Example 5.2.32.** Determine the odd primes $p$ such that $-3$ is a quadratic residue of $p$.

It is necessary to find for which values of $p$ the congruence

$$x^2 + 3 \equiv 0 \pmod{p} \tag{5.29}$$

has a solution. By the law of quadratic reciprocity and the properties of the Legendre symbol,

$$\left(\frac{-3}{p}\right) = \left(\frac{-1}{p}\right)\left(\frac{p}{3}\right)(-1)^{(p-1)/2} = \left(\frac{p}{3}\right).$$

So (5.29) has solutions if and only if the congruence

$$x^2 \equiv p \pmod{3},$$

has solutions, and this happens if and only if $p \equiv 1 \pmod{3}$. In conclusion, (5.29) has a solution if and only if $p$ is a prime of the form $3h + 1$.

A first, simple application of the law of quadratic reciprocity lies in a complete proof of Pépin's test 4.4.8. So we prove the following.

**Proposition 5.2.33.** *If the Fermat number $F_n$ is prime, then*

$$3^{(F_n - 1)/2} \equiv -1 \pmod{F_n}. \tag{5.30}$$

PROOF. If $F_n$ is prime, by applying the law of quadratic reciprocity we have

$$\left(\frac{3}{F_n}\right) = \left(\frac{F_n}{3}\right) = \left(\frac{2}{3}\right) = -1.$$

We may now conclude by applying Euler's Criterion 5.2.19 or part (3) of Proposition 5.2.22. □

### 5.2.5 The Jacobi symbol

The law of quadratic reciprocity is not completely sufficient to compute the Legendre symbol $\left(\frac{a}{p}\right)$, with $p$ odd prime and $a$ an arbitrary number. It is convenient to put ourselves in a more general setting, as follows.

First of all, we introduce a generalisation of the Legendre symbol, the *Jacobi symbol* $\left(\frac{a}{n}\right)$, where $a$ is an arbitrary integer and $n$ an odd positive integer. If the factorisation of $n$ is $n = p_1^{\alpha_1} \cdots p_r^{\alpha_r}$, with $p_1, \ldots, p_r$ distinct odd prime numbers, we define the Jacobi symbol $\left(\frac{a}{n}\right)$ as

$$\left(\frac{a}{n}\right) = \left(\frac{a}{p_1}\right)^{\alpha_1} \cdots \left(\frac{a}{p_r}\right)^{\alpha_r}.$$

It is important to remark that if $n$ is not a prime, the fact that $\left(\frac{a}{n}\right) = 1$ does not at all mean that $a$ is a square modulo $n$. For instance

$$\left(\frac{2}{15}\right) = \left(\frac{2}{3}\right) \cdot \left(\frac{2}{5}\right) = (-1) \cdot (-1) = 1, \tag{5.31}$$

but the equation $x^2 \equiv 2 \pmod{15}$ has no solution (see Exercise B5.42). Nevertheless, the Jacobi symbol enjoys some formal properties analogous to those of the Legendre symbol.

**Proposition 5.2.34.** *Let $n$ be a positive odd number and let $a$ and $b$ be integers relatively prime with $n$. Then:*

*(1) if $a \equiv b \pmod{n}$, then $\left(\frac{a}{n}\right) = \left(\frac{b}{n}\right)$;*
*(2) $\left(\frac{ab}{n}\right) = \left(\frac{a}{n}\right)\left(\frac{b}{n}\right)$;*
*(3) $\left(\frac{1}{n}\right) = 1$ e $\left(\frac{-1}{n}\right) = (-1)^{(n-1)/2}$.*

PROOF. We shall prove only part (3), leaving the easy verification of (1) and (2) as an exercise to the reader (see Exercise A5.55).
    If $n = p_1^{\alpha_1} \cdots p_r^{\alpha_r}$, with $p_1, \ldots, p_r$ distinct odd prime numbers, we have

$$\left(\frac{-1}{n}\right) = \left(\frac{-1}{p_1}\right)^{\alpha_1} \cdots \left(\frac{-1}{p_r}\right)^{\alpha_r} = (-1)^{(\alpha_1(p_1-1)+\cdots+\alpha_r(p_r-1))/2}. \tag{5.32}$$

Notice now that, for every odd prime number $p$ and positive integer $\alpha$, we have

$$p^\alpha = (1 + (p-1))^\alpha \equiv 1 + \alpha(p-1) \pmod{4}$$

(see Exercise A5.57). So we have

$$n \equiv (1 + \alpha_1(p_1 - 1)) \cdots (1 + \alpha_r(p_r - 1)) \pmod{4}$$

and, keeping in mind that all of $p_1 - 1, \ldots, p_r - 1$ are even numbers, we have

$$n \equiv 1 + \alpha_1(p_1 - 1) + \cdots + \alpha_r(p_r - 1) \pmod{4},$$

or

$$\frac{n-1}{2} \equiv \frac{\alpha_1(p_1 - 1) + \cdots + \alpha_r(p_r - 1)}{2} \pmod{2}.$$

This, together with Equation (5.32) proves (3).                                    □

Further, the following holds:

**Proposition 5.2.35.** *If $n$ is a positive odd integer, then*

$$\left(\frac{2}{n}\right) = (-1)^{(n^2-1)/8}$$

*and if $m$ is another positive odd integer, then*

$$\left(\frac{n}{m}\right) = \left(\frac{m}{n}\right)$$

*unless $n \equiv m \equiv 3 \pmod 4$, in which case*

$$\left(\frac{n}{m}\right) = -\left(\frac{m}{n}\right).$$

*Equivalently,*

$$\left(\frac{m}{n}\right)\left(\frac{n}{m}\right) = (-1)^{(n-1)(m-1)/4}.$$

PROOF. Define $\epsilon(n) = (-1)^{(n^2-1)/8}$ for all odd positive integers $n$. It is easy to verify (see Exercise A5.58) that $\epsilon$ is a completely multiplicative function in its domain $\mathbb{Z} \setminus 2\mathbb{Z} = \{1, 3, 5, \ldots\}$, that is, for each pair $(n, m)$ of odd positive integers, $\epsilon(nm) = \epsilon(n)\epsilon(m)$.

Thus, if $n$ factors as $n = p_1^{\alpha_1} \cdots p_r^{\alpha_r}$, with $p_1, \ldots, p_r$ distinct odd prime numbers, we have

$$\epsilon(n) = \epsilon(p_1)^{\alpha_1} \cdots \epsilon(p_r)^{\alpha_r}.$$

By applying Corollary 5.2.27, we get $\epsilon(p_i) = \left(\frac{2}{p_i}\right)$, and so $\epsilon(p_i)^{\alpha_i} = \left(\frac{2}{p_i}\right)^{\alpha_i}$, $i = 1, \ldots, r$. Thus,

$$\epsilon(n) = \left(\frac{2}{p_1}\right)^{\alpha_1} \cdots \left(\frac{2}{p_r}\right)^{\alpha_r},$$

which, by definition, is $\left(\frac{2}{n}\right)$. This proves the first part of the proposition.

As regards the second part, notice first that if $\mathrm{GCD}(n, m) \neq 1$, from the definition itself of Legendre and Jacobi symbols, we have $\left(\frac{n}{m}\right) = \left(\frac{m}{n}\right) = 0$. So, suppose that $\mathrm{GCD}(n, m) = 1$. Write $n = p_1 \cdots p_h$ and $m = q_1 \cdots q_k$, with $p_1, \ldots, p_h, q_1, \ldots, q_k$ not necessarily distinct primes. By definition of the Jacobi symbol and the properties of the Legendre symbol, we find

$$\left(\frac{n}{m}\right) = \prod_{i=1}^{h}\prod_{j=1}^{k}\left(\frac{p_i}{q_j}\right), \qquad \left(\frac{m}{n}\right) = \prod_{i=1}^{h}\prod_{j=1}^{k}\left(\frac{q_j}{p_i}\right).$$

Apply now the law of quadratic reciprocity to each of the $hk$ symbols $\left(\frac{p_i}{q_j}\right)$. Let $\ell$ be the number of pairs $(i, j)$ with $1 \leq i \leq h$, $1 \leq j \leq k$, such that both $p_i$ and $q_j$ are congruent to 3 modulo 4. Then we have

$$\left(\frac{n}{m}\right) = (-1)^{\ell}\left(\frac{m}{n}\right).$$

On the other hand, it is clear that $\ell = \ell_1 \ell_2$ where $\ell_1$ [$\ell_2$, respectively] is the number of $is$ [of $js$, resp.] such that $1 \le i \le h$ [$1 \le j \le k$, resp.] and $p_i$ [$q_j$, resp.] is congruent to 3 modulo 4. So we have $n \equiv 3^{\ell_1} \equiv (-1)^{\ell_1} \pmod 4$, and analogously $m \equiv (-1)^{\ell_2} \pmod 4$. In conclusion, we have $\left(\frac{n}{m}\right) = \left(\frac{m}{n}\right)$ unless both $\ell_1$ and $\ell_2$ are odd, that is, unless $n \equiv m \equiv -1 \equiv 3 \pmod 4$, and in this case $\left(\frac{n}{m}\right) = -\left(\frac{m}{n}\right)$. □

This proposition enables us, in principle, to compute any Legendre symbol in a way analogous to what has been done in Example 5.2.30.

**Example 5.2.36.** Compute $\left(\frac{3083}{3911}\right)$.

We have

$$\left(\frac{3083}{3911}\right) = -\left(\frac{3911}{3083}\right)$$

because both 3911 and 3083 are congruent to 3 modulo 4. Dividing 3911 by 3083 the remainder is $828 = 4 \cdot 207$. So,

$$\left(\frac{3083}{3911}\right) = -\left(\frac{4}{3083}\right)\left(\frac{207}{3083}\right).$$

Going on, we find

$$\left(\frac{3083}{3911}\right) = -\left(\frac{207}{3083}\right) = \left(\frac{3083}{207}\right) = \left(\frac{185}{207}\right) = \left(\frac{207}{185}\right) = \left(\frac{2}{185}\right)\left(\frac{11}{185}\right) =$$

$$= \left(\frac{11}{185}\right) = \left(\frac{185}{11}\right) = \left(\frac{9}{11}\right) = \left(\frac{11}{9}\right) = \left(\frac{2}{9}\right) = 1.$$

**Remark 5.2.37.** The law of quadratic reciprocity, which as we have seen holds in general for the Jacobi symbols, together with the formal properties of the latter, enables us to compute the symbol *without having to know the factorisation* of the numbers involved. In conclusion, computing $\left(\frac{n}{m}\right)$, with $n > m$, has the same complexity as computing the greatest common divisor of $n$ and $m$, that is, has complexity $\mathcal{O}(\log^3 n)$.

### 5.2.6 An algorithm to compute square roots

The law of quadratic reciprocity makes it possible to ascertain whether the square root of an integer $a$ modulo a prime $p$ exists or not, but does not give a method to compute it. We conclude this chapter showing an algorithm to compute square roots in $\mathbb{Z}_p$ with $p$ an odd prime. It relies on the knowledge of an integer $n$ that *is not* a quadratic residue modulo $p$.

Let $a$ be an integer such that $\left(\frac{a}{p}\right) = 1$, so we know that there is a solution of the equation $x^2 \equiv a \pmod p$. We want to determine a solution of this equation. We may proceed as follows.

First of all, we may assume $a$ to be positive and smaller than $p$, and we may consider $a$ as an element of $\mathbb{Z}_p$. Write next $p - 1 = 2^r s$ with odd $s$. Compute the remainder $\xi$ of $n^s$ modulo $p$, and consider $\xi$ as an element of $\mathbb{Z}_p$.

**Lemma 5.2.38.** *The number $\xi$ is a primitive $2^r$th root of unity in $\mathbb{Z}_p$.*

PROOF. We begin by verifying that $\xi$ is a $2^r$th root of unity. Indeed, in $\mathbb{Z}_p$ we have

$$\xi^{2^r} = [n]_p^{2^r s} = [n]_p^{p-1} = 1,$$

having denoted by $[n]_p$ the residue class modulo $p$.

Suppose $\xi$ is not a primitive root. Let $\eta$ be a primitive $2^r$th root and let $\xi = \eta^t$, with $2 \leq t < 2^r$. Then $\mathrm{GCD}(t, 2^r) \neq 1$ (see Exercise A3.30), so $t = 2^u$ with $0 < u < r$. So we would have $\xi = \eta^{2^u} = \left(\eta^{2^{u-1}}\right)^2$. But this is not possible, as

$$\left(\frac{\xi}{p}\right) = \left(\frac{n^s}{p}\right) = \left(\frac{n}{p}\right)^s = (-1)^s = -1. \qquad \square$$

**Remark 5.2.39.** Notice that the previous lemma guarantees that $R_{2^r} \subseteq \mathbb{Z}_p$, as $\xi \in \mathbb{Z}_p$ is generator of $R_{2^r}$. So for all positive integers $u < r$ we have $R_{2^u} \subset R_{2^r} \subseteq \mathbb{Z}_p$.

We now compute the remainder $\rho$ of the division of $a^{(s+1)/2}$ by $p$ and, once more, we consider it as an element of $\mathbb{Z}_p$.

**Lemma 5.2.40.** *In $\mathbb{Z}_p$, $(a^{-1}\rho^2)^{2^{r-1}} = 1$.*

PROOF. In $\mathbb{Z}_p$ we have

$$(a^{-1}\rho^2)^{2^{r-1}} = a^{s2^{r-1}} = a^{(p-1)/2} = \left(\frac{a}{p}\right) = 1. \qquad \square$$

Notice that in $R_{2^{r-1}}$ the element $a^{-1}\rho^2$ has an inverse $x$, and $x\rho^2 = a$. As $R_{2^{r-1}} \subset R_{2^r}$, by Lemma 5.2.38 we have $x = \xi^t$, with $0 \leq t < 2^r$. If we prove that $t = 2v$ is even, then we have $a = x\rho^2 = (\xi^v \rho)^2$, so we have found the square root $\xi^v \rho$ of $a$.

In conclusion, we have to find an element $v$ such that $0 \leq v < 2^{r-1}$ and such that $\xi^v \rho$ is a square root of $a$ in $\mathbb{Z}_p$. We write $v = v_0 + 2v_1 + 2^2 v_2 + \cdots + v_{r-2} 2^{r-2}$ in binary form and show that there is an algorithm to compute successively the digits $v_0, v_1, v_2, \ldots, v_{r-2}$.

*Step 1: determining $v_0$.* By Lemma 5.2.40, in $\mathbb{Z}_p$ we have

$$(a^{-1}\rho^2)^{2^{r-2}} = \pm 1.$$

If $(a^{-1}\rho^2)^{2^{r-2}} = 1$, take $v_0 = 0$, else take $v_0 = 1$. In both cases, as $\xi^{2^{r-1}} = -1$, in $\mathbb{Z}_p$ we have

$$(a^{-1}(\xi^{v_0}\rho)^2)^{2^{r-2}} = 1.$$

*Step $h+1$: determining $v_h$, with $0 < h < r-3$.* Suppose we have found $v_0, \ldots, v_{h-1}$ in such a way that in $\mathbb{Z}_p$ we have

$$(a^{-1}(\xi^{v_0+2v_1+\cdots+2^{h-1}v_{h-1}}\rho)^2)^{2^{r-h-1}} = 1.$$

Then in $\mathbb{Z}_p$ we have

$$(a^{-1}(\xi^{v_0+2v_1+\cdots+2^{h-1}v_{h-1}}\rho)^2)^{2^{r-h-2}} = \pm 1.$$

In the case "+1" we take $v_h = 0$, else we take $v_h = 1$. After this choice in $\mathbb{Z}_p$ we have

$$(a^{-1}(\xi^{v_0+2v_1+\cdots+2^h v_h}\rho)^2)^{2^{r-h}} = 1.$$

Going on like this, after $r-1$ steps we find the required root.

**Example 5.2.41.** The simplest case is when $p \equiv 3 \pmod 4$. Then, and only then, we have $p - 1 = 2s$ with $s = (p-1)/2$ odd and, with the notation introduced above, $r = 1$. In this case, by Lemma 5.2.40 we know that $\rho^2 = a$ and so the required root is $\rho$, that is, the residue class modulo $p$ of $a^{(s+1)/2}$, that is, $a^{(p+1)/4}$.

For instance, the square root of 7 in $\mathbb{Z}_{19}$ is the residue class modulo 19 of $7^5 = 16807$, that is, 11.

**Example 5.2.42.** We apply now the algorithm, as an example, to a very simple case, that of finding a square root of $a = 2$ in $\mathbb{Z}_{17}$. Indeed, notice that $\left(\frac{2}{17}\right) = 1$. On the other hand, $\left(\frac{3}{17}\right) = \left(\frac{17}{3}\right) = \left(\frac{2}{3}\right) = -1$. So we may take $n = 3$. Moreover, $a^{-1} = 9$.

Here $r = 4$ e $s = 1$, so $\xi = n = 3$ and $\rho = a = 2$, and it is necessary to compute $v = v_0 + 2v_1 + 4v_2$ such that $\xi^v \rho = 2 \cdot 3^v$ is a square root of 2.

The algorithm tells us that $a^{-1}\rho^2 = a = 2$ is an eighth root of unity. In the first step of the algorithm we have to consider $(a^{-1}\rho^2)^4 = a^4 = 2^4 = 16 \equiv -1 \pmod{17}$. This tells us that $v_0 = 1$ and so that $a^{-1}(\xi \cdot \rho)^2$ is a fourth root of unity. However, $a^{-1}(\xi \cdot \rho)^2 = 9 \cdot (3 \cdot 2)^2 = 1$, and so the algorithm stops here, as we have found the required root, which is $\xi \cdot \rho = 3 \cdot 2 = 6$. This means that $v_1 = v_2 = 0$.

**Remark 5.2.43.** It is easy to compute the complexity of the algorithm just described. Suppose $a$ positive and smaller than $p$. Otherwise, computing its remainder modulo $p$ has complexity $\mathcal{O}(\log a \cdot \log p)$ (see Proposition 2.4.1).

First of all, it is necessary to compute $r$ and $s$. This involves dividing successively, $r < \log_2 p$ times, $p-1$ by 2. Hence, the complexity of this computation is $\mathcal{O}(r \log^2 p)$, that is, $\mathcal{O}(\log^3 p)$.

Netx, we must compute $\xi$, which is the class modulo $p$ of $n^s$. The complexity of this operation is $\mathcal{O}(\log a \log^2 p)$, that is, again, $\mathcal{O}(\log^3 p)$ (see Section 3.3.1). Similarly, computing $\rho$, which is the class modulo $p$ of $a^{(p+1)/2}$, has complexity $\mathcal{O}(\log^3 p)$.

To carry out the $(h+1)$th step of the algorithm, that is to determine $v_h$, it is again necessary to compute a power modulo $p$ of the class modulo $p$ of

$$a^{-1}(\xi^{v_0 + 2v_1 + \cdots + 2^{h-1}v_{h-1}}\rho)^2,$$

which was already computed in the previous step. This has again complexity $\mathcal{O}(\log^3 p)$. Finally, it is necessary to compute the class modulo $p$ of the number $a^{-1}(\xi^{v_0 + 2v_1 + \cdots + 2^h v_h}\rho)^2$, which has again complexity $\mathcal{O}(\log^3 p)$.

As it is necessary to perform $r - 1$ steps to complete the algorithm, and as $r < \log_2 p$, in conclusion the complexity is $\mathcal{O}(\log^4 p)$, so it is polynomial.

**Remark 5.2.44.** The algorithm discussed in this section relies on the knowledge of an integer $n$ that is not a square modulo $p$. Unfortunately there are no known algorithms to compute such an integer. However, it is quite easy to devise a very efficient *probabilistic algorithm* to find such a number.

As this is the first time we are dealing with this notion, let us tell something about it. The algorithms we have discussed so far are called *deterministic*: they *compute exactly* what they are devised for. A probabilistic algorithm to compute, say, a number $x$, on the contrary, consists of a sequence of steps such that in the $n$th step a number $x_n$ is computed, and there exists a sequence of positive real numbers $\{\epsilon_n\}_{n \in \mathbb{N}}$ converging to 0, such that the probability for $x_n$ to be different from *the thing we are looking for* is smaller than $\epsilon_n$.

Thus, a probabilistic algorithm does not give the *certainty* of having computed what we are interested in. However, if the sequence $\{\epsilon_n\}_{n \in \mathbb{N}}$ converges to 0 very quickly, for instance if $\epsilon_n = 1/2^n$, the probability of having found the correct answer is very high after just a few steps. For all intents and purposes, we may consider very difficult to observe an event having a probability smaller than $2^{-100}$.

It is quite easy to devise a polynomial probabilistic algorithm to find a non-square in $\mathbb{Z}_p$. The algorithm consists in choosing *randomly* in the $n$th step a number $x_n$ such that $0 < x_n < p$ and $x_n \neq x_i$, with $0 < i < n$. Compute next $\left(\frac{x_n}{p}\right)$, which can be done in polynomial time (see Remark 5.2.37) and verify whether $x_n$ is a square in $\mathbb{Z}_p$ or not. Keeping in mind that the number of squares is equal to that of non-squares in $\mathbb{Z}_p$ (see Proposition 5.2.16), we observe that the probability that at the $n$th step the algorithm *has not found* a quadratic non-residue is exactly $1/2^n$.

Notice that the same algorithm also finds a quadratic residue modulo $p$.

# Appendix to Chapter 5

# A5 Theoretical exercises

**A5.1.** Given an arbitrary field $\mathbb{K}$ the intersection of all subfields of $\mathbb{K}$ is called the *prime* or *fundamental subfield* of $\mathbb{K}$. Prove that the fundamental subfield of $\mathbb{K}$ is the *smallest* subfield of $\mathbb{K}$, that is to say it is included in any other subfield of $\mathbb{K}$. Verify further that the fundamental subfield of $\mathbb{K}$ is the field $\mathbb{Q}$ if $\mathbb{K}$ has characteristic 0, while it is the field $\mathbb{Z}_p$ if $\mathbb{K}$ has characteristic $p$.

**A5.2.** Prove that the field $\mathbb{Z}_p(x)$ of rational functions over $\mathbb{Z}_p$ is infinite and has characteristic $p$.

**A5.3.** Let $A \subseteq B$ be a field extension and let $b_1, \ldots, b_n$ be elements of $B$. Prove that $A(b_1, \ldots, b_n)$ consists exactly of the *rational expressions* in $b_1, \ldots, b_n$ with co-efficients in $A$, that is to say the elements of $B$ of the form (5.1).

**A5.4.** Let $A, B$ be rings with $A \subset B$ and let $b_1, \ldots, b_n$ be elements of $B$. Let $A[b_1, \ldots, b_n]$ be the smallest subring of $B$ containing $A$ and $b_1, \ldots, b_n$. Prove that $A[b_1, \ldots, b_n]$ consists of all integral rational expressions in $b_1, \ldots, b_n$ with coefficients in $A$.

**A5.5.*** Let $A, B, C$ be fields such that $A \subseteq B \subseteq C$, with $[C : A]$ finite. Prove that $[B : A]$ and $[C : B]$ are finite as well.

**A5.6.*** Verify the claim made in the proof of the theorem about the multiplicativity of degrees: if $\{a_1, \ldots, a_n\}$ is a basis of $B$ as a vector space over $A$ and $\{b_1, \ldots, b_m\}$ is a basis of $C$ as a vector space over $B$, then the elements $a_i b_j$, $i = 1, \ldots, n$, $j = 1, \ldots, m$ form a basis of $C$ as a vector space over $A$.

**A5.7.** Let $A, B$ be rings with $A \subset B$, and fix $b \in B$. Prove that the map $v_b : f(x) \in A[x] \rightarrow f(b) \in A(b)$ is a ring homomorphism.

**A5.8.*** Let $A \subseteq B$ be a field extension, and let $b \in B$ be algebraic over $A$. Prove that the minimal polynomial $f_b(x) \in A[x]$ of $b$ is irreducible or, equivalently, that the ideal $I_b$ generated by it is prime.

**A5.9.** Let $A$ be an integral domain and let $I$ be an ideal of $A$. Prove that $I = A$ if and only if $1 \in I$.

**A5.10.** * Let $A$ be a commutative ring with unity. Prove that $A$ is a field if and only if the only ideals of $A$ are $A$ and $(0)$.

**A5.11.** * Given a field $A$ and an irreducible monic polynomial $f(x) \in A[x]$, define $B = A[x]/(f(x))$. Verify that $B$ is a field and that $a \in A \to a + (f(x)) \in B$ is an injective map. Identify $A$ with its image and set $c = x + (f(x))$; then prove that $f(c) = 0$ and that $f_c(x) = f(x)$.

**A5.12.** Prove Theorem 5.1.5.

**A5.13.** * Let $A \subseteq B$ a field extension and let $b_1, \ldots, b_n$ elements of $B$. Prove that $b_1, \ldots, b_n$ are algebraic over $A$ if and only if $[A(b_1, \ldots, b_n) : A]$ is finite. Prove next that if $b_1, \ldots, b_n$ are algebraic over $A$, then $A(b_1, \ldots, b_n)$ is algebraic over $A$.

**A5.14.** * Let $A, B, C$ be fields such that $A \subseteq B \subseteq C$. Prove that if the extensions $A \subset B$ and $B \subset C$ are algebraic, then the extension $A \subset C$ is algebraic too.

**A5.15.** * Let $A \subseteq B$ be a field extension. Prove that the elements of $B$ algebraic over $A$ form a subfield $C$ of $B$, which is an extension of $A$ called *algebraic closure of $A$ in $B$*. Prove that $C$ is algebraically closed in $B$, that is, its algebraic closure in $B$ is $C$ itself.

**A5.16.** * Prove that if $A \subset B$ is a field extension and if $A$ is countable, then the algebraic closure of $A$ in $B$ is countable (this result is sometimes called *Cantor's theorem*). Deduce that there are countably many algebraic numbers.

**A5.17.** Conclude the proof of Theorem 5.1.13.

**A5.18.** Prove that if $q = p^f$ and $\text{GCD}(n, p) = 1$, then the polynomial $x^n - 1$ has no multiple roots over $\mathbb{F}_q$.

**A5.19.** Let $\mathbb{K}$ be a field, let $f(x) = x^n + a_1 x^{n-1} + \cdots + a_n$ be a monic polynomial of degree $n$ and let $\alpha_1, \ldots, \alpha_n$ be its roots, each repeated as many times as its multiplicity, considered as elements of the splitting field of $f(x)$. Consider the polynomials over $\mathbb{K}$, called *elementary symmetric polynomials*,

$$p_1(x_1, \ldots, x_n) = \sum_{i=1}^{n} x_i, \quad p_2 = \sum_{1 \leq i < j \leq n} x_i x_j, \quad \ldots, \quad p_n = x_1 \cdots x_n;$$

prove that for all $i = 1, \ldots, n$ we have

$$a_i = (-1)^i p_i(\alpha_1, \ldots, \alpha_n).$$

The Exercises A5.20–A5.23 deal with the set $R_n$ of $n$th roots of unity.

**A5.20.** Verify that the set $R_n$ of $n$th roots of unity is a subgroup of the multiplicative group of the splitting field $\mathbb{F}$ of the polynomial $x^n - 1$ over a prime field $\mathbb{K}$.

**A5.21.** Prove that $\prod_{\xi \in R_n} \xi = (-1)^{n+1}$.

**A5.22.** Let $R_n$ be the set of $n$th roots of unity. Prove that $\sum_{\xi \in R_n} \xi = 0$.

**A5.23.** Given two positive integers $n, m$, prove that if $m \mid n$, then $R_m \subseteq R_n$. Prove that the converse holds as well, as long as the characteristic of the field is 0 or does not divide neither $n$ nor $m$.

**A5.24.** * Prove that the element $\xi = \xi_1 \cdots \xi_h$ in the proof of Proposition 5.1.16 has order $n$.

**A5.25.** Prove Proposition 5.1.19.

**A5.26.** Prove that no finite field is algebraically closed.

**A5.27.** Let $A \subseteq B$ be a field extension and let $B$ be algebraically closed. Prove that the algebraic closure of $A$ in $B$ is an algebraic closure of $A$. Deduce that if $A$ is finite or countable (in particular, if $A = \mathbb{Q}$), its algebraic closure is countable.

**A5.28.** Prove that the roots of the polynomial $x^{p^n} - x$ over $\mathbb{Z}_p$, considered in its splitting field, form a field. Deduce that the splitting field of $x^{p^n} - x$ consists exactly of the roots of the polynomial $x^{p^n} - x$.

**A5.29.** Let $\mathbb{F}$ be a finite field of characteristic $p$ and let $f = [\mathbb{F} : \mathbb{Z}_p]$. Prove that every element of $\mathbb{F}$ is a root of the polynomial $x^{p^f} - x$.

**A5.30.** * Let $\mathbb{F}$ be a finite field of order $p^f$ with $p$ prime, and let $\mathbb{F}'$ be a subfield of $\mathbb{F}$ of order $p^{f'}$. Prove that $\mathbb{F}$ is the splitting field over $\mathbb{F}'$ of the polynomial $g_{f,f'}(x)$ such that $x^{p^f} - x = g_{f,f'}(x)(x^{p^{f'}} - x)$ (see Exercise A4.34 and the proof of Theorem 5.1.33).

**A5.31.** * Let $\mathbb{F}$ be a finite field of order $p^f$ with $p$ prime. Prove that, for all positive integers $m$, the polynomial $x^{p^{mf}} - x$ is divisible by all irreducible polynomials over $\mathbb{F}$ of degree $d \mid m$ and by no other polynomial, in particular not by the squares of these polynomials.

**A5.32.** * Let $p$ be a prime number and denote by $n_{d,p^f}$ the number of irreducible monic polynomials of degree $d$ over $\mathbb{F}_{p^f}$. Prove the relation $p^{mf} = \sum_{d \mid m} d\, n_{d,p^f}$. In particular, if $m$ is prime, $n_{m,p^f} = (p^{mf} - p^f)/m$.

**A5.33.** * Set $q = p^f$ and prove that

$$n_{d,q} > \frac{q^d}{\left(1 - \dfrac{1}{q^{-d/2+1}}\right)}.$$

**A5.34.** * Let $\mathbb{F}$ be a finite field of order $p^f$ with $p$ prime, and let $\mathbb{F}'$ be a subfield of order $p^{f'}$. Describe the automorphism group of $\mathbb{F}$ fixing every element of $\mathbb{F}'$.

**A5.35.** * Let $\mathbb{F}$ be a finite field of order $p^f$, $f \geq d$. Let $f(x) \in \mathbb{Z}_p[x]$ be an irreducible monic polynomial of degree $d$ and let $\alpha \in \mathbb{F}$ be one of its roots. Prove that $\alpha^{p^d} = \alpha$ and that $\alpha^{p^i}$, $i = 0, \ldots, d-1$, are all its distinct roots. In particular, every irreducible monic polynomial over $\mathbb{Z}_p$ has distinct roots.

**A5.36.** * Prove that in the ring $\mathbb{K}[x]/(f(x))$ every element can be represented uniquely as a polynomial of degree smaller than $f(x)$. (Hint: by induction on the degree of a representative element.)

**A5.37.** Prove that every non-zero element of the field $\mathbb{F}_9$ can be written as $\alpha^j$ for some integer $j$, where $\alpha$ is an element of $\mathbb{F}_9$ such that $\alpha^2 = \alpha + 1$.

**A5.38.*** Prove that the computational complexity of performing the reduction in row echelon form of a $m \times n$ $(m \leq n)$ matrix with entries in a finite field $\mathbb{F}_q$ using Gaussian elimination (see [13], Cap. 8) is $\mathcal{O}(n^3 \log^3 q)$.

**A5.39.*** Prove that the computational complexity of computing the determinant of a square $n \times n$ matrix with entries in a finite field $\mathbb{F}_q$ is $\mathcal{O}(n^3 \log^3 q)$.

**A5.40.*** Prove that the computational complexity of computing the inverse of a square $s \times s$ matrix with entries in a finite field $\mathbb{F}_q$ is $\mathcal{O}(n^3 \log^3 q)$.

**A5.41.*** Let $f(x), g(x)$ be polynomials of degree at most $d$ over the finite field $\mathbb{F}_q$. Prove that the complexity of computing the greater common divisor of $f(x)$ and $g(x)$ using the Euclidean algorithm is $\mathcal{O}(d^3 \log^3 p)$.

**A5.42.** Let $\mathbb{K}$ be a field of characteristic different from 2. Let $f(x) = ax^2 + bx + c$ be a polynomial of degree two. Prove that the roots of $f(x)$ in the splitting field of $f(x)$ are given by the formula $(-b \pm \sqrt{b^2 - 4ac})/2a$, called *quadratic formula*. Prove that they are in $\mathbb{K}$ if and only if $\Delta = b^2 - 4ac$, called *discriminant* of $f(x)$, is a square in $\mathbb{K}$, that is, if and only if the equation $x^2 = \Delta$ is soluble in $\mathbb{K}$. Prove that $f(x)$ has two distinct roots if and only if $\Delta \neq 0$.

**A5.43.** Prove Lemma 5.2.18 on page 237.

**A5.44.** Prove Proposition 5.2.22 on page 239.

**A5.45.** Prove Corollary 5.2.23 on page 239.

**A5.46.** Let $p$ be an odd prime. Prove that $\sum_{i=1}^{p}\left(\frac{i}{p}\right) = \sum_{i=0}^{p}\left(\frac{i}{p}\right) = \sum_{i=1}^{p-1}\left(\frac{i}{p}\right) = 0$.

In the following exercises we outline a proof of the law of quadratic reciprocity, based on some properties of finite fields, different from the one given in the main text.

**A5.47.** Let $n$ be an odd number. Prove that $n^2 \equiv 1 \pmod 8$.

**A5.48.*** Let $p$ be a prime number. Prove that the field $F_{p^2}$ contains a primitive eighth root $\xi$ of unity.

**A5.49.*** Recall the definition of the function $\epsilon(n) = (-1)^{(n^2-1)/8}$ (for odd $n$) given in the proof of Proposition 5.2.35. Define $G = \sum_{i=0}^{7} \epsilon(i)\xi^i \in F_{p^2}$ (the sum is over odd values of $i$). Prove that $G^2 = 8$ and deduce that $G^p = \left(\frac{2}{p}\right)G$.

**A5.50.*** Following the previous exercise, prove by a direct computation that $G^p = \epsilon(p)G$ and deduce Corollary 5.2.27.

**A5.51.** Let $p$ and $q$ be distinct odd primes. Prove that there exists a positive integer $f$ such that $p^f \equiv 1 \pmod q$. Deduce that the field $\mathbb{F}_{p^f}$ contains a primitive $q$th root $\xi$ of unity.

**A5.52.*** Following the previous exercise, define $G = \sum_{i=0}^{q-1}\left(\frac{i}{q}\right)\xi^i \in F_{p^f}$. Prove by a direct computation that $G^p = \left(\frac{p}{q}\right)G$.

**A5.53.*** Following the previous exercises, prove that $G^2 = (-1)^{(q-1)/2}q$ and deduce that $G \neq 0$.

**A5.54.**\* Following the previous exercise, prove that $G^p = (-1)^{(p-1)(q-1)/4}\left(\frac{q}{p}\right)G$ and deduce the law of quadratic reciprocity.

**A5.55.** Complete the proof of Proposition 5.2.34 on page 246.

**A5.56.** Let $n$ be a positive odd integer number. Prove that the map $a \in \mathbb{Z}_n^* \to \left(\frac{a}{n}\right) \in \{1, -1\} = \mathbb{Z}_3^*$ is well-defined and is a group homomorphism.

**A5.57.** Let $p$ be an odd number. Prove that for all positive integer $\alpha$, we have $p^\alpha = (1 + (p-1))^\alpha \equiv 1 + \alpha(p-1) \pmod 4$.

**A5.58.** For all positive odd integers $n$, set $\epsilon(n) = (-1)^{(n^2-1)/8}$. Verify that for all pairs $(n, m)$ of positive odd integers, we have $\epsilon(nm) = \epsilon(n)\epsilon(m)$.

# B5 Computational exercises

**B5.1.** Is there a field of characteristic 91?

(a)  Yes, because 91 is a prime number.
(b)  No, because 91 is not a prime number.
(c)  No, although 91 is a prime number.
(d)  None of the above.

**B5.2.** Is there a field of characteristic 997?

(a)  Yes, because 997 is a prime number.
(b)  No, because 997 is not a prime number.
(c)  No, although 997 is a prime number.
(d)  None of the above.

**B5.3.** Is there a field consisting of 64 elements?

(a)  Yes, because 64 is a prime power.
(b)  No, because 64 is not a prime number.
(c)  No, because $64 = 8^2$ and 8 is not a prime number.
(d)  None of the above.

**B5.4.** Is there a field consisting of 323 elements?

(a)  Yes, because 323 is a prime number.
(b)  No, because 323 is not a prime number.
(c)  No, although 323 is a prime number.
(d)  None of the above.

**B5.5.** Write down the addition and multiplication tables of the elements of the field $\mathbb{F}_8$.

**B5.6.** Write down the addition and multiplication tables of the elements of the field $\mathbb{F}_{16}$.

**B5.7.** Write down the addition and multiplication tables of the elements of the field $\mathbb{F}_9$.

**B5.8.** Which is the fundamental subfield of $\mathbb{F}_{32}$?

(a) $\mathbb{Z}_2$.
(b) $\mathbb{F}_4$.
(c) $\mathbb{F}_{16}$.
(d) $\mathbb{F}_{32}$.

**B5.9.** Verify that the multiplicative group $\mathbb{F}_4^* = \mathbb{F}_4 \setminus \{0\}$ of the field $\mathbb{F}_4$ is a cyclic group. How many generators has $\mathbb{F}_4^*$?

**B5.10.** Verify that the multiplicative group $\mathbb{F}_8^*$ of the field $\mathbb{F}_8$ is a cyclic group and write down all its generators.

**B5.11.** Verify that the multiplicative group $\mathbb{F}_{16}^*$ of the field $\mathbb{F}_{16}$ is a cyclic group and write down all its generators.

**B5.12.** How many generators has the multiplicative group $\mathbb{F}_9^*$?

(a) 1.
(b) 2.
(c) 3.
(d) 4.

**B5.13.** Which is the degree of $\mathbb{F}_{16}$ as an extension of $\mathbb{Z}_2$?

(a) 1.
(b) 2.
(c) 4.
(d) 8.

**B5.14.** Prove that $\mathbb{F}_4$ is a subfield of $\mathbb{F}_{16}$.

**B5.15.** Prove that (5.4) on page 225 is a factorisation into irreducible polynomials over $\mathbb{Z}$.

**B5.16.** Prove that (5.5) on page 225 is a factorisation into irreducible polynomials over $\mathbb{Z}_2$.

**B5.17.** Prove that (5.6) on page 226 is a factorisation into irreducible polynomials over $\mathbb{Q}$.

**B5.18.** Prove that (5.7) on page 226 is a factorisation into irreducible polynomials over $\mathbb{Z}_2$.

**B5.19.** Prove that (5.8) on page 226 is a factorisation into irreducible polynomials over $\mathbb{Z}_3$.

**B5.20.** How many irreducible polynomials of fourth degree are there in $\mathbb{Z}_2[x]$?

(a) 1.
(b) 2.
(c) 3.
(d) 4.

**B5.21.** Is $x^4 + x^2 + 1$ irreducible over $\mathbb{Z}_2$?

(a) Yes.

(b) No, as it factors into a degree two polynomial and a degree three one.
(c) No, as it factors into four first degree polynomials.
(d) No, as it factors into two degree two polynomials.

**B5.22.** How many irreducible factors has $x^{27} - x$ over $\mathbb{Z}$?

(a) 5.
(b) 7.
(c) 9.
(d) 10.

**B5.23.** How many irreducible factors has $x^{25} - x$ over $\mathbb{Z}_5$?

(a) 8.
(b) 10.
(c) 12.
(d) 15.

**B5.24.** How many degree two irreducible polynomials are there in $\mathbb{Z}_3[x]$?

(a) 2.
(b) 3.
(c) 4.
(d) 6.

**B5.25.** How many irreducible factors has $x^{27} - x$ over $\mathbb{Z}_3$?

(a) 5.
(b) 7.
(c) 9.
(d) 11.

**B5.26.** How many irreducible factors has $x^{27} - x$ over $\mathbb{Z}_9$?

**B5.27.** Is it true that $\mathbb{F}_{25}$ is a subfield of $\mathbb{F}_{125}$?

(a) No.
(b) Yes, as it is its fundamental subfield.
(c) Yes, as 25 divides 125.
(d) None of the above.

**B5.28.** How many proper (i.e., strictly included) subfields has $\mathbb{F}_{125}$, up to isomorphism?

(a) 1.
(b) 2.
(c) 3.
(d) 5.

**B5.29.** Is it true that $\mathbb{F}_9$ is a subfield of $\mathbb{F}_{27}$?

(a) Yes, as it is its fundamental subfield.
(b) Yes, as 9 divides 27.

(c) No.

(d) None of the above.

**B5.30.** How many proper (i.e., strictly included) subfields has $\mathbb{F}_{32}$, up to isomorphism?

(a) 1.

(b) 2.

(c) 3.

(d) 5.

**B5.31.** Write all automorphisms of $\mathbb{F}_4$, $\mathbb{F}_8$ e $\mathbb{F}_{16}$.

**B5.32.** Write all automorphisms of $\mathbb{F}_9$.

**B5.33.** Write all irreducible monic polynomials of degree $d \leq 6$ over $\mathbb{Z}_2$.

**B5.34.** Write all irreducible monic polynomials of degree 3 over $\mathbb{Z}_3$.

**B5.35.** Write all irreducible monic polynomials of degree $d \leq 3$ over $\mathbb{Z}_4$.

**B5.36.** Write all irreducible monic polynomials of degree $d \leq 3$ over $\mathbb{Z}_9$.

**B5.37.** Solve all quadratic equations over $\mathbb{Z}_p$, $p = 2, 3, 5$.

**B5.38.** Solve the congruence of Example 5.2.3 on page 230 by evaluating $x^3 + 2x^2 - 1$ modulo 10 for all $x$ with $-4 \leq x \leq 5$.

**B5.39.** Solve the congruence of Example 5.2.4 on page 231.

**B5.40.** Prove that the congruence $x^3 + 3x + 2 \equiv 0$ (mod 7) has no solutions (see Example 5.2.7 on page 232).

**B5.41.** Prove that the congruence $x^2 \equiv p$ (mod 5) admits solutions if and only if $p \equiv 1$ (mod 5) or $p \equiv -1$ (mod 5).

**B5.42.** Prove that the congruence $x^2 \equiv 2$ (mod 15) has no solutions.

**B5.43.** How many solutions has the congruence $x^2 - 3x - 1 \equiv 0$ (mod 7)?

(a) 0.

(b) 1.

(c) 2.

(d) 3.

**B5.44.** How many solutions has the congruence $x^2 - 3x - 4 \equiv 0$ (mod 14)?

(a) 0.

(b) 1.

(c) 2.

(d) 4.

**B5.45.** How many solutions has the congruence $x^2 - 2x + 1 \equiv 0$ (mod 8)?

(a) 0.

(b) 1.

(c) 2.

(d) 4.

**B5.46.** How many solutions has the congruence $x^3 + 5x - 4 \equiv 0 \pmod{11}$?

(a) 0.
(b) 1.
(c) 2.
(d) 3.

**B5.47.** How many solutions has the congruence $x^4 - x^3 - 1 \equiv 0 \pmod{49}$?

(a) 0.
(b) 1.
(c) 2.
(d) 4.

**B5.48.** Which is the value of the Legendre symbol $\left(\frac{95}{11}\right)$?

(a) 0.
(b) 1.
(c) $-1$.
(d) None of the above.

**B5.49.** Which is the value of the Legendre symbol $\left(\frac{65}{13}\right)$?

(a) 0.
(b) 1.
(c) $-1$.
(d) None of the above.

**B5.50.** Which is the value of the Legendre symbol $\left(\frac{271}{143}\right)$?

(a) 0.
(b) 1.
(c) $-1$.
(d) None of the above.

**B5.51.** Which is the value of the Legendre symbol $\left(\frac{1001}{971}\right)$?

(a) 0.
(b) 1.
(c) $-1$.
(d) None of the above.

**B5.52.** Which is the value of the Jacobi symbol $\left(\frac{41}{35}\right)$?

(a) 0.
(b) 1.
(c) $-1$.
(d) None of the above.

**B5.53.** Which is the value of the Jacobi symbol $\left(\frac{105}{91}\right)$?

(a) 0.
(b) 1.
(c) −1.
(d) None of the above.

**B5.54.** Which is the value of the Jacobi symbol $\left(\frac{1003}{973}\right)$?

(a) 0.
(b) 1.
(c) −1.
(d) None of the above.

## C5 Programming exercises

**C5.1.** Write a program that writes out the addition and multiplication tables of the field $\mathbb{F}_q$, with $q = p^h$, for a prime number $p$ and a positive integer $h$.

**C5.2.** Write a program that finds a generator of the multiplicative group $\mathbb{F}_q^*$ of the field $F_q$.

**C5.3.** Write a program that solves a polynomial congruence (linear or not) by trial and error.

**C5.4.** Write a program that solves a polynomial congruence using the method described in Section 5.2 (see Proposition 5.2.9).

**C5.5.** Write a program that computes the Legendre symbol $\left(\frac{a}{p}\right)$, where $a$ is an integer and $p$ is an odd prime.

**C5.6.** Write a program that computes the Jacobi symbol $\left(\frac{a}{n}\right)$, where $a$ is an integer and $n$ is an odd positive integer.

**C5.7.** Write a program that determines all the irreducible polynomials of degree $d$ over $\mathbb{Z}_p$, for a prime $p$ and a positive integer $d$.

**C5.8.** Write a program that finds a quadratic residue or a quadratic non-residue modulo a prime $p$ using the probabilistic algorithm described in Remark 5.2.44.

**C5.9.** Write a program that finds a square root in $\mathbb{Z}_p$ using the algorithm described in § 5.2.6.

# 6

## Primality and factorisation tests

In this chapter we shall discuss *primality tests* and *factorisation methods*. The former are used to check whether a given number is prime or not, the latter to decompose a given number into prime factors. We shall encounter again a *probabilistic* approach, already mentioned in Chapter 5.

## 6.1 Pseudoprime numbers and probabilistic tests

In Chapter 4 we saw some naive primality tests relying upon Fermat's little theorem and Wilson's theorem (see Section 4.2.2 and Section 4.2.3). Those tests are not computationally efficient, as they are exponential. Nevertheless, as we shall see, it is useful to investigate further the analysis of the test based upon Fermat's little theorem.

### 6.1.1 Pseudoprime numbers

The test based upon Fermat's little theorem 4.2.5 can be used to verify if a given number $n$ *is not* a prime. So, given $n > 2$, which obviously is not an even number, otherwise clearly it is not prime, we begin by applying the test described in Section 4.2.2 by computing, for instance, $2^{n-1} \pmod{n}$. If $2^{n-1} \not\equiv 1 \pmod{n}$ we are sure that $n$ *is not* a prime, as we know, by Fermat's little theorem, that if $n$ is a prime, then for all integers $a$ that are coprime with $n$:

$$a^{n-1} \equiv 1 \pmod{n}. \tag{6.1}$$

But what may we conclude if $2^{n-1} \equiv 1 \pmod{n}$? May we say that $n$ is a prime number? More in general, we may ask: *is it sufficient to find an integer $a$ that is coprime with $n$ and such that* (6.1) *holds, in order to claim that $n$ is prime?*

The answer to this question is *no*, as the following example shows.

**Example 6.1.1.** The number $91 = 7 \cdot 13$ is not a prime, but $3^{90} \equiv 1 \pmod{91}$. Indeed, it suffices to verify that

$$3^{90} \equiv 1 \pmod{7} \quad \text{and} \quad 3^{90} \equiv 1 \pmod{13}. \tag{6.2}$$

Fermat's little theorem guarantees that $3^6 \equiv 1 \pmod{7}$ and, as $90 = 6 \cdot 15$, we immediately get the first part of (6.2). The second one is verified analogously and is left to the reader (see Exercise B6.1).

Notice that, on the other hand, $2^{90} \not\equiv 1 \pmod{91}$ (see Exercise B6.4); hence, if we had not already known 91 not to be a prime, we would have discovered it by applying to it the Fermat test described in § 4.2.2, with respect to the integer $a = 2$.

These remarks justify the following definition.

**Definition 6.1.2.** *If $n$ is an odd, non-prime number and $a$ is an integer relatively prime with $n$, then $n$ is said to be a* pseudoprime in base $a$ *if $a^{n-1} \equiv 1 \pmod{n}$.*

For instance, $n = 91$ is a pseudoprime in base 3 but not in base 2.

An important question is, *how many pseudoprimes are there in a given base $a$?* If we knew them to be finitely many and we were able to compute them all, we could apply Fermat's little theorem to determine an efficient polynomial primality test. Unfortunately, this is not how things go: for instance, there are infinitely many pseudoprimes in base 2 (see Exercise A6.1). Nevertheless, it is interesting to be aware of the following proposition.

**Proposition 6.1.3.** *Let $n > 1$ a non-prime odd number.*

(a) *If $n$ is pseudoprime in the bases $a_1$ and $a_2$ such that $\mathrm{GCD}(a_1, n) = 1$ and $\mathrm{GCD}(a_2, n) = 1$, then $n$ is pseudoprime in the bases $a_1 a_2$ and $a_1 a_2^{-1}$, where $a_2^{-1}$ is the inverse of $a_2$ modulo $n$.*

(b) *If there exists an integer $a$, with $1 < a < n$ and $\mathrm{GCD}(a, n) = 1$, such that $n$ is not a pseudoprime in base $a$, then $n$ is not a pseudoprime in base $b$ for at least one half of the values of $b$ such that $1 < b < n$ and $\mathrm{GCD}(b, n) = 1$.*

PROOF. The proof of (a) is left as an exercise (see Exercise A6.3). We prove now (b).

Consider $a$ as an element of $U(\mathbb{Z}_n)$. Let $P$ the subset of $U(\mathbb{Z}_n)$ consisting of the classes whose remainder $b$ modulo $n$ is such that $n$ is pseudoprime in base $b$. Keeping in mind (a), it is clear that if $b \in P$ then $ab \notin P$. So there is an injective map $f : b \in P \to ab \in U(\mathbb{Z}_n) \setminus P$. So the order of $P$ equals at most the order of $U(\mathbb{Z}_n) \setminus P$. This proves (b). □

### 6.1.2 Probabilistic tests and deterministic tests

The previous proposition is the base for a quite efficient primality test of a completely new kind, that is, a *probabilistic* test. The difference between probabilistic and *deterministic* tests, the ones we have encountered so far, is similar to the difference between probabilistic and deterministic algorithms, discussed in Remark 5.2.44. Indeed, deterministic tests are designed to determine with *certainty* whether a number is prime or not. On the other hand, probabilistic tests $\mathcal{T}$ consist of a sequence of tests $\{\mathcal{T}_m\}_{m\in\mathbb{N}}$, for which there is a sequence $\{\epsilon_m\}_{m\in\mathbb{N}}$ converging to 0 of positive real numbers smaller than 1, such that if a positive integer number $n$ does not pass a test $\mathcal{T}_m$ then *it is not prime*, while the probability that a positive integer number $n$ passes tests $\mathcal{T}_1, \ldots, \mathcal{T}_m$ and is not a prime is smaller than $\epsilon_m$.

So a probabilistic test does not give the *certainty* that a number passing the test is prime. However, especially if the sequence $\{\epsilon_m\}_{m\in\mathbb{N}}$ converges to 0 very quickly, for instance if $\epsilon_m = 1/2^m$, the probability that a number passing the tests $\mathcal{T}_1, \ldots, \mathcal{T}_m$ is not prime is extremely low, even for a small value of $m$, that is, after just a few steps.

It is true that deterministic tests, like the sieve of Eratosthenes or the test based on Wilson's theorem, give in principle a reliable answer to the question whether a given number is prime or not. However, as we have seen, they are often so computationally expensive to be useless. On the contrary, some probabilistic tests are computationally very efficient.

### 6.1.3 A first probabilistic primality test

The probabilistic test $\mathcal{T}$ we are about to discuss works as follows. Fix an odd integer $n > 1$. The test $\mathcal{T}_1$ consists of the following steps:

(1) choose randomly an integer $a_1$ with $1 < a_1 < n$ and compute $\mathrm{GCD}(a_1, n)$;
(2) if $\mathrm{GCD}(a_1, n) > 1$ then $n$ is not prime, and the test is over;
(3) if $\mathrm{GCD}(a_1, n) = 1$ and condition (6.1) with $a = a_1$ does not hold, then $n$ is not prime, and the test is over;
(4) otherwise, apply test $\mathcal{T}_2$ to $n$.

Define next recursively test $\mathcal{T}_m$ as follows:

(1) choose randomly an integer $a_m$ such that $1 < a_m < n$ with $a_m$ different from $a_1, \ldots, a_{m-1}$, and compute $\mathrm{GCD}(a_m, n)$;
(2) if $\mathrm{GCD}(a_m, n) > 1$ then $n$ is not prime, and the test is over;
(3) if $\mathrm{GCD}(a_m, n) = 1$ and condition (6.1) with $a = a_m$ does not hold, then $n$ is not prime, and the test is over;
(4) otherwise, apply test $\mathcal{T}_{m+1}$ to $n$.

Now assume the following hypothesis, which we shall call *hypothesis H*:

> if $n$ is not prime, there is an $a$, with $1 < a < n$, relatively prime with $n$, such that $n$ is not a pseudoprime in base $a$.

If this is verified, then by (b) of Proposition 6.1.3, when we perform test $\mathcal{T}_m$, the probability of $n$ being a pseudoprime in base $a_m$ is smaller than $1/2$. So the probability of $n$ not being a prime but passing tests $\mathcal{T}_1, \ldots, \mathcal{T}_m$ is smaller than $1/2^m$.

**Remark 6.1.4.** Notice that, if $n$ is fixed, the number of tests $\mathcal{T}_m$ we have to perform is finite, equal to $\varphi(n)$. However, if $n$ is very large, we might not want to perform all the $\varphi(n)$ tests of the sequence. Indeed, after the first, say, 50 steps a number $n$ successfully got through we may claim that, with a good degree of probability, $n$ is prime.

Notice that the complexity of each test $\mathcal{T}_m$ is polynomial. Indeed, after having randomly chosen a number $a$ with $1 < a < n$, verifying that $\mathrm{GCD}(a, n) = 1$ has polynomial complexity $\mathcal{O}(\log^3 n)$ (see Proposition 2.5.4). Moreover, the third step of test $\mathcal{T}_m$ has complexity $\mathcal{O}(\log^3 n)$ too (see Section 3.3.1).

In conclusion, if hypothesis $H$ holds and we want to ascertain whether $n$ is prime with a probability greater or equal than $(2^{50} - 1)/2^{50}$, the test's computational cost is $\mathcal{O}(\log^3 n)$.

### 6.1.4 Carmichael numbers

For the above probabilistic test to be meaningful it is necessary that hypothesis $H$ holds. Unfortunately, this hypothesis is not true. However, we shall shortly see that by exploiting ideas that are very similar to those leading to the above test it is possible to devise sound probabilistic primality tests.

Let us give a definition which is useful to elucidate the state of things about hypothesis $H$.

**Definition 6.1.5.** *If $n$ is an non-prime odd number that is pseudoprime in base $a$ for all $a$ such that $1 < a < n$ and $\mathrm{GCD}(a, n) = 1$, then $n$ is said to be a Carmichael number.*

So, if Carmichael numbers exist, hypothesis $H$ does not hold. And such numbers do exist, as we shall see, although not much is known about them. Recently it has been proved that there are infinitely many Carmichael numbers (see [3]).

The following proposition contains a very important information about Carmichael numbers.

**Proposition 6.1.6.** *Let $n > 1$ be a non-prime odd number.*

*(a) If $n$ is divisible by a square greater than 1, then $n$ is not a Carmichael number.*

*(b) If $n$ is not divisible by a square greater than 1, then $n$ is a Carmichael number if and only if $p-1$ divides $n-1$ for every prime factor $p$ of $n$.*

PROOF. We shall prove here just a part of (b): the fact that if $n$ is not divisible by a square greater than 1 and if $p-1$ divides $n-1$ for every prime factor $p$ of $n$, then $n$ is a Carmichael number. The remaining claims will be proved in Section 6.2.

By hypothesis we have $n = p_1 \cdots p_h$ with $p_1, \ldots, p_h$ distinct primes. Let $a$ be an integer such that $1 < a < n$ and $\mathrm{GCD}(a, n) = 1$, so also $\mathrm{GCD}(a, p_i) = 1$, $i = 1, \ldots, h$. By Fermat's little theorem we have $a^{p_i - 1} \equiv 1 \pmod{p_i}$, $i = 1, \ldots, h$. As $p_i - 1 \mid n - 1$, $i = 1, \ldots, h$, we also have $a^{n-1} \equiv 1 \pmod{p_i}$, $i = 1, \ldots, h$; hence $a^{n-1} \equiv 1 \pmod{n}$.  □

For instance, $561 = 3 \cdot 11 \cdot 17$ is a Carmichael number, as 560 is divisible by $3 - 1$, $11 - 1$, $17 - 1$.

**Proposition 6.1.7.** *Every Carmichael number is a product of at least three distinct primes.*

PROOF. By part (a) of Proposition 6.1.6 we know that every Carmichael number is a product of *distinct* primes. Assume by contradiction that $n = pq$ is a Carmichael number that is a product of exactly two distinct primes, with $p < q$. Again by Proposition 6.1.6, part (b), we know that $n - 1 \equiv 0 \pmod{q - 1}$. But

$$n - 1 = p(q - 1 + 1) - 1 = p(q - 1) + p - 1 \equiv p - 1 \pmod{q - 1};$$

on the other hand, $p - 1 \not\equiv 0 \pmod{q - 1}$, as $0 < p - 1 < q - 1$. So we have reached a contradiction.  □

Keeping this result in mind, it is easy to see that 561 is the smallest Carmichael number (see Exercise B6.8).

### 6.1.5 Euler pseudoprimes

Euler's criterion 5.2.19, or equivalently part (3) of Proposition 5.2.22 may also be used as a primality test, to verify that an odd number *is not* prime. Indeed,

> if there is an $a$ such that $\left(\dfrac{a}{n}\right) \not\equiv a^{(n-1)/2} \pmod{n}$, then $n$ is not a prime.

For instance, $n = 15$ is not prime because $2^7 = 128 \equiv 8 \pmod{15}$, while $\left(\frac{2}{15}\right) = 1$ (see (5.31) on page 246).

Analogously to Definition 6.1.2 of pseudoprime numbers, we may give the following definition.

**Definition 6.1.8.** *Let* $b$ *be an integer number. A non-prime odd positive integer* $n$ *such that* $\mathrm{GCD}(n, b) = 1$ *is said to be an* Euler pseudoprime *in base* $b$ *if*

$$b^{(n-1)/2} \equiv \left(\frac{b}{n}\right) \quad (\mathrm{mod}\ n). \tag{6.3}$$

This definition is related to Definition 6.1.2 as follows.

**Lemma 6.1.9.** *If* $n$ *is an Euler pseudoprime in base* $b$*, then* $n$ *is also a pseudoprime in base* $b$*.*

PROOF. By squaring both sides of (6.3), we find $b^{n-1} \equiv \left(\frac{b}{n}\right)^2 = 1 \ (\mathrm{mod}\ n)$.
□

We explicitly remark that the converse of Lemma 6.1.9 does not hold. Indeed, for instance, 91 is a pseudoprime in base 3, but not an Euler pseudoprime in base 3 (see Exercise B6.14).

We may further remark that every odd number is an Euler pseudoprime in base $b = \pm 1$; for this reason, from now on we shall exclude $b = \pm 1$ from the bases of Euler pseudoprimes.

The probabilistic test of Section 6.1.3 relies upon two premises:

- the *hope* – unfounded, as we have seen – of hypothesis $H$ being true;
- part (b) of Proposition 6.1.3, which says that, if hypothesis $H$ holds, then if a number $n$ is not prime, the bases $a$ in which it is pseudoprime, with $1 < a < n$, are at most $\varphi(n)/2$.

In the case of Euler pseudoprimes we are in a better position. Indeed, the analogue of hypothesis $H$ is always verified and the analogue of (b) of Proposition 6.1.3 holds too. More precisely, the following holds.

**Proposition 6.1.10.** *Let* $n$ *be a non-prime odd, positive integer. The positive numbers* $b < n$ *coprime with* $n$*, such that (6.3) holds, that is, such that* $n$ *is an Euler pseudoprime in base* $b$*, are at most half of all positive numbers* $b < n$ *such that* $\mathrm{GCD}(b, n) = 1$*.*

The proof needs several steps. The first one is the following proposition, which is the analogue of (a) of Proposition 6.1.3: its easy proof is left as an exercise to the reader (see Exercise A6.5).

**Proposition 6.1.11.** *Let* $n > 1$ *be a non-prime odd number. If* $n$ *is an Euler pseudoprime in bases* $b_1$ *and* $b_2$ *such that* $\mathrm{GCD}(b_1, n) = \mathrm{GCD}(b_2, n) = 1$*, then* $n$ *is pseudoprime in bases* $b_1 b_2$ *and* $b_1 b_2^{-1}$*, where* $b_2^{-1}$ *is the inverse of* $b_2$ *modulo* $n$*.*

We prove next:

**Lemma 6.1.12.** *Let* $n > 1$ *be an odd number that is not a perfect square. Then there is a positive integer* $b < n$*, relatively prime with* $n$*, such that* $\left(\frac{b}{n}\right) = -1$*.*

PROOF. If $n$ is prime, the claim is obvious, as quadratic non-residues modulo $n$ are known to exist.

If $n$ is not a prime, there exists a prime factor $p$ of $n$ appearing in the factorisation of $n$ with an odd exponent $e$. Write $n = p^e m$. Let $t$ be a quadratic non-residue modulo $p$. By the Chinese remainder theorem, there exists a positive number $b$ relatively prime with $n$ and smaller than $n$ such that

$$b \equiv t \pmod{p}, \quad b \equiv 1 \pmod{m}.$$

So we have

$$\left(\frac{b}{n}\right) = \left(\frac{b}{p}\right)^e \left(\frac{b}{m}\right) = (-1)^e = -1. \qquad \square$$

The key step in the proof of Proposition 6.1.10 is the following proposition which, basically, guarantees that for Euler pseudoprimes there is no analogue of Carmichael numbers:

**Proposition 6.1.13.** *Let $n > 1$ be a non-prime odd number. Then there is a positive integer number $b < n$ that is coprime with $n$ and such that $n$ is not an Euler pseudoprime in base $b$.*

PROOF. If the claim were false, by Lemma 6.1.9, $n$ would be a Carmichael number. By Proposition 6.1.6, then, we have $n = p_1 \cdots p_h$ with $p_1, \ldots, p_h$ distinct primes.

We prove now that for every positive integer $b < n$ coprime with $n$ the following holds:

$$\left(\frac{b}{n}\right) \equiv b^{(n-1)/2} \equiv 1 \pmod{n}. \tag{6.4}$$

This will contradict Lemma 6.1.12, thus proving the present Proposition.

As we are supposing the claim not to be true, $n$ is an Euler pseudoprime with respect to every base $b$, with $b$ a positive integer smaller than $n$ and coprime with $n$. So, for every such integer $b$ (6.3) holds.

Assume the existence of a positive integer $b < n$ coprime with $n$ such that $b^{(n-1)/2} \not\equiv 1 \pmod{n}$; then we must have $b^{(n-1)/2} \equiv -1 \pmod{n}$. By the Chinese remainder theorem, we may find a positive integer $a < n$, coprime with $n$ and such that

$$a \equiv b \pmod{p_1}, \quad a \equiv 1 \pmod{p_2 \cdots p_h}.$$

Then we have

$$a^{(n-1)/2} \equiv b^{(n-1)/2} \equiv -1 \pmod{p_1}$$

and

$$a^{(n-1)/2} \equiv 1 \pmod{p_2 \cdots p_h}.$$

Thus

$$a^{(n-1)/2} \not\equiv \pm 1 \pmod{n},$$

but this yields a contradiction, as (6.3) must hold in particular for $b = a$. $\quad \square$

Now it is easy to conclude the proof of Proposition 6.1.10: it suffices to argue in a pretty similar way to what has been done in the proof of part (b) of Proposition 6.1.3. We leave the details as an exercise for the reader (see Exercise A6.6).

### 6.1.6 The Solovay–Strassen probabilistic primality test

The *Solovay–Strassen probabilistic primality test* $\mathcal{T}$ relies on the notions discussed in the previous section about Euler pseudoprimes. It is pretty analogous to the test described in § 6.1.3, as follows.

Fix an odd integer $n > 1$. The test $\mathcal{T}_1$ consists of the following steps:

1. choose randomly an integer $b_1$ with $1 < b_1 < n$, and compute $\mathrm{GCD}(b_1, n)$;
2. if $\mathrm{GCD}(b_1, n) > 1$ then $n$ is not prime, and the test is over;
3. if $\mathrm{GCD}(b_1, n) = 1$ and condition (6.3) with $b = b_1$ does not hold, then $n$ is not prime, and the test is over;
4. otherwise, apply test $\mathcal{T}_2$ to $n$.

Define next recursively test $\mathcal{T}_m$ as follows:

1. choose randomly an integer $b_m$ such that $1 < b_m < n$ with $b_m$ different from $b_1, \ldots, b_{m-1}$, and compute $\mathrm{GCD}(b_m, n)$;
2. if $\mathrm{GCD}(b_m, n) > 1$ then $n$ is not prime, and the test is over;
3. if $\mathrm{GCD}(b_m, n) = 1$ and condition (6.3) with $b = b_m$ does not hold, then $n$ is not prime, and the test is over;
4. otherwise, apply test $\mathcal{T}_{m+1}$ to $n$.

Keeping in mind Proposition 6.1.10, when we perform test $\mathcal{T}_m$, the probability that $n$ is a pseudoprime in base $b_m$ is smaller than $1/2$. So, the probability that $n$ passes tests $\mathcal{T}_1, \ldots, \mathcal{T}_m$ without being a prime is smaller than $1/2^m$.

**Remark 6.1.14.** Similar comments to those given in Remark 6.1.4 hold here as well. In particular, the complexity of each test $\mathcal{T}_m$ is polynomial. Indeed, having chosen randomly a number $a$ with $1 < a < n$, verifying that $\mathrm{GCD}(a, n) = 1$ has polynomial complexity $\mathcal{O}(\log^3 n)$ (see Proposition 2.5.4). Moreover, the third step of test $\mathcal{T}_m$ also has complexity $\mathcal{O}(\log^3 n)$ (see Section 3.3.1 and Remark 5.2.37).

### 6.1.7 Strong pseudoprimes

Solovay–Strassen test can be improved. This improvement relies on the following remark.

**Remark 6.1.15.** Let $n$ be an odd positive integer and $b < n$ a positive integer such that $GCD(b, n) = 1$. Let $n$ be a pseudoprime in base $b$, that is,

$$b^{n-1} \equiv 1 \pmod{n}.$$

If $n$ is prime, we must have

$$b^{(n-1)/2} \equiv \pm 1 \pmod{n}, \qquad (6.5)$$

as in $\mathbb{Z}_p$ the only square roots of 1 are $\pm 1$. Iterating, if

$$b^{(n-1)/2} \equiv 1 \pmod{n}$$

and if $(n - 1)/2$ is even, we must have

$$b^{(n-1)/4} \equiv \pm 1 \pmod{n}$$

and so forth.

This remark motivates the following definition.

**Definition 6.1.16.** Let $n$ be a non-prime odd positive integer and $b < n$ a positive integer such that $GCD(b, n) = 1$. Set $n = 2^s t + 1$, with odd $t$. The number $n$ is said to be a strong pseudoprime in base $b$ if one of the following conditions holds: either

$$b^t \equiv 1 \pmod{n},$$

or there exists a non-negative integer $r < s$ such that

$$b^{2^r t} \equiv -1 \pmod{n}.$$

**Example 6.1.17.** The number $n = 25$ is a strongpseudoprime for the base $b = 7$. Indeed, $7^2 = 49 \equiv -1 \pmod{25}$ and so $7^{12} = 7^{(n-1)/2} \equiv 1 \pmod{25}$.

It is easy to verify that $n = 2047$ is a strong pseudoprime in base 2: actually, it is possible to verify that this is the smallest strong pseudoprime in base 2 (see [46]). Notice that $(n - 1)/2 = 2046/2 = 1023 = 11 \cdot 93$. Now, $2^{11} = 2048 \equiv 1 \pmod{2047}$, and so $2^{2046} \equiv 1 \pmod{2047}$.

It is clear that a strong pseudoprime in base $b$ is a pseudoprime in base $b$ as well (see Exercise A6.7). But how are strong pseudoprimes and Euler pseudoprimes related? The answer is given by the following two propositions.

**Proposition 6.1.18.** Let $n$ be an odd positive integer and $b < n$ a positive integer such that $GCD(b, n) = 1$. If $n \equiv 3 \pmod{4}$ then $n$ is a strong pseudoprime in base $b$ if and only if $n$ is an Euler pseudoprime in base $b$.

PROOF. In this case, keeping the notation of Definition (6.1.16), as $n \equiv 3 \pmod{4}$, $s$ must be equal to 1. So $n$ is a strong pseudoprime in base $b$ if and only of (6.5) holds.

Now, if $n$ is an Euler pseudoprime in base $b$, then

$$b^{(n-1)/2} \equiv \left(\frac{b}{n}\right) = \pm 1 \pmod{n},$$

that is, (6.5) holds and $n$ is a strong pseudoprime in base $b$. Conversely, let $n$ be a strong pseudoprime in base $b$. As $n \equiv 3 \pmod 4$, we have

$$\left(\frac{\pm 1}{n}\right) = \pm 1$$

and so

$$\left(\frac{b}{n}\right) = \left(\frac{b \cdot b^{2^{(n-3)/4}}}{n}\right) = \left(\frac{b^{(n-1)/2}}{n}\right) = \left(\frac{\pm 1}{n}\right) = \pm 1 \equiv b^{(n-1)/2} \pmod{n},$$

proving that $n$ is an Euler pseudoprime in base $b$.    □

**Proposition 6.1.19.** *Let $n$ be an odd positive integer and $b < n$ a positive integer such that $\mathrm{GCD}(b, n) = 1$. If $n$ is a strong pseudoprime in base $b$ then $n$ is an Euler pseudoprime in base $b$.*

PROOF. Keeping the notation of Definition 6.1.16, consider first the case in which $b^t \equiv 1 \pmod n$. Let $p$ be a prime divisor of $n$. Then we have $b^t \equiv 1 \pmod p$, and so $\mathrm{Gss}(p, b) \mid t$. As $t$ is odd, we find that $\mathrm{Gss}(p, b)$ is too. On the other hand, $\mathrm{Gss}(p, b) \mid \varphi(p) = p - 1$ and so $\mathrm{Gss}(p, b) \mid (p-1)/2$. So $b^{(p-1)/2} \equiv 1 \pmod p$ and Euler's criterion implies that $\left(\frac{b}{p}\right) = 1$. Hence $\left(\frac{b}{n}\right) = 1$. On the other hand, $b^{(n-1)/2} = (b^t)^{2^s} \equiv 1 \pmod n$ and so (6.3) holds, that is $n$ is an Euler pseudoprime in base $b$.

Consider now the case in which there exists a non-negative integer $r < s$ such that $b^{2^r t} \equiv -1 \pmod n$. Let $p$ be a prime divisor of $n$. We have $b^{2^r t} \equiv -1 \pmod p$ and so $b^{2^{r+1} t} \equiv 1 \pmod p$; hence $\mathrm{Gss}(p, b) \mid 2^{r+1} t$ but $\mathrm{Gss}(p, b) \nmid 2^r t$. So $\mathrm{Gss}(p, b) = 2^{r+1} s$ where $s$ is an odd number. As $\mathrm{Gss}(p, b) \mid p - 1$, we have $2^{r+1} \mid p - 1$, that is, $p = 2^{r+1} q + 1$. Hence

$$\left(\frac{b}{p}\right) \equiv b^{\frac{p-1}{2}} = b^{\frac{\mathrm{Gss}(p,b)}{2} \frac{p-1}{\mathrm{Gss}(p,b)}} \equiv (-1)^{\frac{p-1}{\mathrm{Gss}(p,b)}} = (-1)^{\frac{q}{s}} \pmod{p}.$$

In conclusion, as $s$ is odd, we have

$$\left(\frac{b}{p}\right) = (-1)^q. \tag{6.6}$$

Let now $n = p_1^{h_1} \cdots p_m^{h_m}$ be the factorisation of $n$. Set $p_i = 2^{r+1} d_i + 1$, $i = 1, \ldots, m$. Then we have

$$n = \prod_{i=1}^m (2^{r+1} d_i + 1)^{h_i} \equiv \prod_{i=1}^m (1 + 2^{r+1} h_i d_i) \equiv 1 + 2^{r+1} \sum_{i=1}^m h_i d_i \pmod{2^{r+2}}.$$

So

$$2^{s-1}t = (n-1)/2 \equiv 2^r \sum_{i=1}^{m} h_i d_i \quad (\mathrm{mod}\ 2^{r+1});$$

from this follows

$$2^{s-r-1}t \equiv \sum_{i=1}^{m} h_i d_i \quad (\mathrm{mod}\ 2). \tag{6.7}$$

Now, if $r < s - 1$, we have

$$b^{(n-1)/2} = b^{2^{s-1}t} = (b^{2^r t})^{2^{s-r-1}} \equiv (-1)^{2^{s-r-1}} = 1 \quad (\mathrm{mod}\ n).$$

From (6.7) we deduce that $\sum_{i=1}^{m} h_i d_i$ is even and, keeping in mind (6.6), we obtain

$$\left(\frac{b}{n}\right) = \prod_{i=1}^{m} \left(\frac{b}{p_i}\right)^{h_i} = \prod_{i=1}^{m} (-1)^{d_i h_i} = (-1)^{\sum_{i=1}^{m} d_i h_i} = 1.$$

So in this case (6.3) holds, that is, $n$ is an Euler pseudoprime in base $b$.

The case $r = s - 1$ is pretty analogous and is left as an exercise for the reader (see Exercise A6.9). $\qquad \square$

The results of Propositions 6.1.18 and 6.1.19 are described by the following diagram, where PS (or SPS, EPS, respectively) stays for pseudoprime (strong pseudoprime, Euler pseudoprime, respectively):

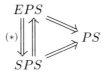

where $(*)$ means that the implication holds only if $n \equiv 3 \ (\mathrm{mod}\ 4)$.

In the light of the previous proposition we may expect strong pseudoprimes to be, in a sense, *even less numerous* than Euler pseudoprimes. This is true, although there still are many of them (see Exercise A6.8).

The first strong pseudoprime in base 2 is 2047. We give the table, originally in [46], with data about the number of pseudoprimes and strong pseudoprimes in base 2. Let $Ps(n)$ the number of pseudoprimes in base 2 smaller than $n$, and $Sps(n)$ the number of strong pseudoprimes in base 2 smaller than $n$. Then:

| $n$ | $Ps(n)$ | $Sps(n)$ |
|---|---|---|
| $10^3$ | 3 | 0 |
| $10^6$ | 245 | 46 |
| $10^9$ | 5597 | 1282 |
| $25 \cdot 10^9$ | 21853 | 4842 |

This table confirms the already mentioned fact that we expect strong pseudoprimes to be far less numerous than pseudoprimes.

This expectation is given a precise meaning by the following proposition, which will be proved in Section 6.2.2:

**Proposition 6.1.20.** *Let $n$ be a non-prime odd positive integer. The positive numbers $b < n$ such that $\mathrm{GCD}(b, n) = 1$ and $n$ is a strong pseudoprime in base $b$ are at most a quarter of all positive numbers $b < n$ such that $\mathrm{GCD}(b, n) = 1$.*

### 6.1.8 The Miller–Rabin probabilistic primality test

Proposition 6.1.20 is the foundation for the *Miller–Rabin probabilistic primality test* $\mathcal{T}$.

Fix an odd integer $n > 1$. Write $n = 2^s t + 1$ with odd $t$. The test $\mathcal{T}_1$ consists of the following steps:

1. choose randomly an integer $b_1$ with $1 < b_1 < n$, and compute $\mathrm{GCD}(b_1, n)$;
2. if $\mathrm{GCD}(b_1, n) > 1$ then $n$ is not prime, and the test is over;
3. if $\mathrm{GCD}(b_1, n) = 1$ compute $b_1^t$ modulo $n$. If $b_1^t \equiv \pm 1 \pmod{n}$, either $n$ is prime or it is a strong pseudoprime in base $b_1$;
4. if $b_1^t \not\equiv \pm 1 \pmod{n}$, compute $b_1^{2t}$ modulo $n$. If $b_1^{2t} \equiv -1 \pmod{n}$, then either $n$ is prime or it is a strong pseudoprime in base $b_1$;
5. if $b_1^{2t} \not\equiv -1 \pmod{n}$ proceed as above. If all the successive powers $b_1^{2^r t}$, for $r = 1, \ldots, s - 1$, are *never* congruent to $-1$ modulo $n$, then, by Remark 6.1.15, $n$ *is not* a prime. Otherwise, $n$ is a strong pseudoprime in base $b_1$.

The test $\mathcal{T}_m$ is defined recursively in a similar way:

1. choose randomly an integer $b_m$ such that $1 < b_m < n$ with $b_m$ different from $b_1, \ldots, b_{m-1}$, and compute $\mathrm{GCD}(b_m, n)$;
2. if $\mathrm{GCD}(b_m, n) > 1$ then $n$ is not prime, and the test is over;
3. if $\mathrm{GCD}(b_m, n) = 1$, compute $b_m^t$ modulo $n$ and proceed as from step (3) of $\mathcal{T}_1$ on. Either $n$ is found to be non-prime or $n$ is a strong pseudoprime with respect to $b_m$.

Keeping in mind Proposition 6.1.20, when we perform the test $\mathcal{T}_m$, the probability of $n$ not being a prime while being a strong pseudoprime in base $b_m$ is smaller than $1/4$. So, the probability of $n$ passing tests $\mathcal{T}_1, \ldots, \mathcal{T}_m$ without being a prime is smaller than $1/4^m$.

**Remark 6.1.21.** Miller–Rabin test is even better than Solovay–Strassen test. In practice, it is not necessary to take too large a value of $m$ to be almost certain that $n$ is prime if it passes tests $\mathcal{T}_1, \ldots, \mathcal{T}_m$. For instance, it has been verified that the only strong pseudoprime in base 2, 3, 5 and 7 smaller than $2.5 \cdot 10^{10}$ is $n = 3215031751$.

It is not difficult to verify that the complexity of Miller–Rabin algorithm is $\mathcal{O}(k \log^4 n)$, where $k$ is the number of times the test is repeated (see Exercise A6.11). Thus, if we fix $k$, we have a probabilistic primality test with polynomial complexity. So we might get the idea of using Miller–Rabin algorithm as a deterministic test, by trying out all the values of $b$ with $1 < b < n$, $\mathrm{GCD}(b, n) = 1$. Unfortunately, the complexity becomes $\mathcal{O}(n \log^4 n)$, that is exponential, and so it is computationally unfeasible.

In order to be able to use Miller–Rabin algorithm to get a polynomial deterministic test, we would have to *know* that if $n > 0$ is a non-prime odd number, then $n$ does not pass Miller–Rabin test for at least one value of $b$ and to have an efficient estimate for such a $b$, that is, it would be necessary to know that $b$ is *small* with respect to $n$, i.e., $b$ must be $\mathcal{O}(\log^h n)$ with $h$ a constant. It is conjectured that such an estimate actually exists and that $b$ is $\mathcal{O}(\log^2 n)$. This is a consequence of the so-called *generalised Riemann hypothesis*.

The classic *Riemann hypothesis* is a conjecture about the *zeros* of Riemann *zeta* function, which is obtained by *analytic continuation* to the complex field of the function $\zeta(s) = \sum_{n=0}^{\infty} 1/n^s$, defined for $s > 1$, considered in this book in Section 4.1.2 (see [59]).

The generalised Riemann hypothesis is an analogous conjecture about the zeros of similar, more general functions, called *Dirichlet L-series*. Unfortunately, neither the classic Riemann hypothesis and even less so the generalised one, have been proved. In fact, their solution is among the greatest open problems in mathematics.

## 6.2 Primitive roots

The notions discussed in this section will help, as we shall see, in dealing computationally with arithmetic modulo $n$. In order to do so, we shall investigate further the group $U(\mathbb{Z}_n)$ for an arbitrary positive integer $n$. This will enable us to prove some results already mentioned and used in the previous sections. In this section we shall often use basic properties of cyclic groups recalled in Exercises A3.23-A3.35, and the notion of Gaussian and its properties (see Exercises A4.30 and B4.20).

Notice that if $n$ is prime, $\mathbb{Z}_n$ is a field and the structure of $U(\mathbb{Z}_n)$ is made clear in Theorem 5.1.30: this group is cyclic of order $n - 1$. Recall further that, for all $n$, $U(\mathbb{Z}_n)$ has order $\varphi(n)$.

**Definition 6.2.1.** *Let $n, r$ be positive integers such that $\mathrm{GCD}(r, n) = 1$. Then $r$ is said to be a primitive root modulo $n$ if $\mathrm{Gss}(n, r) = \varphi(n)$, that is, if $U(\mathbb{Z}_n)$ is a cyclic group and if the residue class of $r$ modulo $n$ is a generator of the group. Then we also say that $r$ is a generator of $U(\mathbb{Z}_n)$.*

**Example 6.2.2.** As already recalled, there is always a primitive root modulo a prime number. As $\mathrm{Gss}(4, 3) = 2 = \varphi(4)$, we have that 3 is a primitive root modulo 4.

However, there are integers for which there are no primitive roots. For instance, there are no primitive roots modulo 8, as

$$\mathrm{Gss}(8, 1) = 1, \qquad \mathrm{Gss}(8, 3) = 2, \qquad \mathrm{Gss}(8, 5) = 2, \qquad \mathrm{Gss}(8, 7) = 2.$$

**Remark 6.2.3.** If $r$ is a primitive root modulo $n$, then there are $\varphi(\varphi(n))$ primitive roots modulo $n$ (see Exercise A3.30). They can be obtained by taking the powers $r^m$, with $\mathrm{GCD}(m, \varphi(n)) = 1$.

Knowing that a primitive root $r$ modulo $n$ exists and effectively finding it is important, as this makes arithmetic modulo $n$ much less computationally expensive. For instance, multiplying two elements of $U(\mathbb{Z}_n)$, both expressed as powers of $r$, just amounts to summing their exponents.

The following theorem gives a complete answer to the basic problem of finding all integers $n$ such that a primitive root modulo $n$ exists.

**Theorem 6.2.4.** *Let $n$ be a positive integer number. A primitive root modulo $n$ exists if and only if $n$ is one of the following numbers: $2, 4, p^h, 2p^h$, with $p$ an odd prime and $h$ a positive integer.*

The proof consists of several steps, and we shall give it by dividing it into lemmas. Clearly, cases $n = 2, 4$ are trivial.

The following three lemmas prove that the integers of the form $p^h$ and $2p^h$, with $p$ an odd prime and $h$ a positive integer, have a primitive root.

**Lemma 6.2.5.** *Let $p$ be an odd prime. If $r$ is a primitive root modulo $p$ and if $r$ is not a primitive root modulo $p^2$, then $r + p$ is a primitive root modulo $p^2$.*

PROOF. Set $m = \mathrm{Gss}(p^2, r)$. By the hypothesis, $m$ is a proper divisor of $\varphi(p^2) = p(p - 1)$. On the other hand, $r^m \equiv 1 \pmod{p^2}$ and so $r^m \equiv 1 \pmod{p}$; hence, $p - 1 = \mathrm{Gss}(p, r) \mid m$. So, $m = p - 1$.

Notice next that, as $r + p \equiv r \pmod{p}$, reasoning in the same way as above we find that either $\mathrm{Gss}(p^2, r + p) = p - 1$ or $\mathrm{Gss}(p^2, r + p) = p(p - 1)$. To conclude, we prove that the first case never happens.

We have

$$(r + p)^{p-1} = r^{p-1} + (p - 1)pr^{p-2} + \binom{p - 1}{2}p^2 r^{p-3} + \cdots,$$

so

$$(r + p)^{p-1} \equiv r^{p-1} - pr^{p-2} \equiv 1 - pr^{p-2} \pmod{p^2}.$$

If $\mathrm{Gss}(p^2, r + p) = p - 1$, then $pr^{p-2} \equiv 0 \pmod{p^2}$, which is false.    □

**Remark 6.2.6.** The previous lemma implies that it is possible to find a common primitive root of $p$ and $p^2$, for every odd prime $p$.

**Lemma 6.2.7.** *Let $p$ be an odd prime and let $h \geq 2$ be an integer. If $r$ is a primitive root modulo both $p$ and $p^2$, then $r$ is a primitive root modulo $p^h$ as well.*

PROOF. The lemma is trivially verified for $h = 2$. Then proceed by induction on $h$, with $h \geq 3$, assuming that $r$ is a primitive root modulo $p^\ell$ with $2 \leq \ell \leq h - 1$ and proving that it is so modulo $p^h$ as well.

Let $m = \mathrm{Gss}(p^h, r)$. We have $m \mid \varphi(p^h) = p^{h-1}(p - 1)$. Reasoning as in the proof of the previous lemma we also find $p - 1 \mid m$. So $m = p^k(p - 1)$, with $0 \leq k \leq h - 1$. To prove the claim, that is, $k = h - 1$, it is sufficient to prove that it is not the case that

$$r^{p^{h-2}(p-1)} \equiv 1 \pmod{p^h}, \tag{6.8}$$

but, on the contrary, one has

$$r^{p^{h-2}(p-1)} \not\equiv 1 \pmod{p^h}. \tag{6.9}$$

We proceed in a way pretty analogous to that of the proof of the previous lemma. Notice that, by the induction hypothesis,

$$r^{\varphi(p^\ell)} = r^{p^{\ell-1}(p-1)} \equiv 1 \pmod{p^\ell}, \tag{6.10}$$

while

$$r^{p^{\ell-2}(p-1)} \not\equiv 1 \pmod{p^\ell} \tag{6.11}$$

for all $\ell$ such that $2 \le \ell \le h-1$. In particular, taking $\ell = h-2$ in (6.10) we have

$$r^{p^{h-3}(p-1)} = 1 + dp^{h-2}, \tag{6.12}$$

while taking $\ell = h-1$ in (6.11) we see that $p \nmid d$. From (6.12) we find

$$r^{p^{h-2}(p-1)} = 1 + p \cdot dp^{h-2} + \binom{p}{2} d^2 p^{2h-4} + \cdots$$

and so (see Exercise A4.9)

$$r^{p^{h-2}(p-1)} \equiv 1 + dp^{h-1} \pmod{p^h}.$$

If (6.9) did not hold, and (6.8) held instead, we would have

$$dp^{h-1} \equiv 0 \pmod{p^h}$$

and so $p \mid d$, yielding a contradiction.  □

**Lemma 6.2.8.** *Let $n = 2p^h$ with $p$ an odd prime and $h$ a positive integer. Then there exists a primitive root modulo $n$.*

PROOF. Let $r$ be a primitive root modulo $p^h$. Notice that $\varphi(n) = \varphi(p^h)$. Let $d$ be a proper divisor of $\varphi(n)$. Then $r^d \not\equiv 1 \pmod{p^h}$. So we cannot have $r^d \not\equiv 1 \pmod{2p^h}$ either, proving that $\text{Gss}(n, r) = \varphi(n)$.  □

We study now the case of the numbers that are powers of two:

**Lemma 6.2.9.** *Let $m$ be an odd positive integer.*

(1) *For all integer $h \ge 3$:*

$$m^{2^{h-2}} = m^{\varphi(2^h)/2} \equiv 1 \pmod{2^h}$$

*so $2^h$ has no primitive roots.*

(2) *There are odd positive integers $m$ such that $\text{Gss}(2^h, m) = 2^{h-2} = \varphi(2^h)/2$; for instance, this is true for $m = 5$.*

PROOF. We prove the first part of the lemma. The claim is true for $h = 3$. Proceed by induction on $h$, with $h \ge 4$, assuming the claim for every integer $k$, with $3 \le k \le h-1$. By induction:

$$m^{2^{h-3}} = 1 + 2^{h-1}d.$$

Squaring:

$$m^{2^{h-2}} = 1 + 2^{2h-2}d^2 + 2^h d \equiv 1 \pmod{2^h}.$$

We prove now the second part of the lemma. We show that, for every integer $h \geq 3$, we have

$$5^{2^{h-3}} \equiv 1 + 2^{h-1} \not\equiv 1 \pmod{2^h}. \tag{6.13}$$

This relation is true for $h = 3$. Proceed by induction on $h$, with $h \geq 4$, assuming the claim for every integer $k$, with $3 \leq k \leq h - 1$. By induction:

$$5^{2^{h-4}} = 1 + 2^{h-2} + 2^{h-1}d,$$

so

$$5^{2^{h-3}} = 1 + 2^{2h-4} + 2^{2h-2}d^2 + 2^{h-1} + 2^h d + 2^{2h-2}d,$$

hence (6.13) follows.    □

**Remark 6.2.10.** The previous lemma completely describes the structure of the group $U(2^h)$, whose order is $\varphi(2^h) = 2^{h-1}$. It is cyclic if $h = 1, 2$. Otherwise, it is a direct product of a cyclic group of order $2^{h-2}$ and of a cyclic group of order 2. Indeed, if $h \geq 3$, in $U(2^h)$ there is the element $-1$ which has order 2, and 5 which has order $2^{h-2}$, and the cyclic groups $\langle 5 \rangle$ and $\langle -1 \rangle$, clearly, intersect only in 1 (see Exercise A6.12), so their direct product is included in $U(\mathbb{Z}_p)$ and is an abelian group of order $2^{h-1}$, hence, it is the whole $U(2^h)$.

To conclude the proof of the Theorem 6.2.4 we prove the following lemma.

**Lemma 6.2.11.** *If the positive integer $n \geq 5$ is not a prime power or a product of 2 and a prime power, then there are no primitive roots modulo $n$.*

PROOF. Let $n = p_1^{h_1} \cdots p_t^{h_t}$ be the prime decomposition of $n$. Assume $r$ to be a primitive root modulo $n$.

As $\mathrm{GCD}(n, r) = 1$, then also $\mathrm{GCD}(p_i, r) = 1$, so

$$r^{\varphi(p_i^{h_i})} \equiv 1 \pmod{p_i^{h_i}}$$

for all $i = 1, \ldots, t$. Set $s = \mathrm{lcm}(\varphi(p_1^{h_1}), \ldots, \varphi(p_t^{h_t}))$ and notice that $s \mid \varphi(p_1^{h_1}) \cdots \varphi(p_t^{h_t}) = \varphi(n)$.

On the other hand,

$$r^s \equiv 1 \pmod{p_i^{h_i}},$$

so

$$r^s \equiv 1 \pmod{n}$$

hence $\varphi(n) = \mathrm{Gss}(n, r) \mid s$. In conclusion, $\mathrm{lcm}(\varphi(p_1^{h_1}), \ldots, \varphi(p_t^{h_t})) = \varphi(p_1^{h_1}) \cdots \varphi(p_t^{h_t})$, which implies that the numbers $\varphi(p_i^{h_i}) = p_i^{h_i-1}(p_i - 1)$ are relatively prime (see Exercise A4.1). This yields that in the factorisation of $n$ no two odd primes can appear. Moreover, if one odd prime appears, then 2 cannot have an exponent greater than 1.    □

We still have to determine the structure of the group $U(\mathbb{Z}_n)$ if it is not cyclic, that is, if $n$ is different from a number of the form $2, 4, p^h, 2p^h$, with $p$ an odd prime and $h$ a positive integer. This is achieved by the following theorem.

**Theorem 6.2.12.** *Let $n$ be a positive integer and let $n = p_1^{h_1} \cdots p_m^{h_m}$ be its prime decomposition. Then the group $U(\mathbb{Z}_n)$ is the direct product of the groups $U(\mathbb{Z}_{p_i^{h_i}})$, $i = 1, \ldots, m$.*

PROOF. From Proposition 3.4.5 follows that the obvious map

$$U(\mathbb{Z}_n) \to U(\mathbb{Z}_{p_1^{h_1}}) \times \cdots \times U(\mathbb{Z}_{p_m^{h_m}})$$

is a group isomorphism. □

Here follows a new definition.

**Definition 6.2.13.** *Let $n$ be a positive integer. A positive integer $s$ is said to be a universal exponent for $n$ if for every positive integer $a < n$ one has $a^s \equiv 1 \pmod{n}$.*

Notice that for all positive integer $n$ there is some universal exponent. For instance, by Euler's Theorem 3.3.11, $\varphi(n)$ is such an exponent. So it makes sense to look for the *least universal exponent* $u(n)$ for $n$. Clearly, $u(n)$ is the least common multiple of the orders of the elements of $U(\mathbb{Z}_n)$ (see Exercise A6.13). Hence:

**Proposition 6.2.14.** *Let $n$ be a positive integer and let $n = 2^h p_1^{h_1} \cdots p_m^{h_m}$ be its prime decomposition. Then*

$$u(n) = \mathrm{lcm}(u(2^h), \varphi(p_1^{h_1}), \ldots, \varphi(p_m^{h_m})),$$

*where $u(2) = 1, u(4) = 2, u(2^h) = 2^{h-2}$, if $h \geq 3$. Moreover, there are elements of $U(\mathbb{Z}_n)$ of order $u(n)$, that is to say, there exist numbers $a$, relatively prime with $n$ such that $\mathrm{Gss}(n, a) = u(n)$.*

PROOF. It is clear that

$$u(n) = \mathrm{lcm}(u(2^h), u(p_1^{h_1}), \ldots, u(p_m^{h_m})).$$

As $U(\mathbb{Z}_{p_i^{h_i}})$ is cyclic, obviously $u(p_i^{h_i}) = \varphi(p_i^{h_i})$, $i = 1, \ldots, m$. The claims about $u(2^h)$ are an immediate consequence of what has been said in Remark 6.2.10.

Choose now $2 = p_0$, $h = h_0$. Identify $U(\mathbb{Z}_n)$ with $U(\mathbb{Z}_{p_0^{h_0}}) \times \cdots \times U(\mathbb{Z}_{p_m^{h_m}})$ by the isomorphism described in the proof of Proposition 6.2.12. Notice that in every group $U(\mathbb{Z}_{p_i^{h_i}})$ there is some element $x_i$ of order $u(p_i^{h_i})$. Clearly, $x = (x_0, \ldots, x_m)$ has order equal to $\mathrm{lcm}(u(2^h), u(p_1^{h_1}), \ldots, u(p_m^{h_m})) = u(n)$. □

Now we are able to complete the:

PROOF OF PROPOSITION 6.1.6. Proof of (a). Assume $n$ to be a Carmichael number. Let $a$ be a number relatively prime with $n$ such that $\mathrm{Gss}(n, a) = u(n)$. As $n$ is a Carmichael number, we have $a^{n-1} \equiv 1 \pmod{n}$, so $u(n) \mid n - 1$.

Let $n = p_1^{h_1} \cdots p_m^{h_m}$ be the prime decomposition of $n$. As $n$ is odd, $p_1, \ldots, p_m$ are odd primes. Moreover, $p_i^{h_i-1}(p_i - 1) \mid u(n) \mid n - 1$, $i = 1, \ldots, m$. This implies that $h_i = 1, i = 1, \ldots, m$, or else we would find an $i = 1, \ldots, m$ such that $p_i \mid n - 1$, yielding a contradiction. This proves (a).

The proof of the part of (b) not already proved is pretty analogous (see Exercise A6.14). □

### 6.2.1 Primitive roots and index

As we already mentioned, if $n$ has a primitive root, arithmetic modulo $n$ is computationally much easier. Let us see why.

Let $r$ be a primitive root modulo $n$. Then $U(\mathbb{Z}_n) = \{1, r, r^2, \ldots, r^{\varphi(n)-1}\}$.

**Definition 6.2.15.** *Let $m$ be an integer relatively prime with $n$, and assume that $n$ has a primitive root $r$. The least non-negative integer $h$ such that $r^h \equiv m$ (mod $n$) is said to be the* index *of $m$ with respect to $r$ and is denoted by* $\operatorname{ind}_r m$.

**Example 6.2.16.** As we already know, a primitive root modulo 14 exists. The number $r = 5$ is such a root. Indeed, $\varphi(14) = 6$ and $5^2 = 11 \equiv -3$ (mod 14), $5^3 \equiv -1$ (mod 14), $5^4 \equiv -5$ (mod 14), $5^5 \equiv 3$ (mod 14), $5^6 \equiv 1$ (mod 14).

As $U(\mathbb{Z}_{14}) = \{1, 3, 5, 9, 11, 13\}$, we have

$$\operatorname{ind}_5 1 = 0, \quad \operatorname{ind}_5 3 = 5, \quad \operatorname{ind}_5 5 = 1, \quad \operatorname{ind}_5 9 = 4, \quad \operatorname{ind}_5 11 = 2, \quad \operatorname{ind}_5 13 = 3.$$

The proof of the following lemma is easy and is left as an exercise for the reader (see Exercise A6.15):

**Lemma 6.2.17.** *Let $r$ be a primitive root modulo $n$. Then:*

(1) $\operatorname{ind}_r 1 = 0$;
(2) *if $m$ is an integer relatively prime with $n$, then $r^h \equiv m$ (mod $n$) if and only if $h \equiv \operatorname{ind}_r m$ (mod $\varphi(n)$);*
(3) *if $m$ is an integer relatively prime with $n$ and if $h$ is an arbitrary integer, then $\operatorname{ind}_r m^h \equiv h \operatorname{ind}_r m$ (mod $\varphi(n)$);*
(4) *if $m, m'$ are integers relatively prime with $n$, then $\operatorname{ind}_r mm' \equiv \operatorname{ind}_r m + \operatorname{ind}_r m'$ (mod $\varphi(n)$).*

Knowing the index makes it easier to solve several problems about congruences. For instance, suppose we want to solve a congruence of the form $ax^h \equiv b$ (mod $n$). If $r$ is a primitive root modulo $n$, the congruence may be rephrased as

$$\operatorname{ind}_r a + h \operatorname{ind}_r x \equiv \operatorname{ind}_r b \pmod{\varphi(n)}$$

which we can solve.

**Example 6.2.18.** Suppose we want to solve the equation $3x^5 \equiv 11$ (mod 14). Keeping in mind Example 6.2.16, if $y = \operatorname{ind}_5 x$, then the congruence becomes $5 + 5y \equiv 2$ (mod 6), which has solution $y \equiv 3$ (mod 6). Thus, the solution modulo 14 of the originary equation is $5^3 = -1$.

The following definition is especially interesting.

**Definition 6.2.19.** *Let $n$ be a positive integer, $m$ an integer relatively prime with $n$, and $h$ a positive integer. The number $m$ is said to be an $h$-ple residue* modulo $n$, *if the equation*

$$x^h \equiv m \pmod{n}. \tag{6.14}$$

*has a solution.*

We have the following result.

**Proposition 6.2.20.** *Let $n$ be a positive integer, $m$ an integer relatively prime with $n$, and $h$ a positive integer. Assume $n$ has a primitive root $r$. Let $d = \text{GCD}(h, \varphi(n))$. Then $m$ is an $h$-ple residue modulo $n$ if and only if*

$$m^{\varphi(n)/d} \equiv 1 \pmod{n}. \tag{6.15}$$

*In this case Equation (6.14) admits exactly $d$ solutions not congruent to each other modulo $n$.*

PROOF. Equation (6.14) can be rephrased as the equation

$$h \operatorname{ind}_r x \equiv \operatorname{ind}_r m \pmod{\varphi(n)}.$$

This equation has solutions, and if so it has $d$ solutions not congruent to each other modulo $\varphi(n)$, yielding $d$ distinct solutions of (6.14) modulo $n$, if and only if $d \mid \operatorname{ind}_r m$, that is, if and only if $\operatorname{ind}_r m = da$ (see Proposition 3.3.4). If this happens, then

$$m^{\varphi(n)/d} \equiv r^{a\varphi(n)} \equiv 1 \pmod{n}.$$

Conversely, if (6.15) holds, by Lemma 6.2.17 we have

$$\operatorname{ind}_r m \cdot \frac{\varphi(n)}{d} \equiv 0 \pmod{\varphi(n)}$$

and so $d \mid \operatorname{ind}_r m$.    □

**Example 6.2.21.** For instance, we ask: is 9 a 4-ple residue modulo 14? We have $\text{GCD}(4, \varphi(14)) = \text{GCD}(4, 6) = 2$. As $9^3 \equiv (-5)^3 \equiv 1 \pmod{14}$ the answer is affirmative and the equation $x^4 \equiv 9 \pmod{14}$ has 2 non-congruent solutions modulo 14. They are found by solving the equation $4 \operatorname{ind}_5 x \equiv 4 \pmod{6}$. It gives $\operatorname{ind}_5 x = 1, 4$ and so $x = 5, 9$ modulo 14.

**Corollary 6.2.22.** *Let $n$ and $h$ be positive integers. Assume $n$ to have a primitive root $r$. Let $d = \text{GCD}(h, \varphi(n))$. Then the equation $x^h \equiv 1 \pmod{n}$ admits exactly $d$ solutions not congruent to each other modulo $n$.*

*Moreover, if $\varphi(n)/d$ is even, and only in this case, the equation $x^h \equiv -1 \pmod{n}$ also admits exactly $d$ solutions not congruent to each other modulo $n$.*

### 6.2.2 More about the Miller–Rabin test

In this section we give the proof of Proposition 6.1.20, which is the key for Miller–Rabin test we have discussed in Section 6.1.8.

PROOF OF PROPOSITION 6.1.20. Notice that if $b$ is a positive integer number such that $n$ is a strong pseudoprime in base $b$ then $n$ is a pseudoprime too in base $b$; hence such values of $b$ are solutions of the equation

$$x^{n-1} \equiv 1 \pmod{n}. \tag{6.16}$$

Let $n = p_1^{h_1} \cdots p_m^{h_m}$ be the factorisation of $n$. By Corollary 6.2.22, the equation

$$x^{n-1} \equiv 1 \pmod{p_i^{h_i}}, \quad i = 1, \ldots, m,$$

has $d_i = \text{GCD}(n-1, p_i^{h_i-1}(p_i-1)) = \text{GCD}(n-1, p_i-1)$ solutions. By the Chinese remainder theorem 3.4.2, Equation (6.16) has $d = \prod_{i=1}^{m} d_i$ solutions.

We examine first the case in which $h_1 > 1$. Set $p_1 = p$ and $h_1 = h$. We have

$$\frac{p^h}{p-1} \geq \frac{p^2}{p-1} = \frac{p^2-1+1}{p-1} = p+1+\frac{1}{p-1} > 4.$$

Then

$$d = \prod_{i=1}^{m} d_i \leq \prod_{i=1}^{m}(p_i-1) < \frac{1}{4}p^h \prod_{i=2}^{m} p_i \leq \frac{n}{4}$$

and this proves the theorem in this case.

So we may assume that $h_i = 1$ for all $i = 1, \ldots, m$. Set

$$n-1 = 2^s t, \qquad p_i - 1 = 2^{s_i} t_i, \quad i = 1, \ldots, m,$$

with $t, t_i$ odd for all $i = 1, \ldots, m$ and $s_1 \leq s_2 \leq \cdots \leq s_m$. Set next

$$\text{GCD}(n-1, p_i-1) = 2^{\sigma_i} \tau_i, \quad i = 1, \ldots, m,$$

where

$$\sigma_i = \min\{s, s_i\}, \qquad \tau_i = \text{GCD}(t, t_i).$$

By Corollary 6.2.22, the equation

$$x^t \equiv 1 \pmod{p_i}, \quad i = 1, \ldots, m,$$

has $\text{GCD}(t, p_i-1) = \text{GCD}(t, t_i) = \tau_i$ solutions. Consider the equation

$$x^{2^r t} \equiv -1 \pmod{p_i}, \quad i = 1, \ldots, m, \tag{6.17}$$

with $r < s$ a non-negative integer. We have

$$\delta_i = \text{GCD}(2^r t, p_i-1) = \text{GCD}(2^r t, 2^{s_i} t_i) = 2^{\sigma_i'} \tau_i, \qquad \text{with } \sigma_i' = \min\{r, s_i\}.$$

Notice that $\varphi(p_i)/\delta_i = (2^{s_i} t_i)/(2^{\sigma_i'} \tau_i)$, and this number is even if and only if $s_i > \sigma_i'$, that is, if and only if $r < s_i$. In this case $\sigma_i' = r$, and Equation (6.17) has exactly $\delta_i = 2^r \tau_i$ solutions (see Corollary 6.2.22), otherwise it has none.

By applying the Chinese remainder theorem we find that the equation

$$x^t \equiv 1 \pmod{n}$$

has $\tau = \tau_1 \cdots \tau_m$ solutions, while the equation

$$x^{2^r t} \equiv -1 \pmod{n},$$

with $r < s$ a non-negative integer, has $\delta = \delta_1 \cdots \delta_m = 2^{mr} \tau$ solutions if $r \leq s_1 - 1 \leq s_i - 1, i = 1, \ldots, m$, otherwise it has none.

In conclusion, the number of integers $b < n$ such that $\text{GCD}(b, n) = 1$ and that $n$ is a strong pseudoprime in base $b$ is

$$N(n) = \tau\left(1 + \sum_{r=1}^{s_1-1} 2^{mr}\right) = \tau\left(1 + \frac{2^{ms_1}-1}{2^m-1}\right).$$

Notice that

$$\varphi(n) = \prod_{i=1}^{m}(p_i - 1) = 2^{s_1 + \cdots + s_m}T,$$

where we set $T = t_1 \cdots t_m$. We have to prove that

$$N(n) \leq \frac{\varphi(n)}{4} \qquad (6.18)$$

and, as $\tau_i \leq t_i, i = 1, \ldots, m$, it suffices to prove that

$$\frac{1}{2^{s_1 + \cdots + s_m}}\left(1 + \frac{2^{ms_1} - 1}{2^m - 1}\right) \leq \frac{1}{4}. \qquad (6.19)$$

As $s_1 \leq s_2 \leq \cdots \leq s_m$, we have

$$\frac{1}{2^{s_1 + \cdots + s_m}}\left(1 + \frac{2^{ms_1} - 1}{2^m - 1}\right) \leq \frac{1}{2^{ms_1}}\left(1 + \frac{2^{ms_1} - 1}{2^m - 1}\right) =$$

$$= \frac{1}{2^{ms_1}} + \frac{1}{2^m - 1} - \frac{1}{2^{ms_1}(2^m - 1)} = \frac{1}{2^m - 1} + \frac{2^m - 2}{2^{ms_1}(2^m - 1)} \leq \frac{2}{2^m - 1}.$$

If $m \geq 4$, then $2/(2^m - 1) \leq 1/4$ and this proves Equation (6.19) in this case. If $m = 3$, then $1/(2^m - 1) + (2^m - 2)/(2^{ms_1}(2^m - 1)) \leq 1/4$, and Equation (6.19) is proved again.

The last case we have to examine is $m = 2$. Here we have

$$\frac{1}{2^{s_1 + s_2}}\left(1 + \frac{2^{2s_1} - 1}{3}\right) = \frac{1}{2^{2s_1 + s_2 - s_1}}\left(1 + \frac{2^{2s_1} - 1}{3}\right) = \frac{1}{2^{s_2 - s_1}}\left(\frac{1}{3} + \frac{1}{3 \cdot 2^{2s_1 - 1}}\right).$$

If $s_2 > s_1$ the last number is smaller than $1/4$ and so Equation (6.19) is proved in this case. Assume finally that $m = 2$, $s_1 = s_2 = \sigma$. It follows from $n = p_1 p_2 = (2^\sigma t_1 + 1)(2^\sigma t_2 + 1)$ that $s \geq \sigma$.

Let $p_1 > p_2$. If $\tau_1 = t_1$ then we would have $p_1 - 1 \mid n - 1$ and $1 \equiv n \equiv p_1 p_2 \equiv p_2$ (mod $p_1 - 1$), which is a contradiction. Then we have $\tau_1 < t_1$ and, as $t_1$ is an odd number, we have $\tau_1 \leq t_1/3$. Thus, $\tau \leq T/3$. Finally, to prove Equation (6.18), notice that

$$N(n) = \tau\left(1 + \frac{2^{2\sigma} - 1}{3}\right) \leq \frac{T 2^{2\sigma}}{6} = \frac{\varphi(n)}{6}. \qquad \square$$

## 6.3 A polynomial deterministic primality test

We describe in this section a recent primality test, due to M. Agrawal, N. Kayal, and N. Saxena [2], and thus called "AKS test". In order to decide whether a number is prime we have described, in the previous sections, several probabilistic algorithms. The reason why those algorithms have been successful is that deterministic algorithms that where known until recently, for instance the sieve of Eratosthenes (see Section 4.1.3), or the tests based on Fermat's little theorem (see Section 4.2.2) or on Wilson's theorem (see Section 4.2.3), have exponential complexity and so are unfeasible for large numbers. The AKS test, on the other hand, is *deterministic* and has *polynomial* complexity. When it was found, it immediately aroused a great interest, as it was all but certain that such an algorithm existed, even if, of course, nobody had proved that it *did not* exist.

The AKS algorithm relies on the following polynomial version of Fermat's little theorem.

**Proposition 6.3.1.** *Let* $a \in \mathbb{Z}$, $n \in \mathbb{N}$, $n \geq 2$ *be numbers such that* $\mathrm{GCD}(a, n) = 1$. *Then* $n$ *is prime if and only if*

$$(x + a)^n \equiv x^n + a \pmod{n}. \tag{6.20}$$

PROOF. Recall that the coefficient of $x^i$ in $(x + a)^n - (x^n + a)$ for $0 < i < n$, is $\binom{n}{i} a^{n-i}$ and, if $n$ is prime, $\binom{n}{i} \equiv 0 \pmod{n}$ (see Exercise A4.9). So, by Fermat's little theorem, (6.20) holds.

Assume now $n$ to be a composite number and $q$ to be a prime dividing $n$. If $\alpha$ is the greatest positive integer such that $q^\alpha \mid n$, then $q^\alpha$ does not divide $\binom{n}{q}$ (see Exercise A6.30) and is relatively prime with $a^{n-q}$. So the coefficient of $x^q$ is not zero modulo $n$, and so $(x+a)^n - (x^n + a)$ is not the zero polynomial in $\mathbb{Z}_n[x]$, that is to say, Equation (6.20) does not hold.    □

The identity (6.20) characterises prime numbers and so it gives a primality test: to verify whether a number $n$ is prime it suffices to choose an $a \in \mathbb{Z}$ such that $\mathrm{GCD}(a, n) = 1$ and to check whether Equation (6.20) holds. However, this test has exponential complexity, as there are $n + 1$ coefficients in the polynomial $(x + a)^n$, and computing all of them may be quite burdensome (see Exercise A2.24).

An idea to make the algorithm more efficient consists in trying to reduce the number of coefficients to be computed. A way to do so is to evaluate both sides of (6.20) modulo a suitable polynomial. For instance, we could choose a polynomial of the form $x^r - 1$ for a suitable, *not too large*, positive integer $r$. In other words, we require for Equation (6.20) to be verified in the quotient ring $\mathbb{Z}_n[x]/(x^r - 1)$, rather that in $\mathbb{Z}_n[x]$: that is, we require that the following equation holds:

$$(x + a)^n = x^n + a \quad \text{in } \mathbb{Z}_n[x]/(x^r - 1). \tag{6.21}$$

Clearly, if $n$ is prime, (6.20) holds, and so (6.21) holds as well. But is the converse true? That is to say, is Equation (6.21) a *sufficient* condition too, and not only a necessary one, for $n$ to be prime? Unfortunately, the answer to this question is negative, as the following simple example shows.

**Example 6.3.2.** We have

$$(x + 3)^4 \equiv x^4 + 3 \quad \text{in } \mathbb{Z}_4[x]/(x^2 - 1). \tag{6.22}$$

Indeed, in $\mathbb{Z}_4[x]$ we have

$$(x + 3)^4 = (x - 1)^4 = x^4 - 4x^3 + 6x^2 - 4x + 1 = x^4 + 2x^2 + 1,$$

so in $\mathbb{Z}_4[x]/(x^2 - 1)$, (6.22) clearly holds. Thus, Equation (6.21) is verified even though 4 is not a prime.

If we intend to carry on our idea of working in $\mathbb{Z}_n[x]/(x^r - 1)$ rather than in $\mathbb{Z}_n[x]$, we have to extend *in a suitable way* Equation (6.20). Luckily, the right extension exists: it involves proving that there is an $r$, not too large with respect to $n$, such that, if (6.21) is verified for *a sufficient number* of $a$s, then $n$ is prime. In order to obtain a polynomial algorithm we have to show that both $r$ and the number of $a$s for which (6.21) must be verified for $n$ to be deemed a prime, are bounded by a *polynomial in* $\log n$.

We show now in detail how it is possible to prove the above. We begin by the following lemma.

**Lemma 6.3.3.** *Let $n$ be a positive integer. There exists an integer $r \leq [16 \log_2^5 n] + 1$ such that $\mathrm{Gss}(r, n) > 4 \log_2^2 n$.*

PROOF. Let $r_1, \ldots, r_t$ be the positive numbers such that $\mathrm{Gss}(r_i, n) \leq 4 \log_2^2 n$ for all $i = 1, \ldots, t$. This means that for all $r_i$, $i = 1, \ldots, t$, we have

$$n^k \equiv 1 \pmod{r_i}, \qquad \text{for some } k = 1, \ldots, [4 \log_2^2 n],$$

that is, every $r_i$, $i = 1, \ldots, t$, divides the product

$$\prod_{i=1}^{[4 \log_2^2 n]} (n^i - 1) < n^{16 \log_2^4 n} \leq 2^{16 \log_2^5 n}. \tag{6.23}$$

If we fix a positive integer $m$, it is easy to prove that, for all $m \geq 7$, we have

$$\mathrm{lcm}(1, 2, \ldots, m) \geq 2^m$$

(see Exercise A6.22). Set $M = [16 \log_2^5 n] + 1$. So, keeping in mind (6.23), we have

$$\mathrm{lcm}(r_1, \ldots, r_t) < 2^M \leq \mathrm{lcm}(1, 2, \ldots, M).$$

If the claim were not true, it would mean that the set $\{r_1, \ldots, r_t\}$ includes the set $\{1, \ldots, M\}$ and so we would have

$$\mathrm{lcm}(1, 2, \ldots, M) \leq \mathrm{lcm}(r_1, \ldots, r_t) < \mathrm{lcm}(1, 2, \ldots, M),$$

a contradiction.                                                                      □

In a short while the following remark will be useful.

**Remark 6.3.4.** Let $r$ be a positive integer and $p$ a prime number not dividing $r$. Consider the polynomial $x^r - 1$ over $\mathbb{Z}_p$ whose roots, in its splitting field, are all $r$th roots of unity which, as we know, form a cyclic group $R_r$ of order $r$ (see Theorem 5.1.16). So there exists $\xi \in R_r$ of order $r$. Consider its minimal polynomial $h(x)$ in $\mathbb{Z}_p[x]$ and let $d$ be its degree. Then $\mathbb{Z}_p[x]/(h(x)) = \mathbb{Z}_p(\xi)$, and this field $F$ has order $p^d$.

Notice that as $F$ contains $\xi$, it also contains all the powers of $\xi$ and so it contains $R_r$; thus $F$ is exactly the splitting field of $x^r - 1$ over $\mathbb{Z}_p$.

We prove now that $d = \mathrm{Gss}(r, p)$, even though this is not strictly necessary for what follows. Set $\mathrm{Gss}(r, p) = n$; then $p^n \equiv 1 \pmod r$. As $h(x)$ divides $x^r - 1$, keeping in mind Exercise A4.36, we have that $h(x)$ divides $x^{p^n - 1} - 1$. So, for all $\eta \in F$ we have $\eta^{p^n - 1} = 1$. In particular, this holds for a generator of $F^*$ (see Theorem 5.1.30). It follows that $p^d - 1 \mid p^n - 1$, and so $d \le n$.

On the other hand, there are in $F$ elements of period $r$: for instance, $\xi$. Thus, $r \mid p^d - 1$, that is, $p^d \equiv 1 \pmod r$, and so $n = \mathrm{Gss}(r, p) \mid d$. Hence $d = n$.

The following terminology will be useful.

**Definition 6.3.5.** *Let $p$ be a prime number and $r$ a positive integer. Given a polynomial $f(x) \in \mathbb{Z}[x]$ and a positive integer $m$, $m$ is said to be* introspective *for $f(x)$ modulo $p$, with respect to $r$, or simply* introspective, *if no confusion is possible, if*

$$(f(x))^m = f(x^m) \quad in \ \mathbb{Z}_p[x]/(x^r - 1).$$

Clearly, a prime $p$ is introspective modulo $p$ itself with respect to any positive integer $r$, due to the freshman's dream (see Exercise A4.10). Moreover, it is not difficult to prove that the product of numbers that are introspective for a polynomial $f(x)$ is still introspective for the same polynomial, and that if a number is introspective for two polynomials then it is introspective for their sum and their product (see Exercise A6.23).

We may now prove the following result, which is fundamental for the AKS algorithm.

**Theorem 6.3.6.** *Let $n > 1$ be an integer that is not a prime power with exponent greater than 1. Let $r < n$ be a positive integer relatively prime with $n$ such that*

$$r \le [16 \log_2^5 n] + 1, \quad \mathrm{Gss}(r, n) > 4 \log_2^2 n.$$

*Assume further that $n$ has no prime factors smaller than or equal to $r$. Then $n$ is prime if and only if for all $a = 1, \ldots, [2\sqrt{\varphi(r)} \log_2 n]$ Equation (6.21) holds, that is,*

$$(x + a)^n = x^n + a \quad in \ \mathbb{Z}_n[x]/(x^r - 1).$$

PROOF. We have already seen that, by Proposition 6.3.1, if $n$ is prime, then (6.21) holds *for all* $a$. So we only have to prove that the condition given in Theorem 6.3.6 is *sufficient* to ensure that $n$ is prime.

Assume by contradiction that $p$ is a prime divisor of $n$: so $p < n/2$ and $p > r$, and so $\mathrm{GCD}(p, r) = 1$.

Set

$$l := [2\sqrt{\varphi(r)} \log_2 n]$$

and notice that

$$l < r - 1. \tag{6.24}$$

Indeed, if formula (6.24) did not hold, we would have $r - 1 \leq 2\sqrt{\varphi(r)} \log_2 n \leq 2\sqrt{r-1} \log_2 n$, and so $\sqrt{r-1} \leq 2 \log_2 n$. Hence $r - 1 \leq 4 \log_2^2 n < \mathrm{Gss}(n,r)$, which is not possible.

By hypothesis, Equation (6.21) is verified for all $a = 1, \ldots, l$. As $p \mid n$, the following relations are verified too:

$$(x + a)^n = x^n + a \quad \text{in } \mathbb{Z}_p[x]/(x^r - 1), \quad \text{for all } a = 1, \ldots, l, \qquad (6.25)$$

that is, $n$ is introspective, modulo $p$ with respect to $r$, for the $l$ polynomials $x + 1, x + 2, \ldots, x + l$.

Consider the following two sets:

$$I := \{n^i \cdot p^j \mid i, j \geq 0\}, \qquad P := \left\{ \prod_{a=1}^{l} (x + a)^{i_a} \mid i_a \geq 0 \right\}.$$

Then *every* number in the set $I$ is introspective for *every* polynomial in the set $P$. Now let us define two groups $G$ and $Q$ connected with the sets $I$ and $P$. Define

$$G := \{x \bmod r \mid x \in I\}.$$

Recalling that $\mathrm{GCD}(n, r) = \mathrm{GCD}(p, r) = 1$, $G$ is the subgroup of $U(\mathbb{Z}_r)$ generated by $n$ and $p$ modulo $r$, that is, $G$ consists of the residue classes modulo $r$ of the elements of $I$. If $|G| = t$, we have

$$\varphi(r) \geq t \geq \mathrm{Gss}(n, r) > 4 \log_2^2 n, \qquad t \geq \mathrm{Gss}(p, r). \qquad (6.26)$$

Consider now the field $F = \mathbb{Z}_p(\xi) = \mathbb{Z}_p[x]/(h(x))$, as in Remark 6.3.4. Denote by $Q$ the subgroup of the multiplicative group $F^*$ generated by the classes of the polynomials $x + 1, x + 2, \ldots, x + l$, that is, the subgroup of $F^*$ generated by the classes in $F = \mathbb{Z}_p[x]/(h(x))$ of the polynomials of the set $P$. Set $q = |Q|$.

Notice that $Q$, as $F^*$ of which it is a subgroup, is a cyclic group, and so it has a generator $\gamma$, which is the class, modulo $p$ and modulo $h(x)$, of a polynomial $G(x) \in \mathbb{Z}[x]$. It is important to remark that

$$m, m' \text{ introspective for } G(x), \ m \equiv m' \pmod{r} \implies m \equiv m' \pmod{q}. \quad (6.27)$$

Indeed, if $m' = m + rv$, we have

$$(G(x))^{m+rv} = G(x^{m+rv}) \quad \text{in } \mathbb{Z}_p[x]/(x^r - 1),$$

hence

$$(G(x))^m (G(x))^{rv} = G(x^m) = (G(x))^m \quad \text{in } \mathbb{Z}_p[x]/(x^r - 1).$$

As $h(x)$ divides $x^r - 1$, we also have

$$(G(x))^m (G(x))^{rv} = (G(x))^m \quad \text{in } F$$

and from here it can be deduced that $\gamma^{rv} = 1$ in $F^*$, and so $q \mid rv = m - m'$.

We shall prove that $q$ verifies simultaneously the following two inequalities:

$$q \geq \binom{t+l-2}{t-1},$$ (6.28)

$$q < \frac{1}{4}n^{2\sqrt{t}}.$$ (6.29)

The formulas (6.28) and (6.29) will be proved later; we see how to conclude the proof of the theorem first. Set $s = [2\sqrt{t}\log_2 n]$. From (6.26) we have $t - 1 \geq s$, and, by (6.28), we have

$$q \geq \binom{s+l-1}{s}.$$

Notice now that $l \geq 2\sqrt{\varphi(r)}\log_2 n \geq 2\sqrt{t}\log_2 n$, which implies that $l \geq s$ and so we have

$$q \geq \binom{2s-1}{s}.$$

On the other hand, for $s \geq 3$, we have

$$\binom{2s-1}{s} = \frac{2s-1}{s-1} \cdot \frac{2s-2}{s-2} \cdot \cdots \cdot \frac{s+2}{2} \cdot (s+1) \geq 2^{s-1}$$

and so

$$q \geq 2^{s-1} \geq 2^{2\sqrt{t}\log_2 n-2} = \frac{1}{4}n^{2\sqrt{t}},$$

contradicting (6.29): the contradiction is a consequence of assuming $n$ not to be prime.

We only have to prove the two inequalities (6.28) and (6.29). To this end, consider the set $P'$ of polynomials in $P$ of the form $\prod_{a=1}^{l}(x+a)^{i_a}$ and whose degree is $\sum_{a=1}^{l} i_a \leq t - 1$. Notice that their number is exactly $\binom{t+l-2}{t-1}$ (see Exercise A1.24). We claim that their classes in $F^*$ are all distinct. From this fact the formula in (6.28) immediately follows.

To prove the above claim, first of all notice that the polynomials in $P'$ are all distinct modulo $p$, that is, in $\mathbb{Z}_p[x]$. This follows from the fact that $x+1, x+2, \ldots, x+l$ are distinct as elements of $\mathbb{Z}_p[x]$, and this in turn is an immediate consequence of the fact that $l < p$, or else by (6.24) we would have $p \leq l < r - 1$, against the hypothesis $p > r$.

So, let $f(x)$, $g(x)$ be two polynomials in $P'$ and assume that their classes in $F$ are not distinct. Set $k(x) = f(x) - g(x)$ and denote by $\bar{k}(x)$ the class of $k(x)$ in $\mathbb{Z}_p[x]$. Then, the polynomial $\bar{k}(x)$ is divisible by $h(x)$, and so it has as a root the $r$th primitive root of unity $\xi$, that is to say, we have

$$\bar{k}(\xi) = 0.$$

As all numbers in $I$ are introspective modulo $p$ with respect to $r$ both for $f(x)$ and $g(x)$, these numbers are introspective for $k(x)$ too, that is, we have

$$(\bar{k}(x))^{n^i \cdot p^j} = \bar{k}(x^{n^i \cdot p^j}) \quad \text{in} \quad \mathbb{Z}_p[x]/(x^r - 1)$$

for every pair of non-negative integers $i, j$. As $h(x)$ divides $x^r - 1$, we also have

$$(\bar{k}(x))^{n^i \cdot p^j} = \bar{k}(x^{n^i \cdot p^j}) \quad \text{in} \quad F = \mathbb{Z}_p[x]/(h(x))$$

for every pair of non-negative integers $i, j$. Consequently, we have

$$\bar{k}(\xi^{n^i \cdot p^j}) = \bar{k}(\xi)^{n^i \cdot p^j} = 0$$

for every pair of non-negative integers $i, j$. When $i$ and $j$ run over the non-negative integers the classes modulo $r$ of the numbers $n^i \cdot p^j$ are as many as the elements of $G$, that is, they are $t$; so the polynomial $\bar{k}(x) \in \mathbb{Z}_p[x]$ has as its $t$ distinct roots $\xi, \xi^2, \ldots, \xi^t$. As $\bar{k}(x)$ has degree smaller than $t$, we have that $\bar{k}(x)$ is the zero polynomial in $\mathbb{Z}_p[x]$. This means that $f(x)$ and $g(x)$ are equal modulo $p$, and as we have seen this implies $f(x) = g(x)$. This concludes the proof of (6.28).

Finally, we prove (6.29). Assume it not to be true, and so

$$q \geq \frac{1}{4} n^{2\sqrt{t}}. \tag{6.30}$$

Consider the elements of the set $I$ of the form $n^i p^j$ with $0 \leq i \leq \sqrt{t}$ and $0 \leq j \leq \sqrt{t}$. The number of such elements is $(\lfloor \sqrt{t} \rfloor + 1)^2 > t$. So in $G$ the classes of these elements are not all distinct. Then there are $i, i', j, j'$ with $0 \leq i, i', j, j' \leq \sqrt{t}$, $i \neq i'$ and $j \neq j'$, such that

$$n^i p^j \equiv n^{i'} p^{j'} \pmod{r}.$$

As $n^i p^j$ and $n^{i'} p^{j'}$ are introspective for the polynomials of $P$, they are also for the polynomial $G(x) \in \mathbb{Z}[x]$ whose class in $F$ generates $Q$; so (6.27) implies

$$n^i p^j \equiv n^{i'} p^{j'} \pmod{q}. \tag{6.31}$$

Suppose $i \geq i'$ and $j \geq j'$. We have $i, j \leq \sqrt{t}$ and so $n^i \leq n^{\sqrt{t}}, p^j \leq (n/2)^{\sqrt{t}}$; then, as $t > 4$ (see (6.26)), we have $1 \leq n^i p^j \leq 1/4 n^{2\sqrt{t}} \leq q$ and analogously $1 \leq n^{i'} p^{j'} \leq q$. Then Equation (6.31) implies $n^i p^j = n^{i'} p^{j'}$, thus $i = i'$, $j = j'$ against the assumptions.

Suppose now that $i \geq i'$ and $j \leq j'$. In this case we have

$$n^{i-i'} \equiv p^{j'-j} \pmod{q}.$$

Arguing as before, we see that this relation implies that $n^{i-i'} = p^{j'-j}$. But in this case $n$, if it were not a prime number, would be a power of $p$, yielding a contradiction. $\qquad \square$

Using these results we can now outline the AKS algorithm.

### The AKS($n$) algorithm

---

Input: an integer $n > 1$;
1. if ($n = m^k$, $m \in \mathbb{N}$, $k \in \mathbb{N}, k > 1$), then return $n$ COMPOSITE;
2. determine the least integer $r$ such that $\mathrm{Gss}(r, n) > 4\log_2^2 n$;
3. if $1 < \mathrm{GCD}(a, n) < n$ for some $a \leq r$, then return $n$ COMPOSITE;
4. if $n \leq r$, output $n$ PRIME;
5. if $(x + a)^n \neq x^n + a$ in $\mathbb{Z}_n[x]/(x^r - 1)$ $\forall\, a = 1, \ldots, \lfloor 2\sqrt{\varphi(r)} \log_2 n \rfloor$,
   then return $n$ COMPOSITE;
6. return $n$ PRIME.

---

Comments about the algorithm:

(1) if $n$ is a power of an integer with exponent greater than 1, that is, $n = m^k$, with $m, k \in \mathbb{N}$ and $k > 1$, this fact can be recognised in polynomial time (see Proposition 6.3.7 below) and clearly in this case $n$ is not prime;

(2) in Lemma 6.3.3 we proved that there is a $r \leq \lceil 16 \log^5 n \rceil$ such that $\mathrm{Gss}(r, n) > 4\log^2 n$;

(3) we are computing a greatest common divisor $r$ times: if one of these is strictly greater than 1 and strictly smaller than $n$, clearly $n$ is composite;

(4) if $n \leq r$ were composite, by the previous remark it would have be identified as such: in fact, in this case it would be sufficient to take as $a$ a non-trivial factor of $n$ and, as $n \leq r$, then $a \leq r$ and so we would have $1 < \mathrm{GCD}(a, n) = a < n$;

(5) see Theorem 6.3.6.

As promised, we prove the following result.

**Proposition 6.3.7.** *Let $n > 1$ be a positive integer. There is a polynomial algorithm of complexity $\mathcal{O}(\log^4 n)$ that recognises if there are positive integers $m, k$, with $k > 1$, such that $n = m^k$.*

PROOF. If there are positive integers $m, k$, with $k > 1$, such that $n = m^k$, then $k = \log_m n < \log_2 n$. So, if $h = \lfloor \log_2 n \rfloor$, we have $2 \leq k \leq h$. Therefore, for all such $k$, we have to verify whether the polynomial $f_k(x) = x^k - n$ has a positive integer root.

Notice that $f_k(x)$ is strictly increasing in the interval $[0, n]$ and that $f_k(0) = -n < 0$, while $f_k(n) = n^k - n > 0$, so $f_k(x)$ has exactly one root in the interval $[0, n]$. We have to verify whether this root is integer or not.

To this end we may proceed as follows. Evaluate the sign of $f_k(n/2)$: we have to compute $n^k - 2^k \cdot n$, which has complexity $\mathcal{O}(k \log n) = \mathcal{O}(\log^2 n)$. If it were the case that $f_k(n/2) = 0$, which does not happen in practice, we would have found the $k$th root of $n$ in $[0, n]$ and the procedure would be over. Assume now that $f_k(n/2)$ is positive, which is what happens in practice. Then evaluate $f_k(x)$ in the midpoint of the interval $[0, n/2]$, that is, in the point

$n/4$. To do so, we have to compute $n^k - 4^k \cdot n$, which has again complexity $\mathcal{O}(\log^2 n)$. If by chance $f_k(n/4) = 0$, we would be through, otherwise repeat this procedure and assume that, in the successive evaluations of $f_k(x)$ in the midpoints of the successive intervals at whose endpoints $f_k(x)$ has values of opposite signs, we never find the $k$th root of $n$ in $[0, n]$. After $h + 1$ steps we have then recursively defined a finite sequence $x_0, \ldots, x_{h+1}$ of points in the interval $[0, n]$ such that:

- for all $i = 1, \ldots, h + 1$, $f_k(x)$ has values with opposite signs in $x_{i-1}$ and in $x_i$;
- $|x_i - x_{i-1}| = n/2^i$.

The computational cost of constructing this sequence is polynomial, as it is $\mathcal{O}(h \log^2 n) = \mathcal{O}(\log^3 n)$.

In particular, we find the $k$th root of $n$ in $[0, n]$, which has to lie in the interval having as endpoints $x_h$ and $x_{h+1}$. This interval has length $n/2^{h+1} < 1$, so it includes at most one integer. If it includes no integers, then $n$ has no integer $k$th roots, otherwise if in this interval there is an integer $m$, by computing $m^k$ we verify at last whether the $k$th root of $n$ is an integer or not.

As we have to repeat this procedure for every $k$ such that $2 \leq k \leq h$, the complexity of the algorithm is $\mathcal{O}(h \log^3 n) = \mathcal{O}(\log^4 n)$.    □

The algorithm AKS gives an effective characterisation of prime numbers and has polynomial running time.

**Theorem 6.3.8.** *Let $n$ be an integer greater than 1. Then the algorithm AKS applied to $n$:*

(1) *outputs PRIME if and only if $n$ is a prime number;*
(2) *has polynomial running time.*

PROOF. The proof of the first claim is contained in the previous remarks.

We prove now that AKS applied to $n$ has polynomial running time. To this end, it suffices to prove that each of the steps of the algorithm is polynomial.

We shall use the following notation:

$$\mathcal{O}^{\sim}(f(n)) := \mathcal{O}(f(n) \cdot p(\log(f(n)))), \quad \text{for some polynomial } p \text{ in } \log(f(n))$$

In particular,

$$\mathcal{O}^{\sim}(\log^k n) = \mathcal{O}(\log^k n \cdot p(\log(\log n))) = \mathcal{O}(\log^{k+\epsilon} n) \quad \text{for all } \epsilon > 0. \quad (6.32)$$

*First step.* By Proposition 6.3.7 this step has complexity $\mathcal{O}(\log^4 n)$.

*Second step.* To verify that, for a *fixed* positive integer $r$, $\mathrm{Gss}(r, n) > 4 \log^2 n$, we have to check that $n^k \neq 1 \pmod{r}$ for all $k \leq 4 \log^2 n$: this requires computing at most $\mathcal{O}(\log^2 n)$ powers modulo $r$, and this has complexity $\mathcal{O}(\log^2 n \log^3 r)$ (see § 3.3.1). Now, by Lemma 6.3.3 we know that we need to test at most $\lceil 16 \log^5 n \rceil$ integers $r$. In conclusion, the complexity of this step is $\mathcal{O}^\sim(\log^7 n)$.

*Third step.* Here we have to compute $r$ times a greater common divisor. Each of these computations takes a time $\mathcal{O}(\log^3 n)$ (see § 2.5.1); so the total complexity is $\mathcal{O}(r \cdot \log^3 n) = \mathcal{O}(\log^8 n)$.

*Fourth step.* As in the previous step.

*Fifth step.* We have to check $\lfloor 2\sqrt{\varphi(r)} \log n \rfloor$ equations. Keeping in mind Exercise A6.26, it is easily seen that each check has complexity $\mathcal{O}(r^2 \log^3 n)$. Thus, the total complexity of this step is

$$\mathcal{O}(r^2 \sqrt{\varphi(r)} \log^4 n) = \mathcal{O}(r^{5/2} \log^4 n) = \mathcal{O}(\log^{16.5} n).$$

The complexity of this step *dominates* all the complexities of the other ones; hence, the complexity of the *whole algorithm* is $\mathcal{O}(\log^{16.5} n)$.

Notice that in [2] a better estimate of this complexity is given, and it has been further improved by other authors, but we shall not go into these details.

$\square$

## 6.4 Factorisation methods

In the previous sections we have seen some primality tests. Suppose we ascertained, by using one of those tests, that a given number $n$ is not prime. As we have already remarked, this tells us nothing about the prime decomposition of $n$.

From now on we shall discuss this problem: knowing that a number is not prime, we are looking for its prime factors.

First of all, when we are about to factor a number, the first thing to be done is to *ensure that it is not a prime*, verifying that it does not pass one of the primality tests seen above. In fact, if the number is prime, most of the factorisation methods we shall shortly see could take an incredibly long time.

As we have already remarked, the question of factoring an integer is computationally very hard. We pointed out this when we discussed the sieve of Eratosthenes (see § 4.1.3), which may be considered to be the first, albeit rather naive, algorithm to factorise integers. As we shall see, all the results we are going to describe provide exponential algorithms, which however in some case work remarkably well.

We face the question of factoring a number in its whole scope when the number is very large, for instance having about one hundred digits. Knowing how such a number decomposes into the product of two primes is like being in possess of a *secret information*. This is the principle on which public-key cryptography relies (we shall return to it in Chapter 7): the public information is the integer, while the secret is its factorisation into two primes. So it is easily understood that the knowledge of large prime numbers is precious in this context. It is well known that the research in this sector is carried out

by professional mathematicians, sometimes funded by agencies dealing with
security systems.

### 6.4.1 Fermat factorisation method

We saw, while discussing the sieve of Eratosthenes, that, in order to factor a
number $n$, it is necessary to ascertain whether it is divisible by the numbers
smaller than or equal to $\sqrt{n}$. However, we also noticed that this method is in
general quite expensive in terms of computational complexity. The following
method due to Fermat is sometimes more efficient. It relies on the following
facts:

1. $n$ may be assumed to be odd;
2. when $n$ is odd, factoring $n$ is *equivalent* to determining two integers $x$ and
   $y$ such that
   $$n = x^2 - y^2.$$

   Indeed, *if* $n = x^2 - y^2$, then $n = (x + y)(x - y)$ is a factorisation of $n$. On
   the other hand, if $n = ab$, then, assuming $a \geq b \geq 1$, we may write

   $$n = \left(\frac{a + b}{2}\right)^2 - \left(\frac{a - b}{2}\right)^2$$

   where $(a + b)/2$ e $(a - b)/2$ are non-negative integers. In fact, as $n$ is odd,
   $a$ and $b$ are odd too, so $a \pm b$ is even;
3. determining $x$ and $y$ such that $n = x^2 - y^2$ is equivalent to determining $x$
   such that $x^2 - n$ is a square, that is, equal to $y^2$.

Fermat factorisation method works as follows: first of all determine the
smallest positive integer $k$ such that $k^2 \geq n$; then successively compute the
differences
$$k^2 - n, \quad (k + 1)^2 - n, \quad (k + 2)^2 - n, \quad \ldots$$

until a value $t \geq k$ is found such that $t^2 - n$ is a square. Notice that this
process *terminates* at most when $t = (n + 1)/2$, as

$$\left(\frac{n + 1}{2}\right)^2 - n = \left(\frac{n - 1}{2}\right)^2,$$

a value that can be obtained only when the number $n$ is prime, so that it has
only the trivial factorisation

$$n = \left(\frac{n + 1}{2} + \frac{n - 1}{2}\right) \cdot \left(\frac{n + 1}{2} - \frac{n - 1}{2}\right) = n \cdot 1.$$

**Example 6.4.1.** Find the prime factors of the number 29591 using Fermat
factorisation method.

In this case $k = 173$ and

$$173^2 - 29591 = 338 \qquad \text{is not a square,}$$
$$174^2 - 29591 = 685 \qquad \text{is not a square,}$$
$$175^2 - 29591 = 1034 \qquad \text{is not a square,}$$
$$176^2 - 29591 = 1385 \qquad \text{is not a square,}$$
$$177^2 - 29591 = 1738 \qquad \text{is not a square,}$$
$$178^2 - 29591 = 2093 \qquad \text{is not a square,}$$
$$179^2 - 29591 = 2450 \qquad \text{is not a square,}$$
$$180^2 - 29591 = 2809 = (53)^2 \qquad \text{is a square.}$$

So, by the last formula

$$29591 = (180 + 53)(180 - 53) = 233 \cdot 127.$$

Now it is easily seen that 233 and 127 are prime numbers, so this is the required factorisation.

It is clear that Fermat factorisation method is efficient in the cases in which $n$ has factors, which can be prime or not, that are close to $\sqrt{n}$, just like the factorisation method based upon the sieve of Eratosthenes is efficient only if $n$ has *small* prime factors. Nonetheless, both methods are equally expensive, that is to say, they are exponential (see § 4.1.3 and Exercise A6.27).

However, the next example shows the remarkable efficiency of Fermat factorisation method when there are (even not prime) factors of $n$ that are *large*, that is, close to $\sqrt{n}$.

**Example 6.4.2.** Factor the number $n = 127433$.

Here $k = 357$. Now

$$357^2 - 127433 = 127449 - 127433 = 16$$

which is a square; hence,

$$127433 = (357 - 4)(357 + 4) = 353 \cdot 361.$$

The number 353 is prime, while $361 = 19^2$. So the prime factorisation of 127433 is

$$127433 = 353 \cdot 19^2.$$

## 6.4.2 Generalisation of Fermat factorisation method

We shall now describe a factorisation method extending Fermat's method described in last section.

Kraitchick in [31] generalised Fermat method as follows: rather than look-ing for two integers $x$ and $y$ such that $x^2 - y^2 = n$, we are requested to find two integers $x$ and $y$ such that

$$x^2 - y^2 \equiv 0 \pmod{n}. \tag{6.33}$$

Clearly, a pair of integers $x$ and $y$ satisfying (6.33) does not ensure that $n$ factors it just says that $n$ *divides* $x^2 - y^2 = (x + y)(x - y)$. Nonetheless, each *prime* divisor of $n$ must be a divisor of either $x - y$ or $x + y$. It may well happen that *all* prime factors of $n$ divide only one of the two numbers $x - y, x + y$: in this case we would have $x \equiv y \pmod{n}$ or $x \equiv -y \pmod{n}$.

But if $x \not\equiv \pm y \pmod{n}$, in order to find a *non-trivial* factor of $n$ it suffices to compute $\mathrm{GCD}(n, x - y)$ or $\mathrm{GCD}(n, x + y)$.

**Example 6.4.3.** Factor with this method the number 559.
Notice that
$$41^2 \equiv 4 \pmod{559}$$

and that $4 = 2^2$. So we found two integers $x = 41$ and $y = 2$ such that $x^2 \equiv y^2$ (mod 559). Determine next $\mathrm{GCD}(41 - 2, 559)$ using the Euclidean algorithm:

$$559 = 39 \cdot 14 + 13,$$
$$39 = 13 \cdot 3 + 0;$$

hence $\mathrm{GCD}(39, 559) = 13$, which is a prime factor of 559. So the factorisation of 559 is
$$559 = 13 \cdot \frac{559}{13} = 13 \cdot 43.$$

**Example 6.4.4.** Factor with this method the number 437.
We have
$$63^2 \equiv 36 = 6^2 \pmod{437}.$$

Moreover, $\mathrm{GCD}(63 - 6, 437) = 19$, so the factorisation of 437 is

$$437 = 19 \cdot 23.$$

Consider now all pairs of numbers $(x, y)$ for which (6.33) holds. In order to get a finite set, we take $|x| < N$ with $N$ a fixed positive integer. Let us ask: which is the probability that for such a pair $x \equiv \pm y \pmod{n}$? Notice that for such pairs the above method does not work.

To give an answer to our question, recall Example 5.2.10 on page 234. It says that if $n$ is odd and not a prime, then there are exactly $2^k$ *square roots* of $y^2$ modulo $n$, where $k$ is the number of distinct primes dividing $n$. Notice that $2^k \geq 4$. So, if we randomly choose $(x, y)$ in our set, it makes sense to suppose that $y$ is a random square root of $x^2$ modulo $n$. So the probability for it to coincide with $\pm x$ is $2/2^k \leq 1/2$.

In conclusion, if $n$ is not a prime, the probability of not finding a non-trivial factor with this method, by randomly choosing the numbers $(x, y)$ for which (6.33) holds, is, at each step, *smaller* than 50%.

This gives rise to a rather efficient factorisation method, with a high probability of success. We shall discuss it in the next section, where we shall try to give an answer to a natural question: how may we find pairs $(x, y)$, and in particular random pairs, satisfying (6.33)? This looks even more problematic than solving the equation $x^2 - y^2 = n$ and we cannot even think about doing it by trial and error. In particular, in Example 6.4.3, how did we find that number 41 having the property that $41^2$ is congruent to a square modulo 559, or, in Example 6.4.4, that number 63 such that $63^2$ is congruent to a square modulo 437? The answer will shortly be given in what follows.

## 6.4.3 The method of factor bases

In order to solve Equation (6.33), we have to find a number $b$ such that $b^2 \bmod n$ is a square.

The point of the method we are going to describe is as follows: take an arbitrary integer $b$, compute $b^2 \bmod n$, and factor this into primes. If $b^2 \bmod n$ is a square, that is, if all the exponents of its factorisation into prime numbers are even, we are done, otherwise iterate this procedure. We shall see that along the way we obtain enough data to ensure that the procedure is successful.

Let us describe the method in detail. To this end it is necessary to give a definition, which justifies the method's name itself.

**Definition 6.4.5.** *A* factor basis *$B$ is an $N$-tuple $(p_1, \ldots, p_N)$ of distinct prime numbers, called the* prime numbers *of the basis.*

*An integer $m$ is said to be a $B$-number modulo an odd positive integer $n$, if in the factorisation of $m \bmod n$ only the primes $p_1, \ldots, p_N$ appear, that is to say, if*

$$m \bmod n = p_1^{\alpha_1} \cdots p_N^{\alpha_N}.$$

*The vector $v(m) = (\alpha_1, \alpha_2, \ldots, \alpha_N) \in \mathbb{N}^N$ is said to be the $B$-vector of $m$ modulo $n$. We may reduce $v(m)$ modulo 2, that is, take the vector $w(m) = (e_1, \ldots, e_N) \in \mathbb{Z}_2^N$, such that $e_i = \alpha_i \bmod 2$, for $i = 1, \ldots, N$. We shall call $w(m)$ the reduced $B$-vector of $m$ modulo $n$.*

Consider again the Examples 6.4.3 and 6.4.4: $41^2$ is a $B$-number modulo 559 with respect to the factorisation basis $B = \{2\}$, while $63^2$ is a $B$-number modulo 437 with respect to the basis $B = \{2, 3\}$.

Let $n$ be the integer to be factored. Having fixed a factorisation basis $B$, the method we are discussing consists in looking for a large enough number $K$ of integers $b_1, b_2, \ldots, b_K$ with $b_1^2, b_2^2, \ldots, b_K^2$ $B$-numbers mod $n$ such that starting with them it is possible to determine a $B$-number $b^2$ that is a square modulo $n$, but in a *non-trivial* way, that is such that $b^2 \equiv c^2 \pmod{n}$ but $b \not\equiv \pm c \pmod{n}$. Notice that we are not asking for the $B$-number to be one of the $B$-numbers $b_1, b_2, \ldots, b_K$, but that it is possible for it *to be determined from them.*

To this end, notice that the set of $B$-numbers is *closed under multiplication*, namely, by multiplying $B$-numbers we still obtain $B$-numbers. Moreover, the set of squares modulo $n$ is closed with respect to multiplication too. So $b$ might be obtained as a product of the numbers $b_1, b_2, \ldots, b_K$ or of some of them.

Let now $b_i^2$, $i = 1, \ldots, K$, be positive $B$-numbers mod $n$. Let $\alpha_{ij}$ be the exponents of the prime factorisation of $b_i^2$ mod $n$ of $B$ and let $e_{ij} := \alpha_{ij}$ mod 2, with $i = 1, \ldots, K$, $j = 1, \ldots, N$. We may consider the matrix $E$ over $\mathbb{Z}_2$, with $K$ rows and $N$ columns,

$$
\begin{pmatrix}
e_{11} & e_{12} & \cdots & e_{1N} \\
e_{21} & e_{22} & \cdots & e_{2N} \\
\vdots & \vdots & \ddots & \vdots \\
e_{K1} & e_{K2} & \cdots & e_{KN}
\end{pmatrix}
$$

its rows being $w(b_1^2), \ldots, w(b_K^2)$.

Assume $K > N$. Then the number of rows of $E$ is strictly larger than the number of columns, so the rows of $E$ are certainly linearly dependent over $\mathbb{Z}_2$, as the rank of $E$ is not greater than $N$. So there is a linear combination of the rows of $E$, with coefficients 0 or 1, not all zero, equal to the zero vector. Such a linear combination can be easily found using Gaussian elimination (see [13], Chapter 8), which over $\mathbb{Z}_2$ only involves row exchanges and sums. So we find a relation of the form

$$
0 = \epsilon_1 w(b_1^2) + \cdots + \epsilon_K w(b_K^2),
$$

where $\epsilon_i = 0, 1, i = 1, \ldots, K$. We may assume that $\epsilon_1 = \cdots = \epsilon_p = 1, \epsilon_{p+1} = \cdots = \epsilon_K = 0$. Then, setting $b = b_1 \cdots b_p$, $b^2 = b_1^2 \cdots b_p^2$ is a square modulo $n$, as required. More precisely, for all $i = 1, \ldots, p$, setting $a_i = b_i^2$ mod $n$, we have

$$
a_i = p_1^{\alpha_{i1}} \cdots p_N^{\alpha_{iN}}.
$$

Then

$$
a := a_1 \cdots a_p = p_1^{\alpha_{11} + \cdots + \alpha_{1p}} \cdots p_N^{\alpha_{N1} + \cdots + \alpha_{Np}}
$$

and $\alpha_{11} + \cdots + \alpha_{1p}, \ldots, \alpha_{N1} + \cdots + \alpha_{Np}$ are even numbers which we shall write as $2\gamma_1, \ldots, 2\gamma_N$, respectively. In conclusion, setting

$$
c = p_1^{\gamma_1} \cdots p_h^{\gamma_N},
$$

we have $b^2 \equiv c^2 \pmod{n}$.

**Remark 6.4.6.** This method is not completely foolproof. In fact, it may happen that $c \equiv \pm b \pmod{n}$, which in particular is the case if the method has been applied in a not judicious way. For instance, if all of $b_1, \ldots, b_K$ have been chosen *too small*, that is, smaller than $\sqrt{n}$, then $b_i^2 < n$ and so $b_i^2$ mod $n = b_i^2$

for all $i = 1, \ldots, K$. Then it is clear that $E$ is the zero matrix and we find a trivial relation.

In any case, if we find a trivial relation, we have to repeat the procedure, *hoping* that the next attempt gives better results. In other words, we have to increase $K$ in order to have *more chances* of finding non trivial relations between the rows of $E$.

On the other hand, if $b_1, \ldots, b_K$ have been chosen in a sufficiently random way, we may expect that, for all $i = 1, \ldots, K$, $b_i \bmod n$ is a random square root of $b_i^2$ modulo $n$, so $c$ is a random square root of $b^2$ modulo $n$. So, by what we said at the end of the previous section, we know that after $k$ steps the probability of finding a trivial relation at each step is smaller than $1/2^k$.

To sum up, in order to factor an integer $n$ with this method, we proceed as follows:

- fix a factor basis $B$ consisting of $N$ prime numbers;
- choose $K > N$ integers such that $b_i^2$ is a $B$-number mod $n$ for every $i$;
- construct the $K \times N$-matrix $E$ having as rows the corresponding $B$-vectors;
- by Gaussian elimination on the rows find a set of rows of $E$ having zero sum;
- the product of the $B$-numbers corresponding to these rows is a $B$-number $b \bmod n$ for which there is a relation $b^2 \equiv c^2 \pmod{n}$;
- if this relation is non-trivial, that is if $b \not\equiv \pm c \pmod{n}$, proceed as in the generalisation of Fermat method to find a factor of $n$.

In practice, how can we choose the factor basis $B$ and the numbers $b_1, \ldots, b_K$?

First of all, notice that the larger the number $N$ of elements of $B$, the higher the probability of finding a factor of $n$, as more numbers are in play and so there is a greater chance of finding non-trivial solutions of (6.33). Analogously, the larger the number $K$, the higher the probability of finding a factor of $n$, as in this case the number of possible relations between rows of $E$ increases. On the other hand, the larger $N$ and $K$, the more computationally expensive the method becomes. So it is necessary to find an efficient way of choosing $B$ and the $B$-numbers $b_i^2$, $i = 1, \ldots, K$.

The first problem lies in factoring the numbers $b_i^2 \bmod n$: they are certainly smaller than $n$, and as such possibly easier to factor than $n$ itself, but how easier? Probably not too much, as these numbers cannot be too small either, as discussed in Remark 6.4.6. In any case, we want to be able to factor them easily. A way to reach this goal consists in choosing as factor basis a set $B$ of sufficiently small primes. In this way, given a $b_i$, to factor $b_i^2 \bmod n$ we have to divide it by all the primes in $B$, a rather undemanding task. In order to obtain $b_1, \ldots, b_K$, take random numbers $b$: if $b^2 \bmod n$ cannot be factored using the primes in $B$, take a different $b$.

In concrete, we may proceed as follows:

(1) choose $B$ first;
(2) depending on the choice of $B$, find next the numbers $b_i$ choosing them randomly and verifying for each of them that every $b_i^2$ mod $n$ is factored by the primes in $B$, until a number $K$ of them is found that is sufficient to apply the above procedure.

This method, as may be easily understood, is not that efficient: it is quite difficult that the randomly chosen numbers $b$ will be factored by the *few* primes in $B$! However, there are other ways.

For instance, we may proceed as follows.

(1) choose a certain number of $b_i'$s, such that $b_i^2$ mod $n$ is small, but not too much (see Remark 6.4.6): for instance, take $b_k = \lfloor \sqrt{nk} \rfloor + 1$, for $k = 1, 2, 3, \ldots$;
(2) factor these numbers, rejecting those having in their factorisation *too large* prime factors, for instance greater than a number $M$ fixed beforehand;
(3) choose the factor basis $B$ starting from the factorisation of the numbers $b_i^2$ mod $n$, including in it, in increasing order, the primes appearing in the factorisation of the numbers $b_i^2$ mod $n$.

This is a more realistic method, but it requires anyhow the preliminary factorisation of *several* quite large numbers, if the method is to work.

In the next examples we shall consider some factorisations we already know, getting them again using this method.

**Example 6.4.7.** Factor 559 using the method of factor bases.
Set $b_k = \lfloor \sqrt{nk} \rfloor + 1$, $k = 1, 2, 3, \ldots$ Depending on the primes appearing in the factorisation of these numbers we shall decide how to choose the factorisation basis $B$. The first values are:

| $b$ | $b^2$ | $b^2$ mod 559 | Prime decomposition of $b^2$ mod 559 |
|---|---|---|---|
| $b_1 = 24$ | 576 | 17 | 17 |
| $b_2 = 34$ | 1156 | 38 | $2 \cdot 19$ |
| $b_3 = 41$ | 1681 | 4 | $2^2$ |

Now we stop, using only the third line, as we have found $b = 41$ such that $b^2 \equiv 2^2$ (mod 559). So we take $B = \{2\}$ and $b = 41$ and $c = 2$. Then $b \not\equiv \pm 2$, so computing $\mathrm{GCD}(41 - 2, 559)$ we obtain the non-trivial factor 13. Thus, $559 = 13 \cdot 43$.

**Example 6.4.8.** Factor 2183 using the method of factor bases.
Again, set $b_k = \lfloor \sqrt{nk} \rfloor + 1$, $k = 1, 2, 3, \ldots$

| $b$ | $b^2$ | $b^2 \bmod 2183$ | Prime decomposition of $b^2 \bmod 2183$ |
|---|---|---|---|
| $b_1 = 47$ | 2209 | 26 | $2 \cdot 13$ |
| $b_2 = 67$ | 4489 | 123 | 123 |
| $b_3 = 81$ | 6561 | 12 | $2^2 \cdot 3$ |
| $b_4 = 94$ | 8836 | 104 | $2^3 \cdot 13$ |
| $b_5 = 105$ | 11025 | 110 | $2 \cdot 5 \cdot 11$ |
| $b_6 = 115$ | 13225 | 127 | 127 |
| $b_7 = 124$ | 15376 | 95 | $5 \cdot 19$ |
| $b_8 = 133$ | 17689 | 225 | $3^2 \cdot 5^2$ |
| $b_9 = 141$ | 19881 | 234 | $2^2 \cdot 3 \cdot 13$ |

Since $b_8$ is a product of two squares, we can find from this the required factorisation: indeed, $133^2 \equiv 15^2 \pmod{2183}$, $133 \not\equiv \pm 15 \pmod{2183}$ and $\mathrm{GCD}(148, 2183) = 37$. Thus $2183 = 37 \times 59$.

However, in order to illustrate how the method works, suppose we did not recognise the squares in $b_8$. This forces us to go on in the algorithm.

Notice that four of the $b_i'$s, and precisely $b_1, b_3, b_4, b_9$, factor with three primes, 2, 3, and 13. So we may take as factor basis $B = \{2, 3, 13\}$. The corresponding matrix $E$ of the exponents modulo 2 is

$$E = \begin{pmatrix} 1 & 0 & 1 \\ 0 & 1 & 0 \\ 1 & 0 & 1 \\ 0 & 1 & 1 \end{pmatrix}.$$

Summing the first and third row we get a row of zeros. These rows correspond to $b_1$ and $b_4$. Taking the element $b = b_1 \cdot b_4$ we get $b^2 \bmod 2183 = 2^4 \cdot 13^2$. So $b^2 \equiv c^2 \pmod{2183}$ with $b = 4418$ and $c = 2^2 \cdot 13 = 52$.

We check now whether $b \equiv \pm c \pmod{2183}$: unfortunately $4418 - 52 = 4366 \equiv 0 \pmod{2183}$. As there is no other way to add the rows so as to get a zero row, we have to add more $B$-numbers, possibly changing $B$ itself. Notice that:

| $b$ | $b^2$ | $b^2 \bmod 2183$ | Prime decomposition of $b^2 \bmod 2183$ |
|---|---|---|---|
| $b_{10} = 48$ | 2183 | 74 | $2 \cdot 37$ |
| $b_{11} = 155$ | 24025 | 12 | $2^2 \cdot 3$ |
| $b_{15} = 181$ | 32761 | 16 | $2^4$ |

We have found a $B$-number, $b_{15} = 181$, and $b_{15}^2$ is a square mod 2183. As $181 \not\equiv \pm 4 \pmod{2183}$, computing $\mathrm{GCD}(181 + 4, 2183) = 37$ yields the factorisation $2183 = 37 \cdot 59$.

In the previous example we have seen that, before finding a $B$-number mod 2183 that is a square and gives a non-trivial relation yielding a factorisation of 2183, some work has been necessary. This suggests that we should look for a more efficient method for finding $b_i'$s such that the numbers $b_i^2 \bmod n$ factor into small enough primes. We shall see in next section how to do so.

**Remark 6.4.9.** We may ask the computational cost of the algorithm described in this section. It is not easy to estimate it. A quite rough estimate may be obtained under some hypotheses with only probabilistic justifications. The estimate for the computational cost of the algorithm is $\mathcal{O}(e^{C\sqrt{\log n \log \log n}})$, where $C$ is a constant. So it is exponential. We do not give here the details of the long and complex derivation of this estimate: the interested reader can find it in [30], pages 148–153.

### 6.4.4 Factorisation and continued fractions

To attack the problem left unsolved at the end of the previous section, the theory of continued fraction comes to our aid (see § 1.5).

Recall the notion of least absolute residue, defined on page 61.

Proposition 1.5.21 on page 61 implies that the least absolute residues of the squares of the numerators of the convergents of the continued fraction of $\sqrt{n}$ are *reasonably small* with respect to $n$. So it is likely that they are easier to factor than $n$. Thus these convergents are good candidates for the choice of numbers $b_1, \ldots, b_K$ in the method of factor bases. As now the integer $LAR(m, n)$ may well be negative, we modify slightly the definition of a *B-number* modulo an odd positive integer $n$ with respect to a factor basis $B$:

**Definition 6.4.10.** *If $B = (p_1, \ldots, p_N)$ is a factor basis in which $p_1, \ldots, p_N$ are distinct prime numbers, a number is said to be a $B$-number modulo an odd positive integer $n$ if in the factorisation of $LAR(m, n)$ only the primes $p_1, \ldots, p_N$ occur, possibly taken with the minus sign, that is, if*

$$LAR(m, n) = \pm p_1^{\alpha_1} \cdots p_h^{\alpha_h}.$$

*For a $B$-number $m$ the $B$-vector $v(m) \in \mathbb{N}$ and the reduced $B$-vector $w(m) \in \mathbb{Z}_2^N$ of $m$ modulo $n$ may be considered, as in Definition 6.4.5.*

**Example 6.4.11.** If $B = \{2, 3\}$, the number 989 is a $B$-number modulo 205. Indeed, $LAR(989, 205) = -36$ and $36 = 2^2 \cdot 3^2$. The $B$-vector of 989 is $(2, 2)$ and the reduced $B$-vector is the zero vector $(0, 0)$.

We now show how to choose a $B$-basis and the numbers $b_1, \ldots, b_K$ to factor a non-square number $n$ using the expression of $\sqrt{n}$ as a continued fraction:

(1) consider the numerators $u_1, \ldots, u_H$ of the first $H$ convergents of the continued fraction of $\sqrt{n}$, which is periodic, as seen in § 1.5.3;
(2) compute next the numbers $LAR(u_i^2, n)$, for $i = 1, \ldots, H$, and factor them. This operation, by Proposition 1.5.21, is distinctly easier than factoring $n$, so we may hope to be able to carry it out;
(3) among the numbers $LAR(u_i^2, n)$, for $i = 1, \ldots, H$, reject the numbers in whose factorisation *too large* prime factors appear, for instance larger than a number $M$ fixed in advance. Let $b_1, \ldots, b_K$ be the numbers among $u_1, \ldots, u_H$ chosen in this way;
(4) choose the factorisation basis $B = (p_1, \ldots, p_N)$ considering the factorisation of the numbers $LAR(b_i^2, n)$, $i = 1, \ldots, K$, including in it, in increasing order, the primes appearing in these factorisations.

Even in the case when $N$ is not too large, so $K$ is not either, starting from $B$ and $b_1, \ldots, b_K$ usually we get in a short time the factorisation of $n$, if $n$ is not prime. Let us see an example of use of this method.

**Example 6.4.12.** Factor the number $n = 2921$.

We need to compute the expression of $\sqrt{2921}$ as a continued fraction, by finding its convergents $C_k = u_k/v_k$ for the first values of $k$:

| $k$ | 1 | 2 | 3 | 4 | 5 |
|---|---|---|---|---|---|
| $u_k$ | 1135 | 1189 | 2324 | 3513 | 5837 |
| $v_k$ | 21 | 22 | 43 | 65 | 108 |

and draw the corresponding list of values $u'_k = u_k \pmod{2921}$ and $a_k = \text{LAR}(u_k^2, 2921)$:

| $k$ | 1 | 2 | 3 | 4 | 5 |
|---|---|---|---|---|---|
| $u'_k$ | 1135 | 1189 | 2324 | 592 | 2916 |
| $a_k$ | 64 | −43 | 47 | −56 | 25 |

Then it makes sense to choose $B = \{2, 5\}$ and to take as $B$-numbers the $a_k$s for $k = 1, 5$, as they are already squares. The method just described guarantees that $b = 1135 \cdot 2916 \equiv 167 \pmod{2921}$ and $c = 2^3 \cdot 5 = 40$ are such that $b^2 \equiv c^2 \pmod{n}$, that is $167^2 \equiv 40^2 \pmod{2921}$. On the other hand, $167 \not\equiv \pm 40 \pmod{2921}$. Thus, $\text{GCD}(167 + 40, 2921) = 23$ is a proper factor of 2921, hence we get the factorisation $2921 = 23 \cdot 127$.

## 6.4.5 The quadratic sieve algorithm

This method was invented by C. Pomerance in 1981. We shall describe it *heuristically*, without giving a proof of why it works. Nor we shall discuss its computational cost which, although exponential as all known factorisation methods, is in some cases more convenient than others. Before more recent techniques, as the number field sieve [33], the quadratic sieve, which is a variation of the method of the factor bases, was the most efficient factorisation algorithm. It is still the fastest way to factor numbers up to about one hundred digits.

Here too, the goal is to have a factor basis $B$ containing small enough numbers and to find a sufficient number of $B$-numbers mod $n$ that are easy to be factored. To this end we may proceed as follows:

- take as $B$ the set of all primes smaller than a suitably chosen integer $M$, and such that $\left(\frac{n}{p}\right) = 1$;
- choose as $b_i$s, as for the originary Fermat factorisation method, the integers $\lfloor \sqrt{n} \rfloor + i$, $i = 1, 2, 3, \ldots, T$, with $T$ suitably fixed in advance too;
- usually both $M$ and $T$ are chosen around $e^{\sqrt{\log n \log \log n}}$ with $M < T < M^2$;
- define next
$$q(b_i) := b_i^2 - n.$$

The $q(b_i)$'s are in the set
$$S := \{b^2 - n \mid \lfloor \sqrt{n} \rfloor + 1 \le b \le \lfloor \sqrt{n} \rfloor + T\}$$

and are our *candidates* for $B$-numbers mod $n$.

Let us see how the algorithm works:

- begin by constructing a table $T$ with the rows corresponding to the integers $q(b_i) = b_i^2 - n$, $i = 1, \ldots, T$;
- fix a prime $p$ in $B$, with $p$ odd. All the operations $(a)$, $(b)$, $(c)$ and $(d)$ that follow refer to this fixed prime $p$;
  $(a)$ solve the congruences

$$x^2 \equiv n \pmod{p^h}, \qquad h = 1, 2, \ldots,$$

  until there are no more solutions $b$ in the interval $I = [\lfloor \sqrt{n} \rfloor + 1, \lfloor \sqrt{n} \rfloor + T]$;
  $(b)$ call $\alpha$ the largest integer such that the equation

$$x^2 \equiv n \pmod{p^\alpha} \tag{6.34}$$

  admits a solution in the interval $I$ and let $\bar{b}_1$ and $\bar{b}_2 \equiv -\bar{b}_1$ be two solutions of (6.34) (see Example 5.2.10). These solution do not necessarily belong to the interval $I$;
  $(c)$ keep building on the table $T$ inserting a 1 near each $q(b_i)$ such that $b_i \equiv \bar{b}_1$ (mod $p$), a 2 (possibly substituting it for a previous 1) near each $q(b_i)$ such that $b_i \equiv \bar{b}_1$ (mod $p^2$), a 3 (possibly substituting it for a previous 2) near each $q(b_i)$ such that $b_i \equiv \bar{b}_1$ (mod $p^3$) and so on. In this way, under the first $p$ we obtain a column in which some integers $h \in \{0, 1, \ldots, \alpha\}$ appear;
  $(d)$ each time a 1 is inserted, or a 1 is increased to 2, or a 2 to 3 and so on, *at the same time divide by $p$ the corresponding $q(b_i)$ and write down the new result*;
  $(e)$ repeat the same operation starting with $\bar{b}_2$;
- repeat the previous operations $(a)$, $(b)$, $(c)$ and $(d)$ for every odd prime in $B$;
- if $p = 2$:
  $(a)$ if $n \not\equiv 1$ (mod 8) insert a 1 near the $q(b_i)$s with odd $b_i$ and at the same time divide $q(b_i)$ by 2;
  $(b)$ if $n \equiv 1$ (mod 8), denoting as in $(b)$ above by $\alpha$ the largest integer such that the equation

$$x^2 \equiv n \pmod{2^\alpha}$$

  admits a solution in the interval $I = [\lfloor \sqrt{n} \rfloor + 1, \lfloor \sqrt{n} \rfloor + T]$, proceed as for odd $p$, but for the fact that we can find four congruent roots modulo $2^\alpha$ rather than two (see Exercise A6.20);
- after concluding the previous operations for all $p$ in $B$, eliminate (hence the name of *sieve*) all the $q(b_i)'s$, except those that became 1 after the division by all the powers of the $ps$ in $B$. After this sieving, the remaining $q(b_i)'s$ are $B$-numbers, with $b_i \in [\lfloor \sqrt{n} \rfloor + 1, \lfloor \sqrt{n} \rfloor + T]$;
- to find a square $q(b_i)$, proceed as in the method of the factor bases, using Gaussian elimination in order to find a zero row.

**Remark 6.4.13.** Before an example we give some remarks about the steps of the algorithm:

- the first check that the integer $n$ to be factored is a quadratic residue with respect to the primes in the factor basis, that is to say, that $\left(\frac{n}{p}\right) = 1$, is done using the properties of Legendre symbol and the law of quadratic reciprocity (see § 5.2.4).

For instance, assume we want to factor $n = 1003$. To construct a factor basis we have to exclude the primes with respect to which $n$ is not a quadratic residue. We have

$$\left(\frac{1003}{3}\right) = \left(\frac{1}{3}\right) = 1 \qquad\qquad \text{so 3 may be included in } B,$$

$$\left(\frac{1003}{5}\right) = \left(\frac{3}{5}\right) = \left(\frac{5}{3}\right) = \left(\frac{2}{3}\right) = -1 \qquad \text{so 5 must be rejected,}$$

$$\left(\frac{1003}{7}\right) = \left(\frac{2}{7}\right) = 1 \qquad\qquad \text{so 7 may be included in } B;$$

- as regards step $(a)$ in the algorithm, we are able to solve congruences modulo odd prime powers with the techniques seen in § 5.2;
- as regards step $(c)$, once we know two solutions $\bar{b}_1$ and $\bar{b}_2 \equiv -\bar{b}_1 \pmod{p^\alpha}$, in order to find the values $b_i \equiv \bar{b}_1$ it suffices to take every $p$-th $b_i$, as they are consecutive integers, then skip to every $p^2$-th one, and so on, with $p$ in $B$;
- when the operations are over with *all* primes in $B$, the only $q(b_i)'$s to be kept are the ones that, after the divisions by the $p$'s in $B$, became 1, because this means that in their factorisation only primes in $B$ appear.

Let us now show an example which, by obvious reasons, will not involve numbers with many digits. So the efficiency of this method cannot be fully appreciated.

**Example 6.4.14.** Factor $n = 56851$ by using the quadratic sieve method.

Take as factor basis $B$ the set of primes $\leq M = 7$, that is,

$$B := (2, 3, 5, 7).$$

Among the odd primes in $B$ we have only to take those such that $n$ is a quadratic residue modulo $p$. As $56851 \equiv 1$ both modulo 3 and modulo 5, and $56851 \equiv 4 \pmod 7$, which implies

$$\left(\frac{56851}{7}\right) = \left(\frac{4}{7}\right) = 1,$$

we get that 56851 is a quadratic residue modulo 3, modulo 5 and modulo 7. So we keep all the primes in $B$.

Now set $b_i = \lfloor \sqrt{n} \rfloor + i$, $i = 1, 2, 3, \ldots, T = 20$.

By choosing $T = 20$ rather than $T = 7^2 = 49$, we are being optimist: to avoid a column of 49 integers $b_i$ we prefer to risk some inefficiency, hoping for the best. Let us see if our optimism will be rewarded.

We have $\lfloor \sqrt{56851} \rfloor = 238$. Compute now $q(b_i) = b_i^2 - n$, $i = 1, \ldots, T = 20$:

| $b$ | $q(b) = b^2 - n$ |
|---|---|
| $b_1 = 239$ | $57121 - 56851 = 270$ |
| $b_2 = 240$ | $57600 - 56851 = 749$ |
| $b_3 = 241$ | $58081 - 56851 = 1230$ |
| $b_4 = 242$ | $58564 - 56851 = 1713$ |
| $b_5 = 243$ | $59049 - 56851 = 2189$ |
| $b_6 = 244$ | $59536 - 56851 = 2685$ |
| $b_7 = 245$ | $60025 - 56851 = 3174$ |
| $b_8 = 246$ | $60516 - 56851 = 3665$ |
| $b_9 = 247$ | $61009 - 56851 = 4158$ |
| $b_{10} = 248$ | $61504 - 56851 = 4653$ |
| $b_{11} = 249$ | $62001 - 56851 = 5150$ |
| $b_{12} = 250$ | $62500 - 56851 = 5649$ |
| $b_{13} = 251$ | $63001 - 56851 = 6150$ |
| $b_{14} = 252$ | $63504 - 56851 = 6653$ |
| $b_{15} = 253$ | $64009 - 56851 = 7158$ |
| $b_{16} = 254$ | $64516 - 56851 = 7665$ |
| $b_{17} = 255$ | $65025 - 56851 = 8174$ |
| $b_{18} = 256$ | $65536 - 56851 = 8685$ |
| $b_{19} = 257$ | $66049 - 56851 = 9198$ |
| $b_{20} = 258$ | $66564 - 56851 = 9713$ |

As we remarked, we shall work separately with each $p \in B$. In the tables we are going to construct we set up the columns for all $p's$, but we shall fill them in only later.

We shall denote by $M_p(I)$ the greatest exponent $h$ of $p$ for which the equation

$$x^2 \equiv n \pmod{p^h} \tag{6.35}$$

admits solutions in the interval $I = [\lfloor \sqrt{n} \rfloor + 1, \lfloor \sqrt{n} \rfloor + 20]$.

Begin with the odd prime 3 in $B$. By solving the congruences $x^2 \equiv n \pmod{p^h}$, $h = 1, 2, \ldots$, we conclude that $M_3(I)$ is 3. The solutions of the equation $x^2 \equiv n \pmod{3^3}$ are $\bar{b}_1 = b_1$ and $\bar{b}_2 \equiv -\bar{b}_1 \pmod{3^3}$.

Begin with $b_1 = 239$, which shall be written in bold in the tables, as it is the starting point for the *skips* for 3. By taking all $b_i \equiv b_1 \pmod 3$, that is, by skipping to every third value starting from $b_1$, we clearly obtain $b_1, b_4, b_7, b_{10}, b_{13}, b_{16}, b_{19}$. Put a 1 near them and at the same time divide the correponding $q(b_i)'s$ ($i = 1, 4, 7, 10, 13, 16, 19$) by 3: we *know* that they are divisible by 3, as these $b_i's$ are congruent to $b_1$ modulo 3.

| $b$ | $q(b) = b^2 - n$ | $p = 2$ | $p = 3$ | $p = 5$ | $p = 7$ |
|---|---|---|---|---|---|
| $\mathbf{b_1 = 239}$ | 90 | | 1 | | |
| $b_2 = 240$ | 749 | | | | |
| $b_3 = 241$ | 1230 | | | | |
| $b_4 = 242$ | 571 | | 1 | | |
| $b_5 = 243$ | 2198 | | | | |
| $b_6 = 244$ | 2685 | | | | |
| $b_7 = 245$ | 1058 | | 1 | | |
| $b_8 = 246$ | 3665 | | | | |
| $b_9 = 247$ | 4158 | | | | |
| $b_{10} = 248$ | 1551 | | 1 | | |
| $b_{11} = 249$ | 5150 | | | | |
| $b_{12} = 250$ | 5649 | | | | |
| $b_{13} = 251$ | 2050 | | 1 | | |
| $b_{14} = 252$ | 6653 | | | | |
| $b_{15} = 253$ | 7158 | | | | |
| $b_{16} = 254$ | 2555 | | 1 | | |
| $b_{17} = 255$ | 8174 | | | | |
| $b_{18} = 256$ | 8685 | | | | |
| $b_{19} = 257$ | 3066 | | 1 | | |
| $b_{20} = 258$ | 9713 | | | | |

Skipping to every $3^2$-th term starting from $b_1$, that is in $b_1, b_{10}$ and $b_{19}$, we change 1 into 2 and divide the corresponding $q(b_i)$ (which shall still be called this way, even if it has already been divided by 3): we are certain that $q(b_{10})$, $q(b_{19})$ are again divisible by 3.

| $b$ | $q(b) = b^2 - n$ | $p = 2$ | $p = 3$ | $p = 5$ | $p = 7$ |
|---|---|---|---|---|---|
| $\mathbf{b_1 = 239}$ | 30 | | 2 | | |
| $b_2 = 240$ | 749 | | | | |
| $b_3 = 241$ | 1230 | | | | |
| $b_4 = 242$ | 571 | | 1 | | |
| $b_5 = 243$ | 2198 | | | | |
| $b_6 = 244$ | 2685 | | | | |
| $b_7 = 245$ | 1058 | | 1 | | |
| $b_8 = 246$ | 3665 | | | | |
| $b_9 = 247$ | 4158 | | | | |
| $b_{10} = 248$ | 517 | | 2 | | |
| $b_{11} = 249$ | 5150 | | | | |
| $b_{12} = 250$ | 5649 | | | | |
| $b_{13} = 251$ | 2050 | | 1 | | |
| $b_{14} = 252$ | 6653 | | | | |
| $b_{15} = 253$ | 7158 | | | | |
| $b_{16} = 254$ | 2555 | | 1 | | |
| $b_{17} = 255$ | 8174 | | | | |
| $b_{18} = 256$ | 8685 | | | | |
| $b_{19} = 257$ | 1022 | | 2 | | |
| $b_{20} = 258$ | 9713 | | | | |

If we now skip to every $3^3$-th term starting from $b_1$ we do not find any $b_i$ in our interval, apart from $b_1$, so change 2 into 3 and divide $q(b_1)$ by 3. We get:

| $b$ | $q(b) = b^2 - n$ | $p = 2$ | $p = 3$ | $p = 5$ | $p = 7$ |
|---|---|---|---|---|---|
| $\mathbf{b_1 = 239}$ | 10 | | 3 | | |
| $b_2 = 240$ | 749 | | | | |
| $b_3 = 241$ | 1230 | | | | |
| $b_4 = 242$ | 571 | | 1 | | |
| $b_5 = 243$ | 2198 | | | | |
| $b_6 = 244$ | 2685 | | | | |
| $b_7 = 245$ | 1058 | | 1 | | |
| $b_8 = 246$ | 3665 | | | | |
| $b_9 = 247$ | 4158 | | | | |
| $b_{10} = 248$ | 517 | | 2 | | |
| $b_{11} = 249$ | 5150 | | | | |
| $b_{12} = 250$ | 5649 | | | | |
| $b_{13} = 251$ | 2050 | | 1 | | |
| $b_{14} = 252$ | 6653 | | | | |
| $b_{15} = 253$ | 7158 | | | | |
| $b_{16} = 254$ | 2555 | | 1 | | |
| $b_{17} = 255$ | 8174 | | | | |
| $b_{18} = 256$ | 8685 | | | | |
| $b_{19} = 257$ | 1022 | | 2 | | |
| $b_{20} = 258$ | 9713 | | | | |

Now repeat everything starting from $-b_1 = -239$, which does not belong to our interval. The first $b_i$ that is congruent to $-239$ modulo 27 is $b_9$, which we write in bold; we now repeat the previous operations on all the $b_i \equiv b_9 \pmod 3$: $b_3$, $b_6$, $b_9$, $b_{12}$, $b_{15}$, $b_{18}$.

| $b$ | $q(b) = b^2 - n$ | $p = 2$ | $p = 3$ | $p = 5$ | $p = 7$ |
|---|---|---|---|---|---|
| $b_1 = \mathbf{239}$ | 10 | | 3 | | |
| $b_2 = 240$ | 749 | | | | |
| $b_3 = 241$ | 410 | | 1 | | |
| $b_4 = 242$ | 571 | | 1 | | |
| $b_5 = 243$ | 2198 | | | | |
| $b_6 = 244$ | 895 | | 1 | | |
| $b_7 = 245$ | 1058 | | 1 | | |
| $b_8 = 246$ | 3665 | | | | |
| $b_9 = \mathbf{247}$ | 1386 | | 1 | | |
| $b_{10} = 248$ | 517 | | 2 | | |
| $b_{11} = 249$ | 5150 | | | | |
| $b_{12} = 250$ | 1883 | | 1 | | |
| $b_{13} = 251$ | 2050 | | 1 | | |
| $b_{14} = 252$ | 6653 | | | | |
| $b_{15} = 253$ | 2386 | | 1 | | |
| $b_{16} = 254$ | 2555 | | 1 | | |
| $b_{17} = 255$ | 8174 | | | | |
| $b_{18} = 256$ | 2895 | | 1 | | |
| $b_{19} = 257$ | 1022 | | 2 | | |
| $b_{20} = 258$ | 9713 | | | | |

We inserted a 1 near $b_3$, $b_6$, $b_9$, $b_{12}$, $b_{15}$, $b_{18}$ and divided the corresponding $q(b_i)$'s by 3: we *knew* that they were divisible by 3, as these $b_i'$s are congruent to $b_9$ modulo 3.

Skip now to every 9-th term, again starting from $b_9$, and change 1 into 2 for $b_9$ and $b_{18}$, and divide $q(b_9)$ and $q(b_{18})$ by 3. We obtain:

| $b$ | $q(b) = b^2 - n$ | $p = 2$ | $p = 3$ | $p = 5$ | $p = 7$ |
|---|---|---|---|---|---|
| $b_1 = \mathbf{239}$ | 10 | | 3 | | |
| $b_2 = 240$ | 749 | | | | |
| $b_3 = 241$ | 410 | | 1 | | |
| $b_4 = 242$ | 571 | | 1 | | |
| $b_5 = 243$ | 2198 | | | | |
| $b_6 = 244$ | 895 | | 1 | | |
| $b_7 = 245$ | 1058 | | 1 | | |
| $b_8 = 246$ | 3665 | | | | |
| $b_9 = \mathbf{247}$ | 462 | | 2 | | |
| $b_{10} = 248$ | 517 | | 2 | | |
| $b_{11} = 249$ | 5150 | | | | |
| $b_{12} = 250$ | 1883 | | 1 | | |
| $b_{13} = 251$ | 2050 | | 1 | | |
| $b_{14} = 252$ | 6653 | | | | |
| $b_{15} = 253$ | 2386 | | 1 | | |
| $b_{16} = 254$ | 2555 | | 1 | | |
| $b_{17} = 255$ | 8174 | | | | |
| $b_{18} = 256$ | 965 | | 2 | | |
| $b_{19} = 257$ | 1022 | | 2 | | |
| $b_{20} = 258$ | 9713 | | | | |

and finally:

| $b$ | $q(b) = b^2 - n$ | $p = 2$ | $p = 3$ | $p = 5$ | $p = 7$ |
|---|---|---|---|---|---|
| $\mathbf{b_1 = 239}$ | 10 | | 3 | | |
| $b_2 = 240$ | 749 | | | | |
| $b_3 = 241$ | 410 | | 1 | | |
| $b_4 = 242$ | 571 | | 1 | | |
| $b_5 = 243$ | 2198 | | | | |
| $b_6 = 244$ | 895 | | 1 | | |
| $b_7 = 245$ | 1058 | | 1 | | |
| $b_8 = 246$ | 3665 | | | | |
| $\mathbf{b_9 = 247}$ | 154 | | 3 | | |
| $b_{10} = 248$ | 517 | | 2 | | |
| $b_{11} = 249$ | 5150 | | | | |
| $b_{12} = 250$ | 1883 | | 1 | | |
| $b_{13} = 251$ | 2050 | | 1 | | |
| $b_{14} = 252$ | 6653 | | | | |
| $b_{15} = 253$ | 2386 | | 1 | | |
| $b_{16} = 254$ | 2555 | | 1 | | |
| $b_{17} = 255$ | 8174 | | | | |
| $b_{18} = 256$ | 965 | | 2 | | |
| $b_{19} = 257$ | 1022 | | 2 | | |
| $b_{20} = 258$ | 9713 | | | | |

So the case $p = 3$ is over. Consider now $p = 5$. We find $M_5(I) = 2$.

Starting from $b_{11}$, which is a solution of $b^2 \equiv n \pmod{5^2}$ and which we shall write as usual in bold, the $b_i'$s congruent to $b_{11} \pmod 5$ are $b_1, b_6, b_{16}$.

Take next the other solution $-b_{11} = -249 \pmod{5^2}$. The solution congruent mod $5^2$ that lies in the interval $I$ is $b_{13} = 251$: we shall write it in bold. The $b_i'$s congruent to $b_{13}$ modulo 5 lying in the interval are $b_3, b_8, b_{13}, b_{18}$. By performing the usual operations we obtain the following table:

| $b$ | $q(b) = b^2 - n$ | $p = 2$ | $p = 3$ | $p = 5$ | $p = 7$ |
|---|---|---|---|---|---|
| $b_1 = 239$ | 2 | | 3 | 1 | |
| $b_2 = 240$ | 749 | | | | |
| $b_3 = 241$ | 82 | | 1 | 1 | |
| $b_4 = 242$ | 571 | | 1 | | |
| $b_5 = 243$ | 2198 | | | | |
| $b_6 = 244$ | 179 | | 1 | 1 | |
| $b_7 = 245$ | 1058 | | 1 | | |
| $b_8 = 246$ | 733 | | | 1 | |
| $b_9 = 247$ | 154 | | 3 | | |
| $b_{10} = 248$ | 517 | | 2 | | |
| $\mathbf{b_{11} = 249}$ | 206 | | | 2 | |
| $b_{12} = 250$ | 1883 | | 1 | | |
| $\mathbf{b_{13} = 251}$ | 82 | | 1 | 2 | |
| $b_{14} = 252$ | 6653 | | | | |
| $b_{15} = 253$ | 2386 | | 1 | | |
| $b_{16} = 254$ | 511 | | 1 | 1 | |
| $b_{17} = 255$ | 8174 | | | | |
| $b_{18} = 256$ | 193 | | 2 | 1 | |
| $b_{19} = 257$ | 1022 | | 2 | | |
| $b_{20} = 258$ | 9713 | | | | |

Consider now the prime $p = 7$ in $B$ and repeat the same procedure. We find $M_7(I) = 1$.

The first solution of $x^2 \equiv n \pmod 7$ in the interval $I$ is $b_2$: write it in bold. The $b_i$s that are congruent to $b_2$ modulo 7 are $b_2, b_9, b_{16}$. The other solution $-b_2 = -240$ is congruent to $b_5$ modulo 7: write down it in bold. All the $b$'s congruent modulo 7 in the interval are $b_5, b_{12}, b_{19}$. In conclusion we obtain the Table 6.1 on page 309.

We performed the operations with all the odd primes in $B$. We still have the prime $p = 2$. As $56851 \not\equiv 1 \pmod 8$, the column under the prime $p = 2$ is completed by inserting a 1 near the odd $b$'s, as in Table 6.2 on page 310.

There is a single $b_i$ such that the corresponding $q(b_i)$ became 1. This means that $b_1$ is the only B-number in $I$. As $q(b_1)$ is not a square, we are forced to enlarge the interval $I$ to a larger interval $I'$, that is to say, we have to add more $b_i'$s. It appears we have been too optimist when choosing $T = 20$. We enlarge the interval $I$ by setting $T = 40$ rather than $T = 20$, adding $b_{21}, \ldots, b_{40}$. From the previous elements $q(b_i)$, $i = 1, \ldots, 20$, we can only keep the unique B-number, discarding all the others which are useless, unless we enlarge the basis $B$.

Now, however, we have to study again the congruences

$$x^2 \equiv n \pmod{p^h}, \qquad \text{for } p \in B \text{ and the different values of } h.$$

In fact, it might well happen that $M_p(I')$ increases, the interval being different. We find indeed that $b_{36} = 274$ is such that $q(b_{36})$ is a multiple of $3^6$.

We insert $b_{36}$ in bold: for the first $p = 3$ the value $b_{36}$ will be the new starting point. As regards $p = 5$, instead, the previous numbers remain, while $M_7(I') = 2$, corresponding to $b_{40}$. But we are in for a surprise, which luckily saves us from further computations. In fact, from the table

**Table 6.1.** Procedure applied to $p = 7$

| $b$ | $q(b) = b^2 - n$ | $p = 2$ | $p = 3$ | $p = 5$ | $p = 7$ |
|---|---|---|---|---|---|
| $b_1 = 239$ | 2 | | 3 | 1 | |
| **$b_2 = 240$** | 107 | | | | 1 |
| $b_3 = 241$ | 82 | 1 | 1 | | |
| $b_4 = 242$ | 571 | 1 | | | |
| **$b_5 = 243$** | 314 | | | | 1 |
| $b_6 = 244$ | 179 | 1 | 1 | | |
| $b_7 = 245$ | 1058 | 1 | | | |
| $b_8 = 246$ | 733 | | | 1 | |
| $b_9 = 247$ | 22 | 3 | | 1 | |
| $b_{10} = 248$ | 517 | 2 | | | |
| $b_{11} = 249$ | 206 | | | 2 | |
| $b_{12} = 250$ | 269 | 1 | | | 1 |
| $b_{13} = 251$ | 82 | 1 | | 2 | |
| $b_{14} = 252$ | 6653 | | | | |
| $b_{15} = 253$ | 2386 | 1 | | | |
| $b_{16} = 254$ | 73 | 1 | 1 | | 1 |
| $b_{17} = 255$ | 8174 | | | | |
| $b_{18} = 256$ | 193 | 2 | | 1 | |
| $b_{19} = 257$ | 146 | 2 | | | 1 |
| $b_{20} = 258$ | 9713 | | | | |

| $b$ | $q(b) = b^2 - n$ | post-divisioni | $p = 2$ | $p = 3$ | $p = 5$ | $p = 7$ |
|---|---|---|---|---|---|---|
| $b_1 = 239$ | 270 | 1 | 1 | 3 | 1 | |
| **$b_{36} = 274$** | 18225 | 1 | | 6 | 2 | |

we read that $q(b_{36})$ is a $B$-number and is a square, being equal to $3^6 \cdot 5^2$. Notice that it might have happened that no $q(b_i)$ had these properties. In this case we should either further widen the interval $I'$ or enlarge the factor basis. Luckily, it was right as it were.

In conclusion, with the old simple Fermat factorisation method, we have

$$n = 56851 = (274)^2 - (3^3 \cdot 5)^2 = (274 - 135)(274 + 135) = 139 \cdot 409.$$

### 6.4.6 The $\rho$ method

The factorisation methods described in the previous sections are essentially more or less refined variations of the simple Fermat factorisation method expounded in § 6.4.1. The factorisation method we describe now relies on a completely different idea.

Assume we know that a number $n$ is composite. To decompose it into primes we proceed as follows, working in $\mathbb{Z}_n$:

- in the first step we choose an easily computable map from $\mathbb{Z}_n$ to itself, for instance a non-linear polynomial $f(x)$ with integer coefficients that does not induce a bijection of $\mathbb{Z}_n$ to itself: most of the times, a good choice is the polynomial $f(x) = x^2 + 1$;

- choose next an arbitrary element $x_0$ of $\mathbb{Z}_n$. Construct the following sequence $\{x_1, x_2, \ldots, x_j, \ldots\}$ of integers modulo $n$:

$$x_1 = f(x_0),$$
$$x_2 = f(x_1) = f(f(x_0)) = f^2(x_0),$$
$$\vdots$$
$$x_j = f(x_{j-1}) = f^j(x_0),$$

and so on. Notice that this sequence has a finite number of distinct elements, as $\mathbb{Z}_n$ is finite. So there are positive integers $k$ and $h$ with $x_k \equiv x_h \pmod{n}$ and $k > h$. If $h$ is the least positive integer such that this happens, it is clear that the meaningful elements of the sequence $\{x_n\}_{n \in \mathbb{N}}$ are $x_1, \ldots, x_{k-1}$, as $x_{k+i} = x_{h+i}$ for all $i \in \mathbb{N}$;

- for every pair $x_i$ and $x_j$ compute $\mathrm{GCD}(x_i - x_j, n)$. If this greatest common divisor is not equal to 1 nor to $n$, we have found a proper divisor of $n$.

**Example 6.4.15.** Assume we want to factor the number 111.
Choose the polynomial $f(x) = x^2 + 1$.
Set $x_0 = 1$. We have

$$x_0 = 1, \qquad x_1 = 2, \qquad\qquad\qquad x_2 = 5,$$
$$x_3 = 26, \qquad x_4 = 677 \equiv 11 \pmod{111}, \qquad x_5 = 122 \equiv 11 \pmod{111}.$$

**Table 6.2.** Procedure applied to $p = 2$

| $b$ | $q(b) = b^2 - n$ | $p = 2$ | $p = 3$ | $p = 5$ | $p = 7$ |
|---|---|---|---|---|---|
| $b_1 = 239$ | 1 | 1 | 3 | 1 | |
| $b_2 = 240$ | 107 | | | | 1 |
| $b_3 = 241$ | 41 | 1 | 1 | 1 | |
| $b_4 = 242$ | 571 | | 1 | | |
| $b_5 = 243$ | 157 | 1 | | | 1 |
| $b_6 = 244$ | 179 | | 1 | 1 | |
| $b_7 = 245$ | 529 | 1 | 1 | | |
| $b_8 = 246$ | 733 | | | 1 | |
| $b_9 = 247$ | 11 | 1 | 3 | | 1 |
| $b_{10} = 248$ | 517 | | 2 | | |
| $b_{11} = 249$ | 103 | 1 | | 2 | |
| $b_{12} = 250$ | 269 | | 1 | | 1 |
| $b_{13} = 251$ | 41 | 1 | 1 | 2 | |
| $b_{14} = 252$ | 6653 | | | | |
| $b_{15} = 253$ | 1193 | 1 | 1 | | |
| $b_{16} = 254$ | 73 | | 1 | 1 | 1 |
| $b_{17} = 255$ | 4087 | 1 | | | |
| $b_{18} = 256$ | 193 | | 2 | 1 | |
| $b_{19} = 257$ | 73 | 1 | 2 | | 1 |
| $b_{20} = 258$ | 9713 | | | | |

We have found the same class, so we stop. Let us compute the differences and the GCD($x_i - x_j, 111$):

$$x_1 - x_0 \equiv 1$$

$$x_2 - x_0 \equiv 4, \qquad\qquad \text{GCD}(4, 111) = 1,$$

$$x_3 - x_0 \equiv 25, \qquad\qquad \text{GCD}(25, 111) = 1,$$

$$x_4 - x_0 \equiv 10, \qquad\qquad \text{GCD}(10, 111) = 1,$$

$$x_2 - x_1 = 3, \qquad\qquad \text{GCD}(3, 111) = 3.$$

We found a proper divisor of 111 and we may conclude $111 = 3 \cdot 37$. It is obvious that for small numbers we do not actually need these algorithms; it is clear, in fact, that 111 is divisible by 3.

When the number $n$ is large, computing *all* the GCD($x_i - x_j, n$), for all $j < i$, may be very expensive. Here follows a variation of the algorithm which improves its efficiency.

### 6.4.7 Variation of $\rho$ method

We keep the above notation. If $i$ and $j$ satisfy $x_i \equiv x_j \pmod{r}$ with $r$ a proper divisor of $n$, then it will also be the case that $x_{i'} \equiv x_{j'} \pmod{r}$ for all $i'$ and $j'$ such that $i - j = i' - j'$. Indeed, if $i - j = i' - j'$, setting $i' = i + t$, $j' = j + t$ and keeping in mind that $f^t(x_i) = x_{i+t}$, it suffices to apply $t$ times the polynomial $f$ to both sides of congruence $x_i \equiv x_j \pmod{r}$, to get $x_{i'} \equiv x_{j'} \pmod{r}$.

Compute again $x_k$ for every $k$ and then proceed as follows. Assume $k$ has length $h + 1$ in base 2, that is, $2^h \le k < 2^{h+1}$. Compute $GCD(x_k - x_j, n)$, for $j = 2^h - 1$. If this gives a proper factor of $n$ we stop, otherwise we consider $k + 1$ next.

The interest of this method lies in the fact that, for each $k$, we have to compute a single greatest common divisor. The disadvantage, of course, is that we might fail to detect a proper factor of $n$ the first time it might appear: that is, the first time we encounter a pair $(k_0, j_0)$ with $j_0 < k_0$ such that $GCD(x_{k_0} - x_{j_0}, n) = r > 1$.

However, *after a while*, we are sure to find the factor we are looking for. Indeed, if $k_0$ has length $h+1$ in base 2, set $j = 2^{h+1} - 1$ and let $k = j + (k_0 - j_0)$. By what has been said, we have $GCD(x_k - x_j, n) > 1$. Notice that $k < 2^{h+2} = 4 \cdot 2^h \le 4k_0$, so we shall not have to wait long, when compared with the standard $\rho$ method described above, before getting a proper factor of $n$.

**Example 6.4.16.** Factor $n - 3713$ using the polynomial $f(x) = x^2 + 1$ and starting from $x_0 = 1$ with the improved $\rho$ method.

Here are the steps of the procedure (checking the computations is left to the reader):

$$x_1 = f(x_0) = 2, \qquad\qquad \text{GCD}(x_1 - x_0, n) = 1,$$
$$x_2 = f(x_1) = 5, \qquad\qquad \text{GCD}(x_2 - x_1, n) = 1,$$
$$x_3 = f(x_2) = 26, \qquad\qquad \text{GCD}(x_3 - x_1, n) = 1,$$
$$x_4 = f(x_3) = 677, \qquad\qquad \text{GCD}(x_4 - x_3, n) = 1,$$
$$x_5 = f(x_4) \equiv 1631 \pmod{n}, \qquad \text{GCD}(x_5 - x_3, n) = 1,$$
$$x_6 = f(x_5) \equiv 1654 \pmod{n}, \qquad \text{GCD}(x_6 - x_3, n) = 1,$$
$$x_7 = f(x_6) \equiv 2949 \pmod{n}, \qquad \text{GCD}(x_7 - x_3, n) = 79$$

obtaining the factorisation $3713 = 79 \cdot 47$. The reader will fruitfully compare this method with the standard $\rho$ method (see Exercises B6.37 and B6.38).

As to the computational complexity of the $\rho$ method, here follows the main result. Notice the probabilistic quality and the exponential nature of the method.

**Theorem 6.4.17.** *Let $n$ be a composite odd positive integer. Assume $x_0$ and $f(x)$ have been chosen in a sufficiently general way. For every real number $\lambda > 0$ the probability that the $\rho$ method does not yield a proper factor of $n$ in $\mathcal{O}(\sqrt[4]{n}\log^3 n)$ bit operations is smaller than $e^{-\lambda}$.*

We give now a sketch of the proof.

Let $X$ be a set of order $k$. Let $F(X)$ be set of all maps $f : X \to X$. Notice that $F(X)$ has order $k^k$ (see Exercise A1.20).

Let $f : X \to X$ be a map. Let $x \in X$ and define recursively a sequence $\{x_n^{f,x}\}_{n\in\mathbb{N}}$ of elements of $X$ as follows:

$$x_0^{f,x} = x, \quad x_{i+1}^{f,x} = f(x_i^{f,x}), \quad i \geq 0.$$

Let $m$ a positive integer number. We shall denote by $G_m(X)$ the subset of $X \times F(X)$ consisting of all pairs $(x, f)$ such that $x_0^{f,x}, \dots, x_m^{f,x}$ are all distinct. Denote by $g_m(X)$ the order of the set $G_m(X)$. Clearly, $g_m(X) = 0$ if $m > k$. The reader can verify as an exercise (see Exercise A6.28) that, if $m \leq k$, we have

$$g_m(X) = k^{k-m} \prod_{i=0}^{m}(k - i). \tag{6.36}$$

**Lemma 6.4.18.** *Let $\alpha$ be a positive real number. Set $a = 1 + \lfloor\sqrt{2\alpha k}\rfloor$. We have*

$$e^\alpha g_a(X) < k^{k+1},$$

*that is to say, the ratio between the order of $G_a(X)$ and the order of $X \times F(X)$ is smaller than $e^{-\alpha}$.*

PROOF. That ratio is

$$\frac{g_a(X)}{k^{k+1}} \leq k^{-a-1}\prod_{i=0}^{a}(k - i) = \prod_{i=1}^{a}\left(1 - \frac{i}{k}\right).$$

Keeping in mind that $\log(1 - x) < -x$ if $0 < x < 1$ (see Exercise A6.29), we have

$$\log\left(\prod_{i=1}^{a}\left(1 - \frac{i}{k}\right)\right) < -\sum_{i=1}^{a}\frac{i}{k} = \frac{-a(a+1)}{2k} < \frac{-a^2}{2k} < -\alpha,$$

as was to be proved.    $\square$

This lemma gives a crucial information about the *possible duration* of the $\rho$ method, yielding the following proof.

PROOF OF THEOREM 6.4.17. Remember that computing $m \bmod n$ has complexity $\mathcal{O}(\log^2 n)$ and computing a greatest common divisor $\mathrm{GCD}(m, n)$ with $0 < m < n$ has complexity $\mathcal{O}(\log^3 n)$.

Fix a proper divisor $r$ of $n$, with $r < \sqrt{n}$. If $k_0$ is the least index such that there exists a $j_0 < k_0$ with $x_{k_0} \equiv x_{j_0} \pmod{r}$, then the $\rho$ method, improved as described in this section, finds a divisor of $n$ in the $k$-th step, with $k < 4k_0$. So the complexity of the algorithm is $\mathcal{O}(k_0 \log^3 n)$. On the other hand, by Lemma 6.4.18 the probability that $k_0 > 1 + \lfloor \sqrt{2\lambda r} \rfloor$ is smaller than $e^{-\lambda}$. If, instead, $k_0 \leq 1 + \lfloor \sqrt{2\lambda r} \rfloor$, then the complexity of the algorithm is $\mathcal{O}(\sqrt{2\lambda r} \log^3 n) = \mathcal{O}(\sqrt[4]{n} \log^3 n)$.    □

**Remark 6.4.19.** In the preceding proof a fundamental role is played by the hypothesis that $x_0$ and $f(x)$ are chosen in a *sufficiently general* way, namely that the pair $(x_0, f)$ is sufficiently general in $X \times F(X)$. If this is not the case, the probability computation we carried out in the proof does not work. The assumption can be translated in the fact that we may choose an easily computable function $f : \mathbb{Z}_n \to \mathbb{Z}_n$, for instance a polynomial that both has this property and is sufficiently general as said. This has not been fully proved, but practical knowledge suggests that several polynomials, among which $x^2 + 1$, behave well in this regard.

# Appendix to Chapter 6

# A6 Theoretical exercises

**A6.1.*** Prove that if $n$ is a pseudoprime in base 2, then $m = 2^n - 1$ is a pseudoprime in base 2 too. Deduce that there exist infinitely many pseudoprimes in base 2.

**A6.2.** Prove that if $n$ is a pseudoprime in base $b$ and if $b - 1$ is coprime with $n$, then $(b^n - 1)/(b - 1)$ is pseudoprime in base $b$. Deduce that there exist infinitely many pseudoprimes in base 3 and in base 5.

**A6.3.** Prove part (a) of Proposition 6.1.3.

**A6.4.** Which properties do Carmichael numbers possess?

(a) They are prime numbers.
(b) They are product of at most two distinct primes.
(c) There are finitely many of them.
(d) None of the above.

**A6.5.** Prove Proposition 6.1.11.

**A6.6.** Conclude the proof of Proposition 6.1.10.

**A6.7.** Prove that if $n$ is a strong pseudoprime in base $b$, then $n$ is a pseudoprime in base $b$ as well.

**A6.8.*** Prove that if $n$ is pseudoprime in base 2 then $m = 2^n - 1$ is a strong pseudoprime in base 2. Deduce that there are infinitely many strong pseudoprimes in base 2.

**A6.9.**\* Conclude the proof of Proposition 6.1.19 discussing the case $r = s - 1$.

**A6.10.** Prove that, if $n$ is an Euler pseudoprime in base $b$ and if $\left(\frac{b}{n}\right) = -1$, then $n$ is a strong pseudoprime in base $b$.

**A6.11.**\* Verify that the complexity of the Miller–Rabin algorithm is $\mathcal{O}(k \log^4 n)$, where $k$ is the number of times the test is repeated.

**A6.12.**\* Show that, if $h \geq 3$, in the group $U(2^h)$ the cyclic groups $\langle 5 \rangle$ and $\langle -1 \rangle$ have the single element 1 in common.

**A6.13.**\* Let $G$ be a direct product of cyclic groups of orders $g_1, \ldots, g_n$. Prove that the least $g$ such that $x^g = 1$ for all $x \in G$ is the least common multiple of $g_1, \ldots, g_n$.

**A6.14.** Complete the proof of part (b) of Proposition 6.1.6 (see 277).

**A6.15.** Prove Lemma 6.2.17.

**A6.16.**\* Let $n$ be a positive integer. Prove that $\prod_{0 < m < n, \mathrm{GCD}(n,m)=1} m \equiv 1 \pmod{n}$ unless $n$ is one of the numbers $2, 4, p^h, 2p^h$, with $p$ an odd prime and $h$ a positive integer; in this case, $\prod_{0 < m < n, \mathrm{GCD}(n,m)=1} m \equiv -1 \pmod{n}$. This result is due to Gauss.

**A6.17.** Let $n = 2^l p_1^{l_1} \cdots p_s^{l_s}$ be a positive integer with its factorisation. Consider the equation $x^h \equiv m \pmod{n}$, where $h$ is a positive integer and $m$ is an integer coprime with $n$. Prove that this equation admits solutions if and only if the equations $x^h \equiv m \pmod{2^l}$, $x^h \equiv m \pmod{p_i^{l_i}}$ $(i = 1, \ldots, s)$ do, and that the number $d$ of its solutions equals the product of the numbers of solutions of the latter equations.

**A6.18.** Consider the equation $x^h \equiv m \pmod{p^l}$, where either $h$ is an integer, $p$ an odd prime and $l$ an arbitrary positive integer, or $p = 2$ and $1 \leq l \leq 2$, and moreover $p \nmid m$ and $p \nmid h$. Prove that this equation admits solutions if and only if the equation $x^h \equiv m \pmod{p}$ does, and that they have the same number of solutions $d = \mathrm{GCD}(h, p - 1)$.

**A6.19.**\* Consider the equation $x^h \equiv m \pmod{2^l}$, where $h$ is an integer, $l \geq 3$ is a positive integer and $m$ is odd. Prove that if $h$ is odd the equation admits exactly one solution. Prove next that if $h$ is even, setting $d = \mathrm{GCD}(h, 2^{l-2})$, the equation admits solutions if and only if $m^{2^{l-2}/d} \equiv 1 \pmod{2^l}$ and $m \equiv 1 \pmod{4}$, and in this case it has exactly $2d$ solutions.

**A6.20.** Prove that, for $p = 2$ and $n$ odd, the congruence equation (6.34) has at most four solutions.

**A6.21.**\* Keeping in mind the structure theorems about $U(\mathbb{Z}_n)$, extend the notion of index to the case where there are no primitive roots modulo $n$. For instance, extend Proposition 6.2.20.

**A6.22.** Prove that, for all positive integer $m$, $\mathrm{lcm}(1, 2, \ldots, m) \geq 2^m$.

**A6.23.**\* Prove that a product of numbers that are introspective for a polynomial $f(x)$ is introspective for the same polynomial and that if a number is introspective for two polynomials, then it is introspective for their sum and their product.

**A6.24.**\* Verify Equation (6.32).

**A6.25.**\* Prove that there is an algorithm of complexity $\mathcal{O}^{\sim}(\log^4 n)$ that determines whether a positive integer $n$ is a power of another positive integer $m$. (Recall the definition of $\mathcal{O}^{\sim}$, introduced in the proof of Theorem 6.3.8)

**A6.26.**\* Verify that there is an algorithm of complexity $\mathcal{O}(r^2 \log^3 n)$ that computes $(x + a)^n$ modulo $x^r - 1$ and modulo $n$.

**A6.27.** Verify that the Fermat factorisation method has exponential complexity.

**A6.28.**\* Verify Equation (6.36) on page 312.

**A6.29.** Prove that $\log(1 - x) < -x$ se $0 < x < 1$.

**A6.30.** Let $n$ be a positive integer and let $q$ be a prime number dividing $n$. Prove that, if $\alpha$ is the greatest positive integer such that $q^\alpha \mid n$, then $q^\alpha$ does not divide $\binom{n}{q}$.

**A6.31.** Assume that an integer number is composite and its factors are small. Which factorisation method is the most convenient?

(a) The $\rho$ method.
(b) Fermat method.
(c) The sieve of Eratosthenes.
(d) The quadratic sieve.

# B6 Computational exercises

**B6.1.** Verify that $3^{90} \equiv 1 \pmod{13}$.

**B6.2.** Determine all the bases $b$ for which 14 is a pseudoprime.

**B6.3.** Determine the least positive integer $n$ that is a pseudoprime in base 3.

**B6.4.** Which of the following equals $2^{90} \pmod{91}$?

(a) 1.
(b) 57.
(c) 64.
(d) None of the above.

**B6.5.** Is the number 179 a pseudoprime in base 2?

(a) Yes.
(b) No, because $2^{178} \not\equiv 1 \pmod{179}$.
(c) No.
(d) None of the above.

**B6.6.** Is the number 341 a pseudoprime in base 2?

(a) Yes.
(b) No, because $2^{340} \not\equiv 1 \pmod{341}$.
(c) No.
(d) None of the above.

**B6.7.** Is the number 561 a pseudoprime in base 2?

(a) Yes.
(b) No, because $2^{560} \not\equiv 1 \pmod{561}$.
(c) No, because 561 is prime.
(d) None of the above.

**B6.8.** Prove that 561 is the smallest Carmichael number.

**B6.9.** Are there Carmichael numbers of the form $5p$ with $p$ a prime number?

**B6.10.** Are there Carmichael numbers of the form $15p$ with $p$ a prime number?

**B6.11.** Find a Carmichael number that is divisible by 91.

**B6.12.** Show that 15841 is a Carmichael number.

**B6.13.** Show that 6601 is a Carmichael number.

**B6.14.** Show that the integer number 91 is not an Euler pseudoprime in base 3.

**B6.15.** Show that 15841 is an Euler pseudoprime in base 2.

**B6.16.** Find all $b$s such that 15 is an Euler pseudoprime in base $b$.

**B6.17.** Show that 15841 is a strong pseudoprime in base 2.

**B6.18.** Show that 65 is a strong pseudoprime in base 8.

**B6.19.** Find all $b$s such that 15 is a strong pseudoprime in base $b$.

**B6.20.** Find a primitive root modulo 9.

**B6.21.** Find a primitive root modulo 49.

**B6.22.** Find a primitive root modulo 81.

**B6.23.** Determine modulo which of the integers $n = 2, 5, 35, 14, 25, 121$ there is a primitive root.

**B6.24.** Describe the structure of the group $U(\mathbb{Z}_{15})$.

**B6.25.** Describe the structure of the group $U(\mathbb{Z}_{16})$.

**B6.26.** Describe the structure of the group $U(\mathbb{Z}_{17})$.

**B6.27.** Describe the structure of the group $U(\mathbb{Z}_{18})$.

**B6.28.** Determine $|U(15)|, |U(16)|, |U(17)|, |U(18)|$.

**B6.29.** Determine $\text{ind}_5\, 7$ modulo 9.

**B6.30.** Study the equation $x^3 \equiv 5 \pmod{9}$.

**B6.31.** Study the equation $x^6 \equiv 7 \pmod{25}$.

**B6.32.** Study the equation $x^4 \equiv 3 \pmod{14}$.

**B6.33.** Find how many solutions has, at most, the equation $x^5 \equiv 3 \pmod{21}$.

**B6.34.** Factor into primes the number 86989 using Fermat's method. How many steps are necessary to find a factor of this number?

(a) 1.
(b) 3.
(c) 5.
(d) None of the above.

**B6.35.** Factor into primes the number 141553 using Fermat method. How many steps are necessary to find a factor of this number?

(a) 1.
(b) 3.
(c) 5.
(d) None of the above.

**B6.36.** Factor using the method of factor bases the number 906113.

**B6.37.** Factor 589 using the standard $\rho$ method, using the polynomial $f(x) = x^2 + 1$ and starting from $x_0 = 2$.

**B6.38.** Factor 589 using the improved $\rho$ method, using the polynomial $f(x) = x^2 + 1$ and starting from $x_0 = 2$.

# C6 Programming exercises

**C6.1.** Write a program that, given a non-prime positive integer $n$, finds all the bases for which it is not a pseudoprime.

**C6.2.** Write a program that, given a positive integer $b$, finds the least $n$ that is a pseudoprime in base $b$.

**C6.3.** Write a program that finds all Carmichael numbers smaller than 10,000.

**C6.4.** Write a program that, given a non-prime positive integer $n$, finds all the bases for which it is not an Euler pseudoprime.

**C6.5.** Write a program that, given a positive integer $b$, finds the least $n$ that is an Euler pseudoprime in base $b$.

**C6.6.** Write a program that performs the probabilistic primality test described in § 6.1.3.

**C6.7.** Write a program that performs the Solovay–Strassen probabilistic primality test.

**C6.8.** Write a program that performs Miller–Rabin primality test.

**C6.9.** Write a program that determines whether a given integer is a $k$-th power of another integer.

**C6.10.** Write a program that performs the AKS test.

**C6.11.** Write a program that, for every positive integer $n < 100,000$ finds whether a primitive root modulo $n$ exists and, if it does, finds such a root.

**C6.12.** Write a program that decides whether a given integer $m$ is an $h$-tuple residue modulo $n$, when a primitive root modulo $n$ exists.

**C6.13.** Write a program that factors an integer using Fermat factorisation method described in § 6.4.2.

**C6.14.** Write a program that factors an integer using the method of factor bases described in § 6.4.3.

**C6.15.** Write a program that factors an integer using the improved method of factor bases described in § 6.4.4.

**C6.16.** Write a program that factors an integer using the quadratic sieve method described in § 6.4.5.

**C6.17.** Write a program that factors an integer using the $\rho$ method as described in § 6.4.6 or in § 6.4.7.

# 7

## Secrets. . . and lies

How can we transmit information in such a way that only authorised persons can understand it? How can we be sure that the information we transmit reaches its destination without being altered? Moreover, how can we be sure of the origin of a message, and so trust its content? In this chapter we shall deal with these problems. We shall first examine the earliest *classic* cryptographic methods, rapidly outlining their development along the centuries, and then we shall discuss the most recent research about *public key* cryptography. For further details about the history and development of cryptography, the reader can have a look at the good popular scientific book [58].

## 7.1 The classic ciphers

Humanity has always felt the need for efficient methods to communicate in a secret and secure way: by this we mean the ability of sending messages that can be easily read by the addressees and cannot possibly be deciphered by unauthorised people. This millennium-old problem is extremely important today, when the advances in electronic communication systems make exchanging information both easier and more vulnerable.

The earliest examples of secret messages appear in the *Histories* by Herodotus, the Greek historian who lived in the 5th century BCE and chronicled the contemporary Greco-Persian wars.

### 7.1.1 The earliest secret messages in history

Herodotus was an extraordinary narrator: he had an unbelievable talent for reporting what he had seen and been told in his travels in Asia Minor, Greece, Africa, Sicily and so on.

In Book VII of the *Histories* he tell how Xerxes, having succeeded his father Darius, after crushing a repellion in Egypt, is about to wage war on Greece: preparations for the expedition are made by creating a formidable army. The pages narrating this preparations are engrossing, with a survey of the army, a detailed portrayal of the costumes and the armours of each of the peoples composing the Persian army, the description of the fleet. Finally, the expedition leaves: but somebody has warned the Greeks of Xerxes's actions and Book VII ends with following passage:

*The Lacedemonians [= Spartans] had been informed before all others that the king was preparing an expedition against Hellas; and thus it happened that they sent to the Oracle at Delphi, where that reply was given them which I reported shortly before this. And they got this information in a strange manner; for Demaratos the son of Ariston after he had fled for refuge to the Medes was not friendly to the Lacedemonians, as I am of opinion and as likelihood suggests supporting my opinion; but it is open to any man to make conjecture whether he did this thing which follows in a friendly spirit or in malicious triumph over them. When Xerxes had resolved to make a campaign against Hellas, Demaratos, being in Susa and having been informed of this, had a desire to report it to the Lacedemonians. Now in no other way was he able to signify it, for there was danger that he should be discovered, but he contrived thus, that is to say, he took a folding tablet and scraped off the wax which was upon it, and then he wrote the design of the king upon the wood of the tablet, and having done so he melted the wax and poured it over the writing, so that the tablet (being carried without writing upon it) might not cause any trouble to be given by the keepers of the road. Then when it had arrived at Lacedemon, the Lacedemonians were not able to make conjecture of the matter; until at last, as I am informed, Gorgo, the daughter of Cleomenes and wife of Leonidas, suggested a plan of which she had herself thought, bidding them scrape the wax and they would find writing upon the wood; and doing as she said they found the writing and read it, and after that they sent notice to the other Hellenes. These things are said to have come to pass in this manner.* (Translation by G.C. Macaulay.)

Moreover, it is well known that in the Battle of Salamis the Greeks, having been informed of Xerxes's expedition thanks to this stratagem and so being ready to confront him, managed in 480 BCE to defeat the Persians. So a secret, *cleverly hidden* message, changed the outcome of a war.

Herodotus himself, in Book V of the *Histories*, tells the story of Histiaios who *desiring to signify to Aristagoras that he should revolt, was not able to do it safely in any other way, because the roads were guarded, but shaved off the hair of the most faithful of his slaves, and having marked his head by pricking it, waited till the hair had grown again; and as soon as it was grown, he sent him away to Miletos, giving him no other charge but this, namely that when he should have arrived at Miletos he should bid Aristagoras shave his hair and look at his head: and the marks, as I have said before, signified revolt. This thing Histiaios was doing, because he was greatly vexed by being detained at Susa. He had great hopes then that if a revolt occurred he would be let go to the sea-coast; but if no change was made at Miletos he had no expectation of ever returning thither again.* (Translation by G.C. Macaulay.)

It is clear that as soon as the enemy suspects the existence of a hidden message, the obvious countermove is to inspect with the greatest care all possible hiding places. In the episodes told by Herodotus, the suspect would be searched until the message hidden under the hair is found, or the tablet would be examined meticulously until the place where the message was written is spotted.

We conclude this historical preamble with a last anecdote (see [58]).

Mary Stuart, Queen of Scots, imprisoned in 1568 by Queen Elizabeth, was a prisoner for 18 years. In 1586 a plot to free her and simultaneously assassinate Queen Elizabeth was organised: the conspirators deemed it necessary for their plan to be approved by the Queen of Scots. To do so, they used hidden and ciphered secret messages. But both the presence of a double-crosser and the deluded certainty of being able to write freely in the messages, in the (mistaken) confidence in the cryptosystem they used being indecipherable, drove Mary to write more than she should have; this gave Queen Elizabeth the proof of her involvement in the plot and led to her death sentence.

These examples and many more show how, mainly during wartime, the need for devices to send messages in such a way that the adversaries could not discover them has been felt for centuries. The most spontaneous way is to hide the message as the above episodes relate: this technique is called *steganography.*

Another way to send a message in such a way that the enemy cannot understand it is obtained by hiding *not the message, but its meaning.* In this case we are dealing with *cryptography.* We are *enciphering* a message so that it can be read by whomever obtains it, but only the actual addressee is able to decipher it, while the enemy cannot, even if he gets hold of it. A first, simple example of ciphering of a message consisted in substituting Greek characters for Latin ones. But perhaps one of the first recorded examples of a ciphered message in the history dates back to Julius Caesar. Thanks to Suetonius's *On the Life of the Caesars* (2nd century CE), we know one of the systems used by Caesar to encipher his messages: he shifted by three positions, with respect to its position in the alphabet, each letter of the message to be sent.

If we denote by lower case letters the 26 letters of the alphabet, each letter of the message (*plaintext*) will be substituted with the letter following it by three positions, which we shall write in upper case: so we get a new message (*ciphertext*). The explicit correspondence between the letters is described in Table 7.1.

The *enciphering* or *encryption* is the rule describing how to pass from one alphabet to the other one, that is, allowing us to rewrite a message so to make it unreadable for those who do not know the rule. For instance, if the message to be sent is

> attack tomorrow          (plaintext),

the result after enciphering is

> DWWDFN WRPRUURZ          (ciphertext).

**Table 7.1.** Cipher used by Caesar

| a | b | c | d | e | f | g | h | i | j | k | l | m | n | o | p | q | r | s | t | u | v | w | x | y | z |
|---|---|---|---|---|---|---|---|---|---|---|---|---|---|---|---|---|---|---|---|---|---|---|---|---|---|
| D | E | F | G | H | I | J | K | L | M | N | O | P | Q | R | S | T | U | V | W | X | Y | Z | A | B | C |

A system like this, in which the cipher alphabet is obtained from the plain alphabet by moving each letter by a fixed number of positions, is called *Caesar cipher*. In English there are altogether 26 possible Caesar ciphers, or rather 25, as clearly if a letter is moved by 26 positions it comes back to its starting point and the ciphered message is equal to the original one. In other words, we may define a bijection between the possible cipher alphabets and the residue classes modulo 26, that is with the integers $n$ such that $0 \leq n \leq 25$. Given such an integer $n$, called *key*, the corresponding cipher alphabet is the one moving the letters of the plaintext by $n$ positions, that is to say, the alphabet obtained by an $n$-position *shift*. Clearly the value $n = 0$ corresponds to the initial alphabet, that is to the plaintext.

If a message that has possibly been enciphered using a Caesar cipher has been intercepted, it suffices, in order to decrypt it, to use the 26, or rather the 25, keys of the possible cipher alphabets. So this enciphering can be sidestepped very easily, especially so if one has good computing instruments, as we have today, while Caesar and his enemies had not.

Refining this principle, we may use as enciphering, rather than just the *shifts*, all possible permutations of the 26 letters. In practice, each permutation of the set $\{0, 1, 2, \ldots, 25\}$, called *key* as above, determines a cipher alphabet, and the other way around: for instance, for the identity permutation it suffices to have 0 correspond to $A$, 1 to $B$, 2 to $C$ and so forth.

But how can we remember the key? We should remember the whole letter sequence, lacking a specific scheme to memorise it. However, there is a good system to generate a permutation of the alphabet that can be easily memorised: it consists in using a key that is itself determined by a *key word* or a *key phrase*, or any letter string we can easily remember. Let us see an example to clarify this method.

**Example 7.1.1.** Assume we have chosen as key phrase the following:

*to be or not to be that is the question.*

First of all, remove the spaces between the words of the key phrase and then the repetitions, obtaining in our case

*tobernhaisqu.*

The cipher alphabet will be constructed by putting in the order, under the plain alphabet, first the letters of the key word modified as above, and then the letters of the plain alphabet not appearing in the key phrase, in the usual alphabetic order. So we get:

```
a b c d e f g h i j k l m n o p q r s t u v w x y z
T O B E R N H A I S Q U C D F G J K L M P V W X Y Z
```

In this way we have associated to the key, that is to the phrase *to be or not to be that is the question*, the cipher alphabet shown. We may verify that the permutation determined by the key word and that determines the alphabet is described by the following table:

| 0 | 1 | 2 | 3 | 4 | 5 | 6 | 7 | 8 | 9 | 10 | 11 | 12 | 13 | 14 | 15 | 16 | 17 | 18 | 19 | 20 | 21 | 22 | 23 | 24 | 25 |
|---|---|---|---|---|---|---|---|---|---|----|----|----|----|----|----|----|----|----|----|----|----|----|----|----|----|
| 19 | 14 | 1 | 4 | 17 | 13 | 7 | 0 | 8 | 18 | 16 | 20 | 2 | 3 | 5 | 6 | 9 | 10 | 11 | 12 | 15 | 21 | 22 | 23 | 24 | 25 |

So, if the message to send were

$$\boxed{\texttt{it is the early bird that gets the worm}},$$

the result after being enciphered in this way would become

$$\boxed{\texttt{IM IL MAR RTKUY OIKE MATM HRML MAR WFKC}}.$$

In general, in this way the number of possible keys, and so of cipher alphabets, increases quickly form 26 (Caesar ciphers) to 26!, the number of all possible permutations of 26 elements (see Exercise A1.11). This number is

$$510909421717094400000,$$

that is, about $51 \cdot 10^{18}$, or more than *fifty billion billion*: if an adversary intercepts the message and suspects this enciphering method has been used he cannot possibly *try to decrypt it by trial and error*. To realise the hugeness of this number it suffices to recall that the Big Bang occurred approximately 15 billion years ago. So whoever has to *send secret messages* may rest easy and relax: nobody can possibly decrypt them! But are things really like this?

Unfortunately, the answer is no: the frequency with which a given letter appears in a text long enough, and other factors depending on the alphabet used can reduce substantially the number of attempts necessary to find the key! We shall return on this in next section.

As regards our ciphers, an enciphering using a single cipher alphabet, as those seen so far, is called *monoalphabetic cipher*. However, we may consider using more than one cipher alphabet. How?

Suppose we want to use $s \in \mathbb{N} \setminus \{0\}$ cipher alphabets. Then, divide the message into $s$-letter blocks and successively encipher the letters in each block with the $s$ alphabets, always using them in the same order. In other words, denoting by $A_i$, $1 \leq i \leq s$, the $s$ alphabets, all the letters that are in the $i$th position of a block will be enciphered with the same alphabet $A_i$. Such a cipher is called *periodic polyalphabetic cipher*. If $s$ is equal or greater than the length of the message we shall simply have what is called an *(aperiodic) polyalphabetic cipher*.

The first example of this kind is apparently due to Leon Battista Alberti, in the second half of 15th century: he proposed the use of *two* cipher alphabets for each message. His idea was improved later by Vigenère in the second half of 16th century. Vigenère proposed that each message should be enciphered using 26 cipher alphabets. The 26 cipher alphabets are shown in table 7.2, where the integer appearing in each line is exactly the shift key giving the cipher alphabet.

To fix ideas, and prepare the mathematical model, we may label the letters of the message using integers as described by table 7.3.

The integers from 0 to 25 are called *numerical equivalents* of the alphabet letters. In this way, as already remarked, we write 0 in the place of $a$, 12 in

**Table 7.2.** Vigenère table

```
      a b c d e f g h i j k l m n o p q r s t u v w x y z
 0    A B C D E F G H I J K L M N O P Q R S T U V W X Y Z
 1    B C D E F G H I J K L M N O P Q R S T U V W X Y Z A
 2    C D E F G H I J K L M N O P Q R S T U V W X Y Z A B
 3    D E F G H I J K L M N O P Q R S T U V W X Y Z A B C
 4    E F G H I J K L M N O P Q R S T U V W X Y Z A B C D
 5    F G H I J K L M N O P Q R S T U V W X Y Z A B C D E
 6    G H I J K L M N O P Q R S T U V W X Y Z A B C D E F
 7    H I J K L M N O P Q R S T U V W X Y Z A B C D E F G
 8    I J K L M N O P Q R S T U V W X Y Z A B C D E F G H
 9    J K L M N O P Q R S T U V W X Y Z A B C D E F G H I
10    K L M N O P Q R S T U V W X Y Z A B C D E F G H I J
11    L M N O P Q R S T U V W X Y Z A B C D E F G H I J K
12    M N O P Q R S T U V W X Y Z A B C D E F G H I J K L
13    N O P Q R S T U V W X Y Z A B C D E F G H I J K L M
14    O P Q R S T U V W X Y Z A B C D E F G H I J K L M N
15    P Q R S T U V W X Y Z A B C D E F G H I J K L M N O
16    Q R S T U V W X Y Z A B C D E F G H I J K L M N O P
17    R S T U V W X Y Z A B C D E F G H I J K L M N O P Q
18    S T U V W X Y Z A B C D E F G H I J K L M N O P Q R
19    T U V W X Y Z A B C D E F G H I J K L M N O P Q R S
20    U V W X Y Z A B C D E F G H I J K L M N O P Q R S T
21    V W X Y Z A B C D E F G H I J K L M N O P Q R S T U
22    W X Y Z A B C D E F G H I J K L M N O P Q R S T U V
23    X Y Z A B C D E F G H I J K L M N O P Q R S T U V W
24    Y Z A B C D E F G H I J K L M N O P Q R S T U V W X
25    Z A B C D E F G H I J K L M N O P Q R S T U V W X Y
```

the place of $m$, and so forth; if necessary, we shall use the notation $a \leftrightarrow 0$, $m \leftrightarrow 12$. We might further use other numbers to denote spaces in the text, commas, diacritics and every other symbol that might be useful to reconstruct more easily the text. For simplicity, we shall not use these signs.

**Table 7.3.** Numerical equivalents of the 26 letters

| a ⟶ 0 | h ⟶ 7 | o ⟶ 14 | u ⟶ 20 |
|---|---|---|---|
| b ⟶ 1 | i ⟶ 8 | p ⟶ 15 | v ⟶ 21 |
| c ⟶ 2 | j ⟶ 9 | q ⟶ 16 | w ⟶ 22 |
| d ⟶ 3 | k ⟶ 10 | r ⟶ 17 | x ⟶ 23 |
| e ⟶ 4 | l ⟶ 11 | s ⟶ 18 | y ⟶ 24 |
| f ⟶ 5 | m ⟶ 12 | t ⟶ 19 | z ⟶ 25 |
| g ⟶ 6 | n ⟶ 13 | | |

How may we remember the sequence of the $s$ alphabets to be used in enciphering our message? *By memorising it using a key word.* We may use a word whose length $s$ represents the period by which the alphabets are repeated.

**Example 7.1.2.** Assume we have chosen the key word `FISH` and we want to encipher the sentence "`shoot now`".

The rules to be followed in the encipher are contained in the key word we have chosen, in the sense that we shall use as cipher alphabets, repeating each in each 4-letter word (`shoo | tnow`), the lines of Vigenère table corresponding successively to the letters $F$, $I$, $S$, $H$. In our example, the lines are the ones numbered 5, 8, 18, and 7. Then the first letter of the message, the $s$, shall be enciphered with the letter that is in the position of $s$ using as cipher alphabet that of line 5, corresponding to the letter $F$ of the key word, up to the second letter $o$ that shall be enciphered with the line corresponding to the letter $H$; then we start again, using the alphabet corresponding to the letter $F$ for the letter $t$, and so on.

We conclude this section with another anecdote (see *Scientific American*, August 1977).

In 1839, Edgar Allan Poe, from the pages of a Philadelphia periodical, asked his readers for cryptograms with monoalphabetic substitutions, and guaranteed he would solve them. Among many other ones, he received the following handwritten cryptogram:

**G**E **J**EASGDXV,
**Z**IJ GL MW LAAM XZY ZMLWHFZEK EJLVDXW KWKE TX LBR ATGH LBMX AANU BAI
**V**SMUKKSS PWNVLWK AGH GNUMK **W**DLNZWEG JNBXVV OAEG ENWBZWMGY MO MLW WNBX MW
AL PNFDCFPKH WZKEX HSSF XKIYAHUL? **M**K NUM YEXDM WBXY SBC HV WYX PHWKGNAMCUK?

The letters in bold correspond to upper case letters.

Having read the message, Poe replied that it consisted of random symbols, not corresponding to any monoalphabetic substitution. More than one hundred years later, in 1975, the mathematician Bryan J. Winkel, and the research chemist Mark Lyster, who took part in the course in cryptography given by by Winkel, decrypted the message. In fact, it was not a monoalphabetic substitution, but did not even consist of random letters. Moreover, there were some errors probably due to the transcription of the text (see Exercise B7.20). The solution is as follows:

*Mr. Alexander,*
*How is it that the messenger arrives here at the same time with the Saturday courier and other Saturday papers when according to the date it is published three days previous? Is the fault with you or the postmasters?*

## 7.2 The analysis of the ciphertext

We have just described some techniques to encipher messages that look to be unbreakable, at least if our only way is to proceed by trial and error and we are not incredibly lucky. Is this all, or is there some other information we can use?

Let us look at things from the point of view of the adversary, who wants to decrypt the message at any cost. We know the message has been enciphered using a monoalphabetic substitution and, as we have remarked, we cannot proceed by trial and error, if we are to solve the problem in an admissible time.

So we have to use other methods, independent of the kind of key that has been used, and so of the cipher alphabet it determines. How may we proceed, having only a page of ciphertext? We subject it to a *text analysis*: depending on the language the text is written in, we take into account its properties. In a language like Italian, where most words end with a vowel, most of the symbols at the end of the words of the ciphertext will be vowels. More in general, much information is given by *frequency analysis*. In each language some letters appear with greater frequency, some more rarely. Linguistic and statistical studies have found the frequency of the 26 letters of English alphabet:

| Letter | % | Letter | % | Letter | % | Letter | % |
|--------|-----|--------|-----|--------|-----|--------|-----|
| a | 7,3 | h | 3,5 | o | 7,4 | u | 2,7 |
| b | 0,9 | i | 7,4 | p | 2,7 | v | 1,3 |
| c | 3,0 | j | 0,2 | q | 0,3 | w | 1,6 |
| d | 4,4 | k | 0,3 | r | 7,7 | x | 0,5 |
| e | 13,0 | l | 3,5 | s | 6,3 | y | 1,9 |
| f | 2,8 | m | 2,5 | t | 9,3 | z | 0,1 |
| g | 1,6 | n | 7,8 | | | | |

So if we know that the text to decrypt is written in English, and if it is long enough, we may determine the frequencies of the letters in the text. Those appearing with the greatest frequency might possibly be *es*, or *ts*. Then we may look for correspondences by trial and error, examining successively the letters with smaller frequency, looking for pieces of the puzzle falling into place. If we get meaningless words, we adjust our tentative assignations. Further information is given by the frequency of double letters, the greater or smaller likelihood that certain letters are found close together, and so on.

**Example 7.2.1.** We get back to the example we saw in the previous section, to see how to use efficiently the techniques just described to decrypt the enciphered message

IM IL MAR RTKUY OIKE MATM HRML MAR WFKC .

We write down the frequency of the letters in the message:

| Letter | Times | Letter | Times | Letter | Times |
|--------|-------|--------|-------|--------|-------|
| A | 3 | J | 0 | S | 0 |
| B | 0 | K | 3 | T | 2 |
| C | 1 | L | 2 | U | 1 |
| D | 0 | M | 6 | V | 0 |
| E | 1 | N | 0 | W | 1 |
| F | 1 | O | 1 | X | 0 |
| G | 0 | P | 0 | Y | 1 |
| H | 1 | Q | 0 | Z | 0 |
| I | 3 | R | 4 | | |

As this is a short message, frequency analysis might be misleading, but in any case we may try to use it. We might also exploit the fact that certain letters are more likely to end a word, or the length of the words, but this attempts are easily foiled by breaking up the message in blocks of the same length, which makes it difficult to reconstruct the single words. However, at this stage we shall use everything we know. So let us analyse our text. The most frequent letters, $M$ and $R$, might correspond to $e$ and $t$. Moreover, the repeating subsequence $MAR$ is likely to be the article *the*. So we may tentatively try the correspondences

$$M = t, \quad A = h, \quad R = e,$$

and for the next most frequent letters of the ciphertext, $I$ and $K$, we might try the next most frequent alphabet letters $n$ and $r$. In this way we obtain:

> `nt nL the eTrUY OIrE thTt HetL the WFrC` .

Clearly, the choice $I = n$ is not promising, and anyway, if the other choices are right, it looks like $I$ must represent a vowel, or the sentence would begin with the "word" $nt$. Trying out $o$, $i$ and $a$, the best choice appears to be $i$. So we get:

> `it iL the eTrUY OIrE thTt HetL the WFrC`

and after some more attempts and frequency analyses we get the solution.

It is clear that this example is not too typical because we are analysing a very short text, but the important thing is to emphasise the fact that the far too many theoretical possibilities to be considered in order to find the key decrease enormously by using data about the language and making some educated guesses and backtracking.

Analysing a Caesar cipher by studying frequencies is clearly even easier.

Vigenère system too can be attacked with a suitably adapted frequency analysis, if its period is known. For instance, if we assume that the period, that is to say the length of the key word is four, then we have to line up the letters of the ciphertext in four columns, as follows:

```
* * * *
* * * *
* * * *
· · · · · ·
· · · · · ·
```

If in the same column there are equal cipher letters, they represent the same letter of the plaintext, as all the letters in the same column are enciphered with the same cipher alphabet.

Coming back to Example 7.1.2, the pattern is

$$\begin{array}{ccccc}
\text{shoo} & \longrightarrow & \text{XPGV} & \longrightarrow & 23\ 15\ 6\ 21 \\
\text{tnow} & & \text{YVGD} & & 24\ 21\ 6\ 3
\end{array}$$

and, as is immediately seen, the letter $G$ (or its numerical equivalent 6) appearing in the third column is repeated and corresponds to the same letter $o$ of the plaintext.

These remarks allow us, with due caution, to use frequency analysis *on each column separately* to reconstruct the key word. Moreover, a German cryptologist who lived at the end of 19th century, F. W. Kasiski, found a method to determine the period of the alphabet, and this partly explains the loss of interest for this kind of ciphers.

**Remark 7.2.2.** Some of these methods can also be applied to ancient inscriptions: clearly the people writing them did not, in general, intend to encipher a message, but for us those texts actually are enciphered messages we have to decrypt. Decrypting an unknown form of writing is something of a magic, as it allows to enter a past world, to get to know a dead civilisation, to call to mind a remote age. The main example are Egyptian hieroglyphics: the most ancient ones date back to fourth millennium BCE. Interest in them was aroused in 16th century when Pope Sixtus V decreed that a new road network should be built in Rome, putting at the crossroads some Egyptian obelisks: confronted with those puzzles, many tried to understand their meaning. The most famous archaeological find with hieroglyphics is undoubtedly the *Rosetta stone*, made of black basalt, discovered in 1799 near Nile's delta and engraved in 196 BCE: it is an inscription regarding a decree by an assembly of priests honouring the Pharaoh and, as is well known, it carries the same text in three versions: hieroglyphic Egyptian, Demotic Egyptian and Greek. The Greek text was easily translated and so it became, in a sense, the *plaintext* against which the two other texts could be compared: it yielded both a great opportunity and an irresistible challenge. J.F. Champollion measured himself against it and, in 1822, solved the mystery. Today the stone is kept in the British Museum in London.

Archaeologists have deciphered several ancient writing systems and languages, but many of them are still undeciphered, among them Etruscan. The most intriguing deciphering has perhaps been that of *Linear B*, a Mediterranean script, dating back to Bronze Age. The renowned English archaeologist Sir Arthur Evans, during his excavations in Crete in 1900, unearthed a great number of clay tablets bearing writings, partly in a script which was called *Linear B*. These tablets made up the archives of Cretan palaces. They were deciphered in 1953 by M. Ventris and J. Chadwick. It would be very long to relate the whole story of the deciphering: suffice it to say that it could form the plot of a thrilling detective novel. Due to the understanding of *Linear B*, the political and social situation of Cretan society has been now reconstructed, at least in its main lines.

We have said above that the purpose of the authors of the inscriptions was not, in general, to encipher them. We kept on the safe side by saying *in general*: indeed, some scholars have recently discovered the presence of cryptographical methods in Egyptian hieroglyphics. Apparently, some of them were enciphered by order of the Pharaohs, using several techniques, among which enciphering by substitution.

We are coming to the end of our digression, which teaches us that history is full of messages to be deciphered. Every discovery unearths a secret whose key was hidden.

Notice that, in trying to understand a message enciphered using a monoalphabetic substitution, we have used quite subtle statistical techniques. Apparently, these cryptanalytical techniques have been invented by Arabs during Middle Ages. A subject like this, in fact, can only appear and flourish within a civilisation possessing an advanced knowledge of mathematics, linguistics, and statistics. Arab civilisation undoubtedly had these traits. The most ancient document describing explicitly the frequency method dates back to 9th century, and is due to Abū Yūsuf ibn Ishāq al-Kindī, known as *the Arab Philosopher*, who in his monograph *On Deciphering Cryptographic Messages* described in detail techniques based on statistics and Arabic phonetics and syntax to be used to decrypt documents.

To complete the historical sketch of cryptography we deal in brief with the machines that, along the centuries, have been devised to put in practice various ciphers.

### 7.2.1 Enciphering machines

The first enciphering machine is the so-called *cipher disc* by Leon Battista Alberti, which is made up of two concentric copper discs, of different diameters, which can rotate one with respect to the other around a central axis. Along the circumferences of the two discs two alphabets are engraved. To encipher a message using Caesar cipher shifting letters by two positions, it suffices to put the *a* of the internal disc, representing the plain alphabet, next to the *C* of the external disc, representing the cipher alphabet (with key $n = 2$). After this simple operation, to encipher the message it suffices, *without any further rotation of the discs*, to read successively the letters on the external disc corresponding to the letters on the internal one. It is a very simple and effective device, which has been in use for several centuries.

The same device can be also used, in quite a natural way, for a polyalphabetic enciphering: changing the position of the second disc means exactly choosing a new alphabet.

As already remarked, this enciphering machine has lived on for several centuries, up to the moment, towards the end of World War I, it has been superseded by the famous *Enigma* machine, invented and constructed by Arthur Scherbius and Richard Ritter, and used until World War II by the German army. In a first form, it consisted of the following three elements, connected by electric wires (see figure 7.1):

- an alphabetic keyboard, to input the plaintext;
- a *scrambler unit*, which is the part actually performing the enciphering;
- a board with as many light bulbs as the letters of the plain alphabet, devised in such a way that the processed electric signal would light the lamp corresponding to the enciphered letter.

The scrambler unit is the system's main part. It is the device actually enciphering the message: it consisted of a thick rubber disc through which a complex network of electric wires passed. For instance, to encipher the letter *a* with the letter *D*, *a* is input on the keyboard: in this way the electric current enters the scrambler, follows the route through the electric wires, and lights the lamp corresponding to the letter *D*.

Later, Scherbius modified the machine, substituting a scrambling rotor for the original scrambler: in this way, the scrambling disc automatically rotated by one twentysixth of a revolution (if the alphabet consisted of 26 letters) after enciphering each letter. So, to encipher the next letter, a different cipher alphabet is used. The rotating scrambler defines 26 cipher alphabets. Further improvements substituted the single scrambler with three scrambling rotors, and introduced a *reflector*, which could reflect the signals processed by the rotors, adding complexity to the machine. In short, it was a very sophisticated enciphering machine, so much so that Scherbius believed that *Enigma* generated unbreakable coded messages. In 1943, during World War II, the English used the *Colossus* computers to decrypt messages generated by *Enigma*. It is very interesting to notice that many of the researchers who collaborated to the breaking of Enigma, among which the mathematician A. Turing, perhaps inspired by the peculiarities of the problem, went on to give fundamental contributions to the development of computer science and artificial intelligence.

## 7.3 Mathematical setting of a cryptosystem

Let us go back to enciphered messages. The science of decrypting messages for which the key is not known is called *cryptanalysis*.

**Fig. 7.1.** Enigma's keyboard and display

So on the one hand there are the *cryptologists*, designing methods to encipher messages in such a way that they cannot be read by unauthorised people, on the other there are the *cryptanalysts*, who try to decrypt messages, looking for weaknesses in the cryptographic system. The interaction of these two subjects, *cryptology* and *cryptanalysis*, which taken together form *cryptography* and deal, from different viewpoints, with the same object, leads, as can be expected, to ever more complex and secure enciphering systems.

In this section we shall show how to give a proper mathematical layout to cryptology-cryptanalysis.

In table 7.4 we give a short glossary of the terms used in cryptography. We further remark that the root *crypt-* derives from Greek *kryptos*, meaning "hidden, secret".

Before going on, some remarks are in order.

- The alphabets used for plaintexts and ciphertexts can be different among them and with respect to the one commonly used in the language. In general, it is convenient to write messages using, rather than letters, integer numbers, which are more suitable to the description of the transformations, that is, the enciphering methods, to be used.
- The transformation procedure, that is, the function describing the passage from plaintext to ciphertext, must be *bijective* if we want to be able to reverse the procedure to decipher the message and find back the original text rather than something else. The crucial thing is that the person who shall have to decipher the message has to be in possession of the key!

**Table 7.4.** Glossary of cryptography

| Lexeme | Meaning |
|---|---|
| Plaintext | Original message to be sent in a secret way, or string of symbols in a given alphabet representing the message or text to be enciphered |
| Ciphertext | Modified, disguised version of the plaintext |
| Encipher, (encrypt) | Convert a plaintext into a ciphertext |
| Decypher, decrypt | Convert a ciphertext into a plaintext |
| Cipher | *Method* used to convert a plaintext into a ciphertext |
| Key | Data determining both a particular enciphering and the correponding deciphering rule, among all the possible ones: in the first case it is called *cipher key*, in the second *decipher key* |
| Cryptology | Science of enciphering messages |
| Cryptanalysis | Science of interpreting enciphered messages |

- Why do we need a key? Is it not sufficient to deal with the enciphering and deciphering transformations? The fact is, once we have perfectioned a system to send enciphered messages, changing often the key offers a greater security without having to modify the whole enciphering system. In other words, we could describe the system as a combination lock, and the key as one of several possible combinations. The lock is the enciphering system we are using, while the key is the combination.
- We are not especially interested here in detailing precisely the set of possible keys. The important thing we want to emphasise is that it has to have a size that is not too small, or else it would be feasible for a cryptanalyst to try out all possible keys for that kind of cipher. For instance, Caesar cipher has a far too small number of keys.

All in all,

- *the task of the cryptologist* is to invent systems to transform a plain message into a cipher message; such systems are called *cryptosystems*;
- *the task of the cryptanalyst* is to oppose this activity, finding ways to interpret enciphered messages, in general without the authorisation of the sender.

In conclusion, a cryptosystem consists of:

- a set $\mathcal{P}$, consisting of the possible plaintexts; a single plaintext shall be denoted by the letter $p$;
- a set $\mathcal{K}$, called key set. We shall denote a key by the letter $k$. Each element $k \in \mathcal{K}$ determines an enciphering transformation $C_k$ and a deciphering transformation $D_k$, inverse of each other. In particular, $D_k \cdot C_k(p) = p$;
- a set $\mathcal{C}$ consisting of the enciphered messages. We shall denote one of these ciphertexts by the letter $c$.

So, given a cryptosystem determined by the triple $(\mathcal{P}, \mathcal{C}, \mathcal{K})$ with

$$\mathcal{P} = \{\text{plaintexts}\}, \qquad \mathcal{C} = \{\text{ciphertexts}\}, \qquad \mathcal{K} = \{\text{keys}\},$$

the communication between two persons, Ariadne and Blanche, is described by the following diagram:

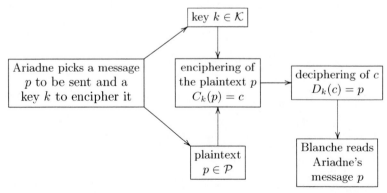

In general the messages, both plain and enciphered ones, are split up into *unitary messages*. A *unitary message* may consist of a *single* letter, or a pair of letters (*digraph*), a triple of letters (*trigraph*), or $s$-letter blocks. The advantage of dividing a message into blocks of a fixed length is in preventing the easy recognition of the beginning and end of the words, making the cryptanalysis based on frequencies more difficult.

Suppose for the time being to have unitary messages consisting of a single letter each, in a given alphabet. To describe mathematically a cryptosystem, the most effective way, as already remarked, is to associate with each symbol of our alphabet an *integer number*. Assume for simplicity the alphabet in which we write our messages to be the English one, and consider its *numerical equivalents* (see Table 7.3 on page 324).

As we are using a 26-letter alphabet, it is natural to perform all mathematical operations on the numerical equivalent of letters modulo 26. In this way 26 is identified with 0, that is with the letter $a$, and so forth. As already suggested, we might use other numbers to denote spaces in the text, commas, diacritics and other symbols which may help in reconstructing more easily the text, but we are not presently interested in them. In general, if the unitary message is an $s$-*letter block* $a_1 a_2 \ldots a_s$, then we *would like* to label the unitary message $a_1 a_2 \ldots a_s$ by the string of integers $x_1 x_2 \ldots x_s$, where $x_i$, with $1 \le i \le s$, is the numerical equivalent of $a_i$. Why are we prevented from doing so? Unfortunately, there is a notational ambiguity which we want to draw attention to.

Indeed, assume we want to transmit a message consisting of 2-letter blocks, using the numerical equivalent of the letters. Then the numerical sequence 114 might correspond to the message *bo*, but also to the message *le*, depending on whether we look at the number 114 as consisting of 1 and 14 or 11 and 4. This is due to having a correspondence with numbers not all consisting of the same number of digits.

When the unitary message consists of more than one letter, in order to avoid ambiguities we may use other correspondences: for instance 2-digit numerical equivalents for the letters or binary numerical equivalents, as described by Table 7.5.

**Table 7.5.** 2-digit and binary numerical equivalents

| a | $\longrightarrow$ 00 = 00000 | j | $\longrightarrow$ 09 = 01001 | s | $\longrightarrow$ 18 = 10010 |
|---|---|---|---|---|---|
| b | $\longrightarrow$ 01 = 00001 | k | $\longrightarrow$ 10 = 01010 | t | $\longrightarrow$ 19 = 10011 |
| c | $\longrightarrow$ 02 = 00010 | l | $\longrightarrow$ 11 = 01011 | u | $\longrightarrow$ 20 = 10100 |
| d | $\longrightarrow$ 03 = 00011 | m | $\longrightarrow$ 12 = 01100 | v | $\longrightarrow$ 21 = 10101 |
| e | $\longrightarrow$ 04 = 00100 | n | $\longrightarrow$ 13 = 01101 | w | $\longrightarrow$ 22 = 10110 |
| f | $\longrightarrow$ 05 = 00101 | o | $\longrightarrow$ 14 = 01110 | x | $\longrightarrow$ 23 = 10111 |
| g | $\longrightarrow$ 06 = 00110 | p | $\longrightarrow$ 15 = 01111 | y | $\longrightarrow$ 24 = 11000 |
| h | $\longrightarrow$ 07 = 00111 | q | $\longrightarrow$ 16 = 10000 | z | $\longrightarrow$ 25 = 11001 |
| i | $\longrightarrow$ 08 = 01000 | r | $\longrightarrow$ 17 = 10001 | | |

**Table 7.6.** ASCII code

| | | | | | | | | | | | | | | | |
|---|---|---|---|---|---|---|---|---|---|---|---|---|---|---|---|
| 32 | | 44 | , | 56 | 8 | 68 | D | 80 | P | 92 | \ | 104 | h | 116 | t |
| 33 | ! | 45 | – | 57 | 9 | 69 | E | 81 | Q | 93 | ] | 105 | i | 117 | u |
| 34 | " | 46 | . | 58 | : | 70 | F | 82 | R | 94 | ^ | 106 | j | 118 | v |
| 35 | # | 47 | / | 59 | ; | 71 | G | 83 | S | 95 | _ | 107 | k | 119 | w |
| 36 | $ | 48 | 0 | 60 | < | 72 | H | 84 | T | 96 | ' | 108 | l | 120 | x |
| 37 | % | 49 | 1 | 61 | = | 73 | I | 85 | U | 97 | a | 109 | m | 121 | y |
| 38 | & | 50 | 2 | 62 | > | 74 | J | 86 | V | 98 | b | 110 | n | 122 | z |
| 39 | ' | 51 | 3 | 63 | ? | 75 | K | 87 | W | 99 | c | 111 | o | 123 | { |
| 40 | ( | 52 | 4 | 64 | @ | 76 | L | 88 | X | 100 | d | 112 | p | 124 | \| |
| 41 | ) | 53 | 5 | 65 | A | 77 | M | 89 | Y | 101 | e | 113 | q | 125 | } |
| 42 | * | 54 | 6 | 66 | B | 78 | N | 90 | Z | 102 | f | 114 | r | 126 | ~ |
| 43 | + | 55 | 7 | 67 | C | 79 | O | 91 | [ | 103 | g | 115 | s | | |

By using the binary or 2-digit numerical correspondence, the ambiguity disappears. For instance, the messages *le* and *bo*, which had the same numerical equivalent, now have different ones:

| | Num. eq. | 2-digit | Binary |
|---|---|---|---|
| le | 114 | 1104 | 0101100100 |
| bo | 114 | 0114 | 0000101110 |

In next section we shall discuss further how to avoid this ambiguity. However, notice that for the sake of simplicity we might keep using the standard numerical correspondence using the simple device of separating with spaces the numbers corresponding to different letters. This is the method we shall use in the simplest examples. For instance, the message $CIAO$ will be transcribed 2 8 0 14 rather than 28014.

Another standard way of associating with each alphabet letter and with each character a number is described, as in the *American Standard Code for Information Interchange* (ASCII), by Table 7.6: it is a usual code used to translate the symbols, more commonly employed when inputting a text into a computer. In Table 7.6 numbers start from 32, as the integers smaller than 32 represent special control characters affecting the operation of the computer.

## 7.4 Some classic ciphers based on modular arithmetic

We shall now describe the mathematical aspects of some of the ciphers seen up to now. For each of them we shall give some examples according to the following pattern:

- fix the length of the unitary message, appending at the end of the message, if necessary, the letter $x$ a number of times sufficient for the whole message to have a length suitable to divide it in blocks of the same length;

- transform the blocks into numerical equivalents following a procedure to be described;
- choose a key $k$ and a corresponding cipher: that is, define the function $C_k$ that determines the cipher on the alphabet in its numerical form;
- determine $D_k = C_k^{-1}$;
- reconstruct the message in the usual alphabet.

**Remark 7.4.1.** Let $N$ be the length of the alphabet. As remarked, usually we take $N = 26$. Once the length $s$ of the unitary message consisting of $s$ letters is fixed, if we call $x_1, \ldots, x_s$ the numerical equivalents of these letters, we describe a procedure that uniquely determines what shall be written taking $x_1, \ldots, x_s$ as starting point. We shall associate with the block $x_1 \ldots x_s$ the number $a$ that in base $N$ is $(x_1 \ldots x_s)_N$. We know that $a \in \{0, \ldots, N^s - 1\}$, that is to say, we may identify the set of unitary messages ($s$-letter blocks) with $\mathbb{Z}_{N^s}$. So it is clear that an enciphering is just an invertible function on, and taking values in, $\mathbb{Z}_{N^s}$. In the previous example, where $N = 26$, we shall have

$$le \longrightarrow 26 \cdot 11 + 4 = 290, \qquad bo \longrightarrow 26 \cdot 1 + 14 = 40.$$

On the other hand, if we begin with the number 40, knowing that $N = 26$, we find that the corresponding plaintext is $bo$ and only this. So we have solved in yet another way the ambiguity issue mentioned in the previous section.

An alternative way of denoting the $s$-letter block $x_1 \ldots x_s$ with no ambiguity is to represent it as an element of $\mathbb{Z}_N^s = \underbrace{\mathbb{Z}_N \times \mathbb{Z}_N \times \cdots \times \mathbb{Z}_N}_{s}$, as each $x_i$ is in $\mathbb{Z}_N$.

This defines an obvious bijection between $\mathbb{Z}_N^s$ and $\mathbb{Z}_{N^s}$ given by

$$\mathbb{Z}_N^s = \underbrace{\mathbb{Z}_N \times \mathbb{Z}_N \times \cdots \times \mathbb{Z}_N}_{s} \to \mathbb{Z}_{N^s}, \tag{7.1}$$

$$(x_1, x_2, \ldots, x_s) \mapsto x_1 + x_2 N + \cdots + x_{s-1} N^{s-2} + x_s N^{s-1} = (x_s \cdots x_1)_N.$$

So, in general, if we split up the message in $s$-letter blocks, on an $N$-letter alphabet, the enciphering function is a bijection

$$f : \mathbb{Z}_N^s \longrightarrow \mathbb{Z}_N^s.$$

A feature of the cipher we are going to describe in this section, the "classic" ciphers, is that the deciphering key $D_k$ is easily computed from the enciphering key $C_k$. In other words, from a computational viewpoint, the knowledge of the deciphering key is *essentially equivalent* to the knowledge of the enciphering key. In public key ciphers we shall describe later, which rely on very different mathematical ideas, it is possible, on the contrary, to divulge the enciphering key without compromising the secrecy of $D_k$. Indeed, in these systems, computing $D_k$ from $C_k$ is so computationally hard to be unfeasible in practice. For these reasons, classic ciphers are called *two-way* or *symmetric*, while public key ciphers are also called *one-way* or *asymmetric*. But more on this later.

Let us now examine systematically some kinds of classic ciphers, all relying on modular arithmetic, which admit as particular instances those considered above. For the sake of simplicity, in the examples we shall use the numerical equivalents, inserting spaces between the numbers to avoid any ambiguity.

### 7.4.1 Affine ciphers

Assume the unitary message consists of a single letter, that is to say, the numerical alphabet of the messages is $P = \mathbb{Z}_{26}$. If $P$ is a letter, we shall also denote by $P$ its numerical equivalent. We shall use the same convention for the letters $C$ in the ciphertext.

Affine ciphers are described by an enciphering function that uses an *affine transformation*, that is a bijection

$$C_k : \mathbb{Z}_{26} \longrightarrow \mathbb{Z}_{26}, \qquad P \longrightarrow (aP + b) \bmod 26,$$

where $a, b \in \mathbb{Z}$ and the pair $k = (a, b)$ represents the *key* of the system. For $C_k$ to be bijective, it is necessary for $a$ to be relatively prime with 26, that is, $\mathrm{GCD}(a, 26) = 1$ (see Exercise A7.3). In this case the congruence $xa \equiv 1 \pmod{26}$ has a unique solution $a'$ modulo 26. Then the inverse deciphering function $D_k$ is

$$D_k = C_k^{-1} : \mathbb{Z}_{26} \longrightarrow \mathbb{Z}_{26}, \qquad C \longrightarrow a'(C - b) \bmod 26.$$

Consider now a key $k = (a, b) \in \mathcal{K}$. Notice that *Caesar* or *translation ciphers* are affine ciphers with $a = 1$.

Clearly, we may assume $0 \le b \le 25$ and $1 \le a \le 25$, taking $a$ and $b$ modulo 26. Recalling that $a$ is relatively prime with 26 if and only if $a$ is an invertible element in $\mathbb{Z}_{26}$, that is, if $a \in U(\mathbb{Z}_{26})$, it follows that the key set is

$$\mathcal{K} = U(\mathbb{Z}_{26}) \times \mathbb{Z}_{26}.$$

How many affine ciphers are there? In other words, how many elements are ther in $\mathcal{K}$?

Recall (see § 3.3 and § 4.2.1) that

$$|U(\mathbb{Z}_{26})| = \varphi(26) = |\{\, 1 \le a \le 25 \,|\, \mathrm{GCD}(a, b) = 1 \,\}| = 12;$$

so there are $12 \cdot 26 = 312$ affine ciphers, including a trivial one, corresponding to $k = (1, 0)$.

In Table 7.7 on page 337 we give a step-by-step example of one of these ciphers, with $a = 7$ and $b = 10$, that is, $k = (7, 10)$. Notice that a Bézout's identity for 7 and 26 is

$$-11 \cdot 7 + 3 \cdot 26 = 1,$$

as can be found by applying the Euclidean algorithm (see § 1.3.3), and so $a' = -11 \equiv 15 \pmod{26}$ and the deciphering function is

$$D_{(7,10)}(C) = 15(C - 10) \bmod 26,$$

that is, $D_{(7,10)} = C_{(15,6)}$, as $-150 \equiv 6 \pmod{26}$.

Suppose we have intercepted a message which is known to be in English and to have been enciphered with this system. How do we decrypt it? That is, how do we find the coefficients $a$ and $b$ of the affine transformation? Once more, with a frequency analysis.

Assume that in the ciphertext the two letters appearing with the least frequency are $R$ and $S$: it stands to reason to guess that these letters correspond, in the plaintext, to $j$ or $z$. If we suppose that $R \leftrightarrow 17$ corresponds to the letter $q \leftrightarrow 16$, and $S \leftrightarrow 18$ corresponds to $z \leftrightarrow 25$, then, according to the relation $C = aP + b$, the following congruence system must hold

$$\begin{cases} 17 \equiv a \cdot 16 + b & (\text{mod } 26), \\ 18 \equiv a \cdot 25 + b & (\text{mod } 26). \end{cases}$$

**Table 7.7.** Affine enciphering with $k = (7, 10)$ modulo 26

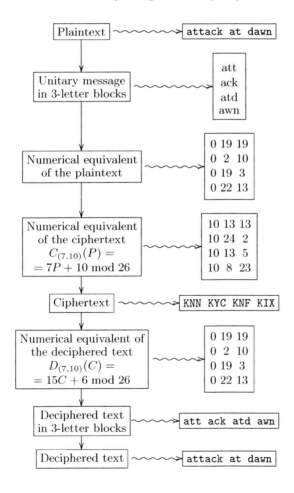

An effective way of describing the system is using matrices. In this way, we may write down the system as

$$A \cdot \begin{pmatrix} a \\ b \end{pmatrix} \equiv \begin{pmatrix} 17 \\ 18 \end{pmatrix} \pmod{26}, \qquad \text{with } A = \begin{pmatrix} 16 & 1 \\ 25 & 1 \end{pmatrix}.$$

To solve the system it is necessary for the matrix $A$ to be invertible modulo 26. In fact, it is easy to prove the following proposition (see Exercise A7.7).

**Proposition 7.4.2.** *Let*

$$A = \begin{pmatrix} a_{11} & a_{12} & \dots & a_{1s} \\ a_{21} & a_{22} & \dots & a_{2s} \\ \vdots & \vdots & \ddots & \vdots \\ a_{s1} & a_{s2} & \dots & a_{ss} \end{pmatrix}$$

*with $a_{ij} \in \mathbb{Z}_N$. The following are equivalent:*

- $\mathrm{GCD}(\det(A), N) = 1$;
- *$A$ is invertible, that is, there is a unique matrix $A^{-1}$ defined over $\mathbb{Z}_N$ such that $A \cdot A^{-1} = A^{-1} \cdot A$ is the identity matrix. The matrix $A^{-1}$ is given by*

$$A^{-1} = \det(A)^{-1} \begin{pmatrix} A_{11} & A_{21} & \dots & A_{s1} \\ A_{12} & A_{22} & \dots & A_{s2} \\ \vdots & \vdots & \ddots & \vdots \\ A_{1s} & A_{2s} & \dots & A_{ss} \end{pmatrix}$$

*where $\det(A)^{-1}$ is the inverse of $\det(A)$ in $\mathbb{Z}_N$ and $A_{ij}$ denotes the cofactor corresponding to the element $a_{ij}$ in $A$;*
- *the map $f : X \in \mathbb{Z}_N^s \to A \cdot X \in \mathbb{Z}_N^s$ is bijective ($X \in \mathbb{Z}_N^s$ is thought of as a column vector of order $s$);*
- *for every column vector $Y \in \mathbb{Z}_N^s$, the system $A \cdot X = Y$ has a unique solution.*

Notice that if $N$ is prime, that is, if we are working in a field, the condition for $A$ to be invertible is $\det(A) \neq 0$. In this case, Proposition 7.4.2 is a well-known result in linear algebra.

Coming back to our example, in which $\det A = -9 \equiv 17 \pmod{26}$ and $\mathrm{GCD}(17, 26) = 1$, we have

$$A^{-1} \equiv \begin{pmatrix} 23 & 3 \\ 23 & 4 \end{pmatrix} \pmod{26}$$

is the inverse modulo 26 of the system matrix. So the solutions $a$ and $b$ are immediately found as follows:

$$\begin{pmatrix} a \\ b \end{pmatrix} \equiv A^{-1} \cdot \begin{pmatrix} 17 \\ 18 \end{pmatrix} \equiv \begin{pmatrix} 3 \\ 21 \end{pmatrix} \pmod{26}.$$

In this particular case the simplest way of solving the original system would have been to subtract the second equation from the first one, immediately finding $a \equiv 3 \pmod{26}$ and then $b \equiv 21 \pmod{26}$. But we have given the general solution method for this kind of problems.

**Remark 7.4.3.** A word of caution is necessary about the cryptanalytical method to find $a$ and $b$ just described. It leads to linear congruence systems of the form

$$A \cdot \begin{pmatrix} a \\ b \end{pmatrix} \equiv \begin{pmatrix} \alpha \\ \beta \end{pmatrix} \pmod{m},$$

where $\alpha, \beta$ and the square $2 \times 2$ matrix $A$ are known. If the determinant of $A$ is invertible modulo $m$, then $A$ has an inverse modulo $m$ and the system admits a unique solution $(a, b)$, given by

$$\begin{pmatrix} a \\ b \end{pmatrix} \equiv A^{-1} \cdot \begin{pmatrix} \alpha \\ \beta \end{pmatrix} \pmod{m}.$$

The reader is encouraged to verify this claim and to extend it to systems in several unknowns (see Proposition 7.4.2 and Exercise A7.8).

If, on the other hand, the determinant of $A$ is not invertible modulo $m$, the system might have no solutions (see Exercise A7.4) or more than one solution (see Exercise A7.5). In the first case this means that our guesses about frequency analysis are certainly wrong and we shall make different cryptanalytical attempts.

If there is more than one solution, we might try out each one of them and check whether it works. For instance, if the determinant of $A$ is not invertible modulo $m$ but is invertible modulo a prime number $q$ divisor of $m$, we might solve the problem modulo $q$. This shall give a unique solution modulo $q$, but several solutions modulo $m$, each one to be analysed to check whether it works (see Exercise A7.6).

We conclude with an example, in the context of Caesar ciphers, which uses an enciphering in $\mathbb{Z}_N^s$, leaving as an exercise an example that relies on the identification of $\mathbb{Z}_N^s$ with $\mathbb{Z}_{N^s}$ described by (7.1).

**Example 7.4.4.** Assume the numerical alphabet of the unitary messages to be represented by $\mathcal{P} = \mathbb{Z}_{26}^s$ with $s$ a fixed integer, that is, the unitary messages to consist of an $s$-letter block. Given the key $k \in \mathcal{K} = \{1, 2, \ldots, 25\}$, as we have seen, the enciphering function is

$$C_k : \mathbb{Z}_{26}^s \longrightarrow \mathbb{Z}_{26}^s, \qquad p \longrightarrow p + \mathbf{k} \bmod 26,$$

where $p$ is the unitary message having numerical entries $p_1 \ldots p_s$, and $\mathbf{k} = (k, \ldots, k)$ is the element of $\mathbb{Z}_{26}^s$ having all entries equal to $k$. By mod 26, it is meant that each entry of an element of $\mathbb{Z}_{26}^s$ is computed modulo 26. Notice that if we take as our key a *vector* $(k_1, \ldots, k_s) \in \mathbb{Z}_{26}^s$, where $k_1, \ldots, k_s$ are *not* all equal, then we construct a polyalphabetic cipher; we shall deal with it again later.

Table 7.8 on page 340 shows an example with $k = 5$.

If, instead, we identify an element $(p_1, \ldots, p_s)$ of $\mathbb{Z}_{26}^s$ with the number written in base 26 as

$$p = p_1 + p_2 \cdot 26 + p_3 \cdot 26^2 + \cdots + p_s \cdot 26^{s-1},$$

where, clearly, $p_1, \ldots, p_s$ are in $\{0, 1, \ldots, 25\}$, we may, as remarked above, identify $\mathbb{Z}_{26}^s$ with $\mathbb{Z}_{26^s}$ via the map

$$(p_1, \ldots, p_s) \in \mathbb{Z}_{26}^s \longrightarrow p = p_1 + p_2 \cdot 26 + \cdots + p_s \cdot 26^{s-1} \in \mathbb{Z}_{26^s}.$$

So we have $\mathcal{P} = \mathbb{Z}_{26^s}$ and $\mathcal{K} = \mathbb{Z}_{26^s}$. Exercise B7.25 uses this different enciphering, with key $k = 100$.

**Table 7.8.** Translation enciphering with $k = 5$

| Plaintext | ⟿ | attack today |

| Message in 4-letter blocks | ⟿ | atta ckto dayx |

| Numerical equivalent of the plaintext | ⟿ | 0 19 19  0<br>2 10 19 14<br>3  0  24 23 |

| Numerical equivalent of the ciphertext $C_5(P) = P + 5 \bmod 26$ | ⟿ | 5 24 24  5<br>7 15 24 19<br>8  5  3  2 |

| Ciphertext | ⟿ | FYYF HPYT IFDC |

| Numerical equivalent of the deciphered text $D_5(C) = C - 5 \bmod 26$ | ⟿ | 0 19 19 0<br>2 10 19 14<br>3  0  24 23 |

| Deciphered text in 4-letter blocks | ⟿ | atta ckto dayx |

| Deciphered text | ⟿ | attack today |

## 7.4.2 Matrix or Hill ciphers

These ciphers split up the text into blocks of length $s$, translate each letter of the block into its numerical equivalent and then apply an enciphering function, defined on the blocks, of the form

$$c = Ap + b \bmod 26, \tag{7.2}$$

where $A$ is a square $s \times s$ matrix, $b$ is a fixed column vector of length $s$, and $p$ and $c$ are the column vectors corresponding numerically to plaintext $p$ and ciphertext $c$. Moreover, if we want to be able to decipher the message, that is, have a bijective enciphering function, it is necessary for the matrix $A$ to be invertible modulo 26. The map

$$C_{(A,b)} : \mathbb{Z}_{26}^s \to \mathbb{Z}_{26}^s, \qquad p \to Ap + b \bmod 26$$

is also called an *affine transformation* defined by the *key* $k = (A, b)$.

Notice that in the case $s = 1$ we find again the affine ciphers described above.

**Example 7.4.5.** We conclude giving an example of affine transformation defined by the key

$$k = (A, b), \quad \text{with } A = \begin{pmatrix} 1 & 2 \\ 4 & 3 \end{pmatrix}, \quad b = 0 = \begin{pmatrix} 0 \\ 0 \end{pmatrix}.$$

So the enciphering function is

$$C_{(A,0)}\begin{pmatrix} p_1 \\ p_2 \end{pmatrix} = \begin{pmatrix} 1 & 2 \\ 4 & 3 \end{pmatrix} \cdot \begin{pmatrix} p_1 \\ p_2 \end{pmatrix} = \begin{pmatrix} p_1 + 2p_2 \\ 4p_1 + 3p_2 \end{pmatrix} \mod 26.$$

Notice that $A$ is invertible modulo 26, as $\det(A) = -5 \equiv 21 \pmod{26}$ is relatively prime with 26. So we may compute the inverse modulo 26 of $A$ (see Proposition 7.4.2). The inverse of 21 modulo 26 is 5, as Bézout's identity found with the Euclidean algorithm is

$$5 \cdot 21 - 4 \cdot 26 = 1.$$

So the inverse modulo 26 of $A$ is

$$A^{-1} = 5 \cdot \begin{pmatrix} 3 & -2 \\ -4 & 1 \end{pmatrix} = \begin{pmatrix} 15 & -10 \\ -20 & 5 \end{pmatrix} \equiv \begin{pmatrix} 15 & 16 \\ 6 & 5 \end{pmatrix} \pmod{26}$$

and the deciphering function is

$$D_{(A,0)}\begin{pmatrix} c_1 \\ c_2 \end{pmatrix} = \begin{pmatrix} 15 & 16 \\ 6 & 5 \end{pmatrix} \cdot \begin{pmatrix} c_1 \\ c_2 \end{pmatrix} = \begin{pmatrix} 15c_1 + 16c_2 \\ 6c_1 + 5c_2 \end{pmatrix} \mod 26.$$

In Table 7.9 we show the complete procedure to encipher with the key given the plaintext true.

## 7.5 The basic idea of public key cryptography

In this section we are going to continue the description of some cryptographic systems, outlining the genesis of the so-called public key systems. Later we shall illustrate specific cryptographic systems of this kind, like the system based on the knapsack problem and the $RSA$ system: the security of the former relies on the difficulty of some combinatorial problems, while that of the latter on the difficulty of factoring large numbers.

In the ciphers described so far the deciphering procedure is not difficult, once the enciphering method, and so the key, are known. In fact, in those cases the deciphering function is, in a way, *symmetric* with respect to the enciphering function: it is, both computationally and logically, a function of the same kind. In particular, all classic cryptosystems concern the exchange of messages between *two* users and rely on exchanging a *key* which, basically, enables both enciphering and deciphering.

In an age like the present one, when most information is transmitted by telephone or electronic mail or radio, every sent message, as well as every sent key, is susceptible to being easily eavesdropped. Moreover, it is necessary to make it possible to communicate for users who have never met and so have not had, in principle, the opportunity of exchanging private enciphering keys. So it is indispensable to find new, and more secure, ways of enciphering messages. This is the goal of public key cryptography.

**Table 7.9.** Matrix cipher as in Example 7.4.5

A public key cipher is a cipher that allows both the method employed and the enciphering key to be made public - hence the name of *public key cipher* - without revealing how to decipher the messages. In other words, in these systems, to be able to compute in a *reasonably short* time the deciphering transformation, which is the inverse of the enciphering one, it is necessary to be in possession of a further piece of information, besides the public ones. So this information is kept secret and without it the complexity of the deciphering is enough to make it unfeasible: in essence, to decipher without further information would require a time exceedingly long with respect to the time required to encipher.

**Remark 7.5.1.** For an example which illustrates quite well the fact that being able to do something does not imply being able to perform the inverse operation, consider the telephone directory of a big city. It is easy to look up the telephone number of a certain person, but it might be impossible, that is to say, it might take too long with respect to the available time, to trace a person from his number.

From a mathematical viewpoint, carrying out this idea relies on the notion of *one-way function*.

We shall call a function $f : S \to T$ from a set $S$ to a set $T$ *one-way* if it can be computed easily (for instance, because it is computed in polynomial time), but, having chosen a random $y \in f(S)$, it is computationally much harder, and impossible in practice (for instance because it takes an exponential time), to find an $x \in S$ such that $y = f(x)$.

This notion may appear quite vague, as it uses terms as "easy", "chosen a random $y \in f(S)$", or "impossible in practice", which have not the mathematical rigour of a definition. Nevertheless, we believe that it gives a sufficiently clear idea of the meaning of a one-way function.

**Example 7.5.2.** Consider a finite group $G$ of order $n$ and an element $b \in G$. Set
$$S = \mathbb{Z}_n = \{0, 1, \ldots, n - 1\};$$
then we may consider the *exponential function*
$$f : S \to G, \quad f(x) = b^x.$$

If $y = f(x)$, we call $x$ a *discrete logarithm* of $y$ over $G$ in base $b$ and denote it by the symbol $\log_b y$. When $G$ is the multiplicative group $\mathbb{F}_q^*$ of a finite field $\mathbb{F}_q$, if $b \in \mathbb{F}_q^*$ is one of its generators, then $f$ is bijective and its inverse function is called *discrete logarithm* over $\mathbb{F}_q$ in base $b$. In this case, computing $f$ requires a polynomial time (see Proposition 5.1.44). On the other hand, all known algorithms to compute discrete logarithms are exponential and it is conjectured that there are no polynomial ones. So the exponential function over $\mathbb{F}_q$ can be regarded as a one-way function. However, it must be remarked that some algorithms to compute discrete logarithms, one example of which we shall shortly illustrate with the so-called *Baby step–giant step* algorithm, are, in particular cases, quite effective.

In general, in cryptography it is interesting to consider those groups $G$ for which, as for $\mathbb{F}_q^*$, computing powers is computationally easy (for instance, requiring polynomial time), while computing discrete logarithms is computationally far harder (for instance, exponential). This may yield one-way functions.

**Example 7.5.3.** Let $\mathbb{F}_q$ be a *large enough* finite field, that is of order $q = p^f$ with $p$ a large prime number, and let $f(x) \in \mathbb{F}_q[x]$ be a polynomial such that the corresponding polynomial function $f : \mathbb{F}_q \to \mathbb{F}_q$ is injective. As we are well aware, the function $f$ can be computed in polynomial time, while the inverse function $f^{-1}$ may be quite hard to compute in practice. It is conjectured that such a polynomial function is in many cases a one-way function.

The next example describes another kind of function $f$ which is computed more easily than $f^{-1}$, but in which, unlike exponential and logarithmic functions, both are computed in polynomial time.

**Example 7.5.4.** We may consider enciphering a message $X \in \mathbb{Z}_N^s$ by multiplying it on the left by a square matrix $A$ of order $s$. We have

$$f : X \in \mathbb{Z}_N^s \to A \cdot X \in \mathbb{Z}_N^s.$$

Computing $Y = f(X) = A \cdot X$ requires about $s^2$ operations (see Exercise A2.19). Computing $X = f^{-1}(Y) = A^{-1}Y$, on the contrary, has a much greater computational cost, as it requires inverting a square matrix of order $s$, which implies about $s^3$ operations (see Exercise A2.22).

The existence of one-way functions has not yet been rigorously proved. However, there are many good candidates, like the exponential functions on finite fields, mentioned above. In practice, we are interested in a specific kind of one-way functions that can be defined in a vague but sufficiently eloquent way, as follows:

**Definition 7.5.5.** *A one-way function $f : S \to T$ is said to be a* trapdoor *function if with some further information it becomes computationally feasible to find, for all $y \in f(S)$, an element $x \in S$ such that $f(x) = y$.*

Public key techniques make use of functions of this kind, in the sense that, basically, they are used as enciphering functions. We shall shortly show two examples which shall illustrate this basic idea, which has remained so far quite indeterminate.

### 7.5.1 An algorithm to compute discrete logarithms

We devote a short section to describe an algorithm, the so-called *Baby step–giant step algorithm*, to compute discrete logarithms over the field $\mathbb{Z}_p$ with $p$ a prime number. So we work in $\mathbb{Z}_p^*$, whose elements shall be identified with $1, 2, \ldots, p-1$. Let $g$ be a generator of $\mathbb{Z}_p^*$. We want to determine the discrete logarithm $x$ of $y \in \mathbb{Z}_p$ in base $g$. We proceed as follows:

- **baby steps**: set $n$ equal to the least integer greater than $\sqrt{p}$ and compute the values $g^i \in \{1, \ldots, p-1\}$, for all $i \in \{0, \ldots, n-1\}$, inserting them in a list to be kept in the memory;
- **giant steps**: compute $g^n$ and then $g^{-n}$ and successively $yg^{-n}, yg^{-2n}, yg^{-3n}, \ldots,$ $yg^{-n^2}$. After each of these computations, compare the result with the numbers in the list created in the first step. As soon as we obtain an equality of the form $yg^{-jn} = g^i$, we have found the logarithm $x = jn + i$.

First of all, notice that the algorithm terminates and gives the desired logarithm. In fact, the logarithm exists and is a number $x$ in $\{1, \ldots, p-1\}$. Then, dividing $x$ by $n$ we have $x = nj + i$, with $0 \le i \le n-1$. On the other hand, as $x < p < n^2$ we also have $1 \le j \le n$.

We estimate next the complexity of the algorithm. There are $n$ *baby steps*, each having complexity $\mathcal{O}(\log n)$. So the total complexity of the *baby steps* is $\mathcal{O}(n \log n)$. Notice that the *baby steps* can be thought of as a kind of *precomputation*, in the sense that they are performed just once, independently of the number $y$ of which we are computing the logarithm.

Reasoning in a similar way, we see that the *giant steps* too, which do depend on the number whose logarithm we are computing, have complexity $\mathcal{O}(n \log n)$; so this is the complexity of the whole algorithm. It is exponential in $n$. There are further issues with this algorithm:

- the algorithm needs a large amount of memory if $p$ is large, as a list consisting of $[\sqrt{p}] + 1$ integers must be kept in memory;
- moreover, comparing the numbers in this list and the numbers computed in the *giant steps* has a computational cost, even if we have neglected it so far. A possible way to perform this comparison consists in dividing a number by the other one and checking whether the quotient is greater than zero or not. Clearly, the algorithm leads us to carry out $n^2$ comparisons, so all of them taken together yield an exponential complexity.

**Example 7.5.6.** Let us illustrate the above by means of a very simple example. Take $p = 11$ and $g = 2$, which is a generator of $\mathbb{Z}_{11}^*$. In this case we have $3 < \sqrt{11} < 4$, so $n = 4$. The baby steps yield the list

$$2^0 = 1, \qquad 2^1 = 2, \qquad 2^2 = 4, \qquad 2^3 = 8. \tag{7.3}$$

Compute next $2^4$, which equals 5 modulo 11, while its inverse is 9.

Suppose we want to compute the logarithm $\log_2 6$. So we perform the giant steps. In the first step we compute $6 \cdot 2^{-4}$, which equals 10 modulo 11. As this number is not included in the list (7.3), we have not found the required logarithm. Perform another giant step, computing $6 \cdot 2^{-8}$, which, modulo 11, equals 2. So we find the relation $6 \cdot 2^{-8} = 2$, or $6 = 2^9 \pmod{11}$, that is, $\log_2 6 = 9$.

For other algorithms to compute discrete logarithms, see [30], Ch. IV, or [44].

## 7.6 The knapsack problem and its applications to cryptography

Suppose we are about to leave for an excursion. We have to pack our knapsack and we want to make maximum use of the available space. We have a number, say $n$, of different objects, having volume $v_1, v_2, \ldots, v_n$; we know that the knapsack contains a volume $V$, and we want to carry the greatest possible load. How do we find it? We are looking for a subset $J \subseteq \{1, 2, \ldots, n\}$ such that

$$V = \sum_{j \in J} v_j. \tag{7.4}$$

This scheme may be applied to several similar problems. Assume we have to pay 2 Euros and have at our disposal 2-, 5-, and 10-cent coins: how can we pay with the least number of coins? Or with the largest number, so as to rid ourselves of as many coins as possible? Moreover, how many possible ways are there of paying?

Let us turn back to the knapsack problem and describe a cryptographic system relying on it, devised by Merkle and Hellman in 1978.

First of all, rephrase the problem as follows. Given $n$ positive integers $a_1, a_2, \ldots, a_n$ and a positive integer $m$, can we find $n$ integers $x_1, x_2, \ldots, x_n$ with $x_i \in \{0, 1\}$ so that our integer $m$ can be written as

$$m = a_1 x_1 + a_2 x_2 + \cdots + a_n x_n? \tag{7.5}$$

In other words, is it possible to write $m$ as a sum of *some* of the $a_i$s? It is not always possible and a solution, if it exists, may or may not be unique. The three following examples demonstrate the different possible cases.

**Example 7.6.1.** Set $n = 5$, $(a_1, a_2, \ldots, a_5) = (2, 7, 8, 11, 12)$ and $m = 21$. It is immediate to see that $21 = 2 + 8 + 11$ and $21 = 2 + 7 + 12$; so we have two solutions, and they are the only ones. In explicit form, the first solution is $x_1 = x_3 = x_4 = 1$ and $x_2 = x_5 = 0$, while the second one is $x_1 = x_2 = x_5 = 1$ and $x_3 = x_4 = 0$.

**Example 7.6.2.** Consider now the same $a_i$s as before, but $m = 1$. As $m$ is smaller than each of the $a_i$s, it is not possible to write $m$ as a combination of the $a_i$s with coefficients 0 or 1.

**Example 7.6.3.** If $a_i = 2^{i-1}$, for $i = 1, \ldots, n$, solving the knapsack problem means finding *the binary representation of $m$* which, as we know, exists and is unique.

In principle, in order to find the solution, if it exists, it suffices to consider all the sums of the form (7.4) with $J$ ranging among all the subsets of $\{1, \ldots, n\}$. As is well known (see Exercise A1.22), there are $2^n$ such subsets, including the empty set.

If $n$ is small, this kind of inspection can be carried out, but if $n$ is large, it is computationally unfeasible, as it is likely to require an exponential algorithm. In general, in fact, no algorithm to solve the knapsack problem is known, apart from trying out all the possibilities.

We may ask if there are *integer* solutions $x_i \in \mathbb{N}$ of Equation (7.5). However, this general formulation of the knapsack problem is beyond the scope of this text.

**Remark 7.6.4.** The knapsack problem is known to belong to a category of very hard problems, the so-called $\mathcal{NP}$-problems, for which it is conjectured that *no algorithm giving the solution in polynomial time exists*.

More in detail, let $\mathcal{P}$ be the class of problems $P$ for which a deterministic algorithm that solves $P$ in polynomial time exists. We have seen so far several examples of problems lying in class $\mathcal{P}$: for instance, the problem of finding the greatest common divisor of two integer numbers, or that of recognising whether a number is prime.

A problem $P$ is said to belong to class $\mathcal{NP}$ if there are algorithms - not necessarily polynomial ones - solving it and if it is possible to verify whether given data solve the problem or not, using a polynomial deterministic algorithm.

For instance, the problem of factoring an integer number $n$ is of this kind. The sieve of Eratosthenes is a non-polynomial algorithm solving the problem, while, given a number $m$, we can verify in polynomial time, using the Euclidean algorithm, whether $m$ divides $n$ or not.

It is easy to see that the knapsack problem is too in $\mathcal{NP}$.

Clearly, $\mathcal{P} \subseteq \mathcal{NP}$; the main conjecture in complexity theory states that $\mathcal{P} \neq \mathcal{NP}$.

A problem $P$ in $\mathcal{NP}$ is said to be $\mathcal{NP}$-complete if, for every other problem $Q$ in $\mathcal{NP}$, there is a polynomial deterministic algorithm that reduces solving $Q$ to solving $P$. Clearly, if $P$ is $\mathcal{NP}$-complete and if there were a polynomial deterministic algorithm that solves $P$, then every problem in $\mathcal{NP}$ would also be in $\mathcal{P}$. So, if the main conjecture in complexity theory is true, there are no polynomial deterministic algorithms that solve $\mathcal{NP}$-complete problems. These problems are, basically, the *most computationally difficult* problems in $\mathcal{NP}$.

As we said at the beginning of this remark, the knapsack problem is known to be $\mathcal{NP}$-complete (see [23]).

A special case of our problem is the one in which the sequence $a_1, a_2, \ldots,$ $a_n$ is *superincreasing*, in the sense of the following definition.

**Definition 7.6.5.** *A sequence of $n$ positive integers $a_1, a_2, \ldots, a_n$ is super-increasing if the following inequalities hold*

$$a_1 < a_2,$$
$$a_1 + a_2 < a_3,$$
$$a_1 + a_2 + a_3 < a_4,$$
$$\vdots$$
$$a_1 + a_2 + \cdots + a_{n-1} < a_n.$$

Is there a solution to the knapsack problem in this case? The answer is not always in the affirmative but, if a solution exists, then it is unique and can be found in polynomial time. Indeed, to find the value $x_1, \ldots, x_n$ such that $m = \sum_{i=1}^{n} x_i a_i$ with $x_i \in \{0, 1\}$ the following algorithm may be used

As a first thing, determine $x_n$, by noticing that necessarily:

$$x_n = \begin{cases} 1 & \text{if } m \geq a_n, \\ 0 & \text{if } m < a_n. \end{cases}$$

To determine $x_{n-1}$, do the same, substituting $m - x_n a_n$ for $m$. In other words, we look for a solution to the knapsack problem by trying to express $m - x_n a_n$ as

$$m - a_n x_n = a_1 x_1 + a_2 x_2 + \cdots + a_{n-1} x_{n-1}.$$

Notice that, clearly, the sequence $a_1, a_2, \ldots, a_{n-1}$ is superincreasing too. So we have

$$x_{n-1} = \begin{cases} 1 & \text{if } m - x_n a_n \geq a_{n-1}, \\ 0 & \text{if } m - x_n a_n < a_{n-1}. \end{cases}$$

In general, having found $x_n, \ldots, x_{j+1}$, we shall set

$$
x_j = \begin{cases}
1 & \text{if } m - \sum_{i=j+1}^{n} x_i a_i \geq a_j, \\
0 & \text{if } m - \sum_{i=j+1}^{n} x_i a_i < a_j.
\end{cases}
$$

It is clear that if $m - \sum_{i=j+1}^{n} x_i a_i = 0$, we have found the solution, which is clearly unique. If on the other hand $m - \sum_{i=j+1}^{n} x_i a_i > 0$ but $a_j, \ldots, a_1$ are greater than $m - \sum_{i=j+1}^{n} x_i a_i$, then no solution exists.

It is not hard to see that the algorithm just described is of polynomial type (see Exercise A7.9).

Let us illustrate with an example this algorithm.

**Example 7.6.6.** Consider the superincreasing sequence $(1, 4, 6, 13, 25)$. How do we compute the solution when $m = 26$ according to the algorithm?

As the rightmost term of our sequence is 25 and $25 < 26$, then we have to choose $x_5 = 1$. Now we carry on the procedure with $26 - 1 \cdot 25 = 1 < 13$, so we set $x_4 = 0$. As $26 - 1 \cdot 25 - 0 \cdot 13 = 1 < 6$, then $x_3 = 0$, hence again $x_2 = 0$. The last step, applied to $26 - 1 \cdot 25 - 0 \cdot 13 - 0 \cdot 6 - 0 \cdot 4 = 1 = a_1$ gives $x_1 = 1$ and so the solution. In fact, we have

$$
26 = 1 \cdot 1 + 0 \cdot 4 + 0 \cdot 6 + 0 \cdot 13 + 1 \cdot 25.
$$

Notice that if the sequence were $(2, 3, 6, 13, 25)$, there would have been no solution, as in the last step we should have written 1 as $x_1 \cdot 2$, which is not possible. In other words, the solution does not exist because $a_1 = 2 > 26 - 1 \cdot 25 - 0 \cdot 13 - 0 \cdot 6 - 0 \cdot 3$.

Let us see now how to construct a *cipher* related to the knapsack problem.

### 7.6.1 Public key cipher based on the knapsack problem, or Merkle–Hellman cipher

The cipher consists of the following steps.

- Each user $X$ chooses a superincreasing sequence $a_1, a_2, \ldots, a_N$ of a fixed length $N$, an integer $m$ such that $m > 2a_N$, and an integer $w$ relatively prime with $m$. These data are kept *secret*.
- User $X$ computes the transformed sequence

$$
b_j = w a_j \bmod m, \qquad \text{for } j = 1, \ldots, N.
$$

The sequence $b_1, b_2, \ldots, b_N$ is made *public* by $X$ and is the *enciphering key*.

- A user $Y$ may send a message $p$ to $X$ acting as follows. First of all, he transforms each letter of the text into its binary equivalent using Table 7.5 on page 333. Next, he splits up the resulting sequence of 0s and 1s into blocks of length $N$, adding at the end, if necessary, a number of 1s so as to have blocks all of the same length $N$. For each block, say $p = x_1 x_2 \ldots x_N$, the user $Y$ applies the transformation to encipher $p \rightarrow c = b_1 x_1 + \cdots + b_N x_N$, and sends the enciphered text $c$ to $X$.
- How has $X$ to proceed to decipher $Y$'s message? As a first thing, $X$ computes the *deciphering key*, that is, $k_d = (m, \bar{w})$ with $w\bar{w} = 1 \pmod m$. Next, he computes

$$v = \bar{w}c \bmod m = \bar{w} \cdot (b_1 x_1 + \cdots + b_N x_N) \bmod m. \qquad (7.6)$$

By definition of $\bar{w}$ and of the $b_i$s, we have

$$v = (\bar{w} b_1 x_1 + \cdots + \bar{w} b_N x_N) \equiv \sum_{i=1}^{N} x_i a_i \pmod m. \qquad (7.7)$$

As the sequence $a_1, a_2, \ldots, a_N$ is superincreasing, we have

$$a_1 + \cdots + a_{N-1} + a_N < a_N + a_N = 2a_N < m$$

and so $v = \sum_{i=1}^{N} x_i a_i$.

- Now, $X$ knows the integer $\sum_{i=1}^{N} x_i a_i$ and the superincreasing sequence $a_1$, $a_2$, ..., $a_N$; from them he has to reconstruct the integers $x_1$, $x_2$, ..., $x_N$. This is easily done, in polynomial time, by using the knapsack algorithm for superincreasing sequences.

**Remark 7.6.7.** Choosing the data in a sufficiently general way, *the sequence $b_1$, ..., $b_N$ is no more superincreasing*, and decrypting illegitimately the message starting with $c = b_1 x_1 + \cdots + b_N x_N$ is a computationally hard problem.

Actually, in 1982 Shamir [52] found a polynomial algorithm that allows one to decrypt the message. The main remark by Shamir is that the knapsack problem to be solved for a sequence of the form $b_1, \ldots, b_N$ is not completely general. Indeed $b_1, \ldots, b_N$, even if it is not a superincreasing sequence, may be obtained from a superincreasing sequence $a_1, \ldots, a_N$ by means of a very simple transformation.

For these reasons, the cipher just described cannot be considered secure. There are several ways to get around this problem, and quite recently some variations of Merkle–Hellman cipher have been found, that have not yet succumbed to cryptanalysts' attacks. However, the description of these variations goes beyond the scope of this book.

In Table 7.10 on page 350 we show an example of the use of this cipher.

## 7.7 The *RSA* system

In this section we describe a public key cryptosystem devised by W. Diffie and M. E. Hellman [19], but commonly called *RSA system*, from the names

**Table 7.10.** Knapsack problem cipher

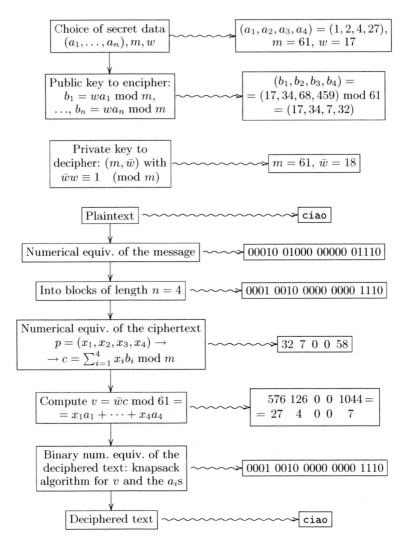

of those who first implemented it: L. M. Adleman, R. L. Rivest, A. Shamir at M.I.T. (Massachusetts Institute of Technology) [1]. This system, as already remarked, uses a public key that allows enciphering a message but not deciphering it. Each user divulges his enciphering key, so anybody may securely communicate with him. This happens using an enciphering method which is a trapdoor function. The user who divulges the enciphering key *keeps secret an additional piece of information*, by which he alone will be able, by inverting the trapdoor function, to decipher the messages he will receive.

Applications of systems of this kind are innumerable: sending enciphered messages among several users, digital authentication of signatures, access to secure archives or databases or simply services such as credit cards, pay-per-view television programmes, and so forth. We shall not enter into the technical details, leaving the reader with the task of thinking about how the systems we are going to describe can be applied to these situations.

### 7.7.1 Accessing the *RSA* system

Suppose we want to use the *RSA* system to exchange messages that are to be read only by the intended addressees and not by eavesdroppers. Then we have to join the system, divulging the *enciphering key*, which is a pair of positive integers $(n, e)$, where $n$ is the product of two large prime numbers $p$ and $q$ we only know, and $e$ must be relatively prime with $\varphi(n) = (p - 1)(q - 1)$, that is, $\mathrm{GCD}(e, \varphi(n)) = 1$, or $\mathrm{GCD}(e, p - 1) = GCD(e, q - 1) = 1$. This pair of integers $(n, e)$ we divulge is kept in a publicly accessible directory.

**Remark 7.7.1.** How do we proceed *in practice* to find two *large* prime numbers having, for instance, 100 decimal digits? We generate a *random* 100-digit odd number $m$. *Random generation* of numbers is an interesting topic in mathematics and computer science, upon which we cannot dwell here. Here it suffices to know that there are programs that generate random numbers.

Apply next to $m$ a primality test. If $m$ passes it, then we have found a prime. Otherwise, we apply the primality test to $m + 2$. If $m + 2$ is not prime either, we test $m + 4$, and so on, until a prime number is found. Recall that, by the prime number theorem, the number $\pi(m)$ of prime numbers smaller than $m$ is of the same order as $m/\log m$ (see page 155). A probabilistic rephrasing of the same theorem states that *the frequency with which prime numbers appear near $m$ is $1/\log m$*. So we may expect to have to perform $\mathcal{O}(\log m)$ primality tests before finding the first prime number larger than $m$. So the number of tests to be performed is polynomial, and so it is feasible. On the other hand, the computational cost of the test itself is usually high.

Returning to the *RSA* system, each user $U$ will act in the same way, that is to say, divulging a pair of integers $(n_U, e_U)$ verifying the same conditions: $n_U$ has to be the product of two prime numbers $p_U$ and $q_U$ that must be large and have to be kept secret, only known to user $U$, while the second number has to be chosen by $U$ in such a way that $\mathrm{GCD}(e_U, p_U - 1) = 1$ and $\mathrm{GCD}(e_U, q_U - 1) = 1$.

We emphasise the fact that *the pair $(n_U, e_U)$ is publicly known*, that is, every user who so desires may look it up, while *the factorisation of $n_U$ is not public and is only known to $U$*. To see how the *RSA* system works, let us consider an example in detail. The general scheme of the procedure is described in Table 7.12 on page 360.

**Example 7.7.2.** In the public directory, next to the name of each person, their enciphering key will be shown, that is, the pair of integers the user has chosen: for instance,

$$\begin{aligned}
\text{Ariadne divulges } A &= (77,\ 13),\\
\text{Beatrix divulges } B &= (1003,\ 3),\\
\text{Charles divulges } C &= (247,\ 5),\\
\text{David divulges } D &= (703,\ 7).
\end{aligned} \qquad (7.8)$$

The numbers have been chosen by the users according the requisites, as

$$\begin{aligned}
n_A &= 77 = 7 \cdot 11, & \text{GCD}(13,6) &= \text{GCD}(13,10) = 1,\\
n_B &= 1003 = 17 \cdot 59, & \text{GCD}(3,16) &= \text{GCD}(3,58) = 1,\\
n_C &= 247 = 13 \cdot 19, & \text{GCD}(5,12) &= \text{GCD}(5,18) = 1,\\
n_P &= 703 = 19 \cdot 37, & \text{GCD}(7,18) &= \text{GCD}(7,36) = 1.
\end{aligned}$$

Notice that in this example we have chosen small numbers, for which it is easy to find the two prime numbers $p_U$ and $q_U$ such that $n_U = p_U q_U$. In general, user $U$, to be safe, shall use an integer $n_U$ that is the product of two primes of about 100 decimal digits, so the number $n_U$ that is their product and that will be divulged, will have about 200 digits. Notice that in order to implement these kinds of ciphers there is a *crucial* need for many *large* prime numbers. By the way, this fact largely justifies the research about primality tests, as well as the hunt for ever larger prime numbers, which is often covered even by the media. Returning to our example, let us see the next steps, after the publication of the key chosen by each of the users.

### 7.7.2 Sending a message enciphered with the *RSA* system

User $A$, Ariadne, has received from user $B$, Beatrix, a message saying: *Which course do you prefer?* She wants to answer:

$$\boxed{\text{algebra}}.$$

To send her answer to Beatrix, Ariadne will have to proceed as follows.

(1) As a first thing, she transforms each letter of the message into an integer using the 2-digit numerical equivalence, as in Table 7.5 on page 333. Indeed, for the enciphering we are about to describe we shall need the letter-number correspondence assigning two digits to each number associated with a letter. The number sequence corresponding to our message will be

$$\boxed{00\ 11\ 06\ 04\ 01\ 17\ 00}.$$

(2) Next, Ariadne looks up in the official directory (7.8) the pair of numbers $(n_B, e_B)$ corresponding to Beatrix. Presently she just needs the first number, $n_B = 1003$.

(3) Now she has to split up the message to be sent into unitary messages *so that the integer associated with each unitary message is smaller than*

$n_B = 1003$ *and relatively prime with* 1003. She notices that, having split up the message *algebra* into 2-letter blocks (or digraphs) as follows

$$\boxed{\text{al}} \;\; \boxed{\text{ge}} \;\; \boxed{\text{br}} \;\; \boxed{\text{ax}} ,$$

the numbers corresponding with the unitary messages are

$$\boxed{0011} \;\; \boxed{0604} \;\; \boxed{0117} \;\; \boxed{0023} ,$$

that is, the numbers 11, 604, 117, and 23, respectively, which are smaller than 1003 and relatively prime with 1003. Notice that, to find the GCD between these numbers and 1003, Ariadne uses the Euclidean algorithm, as she does not know the prime decomposition of $n_B$.

Notice further that Ariadne added the letter $x$ at the end of the last unitary message, to make it a digraph. If she had split up the message into 3- rather than 2-letter blocks, she would have found some unitary messages corresponding to integers greater than $n_B$:

$$alg \longrightarrow 606 < 1003,$$
$$ebr \longrightarrow 40117 > 1003,$$
$$axx \longrightarrow 2323 > 1003,$$

which would not have satisfied our requisites.

So the segments of the plaintext will be represented by the following numbers:

$$\boxed{P_1 = 11,} \;\; \boxed{P_2 = 604,} \;\; \boxed{P_3 = 117,} \;\; \boxed{P_4 = 23.}$$

If the numbers $P_i$ were not relatively prime with 1003, we might nevertheless proceed in a way not dissimilar from the one we are about to describe; we are not dwelling on the differences, which the reader may study in Exercise A7.12.

After these operations, we may assume without loss of generality that each unitary message $P_i$ meets the following two conditions:

$$\boxed{P_i < n_B, \qquad \text{GCD}(P_i, n_B) = 1.}$$

(4) Now the actual enciphering of the message to Beatrix begins, in such a way that the latter *may decipher it and be the only person able to do so in a reasonable time*. To encipher the message to be sent to Beatrix, Ariadne raises each $P_i$ to the $e_B$th power, $e_B$ being the second element of the pair associated with Beatrix.

So the enciphering function is

$$C_B : \mathcal{P} \longrightarrow \mathcal{C}, \qquad P \longrightarrow C = P^{e_B} \bmod n_B.$$

Thus, the enciphered message Ariadne will have to send will consist of the following numbers:

$$C_1 = P_1^{e_B} \bmod n_B = 11^3 \bmod 1003 = 328,$$
$$C_2 = P_2^{e_B} \bmod n_B = 604^3 \bmod 1003 = 797,$$
$$C_3 = P_3^{e_B} \bmod n_B = 117^3 \bmod 1003 = 825,$$
$$C_4 = P_4^{e_B} \bmod n_B = 23^3 \bmod 1003 = 131.$$

Then Beatrix receives the following message:

$$\boxed{C_1 = 328,} \quad \boxed{C_2 = 797,} \quad \boxed{C_3 = 825,} \quad \boxed{C_4 = 131,}$$

consisting of the unitary messages $C_i$, $i = 1, \ldots, 4$. Now Ariadne has carried out her task: she has sent Beatrix the enciphered message. Now Beatrix will have to decipher it.

### 7.7.3 Deciphering a message enciphered with the *RSA* system

Beatrix has received the message

$$\boxed{C_1 = 328,} \boxed{C_2 = 797,} \boxed{C_3 = 825,} \boxed{C_4 = 131,}$$

and has to decipher it, that is, for every $C_i$ has to find the original message $P_i$, knowing that
$$C_i = P_i^{e_B} \bmod n_B.$$
So Beatrix has to determine the deciphering function

$$D_B : \mathcal{C} \longrightarrow \mathcal{P}$$

such that $D_B(E_B(P_i)) = P_i$ for all $i$. How can she find it? A priori it would seem that, to solve this problem, Beatrix would have to find a discrete logarithm which, as we have remarked, is computationally quite hard. However, we have already remarked at the beginning that Beatrix actually has *an additional piece of information enabling her to decipher the message she has received* without difficulty. Let us see which piece of information she has and how she uses it. First of all, Beatrix determines $d_B$, with

$$1 \le d_B < \varphi(n_B) = (p_B - 1)(q_B - 1),$$

such that $d_B$ is a solution of the following congruence

$$\boxed{e_B d_B \equiv 1 \quad (\bmod \ \varphi(n_B))}. \tag{7.9}$$

Such a solution exists and is unique, as $\mathrm{GCD}(e_B, \varphi(n_B)) = 1$.

   In our case, $n_B = 1003$, and Beatrix knows $\varphi(1003)$ because she knows that $1003 = 17 \cdot 59$. So $\varphi(1003) = 16 \cdot 58 = 928$. Then the solution $d_B$ of the congruence (7.9), that is of $3x \equiv 1 \pmod{928}$, is

$$\boxed{d_B = 619}.$$

This number is truly to be framed, because, as we shall shortly see, Beatrix is the only person who can decipher the message because she is the only one who, knowing the factorisation of 1003, is able to compute its Euler function and so $d_B$, which is Beatrix's *private key* to decipher the messages sent to her. Let us what she does to decipher the messages $C_i$.

Beatrix raises each $C_i$ to the power $d_B = 619$, that is, computes

$$328^{619}, \quad 797^{619}, \quad 825^{619}, \quad 131^{619}.$$

The exponent 619 is large, but we know how to proceed in situations like this (see § 3.3.1). The number 619 is written in base 2, that is

$$619 = (1001101011)_2 = 512 + 64 + 32 + 8 + 2 + 1.$$

So we have

$$C_i^{619} = C_i^{512} \cdot C_i^{64} \cdot C_i^{32} \cdot C_i^{8} \cdot C_i^{2} \cdot C_i^{1}$$

and the powers are easily computed according to the following table, for $C_1 = 328$:

| $k$ | $C_1^k \bmod 1003$ |
|---|---|
| 1 | $\boxed{328}$ |
| 2 | $328^2 \bmod 1003 = \boxed{263}$ |
| $2^2=4$ | $263^2 \bmod 1003 = 965$ |
| $2^3 = 8$ | $965^2 \bmod 1003 = \boxed{441}$ |
| $2^4 = 16$ | $441^2 \bmod 1003 = 902$ |
| $2^5 = 32$ | $902^2 \bmod 1003 = \boxed{171}$ |
| $2^6 = 64$ | $171^2 \bmod 1003 = \boxed{154}$ |
| $2^7 = 128$ | $154^2 \bmod 1003 = 647$ |
| $2^8 = 256$ | $647^2 \bmod 1003 = 358$ |
| $2^9 = 512$ | $358^2 \bmod 1003 = \boxed{783}$ |

where we have framed the factors to be multiplied. From the table we may easily see that

$$328^{619} = 328^{512} \cdot 328^{64} \cdot 328^{32} \cdot 328^{8} \cdot 328^{2} \cdot 328^{1} \equiv$$
$$\equiv 783 \cdot 154 \cdot 171 \cdot 441 \cdot 263 \cdot 328 \equiv 11 \quad (\bmod \ 1003).$$

Notice that, by raising $C_1 = 328$ to the exponent $d_B = 619$, we have obtained the number $11 = P_1$, which is the number corresponding to the first part of the original message.

We do the same for the three other message segments $C_2$, $C_3$ and $C_4$. So we get Table 7.11 on page 356 where, as above, we have framed the factors to be multiplied. In conclusion,

**Table 7.11.** Powers of the numbers $C_i$

| $k$ | $C_2^k \bmod 1003$ | $C_3^k \bmod 1003$ | $C_4^k \bmod 1003$ |
|---|---|---|---|
| 1 | $\boxed{797}$ | $\boxed{825}$ | $\boxed{131}$ |
| 2 | $797^2 \equiv \boxed{310}$ | $825^2 \equiv \boxed{591}$ | $131^2 \equiv \boxed{110}$ |
| 4 | $310^2 \equiv 815$ | $591^2 \equiv 237$ | $110^2 \equiv 64$ |
| 8 | $815^2 \equiv \boxed{239}$ | $237^2 \equiv \boxed{1}$ | $64^2 \equiv \boxed{84}$ |
| 16 | $239^2 \equiv 953$ | 1 | $84^2 \equiv 35$ |
| 32 | $953^2 \equiv \boxed{494}$ | $\boxed{1}$ | $35^2 \equiv \boxed{222}$ |
| 64 | $494^2 \equiv \boxed{307}$ | $\boxed{1}$ | $222^2 \equiv \boxed{137}$ |
| 128 | $307^2 \equiv 970$ | 1 | $137^2 \equiv 715$ |
| 256 | $970^2 \equiv 86$ | 1 | $715^2 \equiv 698$ |
| 512 | $86^2 \equiv \boxed{375}$ | $\boxed{1}$ | $698^2 \equiv \boxed{749}$ |

$$328^{619} \bmod 1003 = 11,$$
$$797^{619} \bmod 1003 = 604,$$
$$825^{619} \bmod 1003 = 117,$$
$$131^{619} \bmod 1003 = 23.$$

These are to be seen as 4-digit numbers:

$$0011, \quad 0604, \quad 0117, \quad 0023,$$

and correspond to the four original message segments, which were digraphs. So we have to split each of them into two parts, each of which represents a letter. So Beatrix, using Table 7.5 on page 333, finds

$$00\ 11\ 06\ 04\ 01\ 17\ 00\ 23$$
$$a\quad l\quad g\quad e\quad b\quad r\quad a\quad x$$

and gets to know the course Ariadne likes best.

With did this work? That is, why raising $C_i$ to the exponent $d_B$ we get back $P_i$ such that $C_i = P_i^{e_B}$? In other words, why is

$$D_B : \mathcal{C} \longrightarrow \mathcal{P}, \qquad C_i \longrightarrow C_i^{d_B}$$

the deciphering function? Here follows the reason.

### 7.7.4 Why did it work?

First of all, notice that the congruence (7.9) has exactly one solution modulo $\varphi(n_B)$, because the coefficient $e_B$ is such that

$$\mathrm{GCD}(e_B, p_B - 1) = 1, \qquad \mathrm{GCD}(e_B, q_B - 1) = 1,$$

so also $\mathrm{GCD}(e_B, (p_B - 1)(q_B - 1)) = 1$.

**Remark 7.7.3.** Beatrix is the only person able to solve congruence (7.9), because she is the only one to know the Euler function $\varphi(n_B) = (p_B - 1)(q_B - 1)$, as she knows the prime factors $p_B$ and $q_B$ of $n_B$. In fact, notice that, as $p_B$ and $q_B$ are *large* prime numbers, factoring $n_B$ normally takes a very long time. So, in fact, Beatrix is the only one to know this factorisation.

Actually, one might doubt that knowing $\varphi(n_B)$ is *equivalent* to knowing the prime factors $p_B$ and $q_B$. Of course, whoever knows these factors knows $\varphi(n_B)$ too. But, is it possible to know $\varphi(n_B)$ without knowing $p_B$ and $q_B$? The answer is no: if $\varphi(n_B)$ is known, then $p_B$ and $q_B$ can be reconstructed immediately, in polynomial time. The easy proof is left as an exercise (see Exercise A7.10).

Notice now that $d_B$ is actually the *private key* allowing Beatrix to decipher the message. Indeed, setting $P = P_i$ and $C = C_i$, we have

$$P \equiv C^{d_B} \pmod{n_B},$$

as

$$C^{d_B} \equiv (P^{e_B})^{d_B} = P^{e_B d_B} \pmod{n_B}.$$

On the other hand, $e_B d_B \equiv 1 \pmod{\varphi(n_B)}$ implies that $e_B d_B - 1$ is a multiple of $\varphi(n_B)$, that is $e_B d_B = 1 + \varphi(n_B)k$ for some $k$. So,

$$P^{e_B d_B} = P^{1 + \varphi(n_B) \cdot k} = P \cdot (P^{\varphi(n_B)})^k.$$

As $\mathrm{GCD}(P, n_B) = 1$, by Euler's theorem we have $P^{\varphi(n_B)} \equiv 1 \pmod{n_B}$; hence

$$P^{e_B d_B} \equiv P \pmod{n_B}.$$

So,

$$P \equiv C^{d_B} \pmod{n_B}.$$

As Ariadne has chosen $P < n_B$, there is no ambiguity in determining the number congruent to $C^{d_B}$ modulo $n_B$: it is the only such number between 0 and $n_B - 1$. Once this $P$ is found, Beatrix can read Ariadne's message.

**Remark 7.7.4.** We have said that the unitary message has to be smaller than $n_B$. We have just explained the reason of this request. An example will illustrate the need for it. Consider the message

no

to be sent to Ariadne, whose pair is $(n_A = 77, e_A = 13)$. We opt to consider the whole word *no* as unitary message (digraph). We proceed as above:

(1) transform the message into a number by associating to each letter its numerical equivalent. The associated number is found to be

1314;

notice that 1314 is greater than $n_A = 77$;

(2) raise 1314 to the power $e_A = 13$; we have

$$C_1 = 1314^{13} \equiv 26 \pmod{77};$$

(3) Ariadne receives the message

26.

To decipher this message Ariadne uses her private key, which is $d_A = 37$, the solution of the congruence

$$13d_A \equiv 1 \pmod{\varphi(77)}, \quad \text{that is} \quad 13d_A \equiv 1 \pmod{60}.$$

Raising 26 to the power 37, Ariadne gets 5 (mod 77), which she interprets as

f

so she cannot reconstruct the message she was sent. Also notice that 1314 (mod 77) = 5.

So, if we do not request for *the unitary message to be smaller than $n_A$*, that is, than the first element of the pair of numbers published by the addressee, it becomes *impossible* to define the deciphering transformation.

**Remark 7.7.5.** We have seen that, in order to send the message *algebra* to Beatrix, Ariadne split it up into digraphs. She had to do so by trial and error, verifying all the numbers corresponding to the single digraphs to be smaller than $n_B$ and relatively prime with it.

However, there is a better way of choosing how to split up the message: rather that splitting the original message *algebra*, it is more convenient to split the *numerical* message obtained by associating a 2-digit number with each letter. In this way there is a natural way of splitting it, as follows.

After transforming the message into a sequence of 2-digit numbers, consider the number consisting of the sequence of all the digits, which will be called *numerical message*. Split it up into $k$-digit blocks, where

$$\boxed{k = (\text{number of digits of } n_B) - 1}.$$

In this way, *without even having to examine the message*, each unitary numerical message is smaller than $n_B$. What's more, everybody concerned knows $n_B$ and knows that the sender will split up the message into blocks like this.

Let us illustrate this new method with an example.

**Example 7.7.6.** Suppose Ariadne, user $A$, sends Beatrix, user $B$, the message

$$\boxed{\text{come here}}$$

Then:

(1) first of all Ariadne transforms the message, ignoring spaces, into the sequence of 2-digits numbers

$$\boxed{02 \quad 14 \quad 12 \quad 04 \quad 07 \quad 04 \quad 17 \quad 04}$$

which will be written as

0214120407041704.

There is no ambiguity, as we know that the number associated with each letter of the original message consists of two digits. This is the *numerical message*;

(2) as the number $n_B = 1003$ has 4 digits, $A$ splits the numerical message into blocks of length $4 - 1 = 3$, that is, into trigraphs, as follows:

$$\boxed{021 \quad 412 \quad 040 \quad 704 \quad 170 \quad 423}.$$

Notice that $A$ has added 23, corresponding to the letter $x$, at the end of the message, so all unitary numerical messages consist of three digits. In this way, she has split the numerical message into unitary numerical messages that are trigraphs, and each unitary message is certainly smaller than $n_B = 1003$. Notice that this partitioning is not a partition of the original message *come here*, as the 3-digit unitary numerical blocks do not correspond to any letter group. As we shall see, this will not create any difficulty.

We still have to check that each $P_i$ is relatively prime with 1003. It is easily verified that the only exception is 170. However, we have no reason to worry: keeping in mind Exercise A7.12, we may go on.

So, the unitary numerical messages are

$$P_1 = 21, \quad P_2 = 412, \quad P_3 = 40, \quad P_4 = 704, \quad P_5 = 170, \quad P_6 = 423.$$

Notice that the operations carried out so far do not amount to any enciphering, as transforming a message into a numerical sequence is a standard operation, and so is partitioning it into 3-digit blocks according to the value $n_B$, which is known to everybody;

(3) the enciphered message will be represented by the following numbers:

$$C_1 = P_1^{e_B} \bmod n_B = 21^3 \bmod 1003 = 234,$$
$$C_2 = P_2^{e_B} \bmod n_B = 412^3 \bmod 1003 = 353,$$
$$C_3 = P_3^{e_B} \bmod n_B = 40^3 \bmod 1003 = 811,$$
$$C_4 = P_4^{e_B} \bmod n_B = 704^3 \bmod 1003 = 54,$$
$$C_5 = P_5^{e_B} \bmod n_B = 170^3 \bmod 1003 = 306,$$
$$C_6 = P_6^{e_B} \bmod n_B = 423^3 \bmod 1003 = 587.$$

So Beatrix receives the following sequence of unitary messages:

$$\boxed{C_1 = 234, \; C_2 = 353, \; C_3 = 811, \; C_4 = 54, \; C_5 = 306, \; C_6 = 587.}$$

To decipher it, she raises each $C_i$ to her private key $d_B = 619$, that is, she computes

$$234^{619}, \; 353^{619}, \; 811^{619}, \; 54^{619}, \; 306^{619}, \; 587^{619}.$$

In this way, Beatrix finds

$$234^{619} \bmod 1003 = 21, \qquad\qquad 353^{619} \bmod 1003 = 412,$$
$$811^{619} \bmod 1003 = 40, \qquad\qquad 54^{619} \bmod 1003 = 704,$$
$$306^{619} \bmod 1003 = 170, \qquad\qquad 587^{619} \bmod 1003 = 423.$$

This 3-digit blocks (completed with a leading zero where necessary), regrouped in twos, give

$$02, \; 14, \; 12, \; 04, \; 07, \; 04, \; 17, \; 04, \; 23$$

and now Beatrix can read the message *comeherex*, which she understand as *come here*. This example is shown in Table 7.12 on page 360.

**Table 7.12.** Example of use of $RSA$ system

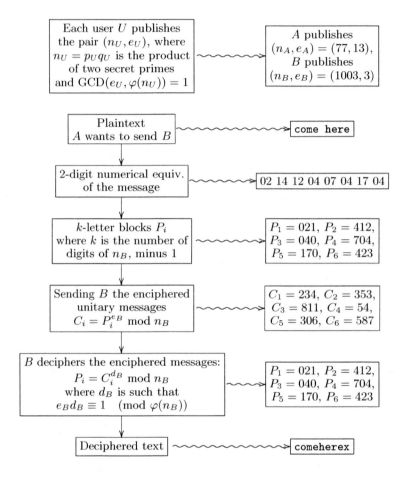

| Each user $U$ publishes the pair $(n_U, e_U)$, where $n_U = p_U q_U$ is the product of two secret primes and GCD$(e_U, \varphi(n_U)) = 1$ | $A$ publishes $(n_A, e_A) = (77, 13)$, $B$ publishes $(n_B, e_B) = (1003, 3)$ |

| Plaintext $A$ wants to send $B$ | come here |

| 2-digit numerical equiv. of the message | 02 14 12 04 07 04 17 04 |

| $k$-letter blocks $P_i$ where $k$ is the number of digits of $n_B$, minus 1 | $P_1 = 021,\ P_2 = 412,$ $P_3 = 040,\ P_4 = 704,$ $P_5 = 170,\ P_6 = 423$ |

| Sending $B$ the enciphered unitary messages $C_i = P_i^{e_B} \bmod n_B$ | $C_1 = 234,\ C_2 = 353,$ $C_3 = 811,\ C_4 = 54,$ $C_5 = 306,\ C_6 = 587$ |

| $B$ deciphers the enciphered messages: $P_i = C_i^{d_B} \bmod n_B$ where $d_B$ is such that $e_B d_B \equiv 1 \pmod{\varphi(n_B)}$ | $P_1 = 021,\ P_2 = 412,$ $P_3 = 040,\ P_4 = 704,$ $P_5 = 170,\ P_6 = 423$ |

| Deciphered text | comeherex |

## 7.7.5 Authentication of signatures with the $RSA$ system

The $RSA$ system allows one to solve of an important problem which is more and more relevant in this era of telecommunications: the problem of *digitally authenticating a signature*.

If Beatrix receives a message from a person signing herself Ariadne, how can she be sure the sender was actually Ariadne? The certainty may be achieved as follows.

Ariadne writes her message $P_1$, putting at the end her signature $F$; to authenticate the signature, Ariadne adds after the message $P_1$ the message

$$P_2 = F^{d_A} \bmod n_A,$$

where $d_A$ is her *private key*, that is, the key known only to her, because only she knows the factorisation of her public key $n_A$. Then she sends Beatrix the message $P$ consisting of the two messages $P_1$ and $P_2$ as usual, that is, raising $P_1$ and $P_2$ to the power $e_B$ and reducing them modulo $n_B$.

On receiving the message, Beatrix reads it using her private key $d_B$. Deciphering message $P_1$, she learns that the message was sent by Ariadne, because the message is signed with Ariadne's signature $F$. But was really Ariadne, and not someone else, who used that signature? Here the section $P_2$ of the message gives an answer. Indeed, it consists of some undecipherable characters, which nevertheless contain the *proof of the authenticity of the signature*.

Now Beatrix, to verify the authenticity, has to proceed as follows. To decipher $P_2$ she cannot use *her private key* $d_B$, which would be useless, as the original message $F$ was enciphered raising it not to the power $e_B$ but to $d_A$. Instead, Beatrix uses Ariadne's public key $e_A$. In this way *she obtains Ariadne's signature* $F$, because

$$P_2^{\,e_A} \equiv \left(F^{d_A}\right)^{e_A} = F^{d_A e_A} \equiv F \pmod{n_A}.$$

This signature has to be authentic, as only Ariadne knows her private key. If what appeared were not Ariadne's signature $F$, the message would have been a fake. Basically, to authenticate a signature, the *sender* uses *her private key*, rather than the *addressee*.

**Example 7.7.7.** Recall that Ariadne published the pair $(n_A = 77, e_A = 13)$. As $77 = 11 \cdot 7$, Ariadne's private key is $d_A = 37$, because $37 \cdot 13 \equiv 1 \pmod{60}$, where $60 = \varphi(77)$. In sending a message to Beatrix, Ariadne authenticates her signature, which we assume to be $F = 5$, raising 5 to the power $d_A$ modulo $n_A$:

$$5^{37} \bmod 77 = 47.$$

Beatrix verifies the authenticity of the signature by raising 47 to the power $e_A = 13$ modulo $n_A$:
$$47^{13} \bmod 77 = 5,$$

that is, she gets again Ariadne's signature. So Beatrix is sure that the message's author is Ariadne.

The previous example is summarised in Table 7.13.

**Table 7.13.** Authentication of a signature with the *RSA* system

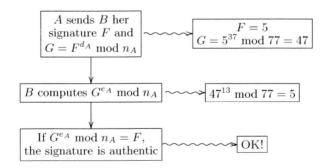

## 7.7.6 A remark about the security of *RSA* system

The security of *RSA* system lies in the fact that, as already emphasised several times in earlier chapters, so far there is no efficient algorithm to factor large numbers. If $A$ sends $B$ a message $C$, an unauthorised eavesdropper who tried to decrypt it should be able to find the factorisation of $n_B$. To find it, when $n_B$ is the product of two 60-digit primes, even using the most advanced algorithms and the fastest computers, would require several months, if not years. The situation is even more unfeasible if we choose primes with 100 or more digits: in this case factoring $n$ is, in practice, impossible. However, if this is true *in general*, it is not *always* so, as the following episode shows.

In August 1977 the three inventors of the public key *RSA* cryptosystem, Rivest, Shamir and Adleman, at MIT, challenged from Martin Gardner's column *Mathematical Games* the readers of *Scientific American* to decrypt a message corresponding to a 129-digit number, an operation they believed to require billions years. They offered a reward of $100 to whomever found the solution.

We are not going to give all the details of Rivest, Shamir and Adleman's problem. Suffice it to say that, in order to decrypt their original message it was necessary to factor the number

$$N = 114381625757888867669235779976146612010218296721242362562561 84293$$
$$5706935245733897830597123563958705058989075147599290026879543541,$$

which had been published together with the number $e = 9007$. Basically, $(N, e)$ was the *public key* of Rivest, Shamir and Adleman.

To ensure that the message came from the MIT team, the following *digital signature* was added, using the private key of the algorithm, that is the number $d$ such that $ed \equiv 1 \pmod{N}$:

$$167178611503808442460152713891683982454369010323583112178350 3844$$
$$6929062655448792237114490509578608655662496577974840004057020373.$$

Raising this number to the power 9007, then reducing it modulo $N$, one obtained the number

$$0609181920001915122205180023091419001$$
$$5140500082114041805040004151212011819$$

corresponding, in Rivest, Shamir and Adleman's cipher, to the sentence:

*First solver wins one hundred dollars,*

which guaranteed that the message really came from MIT.

Seventeen years later, the Dutch mathematician Arjen K. Lenstra, together with a team of hundreds, in just 8 months managed to find the solution. The technique used in tackling the problem is the so-called *multiple polynomial quadratic sieve*, a technique that allows to split up the task into several smaller subtasks. Using this sieve, the possible factors are found among millions of candidates. To organise the work, Lenstra needed hundreds of collaborators all over the world and involved thousands of computers, the whole enterprise being coordinated via the Internet.

The results of this collective effort were sent Lenstra: two days' computations with a supercomputer produced a 64-digit and a 65-digit factor. This allowed Lenstra to decrypt the message by Rivest, Shamir and Adleman.

Are you curious to know what the message said? It said:

*the magic words are squeamish ossifrage.*

The three scientists themselves said that it was a meaningless sentence: they would never have supposed, when they wrote it, that it would someday emerge. What had seemed an impossible challenge, seventeen years later turned out to be within the grasp of the most advanced researchers. The conclusion is that cryptography is still, in several regards, an *experimental science*. It still relies on several conjectures, such as Diffie–Hellman hypothesis, we shall deal with shortly, which might be, if not completely contradicted, at least quite diminished when new algorithms are invented that, at least in many cases, do a good work to elude them. So, when there are no *theorems* telling us whether a given cryptographic procedure is secure, it is convenient to be careful rather than doing as if it were certainly so. A system that today is believed to be secure might not be so tomorrow, as we shall see in Chapter 9.

## 7.8 Variants of *RSA* system and beyond

We are now going to describe some cryptosystems, the first of which is a variant of *RSA* system. Its security relies on the problem of computing discrete logarithms.

### 7.8.1 Exchanging private keys

The *RSA* system, or rather a slight modification of it, allows two users to exchange a private key with which, *independently of the public key system*, they can exchange enciphered messages using one of the classic methods discussed at the beginning of this chapter.

Let us modify the *RSA* system as follows. Choose a very large prime number $p$, which is divulged, and work in the field $\mathbb{Z}_p$. Actually, we might work in any finite field $\mathbb{F}_q$, but we shall limit ourselves to $p$-element fields. Choose next a non-zero element $g \in \mathbb{Z}_p$, which is divulged too. The most convenient choice, to use in the best way the system's resources, would be to choose as $g$ a generator of the multiplicative group of the field $\mathbb{Z}_p$. However, this is not strictly necessary.

Moreover, each user $U$ chooses his *private key* $e_U$, which is a positive number smaller than $p - 1$, and divulges $g^{e_U} \in \mathbb{Z}_p$, that is, the positive number $X_U = g^{e_U} \bmod p$. Notice that, from $X_U$, it is not possible to reconstruct in a reasonable time $U$'s private key $e_U$. Indeed, this would imply finding a *discrete logarithm* which, as we know, requires in general an exponential time.

Let us see now how two users $A$ and $B$ may proceed to exchange a private key. There is a very simple method: $A$ and $B$ may agree to use as a private key $g^{e_A e_B} \in \mathbb{Z}_p$, that is, the number

$$X_{AB} = g^{e_A e_B} \bmod p.$$

Indeed, both $A$ and $B$ can compute $X_{AB}$ in polynomial time. For instance, $A$ knows $X_B$, which is public. Moreover, she knows her private key $e_A$. So she computes

$X_B^{e_A} = (g^{e_B})^{e_A} \bmod p = X_{AB}$. Similarly, $B$ knows $X_A$ and $e_B$, so can compute
$X_A^{e_B} = (g^{e_A})^{e_B} \bmod p = X_{AB}$.

On the other hand, an *eavesdropper* $C$ will find it hard to compute $X_{AB}$. In fact, he knows $X_A$ and $X_B$, but how can he reconstruct $X_{AB}$ from them? In order to do so, he probably should find first $e_A$ and $e_B$, and compute next $g^{e_A e_B}$, which would finally allow him to figure out $X_{AB}$. But in order to find $e_A$ and $e_B$, $C$ should compute some discrete logarithms, which is computationally unfeasible.

But, are we certain that to compute $X_{AB}$ knowing $X_A$ and $X_B$ it is necessary to compute some discrete logarithms? In other words, are we sure that in order to compute $g^{e_A e_B}$ knowing $g^{e_A}$ and $g^{e_B}$ it is necessary to know $e_A$ and $e_B$? So far, nobody has proved nor disproved this fact. Nevertheless, it is conjectured that the complexity of computing $g^{e_A e_B}$ knowing $g^{e_A}$ and $g^{e_B}$ is equal to that of finding discrete logarithms: this is the so-called *Diffie–Hellman hypothesis*. On this hypothesis the security of this method of exchange of private keys is based.

**Example 7.8.1.** Assume $p = 19$ and $g = 2$ have been divulged. Let $e_A = 16$ and $e_B = 11$ be the private keys of $A$ and $B$, respectively. Then $A$ and $B$ *publish* the values

$$X_A = 2^{16} \bmod 19 = 5, \qquad X_B = 2^{11} \bmod 19 = 15,$$

respectively. The common key $A$ and $B$ will use to exchange messages is $X_{AB} = g^{e_A e_B} \bmod 19$. $A$ will compute it as follows:

$$X_{AB} = X_B^{e_A} \bmod 19 = 15^{16} \bmod 19 = 6.$$

Clearly $B$ gets the same result by computing

$$X_A^{e_B} \bmod 19 = 5^{11} \bmod 19 = 6.$$

Notice that $A$ and $B$ can now use the key 6 to exchange messages enciphered, for instance, using a Caesar cipher operating on 26 letters: 6 might be the enciphering key, that is, the number of positions the letters are shifted in Caesar cipher.

### 7.8.2 ElGamal cryptosystem

Fix a large finite field $\mathbb{F}_q$ (we may well take $\mathbb{Z}_p$, for a large $p$) and an element $g \in \mathbb{F}_q^*$ (preferably, but not necessarily, a generator of $\mathbb{F}_q^*$). We shall assume that the numerical equivalents of the messages are in $\mathbb{F}_q$.

Each user $A$ has a public key and a private key: the private key is an integer $a = a_A$, randomly chosen by $A$ ($0 < a < q - 1$), while the public key is $g^a \in \mathbb{F}_q$.

Assume $B$ wants to send $A$ a message $P$. Then $B$ proceeds as follows:

- $B$ randomly chooses an integer $k < q$;
- he computes $g^k$ in $\mathbb{F}_q$;
- he computes $g^{ak}$ in $\mathbb{F}_q$;
- he multiplies the message $P$ by $g^{ak}$ in $\mathbb{F}_q$;
- he sends $A$ the pair $(y_1 = g^k, y_2 = P \cdot g^{ak})$.

*Notice that in order to compute $g^{ak}$ it is not necessary to know $A$'s private key; it suffices to know $g^a$, as $g^{ak} = (g^a)^k$.*

On receiving the pair $(y_1, y_2)$, $A$, who knows $a$, which is her own private key, can discover the message $P$ by raising $y_1$ to the exponent $a$ and dividing $y_2$ by the result found. Indeed,

$$y_2 \cdot ((y_1)^a)^{-1} = P \cdot g^{ak} \cdot ((g^k)^a)^{-1} = P.$$

Somebody who could solve the discrete logarithm problem could violate the cryptosystem by determining the private key $a$ from the knowledge of $g^a$. In theory, it could be possible to obtain $g^{ak}$ knowing $g^a$ and $g^k$ (and so to arrive at the message $P$), but here too, as already said in § 7.8.1, it is conjectured that solving this kind of problem without solving the problem of computing discrete logarithms is impossible.

### 7.8.3 Zero-knowledge proof: or, persuading that a result is known without revealing its content nor its proof

Suppose Paul has found a very important formula: he wants to persuade a colleague he has found it, *but without giving him any indication about the formula itself nor about the way he has proved it*. Is it possible? This kind of communication is said to be a *zero-knowledge* protocol, that is to say, it is a communication that does not transmit any information that could give away the formula or its proof, but lets the addressee know we actually have it. It looks like an impossible feat. However, we shall see that it is possible. Let us see an example to illustrate how to proceed.

Let $G$ be a finite group with $N$ elements, and let $b$ and $y$ be two elements of $G$. Suppose Paul has found a discrete logarithm for $y$ in base $b$, that is, he has determined a positive integer $x$ such that

$$b^x = y.$$

His friend Sylvia is sceptical: Paul wants to convince her he knows $x$ *without telling her $x$*. Assume Sylvia knows the order $N$ of the group $G$ (the case in which Sylvia does not know $N$ can also be dealt with, but we shall not do so). They may proceed as follows:

(1) Paul generates a random positive integer $e < N$ and sends Sylvia

$$b' = b^e;$$

(2) Sylvia tosses a coin: if it shows heads, Paul must disclose $e$ to Sylvia and she checks whether actually $b' = b^e$;
(3) if the coin shows tails, then Paul must disclose the positive integer $x + e \bmod N$. As $b^x = y$ and $b^e = b'$, we have $b^{x+e} = yb'$. Sylvia will check that the number has the required property (notice that Sylvia knows both $y$ and $b'$).

The three steps are repeated (and so there will be a *new* choice of a random integer $e$, a new coin toss, and so on), until Sylvia is convinced that Paul has actually found the discrete logarithm of $y$.

How can she be convinced? If Paul did not really know the discrete logarithm of $y$ and were cheating, he would be able to *answer just one of the two possible questions*. If the coin comes up as heads he certainly may disclose $e$, but if it comes up as tails, how can he disclose $x + e \bmod N$ without knowing $x$? He might try to elude the problem by sending, in step (1), $b' = b^e/y$ rather than $b^e$: so, if tails shows up, he may reveal $e = (e - x) + x$ (which he can easily do). But in this case he would be exposed if the coin shows heads: indeed, in this case he should reveal $e - x$, and how could he without knowing $x$?

By iterating the procedure a sufficient number of times, sooner or later Sylvia *will be persuaded* that Paul actually knows what he claims to know.

So Paul manages to *prove* Sylvia that he knows the discrete logarithm $x$ of $y$ without explicitly exhibiting $x$, and so his secret remains his own.

### 7.8.4 Historical note

This *challenge* scheme calls to mind the challenges that took place centuries ago. In the 16th century a mathematician's ability was demonstrated through "public challenges": these scientific duels were actual tournaments with witnesses, judges, referees and so on. In these challenges fame and money were at stake. For this reason, the most important discoveries were kept *jealously secret*. So, when a mathematician came in possession of a new discovery, he sent a *cartello di matematica disfida* (public mathematical challenge), in which he claimed to be able to solve a class of problems and proposed in turn some of them, and the "contenders" engaged in proposing and solving such problems.

Among the most famous challenges, there were those between Dal Fior and Tartaglia (Nicolò Fontana): the problems presented by Dal Fior can be reduced to solving equations of the form $x^3 + px = q$, which Dal Fior could solve because their solution was transmitted him by his teacher Scipione Dal Ferro before dying. Tartaglia proposed a series of problems reducible to the solution of equations of the form $x^3 + mx^2 = q$, which he could solve. It happened that, even without knowing the general formula for the equations in possession of Dal Fior, Tartaglia managed to find it in time to solve all the problems, while Dal Fior could not solve any.

Another renowned challenge was the one between Ferrari and Tartaglia in 1548: Tartaglia in 1539 had given Cardano the solution of a class of third degree equations (the *casus non irriducibilis*), making him promise he would not divulge it. In 1545 Cardano published his work "Ars Magna" in which, violating his promises, he gave the formula to solve cubic equations. Tartaglia took offence and Cardano's pupil Ferrari challenged him in another famous confrontation.

## 7.9 Cryptography and elliptic curves

So far we have only described the development of several classic and modern cryptographic methods, all based on algebraic, and mostly arithmetic, ideas. In other words, these methods rely on properties of numbers or their congruence classes.

In this section we are going to discuss some new frontiers recently opened to cryptography, especially for what regards the security and the prevention of cryptanalysis. This is due to the interaction of classic algebra and arithmetic with ideas and notions from geometry, and in particular from the study of certain plane curves called *elliptic curves*.

### 7.9.1 Cryptography in a group

Before going on, we give explicitly a remark that, in an implicit form, we have already mentioned elsewhere in this chapter. To fix ideas, consider the exchange of private keys through the $RSA$ system, described in the previous section. It relies on exponentiation in $\mathbb{Z}_p$, with $p$ a prime number. Its easy execution is due to the fact that exponentiating in $\mathbb{Z}_p$ is computationally easy, that is, requires a polynomial time. Its security, on the other hand, depends on the fact that finding discrete logarithms in $\mathbb{Z}_p$ is apparently much harder computationally. More precisely, as we have seen, the Diffie–Hellman hypothesis is relevant here (see page 364).

On the other hand, the theoretical basis of this cryptographic system works *with no changes* if, rather than in the multiplicative group $\mathbb{Z}_p^*$, we work in any other finite group $G$. Leaving the description of the details of the scheme as an exercise, we just remark that actually, to put in practice the theory, and so to implement a cryptosystem to exchange private keys based on exponentiation in an arbitrary finite group $G$, we must ask that:

- it is possible to perform computations in $G$, that is, it is necessary that $G$ is given not only in a theoretical way, but operatively, in such a way that we can actually *work* with its elements;
- exponentiating in $G$ is easy, that is, requires, for instance, a polynomial computational cost;
- determining discrete logarithms in $G$ is computationally much harder, for instance exponential, and that in $G$ the Diffie–Hellman hypothesis holds, that is, for a randomly chosen element $g \in G$ and for $a, b \in \mathbb{Z}$, computing $g^{ab}$ knowing $g^a$ and $g^b$ has *the same computational difficulty* as determining discrete logarithms in $G$.

For instance, if $\mathbb{F}_q$, $q = p^f$, with $p$ a prime number, is a finite field and $G = \mathbb{F}_q^*$ is its multiplicative group, then $G$ has these properties, as:

- $G$ can be described concretely as an extension of $\mathbb{Z}_p$. Some of the examples in Chapter 5 show how to describe its elements;
- exponentiating is easy in group $G$, that is, it requires a polynomial computational cost (see § 5.1.14);
- just like in $\mathbb{Z}_p$, it is conjectured that determining discrete logarithms in $G$ has at least an exponential cost, and that in $G$ the Diffie–Hellman hypothesis holds: indeed, it is clear that if it holds in $\mathbb{Z}_p$ then it holds in $\mathbb{F}_q^*$ too.

So we may use in cryptography, and it is actually used, the multiplicative group of a finite field $\mathbb{F}_q$ rather than that of $\mathbb{Z}_p$ with $p$ a large prime. This yields remarkable advantages. For instance, we may use fields of the form $\mathbb{F}_{2^n}$ of characteristic 2, which are very suitable for a computational approach because their elements can be described as $n$-tuples of 0s and 1s. Moreover, by choosing a *large* $n$, $\mathbb{F}_{2^n}$ becomes in turn *large* very quickly, removing the need for a large prime $p$ to construct $\mathbb{Z}_p$. Unfortunately, choosing $\mathbb{F}_{2^n}$ makes life easier for cryptanalysis. Indeed, recently, in 1984, D. Coppersmith found efficient algorithms to compute discrete logarithms in these fields (see [14], [44]).

So, which groups may we use to do cryptography? We would like groups *quite similar* to $\mathbb{F}_q^*$, which would make them *familiar-looking* and, most important, *computationally easy to use*. At the same time, we would have many of them, to be able to choose among them, perhaps change them frequently, to avoid too easy a cryptanalysis.

Here geometry lends us a helping hand. Let us explore the ideas that lead to considering elliptic curves.

### 7.9.2 Algebraic curves in a numerical affine plane

Rather than considering specifically $\mathbb{F}_q$, consider an arbitrary field $\mathbb{K}$. So we may define the *numerical affine plane* $\mathbb{A}_\mathbb{K}^2$ with coordinates on this field (see [51]). Basically, this is just $\mathbb{K} \times \mathbb{K}$. This terminology is not surprising, and has already been used before. In fact, just consider the case $\mathbb{K} = \mathbb{R}$, leading to the usual plane $\mathbb{A}_\mathbb{R}^2$ with cartesian coordinates $(x, y)$.

In the affine plane $\mathbb{A}_\mathbb{K}^2$ we may *do geometry* exactly as in the real cartesian plane. For instance, we may consider *algebraic curves*. These are subsets of $\mathbb{A}_\mathbb{K}^2$ *defined* by an equation of the form

$$f(x, y) = 0, \qquad (7.10)$$

where $f(x, y)$ is a polynomial with coefficients in $\mathbb{K}$, which we assume to be non-constant and without repeated factors. The curve defined by (7.10) is the set of points $(u, v) \in \mathbb{A}_\mathbb{K}^2$ such that $f(u, v) = 0$. Clearly, substituting the polynomial $kf(x, y)$ for $f(x, y)$, where $k \in \mathbb{K}^*$, we obtain the same curve.

The curve defined by Equation (7.10) is said to be *irreducible* if the polynomial $f(x, y)$ is irreducible over $\mathbb{K}$. Notice that this notion *depends* on the field $\mathbb{K}$, because, as we know, a polynomial may be irreducible over $\mathbb{K}$ but not over an extension of $\mathbb{K}$.

**Example 7.9.1.** Consider the curve in $\mathbb{A}_\mathbb{R}^2$ having equation $x^2 + y^2 = 0$. It is irreducible because such is the polynomial $x^2 + y^2$ over $\mathbb{R}$. On the contrary, the curve in $\mathbb{A}_\mathbb{C}^2$ with the same equation is reducible because we have $x^2 + y^2 = (x + iy)(x - iy)$ over $\mathbb{C}$.

Notice that the curve in $\mathbb{A}_\mathbb{R}^2$ having equation $x^2 + y^2 = 0$ consists of the single point having coordinates $(0, 0)$. So the definition must be studied carefully: the notion of a curve includes sets which do not always correspond to the intuitive idea of a curve as the reader may picture it!

If $f(x, y)$ has degree $d$, we say that $d$ is the *degree* of the curve of equation (7.10). The curves of degree 1, which are clearly irreducible (see Exercise A7.13), are called *lines*, those of degree 2 *conic* curves, those of degree 3 *cubic*, those of degree 4 *quartic* and so forth.

### 7.9.3 Lines and rational curves

The $x$-*axis*, which has equation $y = 0$, may be identified in a natural way with the field $\mathbb{K}$, as it consists of all points $(x, 0)$, with $x$ ranging in $\mathbb{K}$. An analogous remark can be made about the $y$-*axis*, which has equation $x = 0$.

More in general, every straight line may be easily identified with the field $\mathbb{K}$. Indeed, a line $R$ has equation of the form

$$ax + by + c = 0 \tag{7.11}$$

where $a$ and $b$ are not both equal to zero. Assume $b \neq 0$. Then we may *project* the line $R$ on the $x$-axis, associating with each point $(u, v)$ of $R$ the point $(u, 0)$ of the $x$-axis, which will be identified with $u \in \mathbb{K}$ (see Figure 7.2). This

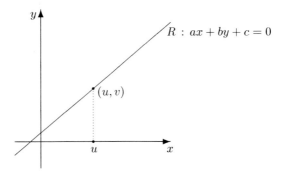

**Fig. 7.2.** Projection of a line $R$ on the $x$-axis

mapping is bijective. Indeed, given $u$, we must have $v = -(au + c)/b$ if the point $(u, v)$ is to lie on $R$. In other words, the projection is given by

$$\pi : (u, v) \in R \rightarrow u \in \mathbb{K}$$

and the inverse mapping is given by

$$\pi^{-1} : u \in \mathbb{K} \rightarrow (u, -(au + c)/b) \in R.$$

Analogously, if $a \neq 0$, the line $R$ of equation (7.11) can be projected on the $y$-axis and the projection is bijective (see Exercise A7.14).

In conclusion, lines are not interesting from our viewpoint: in fact, recall that our goal, in cryptography, is to find groups *different* from $\mathbb{K}^*$.

The idea of projecting a curve on the $x$-axis to study it looks fine. So let us keep it. From this viewpoint, which are the simplest curves after the straight lines? We might answer, for instance, those for which the projection, even if not bijective, is *almost always* so, that is, is bijective but for a finite number of points. For instance, this property is enjoyed by the curves $C$ of equation

$$g(x)y = f(x), \tag{7.12}$$

where $f(x)$, $g(x)$ are polynomials in $x$, with $f(x)$, $g(x)$ different from zero and without common factors. Every curve with these properties is irreducible (see Exercise A7.16). The projection is defined again as

$$\pi : (u,v) \in C \rightarrow u \in \mathbb{K}$$

and the *inverse mapping* is given by

$$\rho : u \in \mathbb{K} \rightarrow \left( u, \frac{f(u)}{g(u)} \right) \in R.$$

Notice that $\rho$ is not defined where $g(x) = 0$, so, strictly speaking, it is not the inverse of $\pi$; however it is its inverse out of finitely many points, the points $u \in \mathbb{K}$ such that $g(u) = 0$. Curves of this kind belong to the class of *rational curves*. These are irreducible curves $C$ defined by an equation of the form (7.10), and such that there are rational functions $\phi(u)$, $\psi(u)$, defined over $\mathbb{K}$ or an algebraic extension of $\mathbb{K}$, such that the rational function $f(\phi(u), \psi(u))$ is the zero function. In other words,

$$x = \phi(u), \quad y = \psi(u)$$

is a so-called *parametric representation* by rational functions of the curve $C$. For instance, the irreducible conic curves are curves of this kind (see Exercise A7.17).

Clearly, rational curves are again too similar to $\mathbb{K}$ to be of interest to us, so we reject them too.

### 7.9.4 Hyperelliptic curves

The next case is given by curves for which the projection on the $x$-axis has no inverse, even after removing a finite number of points. Among these, the simplest case is that of curves for which the preimage of a point under the projection mapping consists in general not of a single point, but of two points. Curves of this kind are called *hyperelliptic*: examples of hyperelliptic curves are given by the irreducible curves $C$ of degree greater than 2 having equation of the form

$$y^2 + yg(x) = f(x), \tag{7.13}$$

where $f(x), g(x) \in \mathbb{K}[x]$. Assume further that $g(x)$ is not the zero polynomial if the characteristic of $\mathbb{K}$ is two: we shall shortly see why this hypothesis is necessary.

If $(u, v)$ is a point on the curve $C$, this means that $v$ is a solution of the equation

$$y^2 + yg(u) = f(u), \tag{7.14}$$

which has *in general* two distinct solutions.

**Remark 7.9.2.** Let us clarify the meaning of the previous claim.

If $\mathbb{K}$ has characteristic different from 2, Equation (7.14) has a single solution if and only if its discriminant is zero, that is, if and only if

$$g(u)^2 + 4f(u) = 0. \tag{7.15}$$

It might happen that Equation (7.15) holds *for all* $(u, v) \in C$. However, if we suppose that the size of the set $C'$ of points $u \in \mathbb{K}$ such that there is a point $(u, v) \in C$ is *large enough*, for instance that its size is greater than the degree of the polynomial $g(x)^2 + 4f(x)$, the factor theorem (see Theorem 1.3.19 and its Corollary 1.3.20) implies that $g(x)^2 + 4f(x)$ is the zero polynomial, and this yields a contradiction, because in this case we would find

$$y^2 + yg(x) - f(x) = \left(y + \frac{g(x)}{2}\right)^2,$$

against the hypothesis that the curve $C$ is irreducible.

If the characteristic of $\mathbb{K}$ is 2, the derivative of the polynomial $y^2 + yg(u) - f(u)$ is $g(u)$. Assume that *for all* $u \in C'$ we have $g(u) = 0$, so Equation (7.14) has a unique solution. Again, assuming that $C'$ has size greater than the degree of $g(x)$, this would imply that $g(x)$ is the zero polynomial, contradicting the hypothesis.

**Example 7.9.3.** Consider the simple case in which $g(x)$ is the zero polynomial, which by hypothesis can happen only if the characteristic of $\mathbb{K}$ is not 2. In this case, if $(u, v)$ is a point on the curve $C$, this means that $v^2 = f(u)$, that is, $f(u)$ is a square in $\mathbb{K}$. So, not only $(u, v)$ lies on $C$ but $(u, -v)$ as well, and these are the only two points of $C$ that project on the point $u \in \mathbb{K}$. They are *symmetric* with respect to the $x$-axis, in the sense that their second coordinates are one the opposite of the other. Of course, if $f(u) = 0$ these points coincide, otherwise they are distinct.

In conclusion, hyperelliptic curves can be thought of as *double coverings* of $\mathbb{K}$. This concept is particularly clear when $\mathbb{K}$ is algebraically closed. In this case, by Remark 7.9.2, if $\mathbb{K}$ has characteristic different from 2, for all $u \in \mathbb{K}$ that are not roots of the polynomial $g(x)^2 + 4f(x)$, we have exactly two points of $C$ *over* the point $(u, 0)$ of the $x$-axis. If $u \in \mathbb{K}$ is a root of $g(x)^2 + 4f(x)$, the unique point $(u, -g(u)/2)$ of $C$ corresponds to it.

If, on the other hand, $\mathbb{K}$ has characteristic 2, for all $u \in \mathbb{K}$ that are not roots of the polynomial $g(x)$, we have exactly two points of $C$ *over* the point $(u, 0)$ of the $x$-axis. If $u \in \mathbb{K}$ is a root of $g(x)$, the unique point $(u, \sqrt{f(u)})$ of $C$ corresponds to it.

Clearly, we may consider curves for which the behaviour of the projection on the $x$-axis is even more complex: for instance, the preimage of a general point of $\mathbb{K}$ may have size greater than two. But we shall not go into these cases because, as already remarked, we want to consider interesting curves which are nevertheless constructible in the easiest possible way.

### 7.9.5 Elliptic curves

So we are left with the problem of finding hyperelliptic curves which also are groups. This may be done, as we shall shortly see, if $f(x)$ and $g(x)$ have the simplest possible form compatible with the hypothesis that the curve has degree greater than 2. Indeed, assume that $g(x) = mx + n$ is of first degree and that $f(x) = x^3 + px^2 + qx + r$. In this case the curve $C$ of equation (7.13) is cubic.

It is important to observe that, with suitable changes of variable, the equation of curve $C$ may be simplified.

**Proposition 7.9.4.** *Let $C$ be the curve of equation*

$$y^2 + y(mx + n) = x^3 + px^2 + qx + r. \qquad (7.16)$$

*It is possible to change coordinates in $\mathbb{A}^2_{\mathbb{K}}$ in such a way that in the new coordinate system*

- *if $\mathbb{K}$ has characteristic different from 2 or 3, $C$ has equation of the form*

$$y^2 = x^3 + ax + b; \qquad (7.17)$$

- *if $\mathbb{K}$ has characteristic 3, $C$ has equation of the form*

$$y^2 = x^3 + ax^2 + bx + c; \qquad (7.18)$$

- *if $\mathbb{K}$ has characteristic 2, $C$ has equation of the form*

$$y^2 + cy = x^3 + ax + b; \qquad (7.19)$$

  *or of the form*

$$y^2 + xy = x^3 + ax^2 + b. \qquad (7.20)$$

PROOF. To begin, assume $\mathbb{K}$ not to have characteristic 2. Change variables as follows:

$$x \to x, \qquad y \to \frac{y - mx - n}{2}.$$

The equation of $C$ becomes of the form (7.16) where $m = n = 0$. This concludes the proof in the case of characteristic 3. If the characteristic is not 3, change again variables as follows:

$$x \to x - \frac{p}{3}, \qquad y \to y. \qquad (7.21)$$

The equation of $C$ becomes now of the form (7.17), concluding the proof if the characteristic is neither 2 nor 3.

Assume now the characteristic of $\mathbb{K}$ to be 2. By performing the change of variable (7.21) the equation of $C$ becomes of the form

$$y^2 + y(mx + n) = x^3 + ax + b.$$

If $m = 0$ the equation is of the form (7.19). Assume then $m \neq 0$. In this case, perform the change of variable

$$x \to m^2 x + \frac{n}{m}, \qquad y \to m^3 y + \frac{m^2 a + n^2}{m^3},$$

obtaining an equation of the form (7.20). $\qquad\qquad\qquad\qquad\qquad\qquad$ □

The equations of the form (7.17), (7.18), (7.19), (7.20) are called canonical equations in *Weierstrass form* of a cubic curve.

Now we shall put ourselves in a *regularity hypothesis*. We shall assume that, if $\mathbb{K}$ has characteristic different from 2, the right-hand side of Equation (7.17) or (7.18) has no multiple roots in the algebraic closure of $\mathbb{K}$. If $\mathbb{K}$ has neither characteristic 2 nor 3, that is, when the equation in Weierstrass form is Equation (7.17), this is equivalent to saying that $27b^2 + 4a^3 \neq 0$ (see Exercise A7.18). If, on the other hand, $\mathbb{K}$ has characteristic 2 and if the equation in Weierstrass form is Equation (7.20), then we shall assume $b \neq 0$. We shall shortly see the meaning of this hypothesis.

Finally, we shall add to $C$ a point $O$ called *point at infinity*, whose meaning is well known to the reader acquainted with projective geometry (see [51]). Next, we shall denote by $E$ the set $C \cup \{O\}$, call $E$ an *elliptic curve*, and say that (7.17), (7.18), (7.19) or (7.20) is its equation.

**Remark 7.9.5.** As is well known, the affine plane $\mathbb{A}^2_{\mathbb{K}}$ can be naturally embedded in the *projective plane* $\mathbb{P}^2_{\mathbb{K}}$, whose points are non-zero ordered triples $[x_0, x_1, x_2]$ of elements of $\mathbb{K}$, up to a multiplication by a constant, that is, $[x_0, x_1, x_2] = [kx_0, kx_1, kx_2]$ for all $k \in \mathbb{K}^*$. Given the point $[x_0, x_1, x_2]$ of $\mathbb{P}^2_{\mathbb{K}}$, $x_0, x_1, x_2$ are said to form a triple of *homogeneous coordinates* of the point. The embedding of $\mathbb{A}^2_{\mathbb{K}}$ in $\mathbb{P}^2_{\mathbb{K}}$ happens as follows:

$$(x, y) \in \mathbb{A}^2_{\mathbb{K}} \to [1, x, y] \in \mathbb{P}^2_{\mathbb{K}}.$$

The complement of $\mathbb{A}^2_{\mathbb{K}}$ in $\mathbb{P}^2_{\mathbb{K}}$ is the set of points satisfying the *equation* $x_0 = 0$, that is, the set of points of the form $[0, a, b]$, called *points at infinity*. This set is called *line at infinity* of the projective plane. If $[x_0, x_1, x_2]$ is not on the line at infinity, its cartesian coordinates in $\mathbb{A}^2_{\mathbb{K}}$ are

$$x = \frac{x_1}{x_0}, \qquad x = \frac{x_2}{x_0}.$$

These are the formulas to pass from homogeneous coordinates to cartesian coordinates.

If we consider a line $R$ in the plane $\mathbb{A}^2_{\mathbb{K}}$, with equation $bx - ay + k = 0$, passing to homogeneous coordinates and multiplying both sides by $x_0$, we find the equation $kx_0 + bx_1 - ax_2 = 0$. Clearly all the solutions to this equation having $x_0 \neq 0$, and they alone, correspond to the points of $\mathbb{A}^2_{\mathbb{K}}$ lying on $R$. On the other hand, by intersecting it with the line at infinity, we obtain the system

$$x_0 = bx_1 - ax_2 = 0,$$

which uniquely determines the point $0 = [0, a, b]$. This leads to the well-known interpretation of the points at infinity: the point $[0, a, b]$ is to be considered as *the common point* of all parallel lines of the plane $\mathbb{A}^2_\mathbb{K}$ having equation of the form $bx - ay + k = 0$ with $k$ ranging in $\mathbb{K}$.

Similarly, considering the curve $C$ of equation (7.17), passing to homogeneous coordinates and multiplying both sides by $x_0^3$, we find the equation

$$x_0 x_2^2 = x_1^3 + a x_0^2 x_1 + b x_0^3. \tag{7.22}$$

All the solutions to this equation having $x_0 \neq 0$, and they alone, correspond to the points of $\mathbb{A}^2_\mathbb{K}$ lying on $C$. On the other hand, intersecting it with the line at infinity, we obtain the system

$$x_0 = x_1 = 0,$$

which uniquely determines the point $0 = [0, 0, 1]$. So it is natural to consider Equation (7.22) as defining the *projective closure* $E$ of $C$. It differs from $C$ only by the point at infinity $O$. We may reason analogously if the curve has equation (7.18), (7.19) or (7.20).

**Remark 7.9.6.** We might wonder whether a curve having equation (7.17), (7.18), (7.19) or (7.20) could itself be rational, and so devoid of interest for our uses. It is not so in the regularity hypothesis we have stipulated: for instance when $\mathbb{K}$ has characteristic different from 2 and 3, the equation is of the form (7.17) and $27b^2 + 4a^3 \neq 0$, while it can be shown that the curve is rational if $27b^2 + 4a^3 = 0$ (see Exercise A7.23). The simplest case is that of the curve of equation $y^2 = x^3$, which has parametric representation

$$x = u^2, \quad y = u^3.$$

To study the matter more in depth, see [56].

### 7.9.6 Group law on elliptic curves

Let us discuss now the group law on an elliptic curve $E$. We shall consider here in detail the case in which $\mathbb{K}$ has characteristic different from 2 or 3 and the equation is of the form (7.17), leaving to the reader as an exercise the analogous discussion of the remaining cases (see Exercises A7.27 and [56], Chapter III, § 2).

The key observation is that *given two points $p = (x_1, y_1)$ and $q = (x_2, y_2)$ of the curve, the line through them intersects the curve in a third point $r = (x_3, y_3)$.* This observation is to be taken with a grain of salt, in the sense we are going to explain.

First of all, we verify it in the case in which $p$ and $q$ are distinct and the line $R$ through them is not *vertical*, that is, has not an equation of the form $x = u$. This means, as already remarked, that $x_1 \neq x_2$.

The equation of $R$ is

$$y = mx + n, \tag{7.23}$$

with

$$m = \frac{y_2 - y_1}{x_2 - x_1}, \qquad n = y_1 - mx_1 \qquad (7.24)$$

(see Exercise A7.19). To find the points of intersection of $R$ with $C$, substitute (7.23) in (7.17), and solve with respect to $x$. So one gets the third degree equation

$$x^3 - (mx + n)^2 + ax + b = 0;$$

clearly $x_1$ and $x_2$ are two of its roots. Let $x_3$ be the third root. As

$$x_1 + x_2 + x_3 = m^2$$

(see Exercise A7.20), we have

$$x_3 = m^2 - x_1 - x_2,$$

which gives the first coordinate of the third point $r$ of intersection of $R$ with $C$. The second coordinate of $r$ is given by

$$y_3 = mx_3 + n.$$

Let us see what happens if the line through two distinct points $p$ and $q$ is the vertical line $x = u$. As we have seen in Example 7.9.3, this means that the two points have coordinates $(u, v)$ and $(u, -v)$, where $\pm v$ are the square roots of $u^3 + au + b$. The intersection of the line of equation $x = u$ with the curve of equation (7.17) in $\mathbb{A}_{\mathbb{K}}^2$ consists only of the points $p$ and $q$. But passing to homogeneous coordinates, the equation of the line becomes $x_1 = ux_0$ and we see that it passes through the point $O$ lying in $E$. So it is natural to regard $O$ as the third intersection point of the line through $p$ and $q$ with the curve.

**Example 7.9.7.** We demonstrate the preceding remarks by examining the real curve of equation

$$y^2 = x^3 - x.$$

The curve corresponds to the union of the graph of the function

$$y = \sqrt{x^3 - x}$$

and of its symmetric with respect to the $x$-axis. Fix a point on the curve, say $p = (2, \sqrt{6})$. Write the equation of a non-vertical line through $p$. It is of the form

$$y - \sqrt{6} = \frac{1}{m}(x - 2), \qquad (7.25)$$

with $m \neq 0$. Intersect this line with the curve, obtaining, besides $p$, two more points, $q$ and $r$. We leave to the reader the task of finding their coordinates as functions of $m$ and of verifying that, as $m$ approaches 0, that is, when the line tends to becoming the vertical line $x = 2$, one of the two points $q$ and $r$ tends to the point $p' = (2, -\sqrt{6})$, symmetric of $p$ with respect to the $x$-axis, which lies indeed on the vertical line $x = 2$, while the other's second coordinate tends to infinity (see Figure 7.3). Basically, this second point tends to infinity and its *limit position*, which may be thought of as infinitely far along the $y$-axis, to which the line of equation (7.25) becomes parallel as $m$ approaches 0, is exactly that of the point $O$ we have added to $C$ in order to get $E$. These heuristic remarks are quite natural and should not sound strange to readers acquainted with projective geometry.

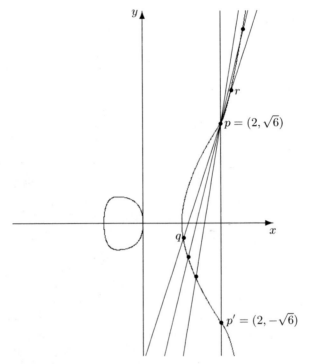

**Fig. 7.3.** Elliptic curve of equation $y^2 = x^3 - x$

Finally, what happens if $p = q$? Here we cannot consider the *line through* $p$ and $q$. Nevertheless, among the infinitely many lines through $p = q$ one is special, with respect to the elliptic curve: the *tangent line* to the curve in $p$, which may be thought of as the line joining $p$ with a point $q$ on the curve that is *so close to $p$ to be undistinguishable from $p$*. This notion is well known in the real case: it is the limit line of the line through $p$ and another point $q$ of the curve, when $q$ approaches $p$.

If a real curve $C$ has equation

$$f(x, y) = 0$$

and if $p = (\xi, \eta)$ is a point of $C$, the condition for the tangent, seen as the above limit, to exist is that in $(\xi, \eta)$ not both partial derivatives of $f(x, y)$ are zero (see [51]), that is, it is not the case that

$$\frac{\partial f}{\partial x}(\xi, \eta) = 0, \qquad \frac{\partial f}{\partial y}(\xi, \eta) = 0. \tag{7.26}$$

So the tangent line to $C$ in $p$ has equation

$$\frac{\partial f}{\partial x}(\xi, \eta)(x - \xi) + \frac{\partial f}{\partial y}(\xi, \eta)(y - \eta) = 0. \tag{7.27}$$

More in general, these notions extend without any difference to the case of a curve on an arbitrary field (see [56], Ch. I, § 1). A point $p = (\xi, \eta)$ of the curve $C$ of equation $f(x, y) = 0$ for which (7.26) hold is said to be *singular*. In it the tangent line does not exist. A non-singular point is said to be *simple* or *smooth*, and in it the tangent line exists and is given by Equation (7.27). A curve having a singular point in $p$ is said to be *singular* in $p$.

**Remark 7.9.8.** A curve of equation $f(x, y) = 0$ is *singular*, that is, has some singular point, if and only if the system

$$f(x, y) = 0, \qquad \frac{\partial f}{\partial x}(x, y) = 0, \qquad \frac{\partial f}{\partial y}(x, y) = 0 \qquad (7.28)$$

admits solutions. Notice that this definition depends on the field $\mathbb{K}$, as it is possible that the system (7.28) has no solutions in $\mathbb{K}$ but has solutions in some extension of $\mathbb{K}$ (see Exercise B7.58).

When a curve is said to be *non-singular*, without specifying the field on which it is considered, the curve is meant to be considered on the algebraic closure of the field $\mathbb{K}$ containing the coefficients of the equation defining the curve.

The reader may easily verify that the regularity hypothesis on page 373 makes sure that the curves $C$ defined by equations in Weierstrass form of the kind (7.17), (7.18), (7.19) or (7.20) are non-singular (see Exercise A7.22). Let us fix our attention, as usual, on the case in which the equation is of the form (7.17). Then, given a point $p = (\xi, \eta)$ of the curve, the tangent line $R_p$ in that point has equation

$$(-3\xi^2 - a)(x - \xi) + 2\eta(y - \eta) = 0.$$

It is vertical if and only if $\eta = 0$, that is on the points in which $C$ intersects the $x$-axis. We leave to the reader the task of verifying that if $\eta \neq 0$, then $R_p$ intersects $C$ in a further point $r$ the coordinates of which may be easily computed (see Exercise A7.25). The point $r$ is to be interpreted as the third intersection with the elliptic curve of the *line through $p$ and $q$* when $p = q = (\xi, \eta)$. Of course, if $p - q = (\xi, 0)$, the tangent line is the vertical line of equation $x = \xi$ and, as we have already seen, it is the point at infinity $O$ that has to be regarded as the third intersection of this line with the curve.

What can be said about the lines through $O$? By our interpretation of points at infinity, they are all the lines that are parallel to the $y$-axis, that is, the vertical lines of equation $x = u$.

So, if $p = (\xi, \eta)$ lies on the curve, the line through $p$ and $O$ is the line of equation $x = \xi$, which, as we know, intersects the curve in the further point $q = (\xi, -\eta)$.

Finally, for reasons we shall not discuss at length (see Exercise A7.26), it can be seen that the line at infinity must be considered as the tangent line to the curve $E$ in $O$ and, as already seen, it intersects the curve only in $O$. So we may say that this line, which must be considered as the line through $p$ and $q$ when $p = q = O$, intersects further the curve in $O$ itself.

In conclusion, we may say that, in the sense made clear above, given two points $p$ and $q$ of $E$, distinct or equal, there is a third point $r$ of $E$ such that $p$, $q$ and $r$ are *collinear*.

So we are very close to defining the group law on $E$, which will be described using an additive notation. One is tempted to define define the *sum* $p + q$ of two points $p$ and $q$ of $E$ as the third point of $E$ that is collinear with $p$ and $q$. But this is not a completely correct idea. We have to take $O$ as the identity element, that is the *zero*, of the group, and define the sum $p + q$ as follows:

- consider first the *third point* $r$ of $E$ collinear with $p$ and $q$;
- define $p + q$ as the *third point* $s$ of $E$ collinear with $r$ and $O$.

In other words, if we consider an elliptic curve defined by an equation of the form (7.17), $p + q$ is the symmetric point with respect to the $x$-axis of the third point $r$ of $E$ collinear with $p$ and $q$ (see Figure 7.4). In this way, the opposite of each point $p$ is exactly its symmetric with respect to the $x$-axis.

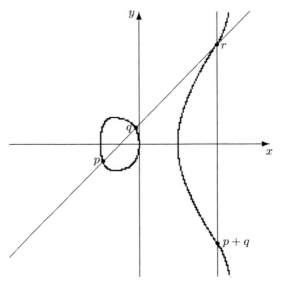

**Fig. 7.4.** Group law on an elliptic curve

Keeping in mind what has been said in this section, as well as in Exercise A7.25, we may compute the coordinates of the point $s = p + q = (x_3, y_3)$ as a function of the points $p = (x_1, y_1)$ and $q = (x_2, y_2)$ of an elliptic curve (see Exercise A7.27). Here we give the details only for the case in which the equation in Weierstrass form is of the kind (7.17).

**Proposition 7.9.9.** *Let $E$ be an elliptic curve on a field $\mathbb{K}$ of characteristic different from 2 and 3, of equation $y^2 = x^3 + ax + b$ with $27b^2 + 4a^3 \neq 0$. Consider the binary operation $+$ in $E$*

$$E \times E \longrightarrow E, \qquad (p, q) \longrightarrow s$$

*defined as follows:*

- *if $p = O$, then $s = q$;*
- *if $p \neq q$ are both different from $O$, $p = (x_1, y_1)$ and $q = (x_2, y_2)$, and if $x_1 = x_2$ and $y_1 = -y_2$, then $s = O$;*
- *if $p \neq q$ are both different from $O$, $p = (x_1, y_1)$ and $q = (x_2, y_2)$ with $x_1 \neq x_2$, then $s \neq O$ and $s = (x_3, y_3)$ with*

$$x_3 = \left( \frac{y_2 - y_1}{x_2 - x_1} \right)^2 - x_1 - x_2, \qquad y_3 = \frac{(y_2 - y_1)}{(x_2 - x_1)}(x_1 - x_3) - y_1;$$

- *if $p = q$ is different from $O$ and $p = (\xi, 0)$, then $s = O$;*
- *if $p = q$ is different from $O$ and $p = (\xi, \eta)$, with $\eta \neq 0$, then $s \neq O$ and $s = (\xi', \eta')$ with*

$$\xi' = \left( \frac{3\xi^2 + a}{2\eta} \right)^2 - 2\xi, \qquad \eta' = \frac{(3\xi^2 + a)}{2\eta}(\xi - \xi') - \eta.$$

*With this operation, $(E, +)$ is an abelian group, whose identity element is the point at infinity $O$ and the opposite of the point $(\xi, \eta)$ of $E$ is the point $(\xi, -\eta)$.*

PROOF. The commutativity of the operation just defined is a simple algebraic computation we leave to the reader as an exercise (see Exercise A7.28). The most delicate check is that the operation is associative: for it see [56], Proposition 2.2. □

**Remark 7.9.10.** Notice that without the hypothesis $27b^2 + 4a^3 \neq 0$, we may have anomalous situations, as the following one. Let $E$ be the real curve of equation

$$y^2 = x^3 - 3x - 2.$$

We have $27 \cdot 4 + 4 \cdot (-3)^3 = 0$. Now it is easy to verify that, if we keep the definition of the sum as given above, for every $(x, y) \in E$ we have

$$(-1, 0) + (x, y) = (-1, 0).$$

### 7.9.7 Elliptic curves over $\mathbb{R}$, $\mathbb{C}$ and $\mathbb{Q}$

We have finally come in possession of a *group for each elliptic curve*, and we may use these groups in cryptography, as we intended. But which ones of these curves should we use? and over which fields?

*Real* elliptic curves possess *infinitely many points*. Indeed, a real elliptic curve $E$ is defined by the equation (7.16). The curve $E$ consists of the graph of the function $y = \sqrt{f(x)}$ and of its symmetric with respect to the $x$-axis. As $x$ approaches $+\infty$, $f(x)$ tends to $+\infty$. Thus $\sqrt{f(x)}$ is defined at least on a half-line, so $E$ has infinitely many points. It is easily verified that $E$ consists of either one or two arcs, depending on the number (one or three) of real roots of $f(x)$ (see Exercise A7.29). In cryptography we need finite groups, so real elliptic curves are useful to draw inspiration from, but cannot be used for our goals.

Elliptic curves over $\mathbb{C}$ are even less useful. $\mathbb{C}$ being algebraically closed, those curves are, as mentioned, *double covers* of $\mathbb{C}$. They can be parametrised using suitable functions, called *elliptic functions*, which are not rational functions, but have properties quite similar to those of trigonometric functions and cannot be expressed in terms of elementary functions. The theory of these functions is very interesting but too complex to allow us more than the briefest mention (see [55]). It is interesting to remark that elliptic curves take their name from these functions.

We may next consider elliptic curves over $\mathbb{Q}$. In this regard, the following fundamental theorem is well known (see [56], pag. 188):

**Theorem 7.9.11 (Mordell–Weil).** *If $E$ is an elliptic curve over $\mathbb{Q}$, then $E$ is a finitely generated abelian group, that is,*

$$E \simeq \mathrm{Tors}(E) \oplus \mathbb{Z}^r,$$

*where $\mathrm{Tors}(E)$ is the torsion subgroup of $E$, that is to say, the subgroup of $E$ consisting of the points of finite order, while $r$ is called rank of $E$.*

The dependence of the rank of an elliptic curve over $\mathbb{Q}$ from its equation is not yet well understood.

**Example 7.9.12.** The point $p = (2,3)$ of the elliptic curve over $\mathbb{Q}$ of equation $y^2 = x^3 + 1$ is a torsion point. Indeed, by using the group law on the curve we find that $2p = p + p = (0,1)$, $4p = (0,-1)$ and so $6p = O$.

Excluding the case in which the rank of a rational elliptic curve $E$ is 0, $E$ is an infinite group too, and so unsuitable for use in cryptography.

*So we are only left with elliptic curves on finite fields $\mathbb{F}_q$.* This is a classical and intriguing subject which has played a central role in last century's mathematics, culminating in the momentous proof by A. Wiles of the well-known so-called Fermat's Last Theorem, which states that the equation $x^n + y^n = z^n$ has no solutions $(x, y, z) \in \mathbb{Z} \times \mathbb{Z} \times \mathbb{Z}$ with $x, y, z$ different from zero, if $n > 2$ (see the popular science book [57] or Wiles's paper [63]; see also Exercises A7.30-A7.35 for the case $n = 2$).

### 7.9.8 Elliptic curves over finite fields

On the road to describe elliptic curves over a finite field $\mathbb{F}_q$, we begin by remarking that such a curve $E$, of equation (7.16), has finitely many points, their number being denoted by $|E/\mathbb{F}_q|$. We also have an estimate

$$|E/\mathbb{F}_q| \leq 2q + 1 \tag{7.29}$$

because, apart from $O$, for all $x \in \mathbb{F}_q$, the equation (7.16) has at most two solutions in $\mathbb{F}_q$.

The estimate (7.29) is very rough: indeed, only one half of the elements of $\mathbb{F}_q$ are squares, so only one half of the elements of $\mathbb{F}_q$ has a square root. To be more precise, we can give the following definition, which is the analogous, for finite fields, of the Legendre symbol:

**Definition 7.9.13.** *The* quadratic character *in $\mathbb{F}_q$ is the function*

$$\chi : u \in \mathbb{F}_q \rightarrow \chi(u) \in \{0, 1, -1\}$$

*defined as follows:*

$$\chi(u) = \begin{cases} 0 & \text{if } u = 0, \\ 1 & \text{if } u \text{ is a square}, \\ -1 & \text{if } u \text{ is not a square.} \end{cases}$$

In particular, $\chi(u) = (\frac{u}{q})$ if $q$ is a prime number.

Notice that the number of solutions of the equation $x^2 = u$ in $\mathbb{F}_q$ equals $1 + \chi(u)$ (see Exercise A7.36). Moreover, $\chi(uv) = \chi(u)\chi(v)$ for every pair $(u, v)$ of elements of $\mathbb{F}_q$ (see Exercise A7.37).

Then, if the characteristic of $\mathbb{F}_q$ is different from 2 and 3 (hence the equation becomes (7.17)), we have

$$|E/\mathbb{F}_q| = 1 + \sum_{x \in \mathbb{F}_q} (1 + \chi(x^3 + ax + b)) = q + 1 + \sum_{x \in \mathbb{F}_q} \chi(x^3 + ax + b).$$

Now, let us reason heuristically: we expect that, for general $a$ and $b$, for a given $x \in \mathbb{F}_q$, $\chi(x^3 + ax + b)$ has the same probability of being equal to 1 or $-1$. That is, for all $x \in \mathbb{F}_q$, computing $\chi(x^3 + ax + b)$ is like tossing a coin to see whether it shows heads or tails.

In probability theory a situation of this kind is called *random walk*. Assume we are on a line, in the coordinate origin. We have a coin and we toss it. If it shows heads, we take a step towards the positive semiaxis, while if it shows tails we take a step towards the negative semiaxis. After $n$ steps, how far may we expect to be from the origin? The answer given by probability theory is that we expect a distance of about $\sqrt{n}$ steps, in one of the two directions (see [29]).

In our case $n = q$, that is, the number of points of $\mathbb{F}_q$. So we expect, by this heuristic argument, that the number $|E/\mathbb{F}_q|$ of points in the elliptic curve $E$ is *on the average* bounded by $q + 1 + \sqrt{q}$. The following result by Hasse (see [56], p. 131) gives us an actual, not just a heuristic, estimate for $|E/\mathbb{F}_q|$. Notice that this estimate is not very far from the previous one.

**Theorem 7.9.14 (Hasse's Theorem).** *If $N = |E/\mathbb{F}_q|$, then*

$$|N - (q + 1)| \le 2\sqrt{q}.$$

Hasse's theorem says that an elliptic curve over $\mathbb{F}_q$ has, after all, *not many more points* than $\mathbb{F}_q$ itself. So elliptic curves over finite fields are actually *not too complicated objects*. Good news for cryptographers!

**Example 7.9.15.** Let us compute now the number of points of the elliptic curve of equation $y^2 = x^3 + x$ over $\mathbb{F}_p$, with $p$ a prime number such that $p \equiv 3$ (mod 4). We have

$$N = p + 1 + \sum_{x \in \mathbb{F}_p} \chi(x^3 + x) = p + 1 + \sum_{x \in \mathbb{F}_p^*} \chi(x^3 + x).$$

But

$$\chi((-x^3) + (-x)) = \chi((-1)(x^3 + x)) = \chi(-1)\chi(x^3 + x) =$$
$$= \left(\frac{-1}{p}\right)\chi(x^3 + x) = -\chi(x^3 + x);$$

hence it follows that $N = p + 1$, as the summands in the sum giving $N$ cancel out in pairs.

A result more precise than Hasse's Theorem is Weil's theorem. It is one of the most important theorems of 20th century mathematics; it led Weil to conjecture more general results which were later proved by Deligne, giving a great boost both to algebraic geometry and number theory.

To state Weil's theorem (see [56], Ch. V, §2), associate with an elliptic curve $E$ defined over $\mathbb{F}_q$ a function called *zeta function* of $E$, denoted by $Z_{E,\mathbb{F}_q}(t)$. If $N_r = |E/\mathbb{F}_{q^r}|$, define

$$Z_{E,\mathbb{F}_q}(t) := e^{\sum_{r=1}^{\infty} N_r t^r / r}.$$

**Theorem 7.9.16 (A. Weil).** *The function $Z_{E,\mathbb{F}_q}(t)$ is rational, of the form*

$$\frac{1 - at + qt^2}{(1 - t)(1 - qt)} \tag{7.30}$$

*and only depends on $E$. More precisely,*

$$a = q + 1 - N_1.$$

Notice that the discriminant $\Delta = a^2 - 4q$ of the numerator of (7.30) is negative by Hasse's theorem, so the latter has two complex conjugate roots $\alpha'$, $\beta'$. It is easy to verify, and we leave it as an exercise to the reader (see Exercise A7.38), that, by setting

$$\alpha = \frac{1}{\alpha'}, \qquad \beta = \frac{1}{\beta'},$$

we have $|\alpha| = |\beta| = \sqrt{q}$.

**Corollary 7.9.17.** *For all $r$ we have*

$$N_r = 1 + q^r - \alpha^r - \beta^r.$$

PROOF. From Weil's theorem it follows that:

$$\sum_{r=1}^{\infty} N_r \frac{t^r}{r} = \log \frac{(1 - \alpha t)(1 - \beta t)}{(1 - t)(1 - qt)} =$$

$$= \log(1 - \alpha t) + \log(1 - \beta t) - \log(1 - t) - \log(1 - qt).$$

Taking derivatives on both sides we find

$$\sum_{r=1}^{\infty} N_r t^{r-1} = -\frac{\alpha}{1 - \alpha t} - \frac{\beta}{1 - \beta t} + \frac{1}{1 - t} + \frac{q}{1 - qt} =$$

$$= -\alpha \sum_{r=0}^{\infty} (\alpha t)^r - \beta \sum_{r=0}^{\infty} (\beta t)^r + \sum_{r=0}^{\infty} t^r + q \sum_{r=0}^{\infty} (qt)^r =$$

$$= \sum_{r=0}^{\infty} (-\alpha^{r+1} - \beta^{r+1} + 1 + q^{r+1}) t^r,$$

hence the corollary immediately follows. $\qquad\square$

Let us see how to apply these results to computing the number of points on an elliptic curve on a finite field.

**Example 7.9.18.** Let us compute the number $N_1$ of points over $\mathbb{F}_2$, and the number $N_r$ of points over any finite extension of degree $r$, of the elliptic curve $E$ of equation $y^2 + y = x^3 + 1$. Computing the number of points of any curve over $\mathbb{F}_2$ is trivial. We may easily proceed by trial and error, keeping in mind that in the affine plane there are exactly 4 points. Then Weil's theorem gives the number of points of the curve over any extension of $\mathbb{F}_2$.

In this case we have $N_1 = 3$, because $E$, apart from the point at infinity $O$, has over $\mathbb{Z}_2$ only the points $(1, 0)$ and $(1, 1)$ (see Exercise B7.66). Then the zeta function is

$$Z(t) = \frac{1 + 2t^2}{(1 - t)(1 - 2t)}.$$

The roots of the numerator are $\pm i/\sqrt{2}$, so

$$N_r = 1 + 2^r - (i\sqrt{2})^r - (-i\sqrt{2})^r = \begin{cases} 1 + 2^r & \text{if } r \text{ is odd,} \\ 1 + 2^r - 2(-2)^{r/2} & \text{if } r \text{ is even.} \end{cases}$$

### 7.9.9 Elliptic curves and cryptography

Let us come back to cryptography. Now we have at our disposal not only the multiplicative groups of the fields $\mathbb{F}_q$, but also the elliptic curves defined on them, and there are many more of them. So, as already mentioned, we have more diversity, and this gives more security to our cryptosystems.

But are elliptic curves on finite fields actually appropriate for cryptography? The answer is essentially in the affirmative. Let us see why, discussing separately the issues already hinted at at the beginning of this section.

- Elliptic curves are given in a concrete way: it suffices to give their equation. This does not always mean that determining their points is easy. Indeed, there is no known polynomial algorithm to generate points on an elliptic curve over $\mathbb{F}_q$. However, there are *probabilistic* polynomial algorithms, which are able to determine points on an elliptic curve with a very high probability (see Exercise A7.39 and A7.40). Let us further mention that there is in fact a polynomial algorithm, due to R. Schoof, that determines the *number* of points of an elliptic curve over $\mathbb{F}_q$, but without determining the points themselves (see [50]).
- Exponentiating on an elliptic curve $E$, which amounts now actually to multiplying a point of $E$ by an integer, has a polynomial computational cost: keeping in mind Proposition 7.9.9, this can be shown in a similar way as in $\mathbb{Z}_p$ or in $\mathbb{F}_q$.
- A. Menezes, T. Okamoto and S. A. Vanstone ([42]) have shown that the problem of discrete logarithms on an elliptic curve is not less hard than on a finite field. It is conjectured that, a fortiori, for an elliptic curve on $\mathbb{F}_q$ the Diffie–Hellman hypothesis holds.

In conclusion, let us sum up how it is possible to exchange a private key using an $RSA$ system relying on the use of an elliptic curve. We choose: a field $\mathbb{F}_q$, an elliptic curve $E$ and a point $p \in E$, which are made public. As we shall see shortly, it is convenient for the system to work best, that $E$ has many points over $\mathbb{F}_q$. To determine such an $E$, the results described above are helpful.

Each user $U$ chooses a key $e_U$, a positive integer, he will keep private. However he publishes $p_U = e_U p$, which is again a point of $E$, computed by $U$ in polynomial time. It is convenient that if $U \neq V$ then $p_U \neq p_V$. To this end, it helps if $p$ has a very large order, far larger than the number of users of the system. So, when the numbers $e_U$ are chosen randomly the points $p_U$ will be different.

If two users $A$ and $B$ want to exchange a private key, they may do so by agreeing about using as their private key the point $p_{AB} = (e_A e_B)p$. They both may determine it in polynomial time, while, due to the difficulty of computing discrete logarithms on $E$ and Diffie–Hellman hypothesis, $p_{AB}$ will be unreachable by anybody else.

Clearly, $A$ and $B$ may use as their private key a number deduced from $p_{AB}$: for instance, one of the coordinates of this point, or their sum, and so on.

We leave to the reader the task of devising a way of applying the general $RSA$ method using elliptic curves (see Exercise A7.41).

## 7.9.10 Pollard's $p-1$ factorisation method

Surprisingly, elliptic curves are not only greatly useful in cryptography, but are also suitable to solving several other problems we have previously discussed, such as primality tests (S. Goldwasser, J. Kilian; A.O.L. Atkin, see [30], Ch. VI) and factorisation (H.W. Lenstra, see [30], l.c.). The ideas are not very different from those described in this book, but elliptic curves make it possible to implement them with an accuracy and a flexibility that make them very effective. We conclude this part by examining a factorisation method (Pollard's $p-1$ method) which does not rely on elliptic curves. Without going into too many details, we mention the fact that the restrictions of this method can be overcome by using elliptic curves, and this yields Lenstra's factorisation mentioned.

Assume we have to factor an integer $N$. The method we are going to describe works when $N$ has a prime factor $p$, to be determined, such that $p-1$ has not too large prime divisors. So we give an a priori estimate of the greatest prime $T$ dividing $p-1$.

We have to find a number $k$ divided by $p-1$. To this end, we may proceed as follows. Let

$$p - 1 = 2^\alpha 3^\beta 5^\gamma \cdots T^\omega$$

be the prime factorisation of $p-1$. As $p$ is not known, the exponents $\alpha$, $\beta$, $\gamma$ etc. are not known either. However, as $2^\alpha \leq p-1 < N$, we have $\alpha < (\log N)/(\log 2)$, and so $\beta < (\log N)/(\log 3)$, etc. Thus, setting

$$k := 2^{\lfloor (\log N)/(\log 2) \rfloor} \cdot 3^{\lfloor (\log N)/(\log 3) \rfloor} \cdots T^{\lfloor (\log N)/(\log T) \rfloor},$$

we have that $p-1$ divides $k$.

Let now $a$ be an integer between 2 and $N-2$, such that $\mathrm{GCD}(a, N) = 1$, so $a$ is relatively prime with all prime factors of $N$. By Fermat's little theorem we have $a^{p-1} \equiv 1 \pmod{p}$ and so, as $p-1 | k$,

$$a^k \equiv 1 \pmod{p}.$$

Compute now $a^k \pmod{N}$ in polynomial time (see § 3.3.1) and simultaneously, using the Euclidean algorithm, compute $d = \mathrm{GCD}(a^k - 1, N)$. Clearly, the first $p$ we are looking for is such that $p | d$, so $d \neq 1$. If $d \neq N$, we have found a (not necessarily prime) factor of $N$ and we are done. Otherwise, if $d = N$, that is, if $N | (a^k - 1)$, modify the choice of the integer $a$ or of the integer $k$ and start over. In practice, take as $k$ a multiple of all integers smaller than or equal to a fixed integer $M$, which is supposed to be greater than all the powers of the prime numbers dividing $p-1$. For instance, we may take $k = M!$.

**Example 7.9.19.** Factor with this method the number 156203. Choose $M = 6$, $k = 6!$, and $a = 2$. Compute $2^k \pmod{156203} = 32219$. Find next

$$\mathrm{GCD}(2^k - 1, 156203) = 181.$$

We have found that 181 (which is a prime) is a factor of 156203 and the factorisation of 156203 is

$$156203 = 181 \cdot 863.$$

**Remark 7.9.20.** While the single steps just described have polynomial cost, the algorithm itself is exponential because, when $d = N$, we have to modify, for instance, our choice of the integer $a$, which may be done in $N$ ways. Moreover, it may happen that for each choice of $a$ we have $N|(a^k - 1)$, and so the algorithm might never give a positive result.

Nevertheless, there are probabilistic reasons for this algorithm to work in some cases. For instance, suppose exactly one of the prime factors of $N$, say $p$, has the property that the prime factors of $p-1$ are bounded by $T$, while for all other factors $q$, $q$ is large with respect to $k$.

In this situation, if $q|(a^k - 1)$ then $a$ is a $k$th root of unity modulo $q$, and the probability of this happening for a random choice of $a$ is $k/q$, because the $k$th roots of unity are at most $k$ in $\mathbb{Z}_q^*$. So this probability is very small, if $q \gg k$. Hence the probability that $N|(a^k - 1)$ is even smaller, and this is exactly the case in which the algorithm has to be repeated.

Which are the limitations of this algorithm? In it, we exploit the structure of the groups $\mathbb{Z}_p^*$, with $p$ ranging among the prime factors of $N$. For a fixed $N$, these groups are fixed and cannot be exchanged for others; and, as we have remarked, the algorithm might not give positive result for any of them. This happens, in particular, if the order $p-1$ of each of these groups has at least a prime factor not bounded by the number $T$ we have chosen at the beginning and which, as seen, determines the number $k$. How may we obtain a larger choice? By using elliptic curves, on which H. W. Lenstra's factorisation method relies (see [35], [30], Ch. VI, § 4). Here is a sketch of the idea: substitute the group $E/\mathbb{Z}_p$ of the points of an elliptic curve $E$ over $\mathbb{Z}_p$ for $\mathbb{Z}_p^*$. The new group, by Hasse's theorem 7.9.14, has order

$$|E/\mathbb{Z}_p| = p + 1 - s, \qquad \text{with } |s| \le 2\sqrt{p}.$$

Different elliptic curves $E$ yield different values of $s$ and we have at our disposal several groups: so it is realistic to expect one of them to have order with small prime factors.

# Appendix to Chapter 7

## A7 Theoretical exercises

**A7.1.** We have seen the frequencies of the different letters in English. Assume we have messages written in two languages, for instance containing an English text and its translation in another language, or the other way around. Explain how to get a frequency table for these messages, assuming we have the tables for the two languages, knowing that each message is written half in one of the languages and half in the other one.

**A7.2.** We have seen how to carry out the cryptanalysis of an enciphered message by using frequency analysis. Exercise C7.2 requests the reader to write a program that computes the frequencies with which the letters appear in a given text. Assume we have a program saying whether a word exists in English language or not: more precisely, on receiving as its input a word, its output is either *true* or *false*, depending on the word being present or not in English lexicon. Describe an algorithm to decrypt a message, enciphered using a monoalphabetic substitution, using these two programs.

**A7.3.** Prove that an affine transformation $f : \mathbb{Z}_{26} \to \mathbb{Z}_{26}$ defined by $f(n) = an + b \bmod 26$, where $a, b \in \mathbb{Z}$, is bijective if and only if $\mathrm{GCD}(a, 26) = 1$.

**A7.4.*** Show by an example that there are systems of two linear congruences modulo 26 in two unknowns having no solutions. (Hint: try a diagonal matrix. Indeed, in this case the system reduces to two independent linear congruences; the system does not admit solutions if and only if one of the two congruences does not.)

**A7.5.*** Show by an example that there are systems of two linear congruences modulo 26 in two unknowns having more than one solution modulo 26. (Hint: try a diagonal matrix.)

**A7.6.** Explain how to carry out the cryptanalysis of an affine cipher, with 2-letter unitary messages, when the system of linear congruences modulo 26 obtained as explained in Remark 7.4.3 on page 339 has more than one solution.

**A7.7.*** Prove Proposition 7.4.2 on page 338.

**A7.8.*** Consider a system of two linear congruences of the form

$$
A \cdot \begin{pmatrix} a_1 \\ a_2 \\ \vdots \\ a_n \end{pmatrix} \equiv \begin{pmatrix} \alpha_1 \\ \alpha_2 \\ \vdots \\ \alpha_n \end{pmatrix} \pmod{m},
$$

where $A$ is a square $n \times n$ matrix with integer coefficients. Prove that if the determinant of $A$ is invertible modulo $m$, then $A$ has an inverse modulo $m$ and the system admits a unique solution in $a_1, a_2, \ldots, a_n$, given by

$$
\begin{pmatrix} a_1 \\ a_2 \\ \vdots \\ a_n \end{pmatrix} \equiv A^{-1} \cdot \begin{pmatrix} \alpha_1 \\ \alpha_2 \\ \vdots \\ \alpha_n \end{pmatrix} \pmod{m}.
$$

**A7.9.** Prove that the computational complexity of the knapsack problem, when $a_1, \ldots, a_n$ is a superincreasing sequence of integers, is polynomial.

**A7.10.*** Assume we know that an integer $n$ is the product of two prime numbers $p$ and $q$. Explain how to find $p$ and $q$ in polynomial time if one knows $\varphi(n)$.

**A7.11.*** Let $n$ be a product of distinct primes. If $d$ and $e$ are positive integers such that $de - 1$ is divisible by $p - 1$ for every prime $p \mid n$, then we have $a^{de} \equiv a \pmod{n}$ for every integer $a$, even if $a$ is not relatively prime with $n$.

**A7.12.** Explain why Exercise A7.11 proves that we have not to worry if in the *RSA* cryptosystem it happens that a unitary message $P_i$ we want to send to the user $B$ is not relatively prime with $n_B$.

**A7.13.** Prove that lines are irreducible curves. (Hint: first degree polynomials are always irreducible.)

**A7.14.** Consider the line $R$ of equation (7.11). Prove that if $a \neq 0$ the projection $(u, v) \in R \to v \in \mathbb{K}$ on the $y$-axis is bijective, and write its inverse.

**A7.15.** Prove that the reducible conics are unions of two lines.

**A7.16.**\* Prove that a curve of equation (7.12) is irreducible.

**A7.17.**\* Prove that an irreducible conic, that is an irreducible curve defined by a degree two equation, is a rational curve, that is admits a parametric representation by rational functions.

**A7.18.**\* Prove that the polynomial $f(x) = x^3 + ax + b$ has no multiple roots if and only if $27b^2 + 4a^3 \neq 0$. (Hint: a root of a polynomial $f(x)$ is a multiple root if and only if it is a root of the derivative of $f(x)$ too.)

**A7.19.** Prove that the line through the points $(x_1, y_1)$ and $(x_2, y_2)$ has equation (7.23) with $m, n$ determined by (7.24). (Hint: the (non vertical) lines through $(x_1, y_1)$ have equation $y - y_1 = m(x - x_1)$ and $m$ can be found by imposing that the point $(x_2, y_2)$ lies on the line.)

**A7.20.** Prove that if $x_1, x_2, x_3$ are the three roots of a degree three monic polynomial, then the degree two coefficient of the polynomial is $-x_1 - x_2 - x_3$ and the constant term is $-x_1 x_2 x_3$. (Hint: the polynomial may be written as $(x - x_1)(x - x_2)(x - x_3)$.)

**A7.21.** Let $\mathbb{K}$ be a field and consider the curve in $\mathbb{A}^2_{\mathbb{K}}$ defined by an equation of the form $y^2 + y(mx + n) = \ell x^3 + px^2 + qx + r$ with $\ell \neq 0$. Prove that, if $\mathbb{K}$ contains the cubic roots of each of its elements, then it is possible to reduce it in Weierstrass canonical form as in Proposition 7.9.4. (Hint: begin by changing variables by $x \to x/\sqrt[3]{\ell}$, $y \to y$, and go on as in the proof of Proposition 7.9.4.)

**A7.22.**\* Prove that a curve defined by an equation in Weierstrass form of the kind (7.17), (7.18), (7.19) or (7.20) is singular in the algebraic closure of $\mathbb{K}$ if and only if it does not verify the regularity hypothesis on page 373.

**A7.23.**\* Prove that if a curve defined by an equation in Weierstrass form of the kind (7.17), (7.18), (7.19) or (7.20) is singular, then it is rational. (Hint: work in the algebraic closure of $\mathbb{K}$ and prove that almost all the lines through a singular point intersect the curve in a unique point out of the singular point.)

**A7.24.** Prove that, if $27b^2 + 4a^3 = 0$, then the curve of equation (7.17) is rational.

**A7.25.**\* Let $p = (\xi, \eta)$ be a point of the elliptic curve (7.17), with $\eta \neq 0$. Determine the intersection point different from $p$ of the tangent line to the curve in $p$. (Hint: obtain $y$ from the equation of the tangent line and substitute it in the equation of the curve; so one finds an equation of degree three in $x$ having a double solution in $x = \xi$; the further solution of the equation is the abscissa of the required point.)

**A7.26.**\* Explain why the line at infinity is to be considered as the tangent line to an elliptic curve in Weierstrass form in the point at infinity. (Hint: write everything in cartesian coordinates $u = x_0/x_3, v = x_1/x_3$ and consider the point $O$ as the origin of a new plane $\mathbb{A}_k^2$.)

**A7.27.**\* Compute the coordinates of the point $p + q = (x_3, y_3)$ in function of the points $p = (x_1, y_1)$ and $q = (x_2, y_2)$ of an elliptic curve in each of the cases (7.17), (7.18), (7.19) and (7.20) (see page 379).

**A7.28.** Prove that the operation of sum of points on an elliptic curve is commutative, that is, $p + q = q + p$. (Hint: carry out the calculations, or consider the geometric definition of the group law.)

**A7.29.** Prove that an elliptic curve $y^2 = f(x)$ defined on the real field consists of a single arc if $f(x)$ has a single real root, while it consists of a closed arc and an open one if $f(x)$ has three real roots. In Figure 7.4 this second case is represented. Give an example of a curve for which the first case is verified. (Hint: study the sign of $f(x)$, since the curve has no point $(x, y)$ if $f(x)$ is negative.)

In the following Exercises A7.30-A7.35 all the solutions $(x, y, z) \in \mathbb{N} \times \mathbb{N} \times \mathbb{N}$ with non-zero $x, y, z$ of the equation $x^2 + y^2 = z^2$ are determined. These solutions are called *Pythagorean triples* as, by Pythagoras's theorem, they are the lengths of the legs and of the hypotenuse of a right triangle. A Pythagorean triple $(x, y, z)$ is said to be *primitive* if $\mathrm{GCD}(x, y, z) = 1$.

**A7.30.** Let $(x, y, z)$ be a Pythagorean triple such that $\mathrm{GCD}(x, y, z) = d$. Prove that $(x/d, y/d, z/d)$ is a primitive Pythagorean triple.

**A7.31.** Let $(x, y, z)$ be a primitive Pythagorean triple. Prove that $\mathrm{GCD}(x, y) = \mathrm{GCD}(x, z) = \mathrm{GCD}(y, z) = 1$.

**A7.32.** Let $(x, y, z)$ be a primitive Pythagorean triple. Prove that one out of $x$ and $y$ is even, and the other is odd.

**A7.33.** Let $r, s$ be positive integers such that $\mathrm{GCD}(r, s) = 1$. Prove that if $rs$ is a square, so are $r$ and $s$.

**A7.34.**\* Let $(x, y, z)$ be a primitive Pythagorean triple with even $y$. Prove that there exist positive integers $n, m$, with $\mathrm{GCD}(m, n) = 1$ and $m > n$, such that

$$x = m^2 - n^2, \qquad y = 2mn, \qquad z = m^2 + n^2. \tag{7.31}$$

**A7.35.** Verify that if $(x, y, z)$ is a triple given by (7.31) with $n, m$ positive integers such that $\mathrm{GCD}(m, n) = 1$ and $m > n$, then $(x, y, z)$ is a primitive Pythagorean triple.

**A7.36.** Prove that the number of solutions of the equation $x^2 = u$ in $\mathbb{F}_q$ is $1 + \chi(u)$, where $\chi(u)$ is the quadratic character of $u$ in $\mathbb{F}_q$.

**A7.37.** Prove that the quadratic character $\chi$ in $\mathbb{F}_q$ satisfies $\chi(uv) = \chi(u)\chi(v)$ for all $u, v \in \mathbb{F}_q$.

**A7.38.** Let $\alpha', \beta'$ be the roots of the numerator of (7.30). Prove that $|\alpha| = |\beta| = \sqrt{q}$ where $\alpha = 1/\alpha', \beta = 1/\beta'$.

In the following two exercises we describe a simple polynomial probabilistic algorithm to determine points over an elliptic curve defined over $\mathbb{Z}_p$.

**A7.39.*** Let $p$ be a prime number. Prove that there is a polynomial probabilistic algorithm that determines an integer $n$ that *is not* a quadratic residue modulo $p$. (Hint: keep in mind the fact that half the elements of $\mathbb{Z}_p$ are not quadratic residues and that computing Jacobi symbols takes a polynomial time.)

**A7.40.*** Let $p > 2$ be a prime number and let $C$ be a hyperelliptic curve of equation $y^2 = f(x)$ defined over $\mathbb{Z}_p$. Prove that there is a polynomial probabilistic algorithm that determines a point of $C$. (Hint: for all $x \in \mathbb{K}$ the probability that $f(x)$ is a square is $1/2$; keep in mind the previous exercise and apply the algorithm of § 5.2.6.)

**A7.41.*** Explain how to use the groups defined over elliptic curves to implement the $RSA$ public-key cryptosystem. (Hint: follow exactly the same steps already seen for the $RSA$ system, substituting the points of an elliptic curve for the integer numbers and the operation of addition of points on the curve for the multiplication among integers.)

**A7.42.*** Explain how to use the groups defined over elliptic curves to implement the method to exchange private keys as described in § 7.8.1.

**A7.43.*** Explain how to use the groups defined over elliptic curves to implement the cryptosystem described in § 7.8.2.

# B7 Computational exercises

**B7.1.** Which is the 2-digit numerical equivalent of `exercise`?

(a) 0423041602081804.
(b) 0423041702091804.
(c) 0423041602091804.
(d) None of the above

**B7.2.** Which string corresponds to the number sequence

$$021418040002001814,$$

if we have used the 2-digit numerical equivalent?

(a) `coseacoso`.
(b) `coseacasa`.
(c) `coseacosa`.
(d) None of the above

**B7.3.** Which is the binary numerical equivalent of `codes`?

(a) 000100110000101001001010.
(b) 000100111000101001001010.
(c) 000100110000011001001010.
(d) 000100111000011001001010.

**B7.4.** To which word does the number sequence

011000010000110100100

correspond if we have used the binary numerical equivalent?

(a) `mine`.
(b) `mien`.
(c) `main`.
(d) `nine`.

**B7.5.** Which is the most frequent consonant in English language texts?

(a) N.
(b) S.
(c) R.
(d) T.

**B7.6.** Write the vowels a, e, i, o, u in decreasing order of their frequency in a long English language text.

(a) A, e, i, o, u.
(b) E, i or o, o or i, a, u.
(c) A, e or o, o or e, i, u.
(d) None of the above.

**B7.7.** Determine the most frequent consonant in the following text: *analysing frequencies often is the key to a successful cryptanalysis of messages enciphered using a monoalphabetic substitution.*

(a) N.
(b) T.
(c) C.
(d) S.

**B7.8.** Analyse the letter frequencies in the following text: *the frequencies with which the letters appear in a short text might be very different from what we would expect* Order the vowels a, e, i, o, u in the order of frequency in this text.

(a) A, e, i, o, u.
(b) E, a, i, o, u.
(c) E, o, i, u, a.
(d) E, i, a, o, u.

**B7.9.** Encipher using Caesar method, shifting each letter forward by three positions, the following message: `i am ready to attack gaul.`

(a) `L DP UHDGB WR DWWDFN JDWO`.
(b) `L DP UIDGB WR DWWDFN JDWO`.
(c) `L DP UHDGB WR DWWDFN JDXO`.
(d) `L DP UIDGB WR DWWDFN JDXO`.

**B7.10.** Suppose a Roman centurion received the following message sent him directly by Caesar:

DOHD NDFZAP HVZ.

After deciphering it, the centurion is very puzzled, as in the plaintext there is a patent Latin grammar error. Which is the plaintext of the message the centurion received?

(a) alea iacta est.
(b) alea iactae est.
(c) alea iactum est.
(d) alea iactus est.

**B7.11.** Consider the plaintext stiff upper lip. Which is the ciphertext in a Caesar cipher with a 13-letter shift?

(a) FGVSS HCCRE YVC.
(b) FGUSS HCCRE YUC.
(c) FGVTT HCCRE YVC.
(d) None of the above.

**B7.12.** Suppose we have received the message XYXCOXCO, known to have been enciphered by a 10 letter shift in a Caesar cipher. Which is the plaintext?

(a) popcorns.
(b) nonsense.
(c) ascience.
(d) None of the above.

**B7.13.** Verify that the cipher with *to be or not to be that is the question* as its key phrase determines the permutation:

0  1  2  3  4   5  6 7 8 9  10 11 12 13 14 15 16  17 18 19 20 21 22 23 24 25
19 14 1 4 17 13 7 0 8 18 16 20  2   3   5   6   9  10 11 12 15 21 22 23 24 25

**B7.14.** We want to encipher a message using a monoalphabetic substitution. Suppose we have chosen as our key phrase *there are more things in heaven and earth.* Which is the enciphering alphabet we get?

(a)
a b c d e f g h i j k l m n o p q r s t u v w x y z
T H E R A M O I N G S V D B C F J K L P Q U W X Y Z

(b)
a b c d e f g h i j k l m n o p q r s t u v w x y z
T H E R A M O T I N G S V D B C F J K P Q U W X Y Z

(c)
a b c d e f g h i j k l m n o p q r s t u v w x y z
T H E R A M O T I N G S D B C F J K L P Q U W X Y Z

(d)
a b c d e f g h i j k l m n o p q r s t u v w x y z
T H E A M O R I N G S D B C F J K L P Q U V W X Y Z

**B7.15.** Consider the monoalphabetic substitution of the previous exercise. Which ciphertext is obtained from the plaintext than are dreamt of in your philosophy?

(a) PITB TKM AKMTDP CO NB YCQK FINVCLCFIY.

(b) `PITB TLA RLATDP CM NB YCQL FINVCLCFIY.`
(c) `PITB TKA RKATDP CM NB YCQK FINVCLCFIY.`
(d) `PITB TKE AKETDP CO NB YCQK FINVCLCFIY.`

**B7.16.** Suppose we are reading a text enciphered using the monoalphabetic substitution given in the two previous exercises. If this text is `LWCKR`, which is the plaintext?

(a) `swear`.
(b) `sword`.
(c) `swore`.
(d) None of the above.

**B7.17.** We want to encipher a message using a monoalphabetic substitution. Suppose we have chosen as our key phrase

*Tyger Tyger, burning bright,*
*in the forests of the night.*

Which is the enciphering alphabet we get from this key phrase?

(a)
| a | b | c | d | e | f | g | h | i | j | k | l | m | n | o | p | q | r | s | t | u | v | w | x | y | z |
|---|---|---|---|---|---|---|---|---|---|---|---|---|---|---|---|---|---|---|---|---|---|---|---|---|---|
| T | Y | G | E | R | B | U | N | I | H | F | O | R | S | A | C | D | J | K | L | M | Q | V | W | X | Z |

(b)
| a | b | c | d | e | f | g | h | i | j | k | l | m | n | o | p | q | r | s | t | u | v | w | x | y | z |
|---|---|---|---|---|---|---|---|---|---|---|---|---|---|---|---|---|---|---|---|---|---|---|---|---|---|
| T | Y | G | E | R | B | U | N | F | I | H | O | S | A | C | D | J | K | L | M | P | Q | V | W | X | Z |

(c)
| a | b | c | d | e | f | g | h | i | j | k | l | m | n | o | p | q | r | s | t | u | v | w | x | y | z |
|---|---|---|---|---|---|---|---|---|---|---|---|---|---|---|---|---|---|---|---|---|---|---|---|---|---|
| T | Y | G | E | R | B | U | N | I | H | F | O | S | A | C | D | J | K | L | M | P | Q | V | W | X | Z |

(d)
| a | b | c | d | e | f | g | h | i | j | k | l | m | n | o | p | q | r | s | t | u | v | w | x | y | z |
|---|---|---|---|---|---|---|---|---|---|---|---|---|---|---|---|---|---|---|---|---|---|---|---|---|---|
| T | Y | G | E | R | B | U | N | F | I | H | O | R | S | A | C | D | J | K | L | M | Q | V | W | X | Z |

**B7.18.** Consider the monoalphabetic substitution determined by the key phrase `Tyger Tyger, burning bright, in the forests of the night`, for which we have determined the enciphering alphabet in Exercise B7.17. Which is the enciphered text we obtain from the plaintext `poem by blake`?

(a) `DCRS YX YOTFR.`
(b) `DCSR XY XOTHR.`
(c) `DCSR YX YOTHR.`
(d) `DCRS XY XOTFR.`

**B7.19.** Suppose we are reading a text enciphered using the monoalphabetic substitution of the previous exercise, that is, using the key phrase `Tyger Tyger, burning bright, in the forests of the night`. If this text is `LIAUTOCAU`, which is the plaintext?

(a) `singasong`.
(b) `longsongs`.
(c) `alongsong`.
(d) None of the above.

**B7.20.** Use the program of exercise C7.3 to encipher the message sent to Edgar Allan Poe, using the key phrase *UNITED STATES*. Check the mistakes that were made in the message as given in the text.

**B7.21.** Let ALGEBRA be the key word chosen to encipher a message using Vigenère method (see pp. 323-325). If the plaintext is **very hard to decrypt**, which is the enciphered text?

(a) VPXC IRRD EU HFTRYAZ.
(b) VPXC IRRD EU HFTRYAZ.
(c) VPXC IRRD EU HFTRYAZ.
(d) VPXC IRRD EU HFTRYAZ.

**B7.22.** Let again ALGEBRA be the key word chosen to encipher a message using Vigenère method. If the ciphertext is **I SGZF JOLGKH OZNE PDISTISPY**, which is the plaintext?

(a) i have solved five exercises.
(b) i have solved four exercises.
(c) i have solved nine exercises.
(d) None of the above.

**B7.23.** Let HARDWORK be the key word chosen to encipher a message using Vigenère method. If the plaintext is **all work and no play makes jack a dull boy**, which is the ciphertext?

(a) HLC ZFKB KUD ES LZRI TABHO XRMR A WXHZ SYF
(b) HLC ZKFB KUD ES LZRI YABHO XRMR A UXHZ SYF
(c) HLC ZFKB KUD ER LZRI YABHO XRMR A UXHZ SYF
(d) HLC ZKFB KUD ER LZRI TABHO XRMR A UXHZ SYF

**B7.24.** Let again HARDWORK be the key word chosen to encipher using Vigenère method. If the ciphertext is **AABHWRRIVUK**, which is the plaintext?

(a) have a day out.
(b) just one more.
(c) too much work.
(d) None of the above.

**B7.25.** Encipher the message **attack today** in 4-letter blocks using the translation method in $\mathcal{P} = \mathbb{Z}_{26}^4$ described in section 7.4.1 using the key $k = 100$ (see Table 7.8 on page 340).

**B7.26.** Consider the plaintext **happy birthday**. Which is the text enciphered using the affine transformation with key $k = (7, 3)$?

(a) ADEEP KHSGAYDP.
(b) ADEEM JHSGAYDM.
(c) ADEEP JHSGAYDP.
(d) ADEEM KHSGAYDM.

**B7.27.** Suppose we receive the message **ZDB AHJ FDSCP**, knowing that it has been enciphered using the affine transformation with key $k = (7, 3)$. Which is the plaintext?

(a) see him later.
(b) saw you early.
(c) see you later.
(d) saw him early.

**B7.28.** Compute, if it exists, the inverse modulo 26 of the matrix

$$\begin{pmatrix} 7 & 3 \\ -5 & -10 \end{pmatrix}.$$

(a) The inverse does not exist because the determinant of the matrix is not relatively prime with 26.

(b) The inverse is $\begin{pmatrix} 12 & -1 \\ 7 & -11 \end{pmatrix}.$

(c) The inverse is $\begin{pmatrix} 11 & 1 \\ 7 & -11 \end{pmatrix}.$

(d) None of the above.

**B7.29.** Compute, if it exists, the inverse modulo 26 of the matrix

$$\begin{pmatrix} 19 & 13 \\ 2 & 11 \end{pmatrix}.$$

(a) The inverse does not exist because the determinant of the matrix is not relatively prime with 26.

(b) The inverse is $\begin{pmatrix} 11 & 13 \\ -2 & -19 \end{pmatrix}.$

(c) The inverse is $\begin{pmatrix} -11 & 13 \\ -2 & 19 \end{pmatrix}.$

(d) None of the above.

**B7.30.** We want to encipher the plaintext computer using an affine transformation of $\mathbb{Z}_{26}^2$ defined by the key $k = (A, b)$, with

$$A = \begin{pmatrix} 7 & 3 \\ -5 & -10 \end{pmatrix}, \qquad b = \begin{pmatrix} 1 \\ 2 \end{pmatrix}.$$

Which is the ciphertext?

(a) FIAAQYCU.
(b) FIABQZCU.
(c) FIABQYCU.
(d) FIABQZCU.

**B7.31.** Consider a text enciphered using the affine transformation given in the previous exercise. If the ciphertext is VOITJOCHGN, which is the plaintext?

(a) twosecrets.
(b) tensecrets.
(c) anysecrets.
(d) sixsecrets.

**B7.32.** Trying to perform the cryptanalysis of a message we have intercepted, we are confronted with the problem of solving the following system of linear congruences:

$$\begin{cases} 19 \equiv 6a + b \pmod{26}, \\ 13 \equiv 3a + b \pmod{26}. \end{cases}$$

How many solutions does this system admit modulo 26?

(a) One.
(b) No one.
(c) Infinitely many.
(d) None of the above.

**B7.33.** Trying to perform the cryptanalysis of a message we have intercepted, we are confronted with the problem of solving the following system of linear congruences:

$$\begin{cases} 11 \equiv 21a + b \pmod{26}, \\ -7 \equiv -5a + b \pmod{26}. \end{cases}$$

How many solutions does this system admit modulo 26?

(a) One.
(b) No one.
(c) Infinitely many.
(d) None of the above.

**B7.34.** Let $S = \{1, \ldots, 15\}$. Consider the function $f : S \to S$, $f(x) = 5^x \bmod 16$. What can be said about $f$?

(a) It is bijective.
(b) It is injective but not surjective.
(c) It is surjective but not injective.
(d) It is neither surjective nor injective.

**B7.35.** Consider the function $f : \mathbb{Z}_4 \to \mathbb{F}_5^*$, $f(x) = 4^x \bmod 5$. What can be said about $f$?

(a) It is bijective.
(b) It is injective but not surjective.
(c) It is surjective but not injective.
(d) It is neither surjective nor injective.

**B7.36.** Consider the function $f : \mathbb{Z}_8 \to \mathbb{F}_9^*$, $f(x) = i^x$, where $i$ is an element of $\mathbb{F}_9$ such that $i^2 = -1$. What can be said about $f$?

(a) It is bijective.
(b) It is injective but not surjective.
(c) It is surjective but not injective.
(d) It is neither surjective nor injective.

**B7.37.** Consider the function $f : \mathbb{Z}_{q-1} \to \mathbb{F}_q^*$, $f(x) = b^x$, where $b$ is an element of $\mathbb{F}_q^*$. What can be said about $f$?

(a) The function $f$ is bijective for all $b$.
(b) There exists at least a value of $b$ such that $f$ is bijective.
(c) There exists exactly one $b$ such that $f$ is bijective.
(d) In general there is no $b$ such that $f$ is bijective.

**B7.38.** Verify that $\bar{2}$ is a generator of $U(\mathbb{Z}_{101})$. Compute the discrete logarithm $\log_2 \bar{3}$.

**B7.39.** Using the Baby step–giant step algorithm determine $\log_2 7$ in base 13. The answer is:

(a) 4.
(b) 5.
(c) 7.
(d) 11.

**B7.40.** Using the Baby step–giant step algorithm determine $\log_3 7$ in base 17. The answer is:

(a) 4.
(b) 11.
(c) 7.
(d) 16.

**B7.41.** Consider the sequence $1, 3, 4, 8, 13, 20$. How many solutions has the knapsack problem for this sequence and $m = 34$?

(a) None.
(b) One.
(c) Two.
(d) Three.

**B7.42.** Consider the sequence $2, 3, 7, 8, 15, 27$. How many solutions has the knapsack problem for this sequence and $m = 35$?

(a) None.
(b) One.
(c) Two.
(d) Three.

**B7.43.** Is the sequence $1, 5, 8, 15, 30, 60$ superincreasing?

(a) Yes.
(b) No, because $2 \cdot 5 > 8$ and $2 \cdot 8 > 15$.
(c) No, because $60 \leq 1 + 5 + 8 + 15 + 30$.
(d) None of the above.

**B7.44.** Is the sequence $1, 2, 5, 9, 17, 45$ superincreasing?

(a) Yes.
(b) No, because $2 \cdot 5 > 9$.
(c) No, because $17 \leq 1 + 2 + 5 + 9$.
(d) None of the above.

**B7.45.** Consider the superincreasing sequence 1, 2, 5, 11, 22, 44, 88. How many solutions has the corresponding knapsack problem for $m = 147$?

(a) None.
(b) The only solution is $x_1 = x_2 = x_4 = x_6 = x_7 = 1$ and $x_3 = x_5 = 0$.
(c) The only solution is $x_3 = x_4 = x_6 = x_7 = 1$ and $x_1 = x_2 = x_4 = 0$.
(d) None of the above.

**B7.46.** Consider the superincreasing sequence 1, 4, 7, 13, 28, 54. How many solutions has the corresponding knapsack problem for $m = 76$?

(a) None.
(b) The only solution is $x_1 = x_3 = x_4 = x_6 = 1$ and $x_2 = x_5 = 0$.
(c) The only solution is $x_1 = x_2 = x_3 = x_6 = 1$ and $x_4 = x_5 = 0$.
(d) None of the above.

**B7.47.** Consider the knapsack problem cipher. We have chosen the superincreasing sequence $1, 3, 6, 12$, $m = 29$ and $w = 10$. Which is the public key we have to publish to have people send us enciphered messages?

(a) It is the sequence $10, 1, 2, 4$.
(b) It is the sequence $10, 2, 4, 8$.
(c) It is the sequence $10, 2, 3, 6$.
(d) None of the above.

**B7.48.** Let $1, 2, 5, 20$ be our superincreasing sequence, $m = 43$ and $w = 25$. Which is the public key we have to publish to have people send us enciphered messages?

(a) It is the sequence $25, 7, 39, 29$.
(b) It is the sequence $25, 7, 37, 27$.
(c) It is the sequence $25, 7, 39, 27$.
(d) None of the above.

**B7.49.** Consider the example of knapsack problem cipher illustrated in Table 7.10 on page 350. If the plaintext to be sent is otto, which is the numerical equivalent of the ciphertext?

(a) 73 17 58 41 58.
(b) 73 17 50 41 58.
(c) 73 34 50 41 58.
(d) 73 34 58 41 58.

**B7.50.** Consider again the example in Table 7.10 on page 350. If the plaintext to be sent is casa, which is the numerical equivalent of the ciphertext?

(a) 32 0 7 34 0.
(b) 32 0 17 34 0.
(c) 32 0 7 0 17.
(d) 32 0 7 17 0.

**B7.51.** To access the $RSA$ system, Blanche wants to publish the enciphering key $(7927, 37)$, but the system does not accept this key. Why?

(a) Because 7927 is too large.
(b) Because 37 is too small.
(c) Because 7927 is not the product of two prime numbers.
(d) Because 37 is not relatively prime with $\varphi(7927)$.

**B7.52.** To access the $RSA$ system, Ariadne wants to publish the enciphering key $(9991, 119)$, but the system does not accept this key. Why?

(a) Because 9991 is too large.
(b) Because 119 is too small.
(c) Because 9991 is not the product of two prime numbers.
(d) Because 119 is not relatively prime with $\varphi(9991)$.

**B7.53.** Consider the user directory (7.8) on page 352. A user of the $RSA$ system wants to send the message **baba** to the user $B$, Beatrix. Which enciphered message does Beatrix receive?

(a) $C_1 = 9$, $C_2 = 9$.
(b) $C_1 = 999$, $C_2 = 9$.
(c) $C_1 = 999$, $C_2 = 999$.
(d) None of the above.

**B7.54.** Another user of the $RSA$ system wants to send Beatrice a message. If the plaintext is **coda**, which enciphered message does Beatrix receive?

(a) $C_1 = 31$, $C_2 = 243$.
(b) $C_1 = 546$, $C_2 = 243$.
(c) $C_1 = 546$, $C_2 = 576$.
(d) $C_1 = 31$, $C_2 = 576$.

**B7.55.** Consider again the user directory (7.8) on page 352. Beatrix has received the following message:

$$C_1 = 31, \quad C_2 = 722, \quad C_3 = 272.$$

Which is the numerical equivalent of the plaintext?

(a) 0214 0308 0204.
(b) 0214 0300 0208.
(c) 0214 0308 0208.
(d) None of the above.

**B7.56.** Beatrix has received the following message:

$$C_1 = 243, \quad C_2 = 722.$$

Which is the numerical equivalent of the plaintext?

(a) 0308 0200.
(b) 0300 0308.
(c) 0200 0308.
(d) None of the above.

**B7.57.** Beatrix has received the following message:

$$C_1 = 546, \quad C_2 = 722, \quad C_3 = 999.$$

Which is the numerical equivalent of the plaintext?

(a) 0314 0308 0208.
(b) 0308 0208 0314.
(c) 0308 0314 0208.
(d) 0314 0208 0308.

**B7.58.** Verify that the real curve of equation $x^3 + xy^2 - x - 3x^2 - 3y^2 + 3 = 0$ is not singular, but it is so as a curve over $\mathbb{C}$.

**B7.59.** Give an example of a curve that is not singular over $\mathbb{Q}$ but is singular over $\mathbb{R}$.

**B7.60.** Consider the elliptic curve over $\mathbb{R}$ of equation $y^2 = x^3 - x$. Let $p = (-1, 0)$ and $q = (2, -\sqrt{6})$. Which are the coordinates of $p + q$?

(a) $p + q$ is the point at infinity $O$.
(b) $p + q = (-1/3, 2\sqrt{6}/9)$.
(c) $p + q = (-1/3, -2\sqrt{6}/9)$.
(d) None of the above.

**B7.61.** Consider the elliptic curve over $\mathbb{R}$ of equation $y^2 = x^3 - x$. Let $p = (1, 0)$ and $q = (2, \sqrt{6})$. Which are the coordinates of $p + q$?

(a) $p + q$ is the point at infinity $O$.
(b) $p + q = (3, 2\sqrt{3})$.
(c) $p + q = (3, -2\sqrt{3})$.
(d) None of the above.

**B7.62.** Consider the elliptic curve over $\mathbb{R}$ of equation $y^2 = x^3 - x$. Let $p = (-1, 0)$. Which are the coordinates of $2p = p + p$ (in the group law on the curve)?

(a) $2p$ is the point at infinity $O$.
(b) $2p = (-1, 0)$.
(c) $2p = (0, 0)$.
(d) None of the above.

**B7.63.** Consider the elliptic curve over $\mathbb{R}$ of equation $y^2 = x^3 - x$. Let $p = (2, -\sqrt{6})$ e $q = (0, 0)$. Which are the coordinates of $p + q$?

(a) $p + q$ is the point at infinity $O$.
(b) $p + q = (-1/2, \sqrt{6}/4)$.
(c) $p + q = (-1/2, -\sqrt{6}/4)$.
(d) None of the above.

**B7.64.** Let $C$ be the elliptic curve of equation $y^2 = x^3 - x$ over the field $\mathbb{F}_7$. Are $p = (1, 0)$ and $q = (-2, -1)$ points of $C$?

(a) Both $p$ and $q$ are points of $C$.
(b) The point $p$ is on $C$, but $q$ is not.

(c) The point $q$ is on $C$, but $p$ is not.
(d) Neither $p$ nor $q$ belong to $C$.

**B7.65.** Let $C$ be the elliptic curve of equation $y^2 = x^3 - x$ over the field $\mathbb{F}_5$. Are $p = (2,1)$ and $q = (-2,2)$ points of $C$?

(a) Both $p$ and $q$ are points of $C$.
(b) The point $p$ is on $C$, but $q$ is not.
(c) The point $q$ is on $C$, but $p$ is not.
(d) Neither $p$ nor $q$ belong to $C$.

**B7.66.** Prove that the elliptic curve of equation $y^2 + y = x^3 + 1$ has three points over $\mathbb{Z}_2$, including the point at infinity.

**B7.67.** How many points has the curve $y^2 = x^3 - x$ over $\mathbb{F}_7$?

(a) 3.
(b) 6.
(c) 8.
(d) 10.

**B7.68.** How many points has the curve $y^2 + y = x^3$ over $\mathbb{F}_8$?

(a) 4.
(b) 5.
(c) 7.
(d) 9.

# C7 Programming exercises

**C7.1.** Write a program that implements any Caesar cipher, that is, given in input a number $n$, with $1 \leq n \leq 25$, and a text, it outputs the text enciphered by shifting each letter by $n$ positions. Then write a program that deciphers a message so enciphered.

**C7.2.** Write a program that, given a text as its input, outputs a frequency table of the letters appearing in the text.

**C7.3.** Write a program that, given as input a key word and a text, outputs the text enciphered using Vigenère method (see pp. 323-325) with the given key word.

**C7.4.** Write a program that, given in input a text enciphered with Vigenère method and the key word, outputs the plaintext.

**C7.5.** Write a program that computes the inverse of a square matrix modulo a positive integer, if it exists.

**C7.6.** Write a program that, given in input a text and two integers $a, b$, outputs the text enciphered with the affine transformation $C_{a,b} : \mathbb{Z}_{26} \to \mathbb{Z}_{26}$, $C_{a,b}(P) = aP + b$ $\pmod{26}$.

**C7.7.** Write a program that, given in input a text, an $s \times s$ square matrix $A$ and a vector $b$ of length $s$, outputs the text enciphered with the affine transformation $C_{A,b} : \mathbb{Z}_{26}^s \to \mathbb{Z}_{26}^s$, $C_{A,b}(p) = Ap + b \pmod{26}$, where $p$ is a vector of length $s$.

**C7.8.** Write a program that computes discrete logarithms using the Baby step–giant step algorithm.

**C7.9.** Write a program that verifies whether an $n$-integer sequence is superincreasing or not.

**C7.10.** Write a program that generates a superincreasing $n$-integer sequence.

**C7.11.** Write a program that, given in input a superincreasing sequence $a_1, \ldots, a_n$ and an integer $m$, outputs the solution to the corresponding knapsack problem, if it exists. (Hint: use the algorithm described in the text.)

**C7.12.** Write a program that, given in input an integer $N$, outputs: (1) a superincreasing sequence $a_1, \ldots, a_N$, an integer $m > 2a_N$ and an integer $w$ relatively prime with $n$ (the *private* data of a user $X$); (2) the sequence $b_j = wa_j \bmod m$ (the *public* key of user $X$).

**C7.13.** Write a program that, given in input a sequence $(b_j)$, constituting the public key in a Merkle–Hellman system, and a plaintext, outputs the text enciphered with the $b_j$s to be sent to the user $X$.

**C7.14.** Write a program that, given in input a ciphertext and the deciphering private key, outputs the plaintext. (Hint: use the program solving the knapsack problem for a superincreasing sequence.)

**C7.15.** Write a program that randomly generates a prime number with a given number of digits. (Hint: use the algorithm described in Remark 7.7.1.)

**C7.16.** Write a program that, given in input a positive integer $N$, outputs a pair of integers $(n, e)$ such that $n$ is the product of two prime numbers $p$, $q$ each having $N$ digits, and $e$ is relatively prime with both $p - 1$ and $q - 1$ (so we may use the pair $(n, e)$ as public key to use an $RSA$ system).

**C7.17.** Write a program that, given in input a plaintext and the public key $(n, e)$ of a user $A$, outputs the ciphertext to be sent to $A$ using the $RSA$ system.

**C7.18.** Write a program that, given in input the ciphertext and the private information $n = pq$ in an $RSA$ system, outputs the deciphered text.

**C7.19.** Write a program that finds points on an elliptic curve on a finite field of characteristic different from 2.

**C7.20.** Write a program that, given in input a prime number $p$ (sufficiently small) and the equation of an elliptic curve over $\mathbb{Z}_p$, computes how many points of the plane lie on the curve. (Hint: proceed by trial and error for all the values in $\mathbb{Z}_p$ of $x$ in the equation in Weierstrass form.)

**C7.21.** Write a program that, given in input the coordinates of two points $p$ and $p'$ of an elliptic curve over a finite field $\mathbb{F}_q$, outputs the coordinates of the point $p + p'$. (Hint: use the equations given in the text.)

**C7.22.** Write a program that, given in input the coordinates of a point $p$ of an elliptic curve defined over a finite field $\mathbb{F}_q$, determines the order of $p$.

**C7.23.** Write a program that factors a number using Pollard's $p - 1$ algorithm.

**C7.24.** Write a program that computes $[\sqrt{n}]$ for an integer $n$.

# Transmitting without... fear of errors

This chapter gives a short introduction to codes and *coding theory*. The reader has nothing to worry about: Justinian the Great, the Eastern Roman Emperor who in 6th century rewrote Roman law has got nothing to do with it! Coding theory studies the mathematical problem of transmitting data through channels in which interference is present. During the transmission, it may happen that the *noise* present in the channel modifies some of the transmitted data, jeopardising the intelligibility of the message on arrival.

So it is necessary to find techniques allowing the detection of the errors and, possibly, their correction, and of course fast *coding* and *decoding* of the data to be sent. It is clear that the cost to pay for this requirement consists in having to transmit additional data: on the one hand, they allow us to verify if an error has occurred, and possibly to correct it, on the other hand, they make the transmission operations more complex. So, for the system to be efficient it is necessary to find the best balance between redundancy and correctness of the information. Coding theory is a quite recent branch of mathematics, living in an important border area close to computer science, an aspect we shall not dwell upon here. It also has significant applications in other sciences, mainly biology: DNA, for instance, is a code. In coding theory we see a convergence of several subtle combinatorial, algebraic, and geometric techniques, partly discussed in previous chapters. We shall try here, without aiming at completeness, and leaving apart several important aspects of the theory, to give a taste of what is happening in this area in which, we are certain, remarkable developments will be seen in the next few years. So this chapter, more than a systematic treatment, is an invitation to further reading; in particular, we recommend the classic references [36], [41], [39], where a rich bibliography will be found. Several more recent texts can also be suggested, such as [6]. A last warning: in this chapter we shall only discuss the so-called *block codes*, which can be more directly studied using the algebraic and geometric techniques we have mentioned. We shall omit a treatment of the *variable-length codes*, which are very important too, both from a theoretical, mathematical viewpoint and for the applications. The interested reader may look them up in the references.

## 8.1 Birthday greetings

If on your birthday you receive a greeting card saying

WARM BIRTKDAY WISHES!

you will be glad, even if you cannot avoid noticing that the sender made a mistake... But let us reflect a moment and ask ourselves: (a) how can we be sure that in the message there is an error? (b) how can we be certain about the number and the position of the errors? (c) why are we sure of being able to correct the error or errors? (d) in other, analogous situations, how could we recognise and correct the errors?

Let us tackle the questions one at a time, starting from (a). English *words*, as in any other language, consist of letters of an *alphabet*. However, not all combinations of letters from the alphabet correspond to words. It is this *redundancy* of the *words in a language* with respect to *all possible words* that makes it possible to recognise errors. For instance, WARM is a word belonging to the English language, while BIRTKDAY is not. So we are sure that in the message there is an error.

As to questions (b) and (c), the above reasoning already suggests part of the answer: certainly, there is an error in the word BIRTKDAY. There could be other ones: we might not notice them because we recognise a mistaken word as another English word. For instance, WARM might have been written in the place of WORM. However we look at the signature and ascertain that the author is a friend who can spell quite well and is not particularly absent-minded. He might well have made a mistake, say in a hurry or through carelessness, but we doubt very much he made two in the same, very short message! As we are certain that he made one in the word BIRTKDAY, we believe that there are no other mistakes. Reasoning in the same way, we also believe that a single letter is wrong in this word, so we read BIRTHDAY rather that BIRTKDAY, and the error is corrected.

As to question (d), it is the more complex one. What if we had received a card saying

HARM BIRTHDAY FISHES

in which all words *are* English words? How could we be certain that there are errors? And how could we correct them? We shall see that in order to be able to come to a conclusion about this last question other elements may be taken in account: for instance, an encouragement to *harm* something would be out of place in a greeting card, our friend is not especially interested in *fishes*, and so on.

In conclusion, we desire to indicate at least three points displayed by our reasoning:

(i) English language, as every language, by the way it is written down, has an inborn high measure of *redundancy*, which is what makes it possible to recognise and correct errors;

(ii) the lower we may assume the probability of each error, the more errors we are able to correct;

(iii) it may happen that, even if we are aware of an error being present in a sentence, we are not be able to correct it with a reasonable degree of certainty.

What has this rambling speech about greeting cards to do with codes, which are the subject of this chapter? There is a very strict link! Coding theory is exactly about this problem. A message is sent through a transmission channel, which might be an optical cable, a satellite link, or any other communication medium. During the transmission the *initially correct* message sustains modifications, due to imperfections in the optical fibre, interference of other signals and so on. We want to recognise if it contains errors and, if so, to be able to correct them. It turns out that it is a problem quite similar to the one of the greeting, so the remarks we made about the latter may help us in giving a correct mathematical formulation to our treatment of the former.

## 8.2 Taking photos in space or tossing coins, we end up at codes

A typical example of a problem solved by coding theory is a quite important one. It was faced by the technicians at the Jet Propulsion Laboratory of the California Institute of Technology at Pasadena, when, at the end of the 1960's, they were working on the Mariner and Voyager space probes designed to approach Mars or Jupiter, photograph these planets and send black and white pictures back to the Earth. How does transmitting one of these pictures work?

Consider a 30 cm × 30 cm photograph. It is divided by a grid consisting of 0.1 mm$^2$ squares, and for each square the corresponding intensity of grey, on a scale, say from 0 to 63, where 0 represents white and 63 black, is sent to the Earth. If numbers are represented in binary form, the probe computer sends a 0 - 1 string of length six for each square. By collecting all these numbers and assigning to each of them its colour, the image can be reconstructed.

Quite easy, isn't it? But what if there are errors, perhaps due to some kind of interference, a natural thing to be expected for such unusual transmissions? For instance, it may happen that there is an error probability $p$, say of the order of $p = 0.001$; that is, we may expect on the average one error for every 1000 transmitted digits. The reader will immediately realise, with simple calculations left to him, that the transmission may be seriously disrupted, and the photograph irreparably altered.

To remedy this problem, it is necessary to introduce in the procedure we are implementing in the probe computer the *redundancy* we have been talking about in the previous section: it allows us to recognise and correct errors.

To simplify the model we are dealing with, without reducing the scope of the problem, consider the following situation. A man in Australia tosses a coin and transmits us, via email, 0 if it comes up as HEADS and 1 if it shows TAILS. But during the transmission there is an error probability $p$ which, as usual, is a number between 0 and 1. For instance, if $p = 0.01$, it happens just once in one hundred transmissions that 1 is transmitted instead of 0 or the other way around.

If we receive 0, we cannot be *certain* that the result was HEADS. However, in this situation, as in the case of Mars photographs, there is no *redundancy*. Let us introduce it, as follows.

We ask the man in Australia to send 00 if the result is HEADS, and 11 if it is TAILS. Assume HEADS shows up. The probability for us to receive 11, that is to say, that two errors have occurred during the transmission, is $p^2 = 0.0001$, very very small! The probability of receiving 01 or 10, that is, that a single error occurs, is $2 \cdot 0.99 \cdot 0.01 = 0.0198$, while the probability of receiving the correct information 00 is $0.99 \cdot 0.99 = 0.9801$. On the other hand, if we receive 01 or 10 we are *certain* that a transmission error has occurred, and we may ask the man in Australia to send again the result. So we are led into error only in one case out of 10000.

If we have at our disposal plenty of time and money to use the communication channel with Australia, we may improve the situation even more. We fix an odd positive integer $n$, and have the man send us a string of $n$ 0's if the result is HEADS, and one of $n$ 1s if the result is TAILS. On arrival, we interpret as HEAD a string containing more 0's than 1's and as TAILS a string containing more 1's than 0's. A simple computation we do not make here shows that the probability of an error becomes

$$P_n = \sum_{0 \leq i < \lfloor n/2 \rfloor} \binom{n}{i} p^{n-i}(1-p)^i$$

and it tends to zero when $n$ approaches infinity (see Exercise A8.1).

The scheme we are using is a *code*, called *repetition code*. Let us expound this notion.

We fix first our attention upon the *set of words* we use in our transmissions. To *write* these words we need an *alphabet*, that is, a set of symbols to be used in our messages. For instance, in the situation discussed above, the alphabet is $\mathbb{Z}_2$, a not unusual choice because transmitting data, as we know, often occurs by transmitting information bits. More in general, we may consider as our alphabet a finite field $\mathbb{F}_q$. We may further assume that the *words* we use consist of a bounded number of alphabet letters, say at most $k$: consider the fact that the longest (non-technical, non-made up) word in English is

ANTIDISESTABLISHMENTARIANISM

which consists of 28 letters. On the other hand, it is not restrictive to assume that all the words we are using have the same length $k$. Indeed, we may use an alphabetic symbol as a *dummy symbol* to be added at the end of the words of length smaller than $k$ to write them with exactly $k$ letters.

So, once we have fixed the *alphabet* $\mathbb{F}_q$, we may assume that the set of words we shall use, that is our *dictionary* is a subset $C$ of $\mathbb{F}_q^k$.

For instance, in the case of the Australian coin tosser, the word set is $C = \mathbb{F}_2$, as the only two words we want to communicate are 0 for HEADS and 1 for TAILS.

But so far there is no space for the *redundancy* mentioned more than once, which is essential to discover whether the message contains errors and, possibly, to correct them. To introduce redundancy we need to represent each word in our dictionary using *more symbols* than strictly necessary to represent it. In order to do so, it suffices to fix another positive integer $n \geq k$ and give, for instance, an injective mapping

$$f : \mathbb{F}_q^k \to \mathbb{F}_q^n,$$

by which we may identify our dictionary $C \subseteq \mathbb{F}_q^k$ with a subset of $C$ of $\mathbb{F}_q^n$.

For instance, for the coin tosser, the mapping $f$ is the one mapping $x \in \mathbb{F}_2$ to $(x, x, \ldots, x) \in \mathbb{F}_2^n$; hence, $C = \mathbb{F}_2$ is identified with the subset $\{(0, 0, \ldots, 0), (1, 1, \ldots, 1)\}$ of $\mathbb{F}_2^n$, which is called a *repetition code*.

In conclusion, we may give the following definition.

**Definition 8.2.1.** *A code* over the alphabet $\mathbb{F}_q$ *is a subset* $C$ *of* $\mathbb{F}_q^n$, *where* $n$ *is a positive integer, called* length *of the code. The number* $M$ *of elements of* $C$ *is called* size *of the code. Codes having as alphabet* $\mathbb{F}_2$ *are called* binary *codes.*

Such a code, in which both the length of the words and the size of the code are fixed, are called *block codes*. It is also possible to consider a more general notion of code, in which length and size may vary. This codes are called *variable-length codes* and we shall not consider them here.

Notice that the code used by the Australian coin tosser, based on a double repetition of each digit, allows the detection of *a single error*. The code in which digits are repeated three times allows the detection of *at most two errors* and so on. Unfortunately, as we have seen, finding an error does not imply *correcting* it. We shall work a bit to obtain codes that also enable us to correct errors.

Moreover, notice that the length $n$ of code $C$ may be seen as a measure of the *cost* of the code. The longer a code, the more time and energy are needed to send messages and so the higher the cost of using it. The size $M$ of the code $C$ may be seen instead as a measure of how *rich* the code is. The more words a code consists of, the larger and more various amount of information it can transmit. A very *rich* code cannot be too *short*. This ratio quality/price may be represented by the number

$$R = \frac{\log_q M}{n}, \tag{8.1}$$

which equals $k/n$ if the original dictionary is $\mathbb{F}_q^k$. The number $R$ is called *information rate* of the code. In principle, one would like it to be as close as possible to 1 for the code not to be too expensive, but it cannot be too close to 1 if the code has to be sufficiently *efficient* as to its capability of detecting and correcting errors.

## 8.3 Error-correcting codes

Let us see now how to improve the efficiency of the codes, enabling them not only to detect errors but also to correct them. At the same time, we want also to minimise the cost of this operation, better than we have done, in a somewhat naive way, in the previous section in the case of the Australian coin tosser.

The inventor of the procedure we are about to describe is R. Hamming (1948), then at the Bell Telephone Laboratories. Consider again a particular example, the one Hamming himself actually examined. It is very similar to that of Mars photographs, mentioned in the previous section. Suppose we want to send messages consisting of numbers from 0 to 15 written in base 2. So they are 4-tuples of 0 and 1 digits. In other words, our code has as its alphabet $\mathbb{F}_2$, and $C = \mathbb{F}_2^4$, so $M = 2^4 = 16$.

When transmitting the 4-tuple $(a_1, a_2, a_3, a_4)$, we do not transmit just it alone, but we transmit instead the 7-tuple

$$(a_1, a_2, a_3, a_4, a_1 + a_3 + a_4, a_1 + a_2 + a_4, a_1 + a_2 + a_3). \tag{8.2}$$

The sums, of course, are computed in $\mathbb{F}_2$. For instance, rather than the 4-tuple $(0, 1, 0, 0)$ we transmit the 7-tuple $(0, 1, 0, 0, 0, 1, 1)$.

Basically, we are constructing a code of length $n = 7$, embedding $C = \mathbb{F}_2^4$ in $\mathbb{F}_2^7$ according to the injective mapping that associates the element (8.2) of $\mathbb{F}_2^7$ to $(a_1, a_2, a_3, a_4) \in \mathbb{F}_2^4$. This code, called *Hamming* $(7, 4)$ *code* has information rate $R = 4/7$: not bad, if compared to the ratio $1/n$ of the coin tosser.

Let us see now what has been gained by *increasing the length* of the code. First of all, notice that, if $(x_1, \ldots, x_7)$ is a word of our code, we have

$$\begin{cases} x_1 + x_3 + x_4 + x_5 = 0, \\ x_1 + x_2 + x_4 + x_6 = 0, \\ x_1 + x_2 + x_3 + x_7 = 0. \end{cases} \tag{8.3}$$

Indeed, for instance, the left-hand side of the first equation is $2x_1 + 2x_3 + 2x_4$, which is equal to 0 in $\mathbb{F}_2$.

We may interpret (8.3) as a linear system of equations over the field $\mathbb{F}_2$. This system is said to be the *parity check* system of the code we are considering. It is easy to understand that every solution of this system is an element of

$\mathbb{F}_2^7$ that is a word of our code (see Exercise A8.2). So if we receive the message $(x_1, \ldots, x_7)$ and it does not verify the system (8.3), we are certain that there is a transmission error. So this is an error-detecting code.

But more can be said. Suppose we know, or regard as *likely enough*, that there is no more than one transmission error. From the form itself of the system (8.3), it is clear that, if there is exactly one error, all possible cases are the following ones:

- if $(x_1, \ldots, x_7)$ does *not verify any* equation of the linear system (8.3), the error is in the first position;
- if $(x_1, \ldots, x_7)$ verifies the first equation of the linear system (8.3), but not the other two, the error is in the second position;
- if $(x_1, \ldots, x_7)$ verifies the second equation of the linear system (8.3), but not the other two, the error is in the third position;
- if $(x_1, \ldots, x_7)$ verifies the third equation of the linear system (8.3), but not the other two, the error is in the fourth position;
- if $(x_1, \ldots, x_7)$ verifies the second and third equation of the linear system (8.3), but not the first one, the error is in the fifth position;
- if $(x_1, \ldots, x_7)$ verifies the first and third equation of the linear system (8.3), but not the second one, the error is in the sixth position;
- if $(x_1, \ldots, x_7)$ verifies the first and second equation of the linear system (8.3), but not the third one, the error is in the seventh position.

In conclusion, if *we know* that there is *a single* error, the system (8.3) also allows us to find *where* the error is, and so to correct it. Indeed, as the code is binary, if we know where an error is, to correct it it suffices to change the symbol lying in the specified place.

**Example 8.3.1.** Suppose we receive the word $p = (0, 1, 0, 1, 0, 1, 0)$, which contains errors, as it does not satisfy linear system (8.3), and assume that there is exactly one error.

As $p$ does not satisfy any of the three equations of linear system (8.3), by the remarks above we know that the error is in the first coordinate, so the correct word is $(1, 1, 0, 1, 0, 1, 0)$.

Why does this code work so well with respect to correcting one error? The crucial remark is that *two distinct words of the code differ in at least three coordinates*. We ask the reader to check this claim as an exercise (see Exercise A8.3).

Let $(x_1, \ldots, x_7)$ be a word of the code which, during the transmission, undergoes a change *in a single coordinate*, for instance the first one, so we receive $(x_1', x_2, \ldots, x_7)$. As every other codeword differs form $(x_1, \ldots, x_7)$ in *at least three* coordinates, it differs from $(x_1', x_2, \ldots, x_7)$ in *at least two* coordinates. Thus, we may correct the error by applying the *maximum likelihood* principle, that is, by assuming that the corrected form of $(x_1', x_2, \ldots, x_7)$ is $(x_1, \ldots, x_7)$ because the latter is the *nearest* codeword, that is to say, the most similar one, being the *only one* differing from $(x_1', x_2, \ldots, x_7)$ in a single coordinate.

The reader will immediately understand that this way of proceeding is completely analogous to the correction of the mistake in the greeting card of the first section.

These remarks suggest an immediate extension of these notions, which will now be formalised.

**Definition 8.3.2.** *Let $x$, $x'$ be two elements of $\mathbb{F}_q^n$. The* Hamming distance *between $x$ and $x'$ is the number of coordinates in which $x$ differs from $x'$ and is denoted by the symbol $d(x, x')$.*

Assume we have a code $C$ of length $n$. The minimum Hamming distance between two words of $C$ is denoted by the symbol $d$ and is called *minimum distance* of $C$.

If a code $C$ which uses a $q$-symbol alphabet has size $M$, length $n$ and minimum distance $d$, it is said to be a *code of type* $(n, M, d)_q$ or a $(n, M, d)_q$-*code*. The numbers $q, n, M, d$ are called the *parameters* of the code.

For instance, the Hamming $(7, 4)$ code is a code of type $(7, 16, 3)_2$.

It is not by chance that we use the term *distance* to designate $d(x, x')$. It satisfies the same formal properties of the distance between points in a plane or in the space, namely:

- *positivity*: for all $x$ and $x'$, $d(x, x') \geq 0$; moreover, $d(x, x') = 0$ if and only if $x = x'$;
- *symmetry*: for all $x$ and $x'$, $d(x, x') = d(x', x)$;
- *triangle inequality*: for all $x$, $x'$ e $x''$,

$$d(x, x') \leq d(x, x'') + d(x'', x').$$

We urge the reader to verify these properties (see Exercise A8.4).

Let us see the connection between the Hamming distance and the problem of error correction. First of all, yet another definition.

**Definition 8.3.3.** *We say that a code $C$* detects *$h$ errors if, whenever in the transmission of a word at most $h$ errors occur, it is possible to find out this fact, as the string obtained does not correspond to any codeword. We say next that a code $C$* corrects *$h$ errors if, whenever in the transmission of a word at most $h$ errors occur, it is possible to find it out, and moreover it is possible to correct the errors by applying the maximum likelihood principle, as there is a single codeword having least Hamming distance from the received string.*

For instance, the Hamming $(7, 4)$ code detects and corrects one error

The connection between error correction and Hamming distance is given by the following theorem, whose proof is left to the reader as an exercise (see Exercise A8.5):

**Theorem 8.3.4.** *Assume a code $C$ has minimum distance $d$. Then:*

*(i) $C$ detects $k$ errors if and only if $d \geq k + 1$;*

*(ii) C corrects k errors if and only if $d \geq 2k + 1$.*

**Remark 8.3.5.** Suppose we are working with a code $C$ having length $n$, which uses a transmission channel having error probability $p < 1/2$. We receive a word $x$ having Hamming distance $d$ from the word $y$. Then the probability of having received $x$ instead of $y$ equals

$$P = p^d(1-p)^{n-d} = \left(\frac{p}{1-p}\right)^d (1-p)^n. \tag{8.4}$$

(see Exercise A8.6). As $p < 1/2$, then $p/(1-p) < 1$ and $(1-p)^n$ does not depend on $d$. Thus, $P$ decreases when $d$ increases. So error correction by the maximum likelihood principle, which coincides with the minimum Hamming distance principle, puts into practice the need for minimising the error probability when decoding.

We conclude this section mentioning a simplified version of a fundamental result, due to C. E. Shannon (1948), which may be said to have originated coding theory. We need some more notation.

Let $C = \{x_1, \ldots, x_M\}$ be a code having length $n$, which does not necessarily detect or correct errors. We shall use the maximum likelihood principle to correct errors. Let $P_i$ be the probability of being wrong by decoding like this when the word $x_i$ had been transmitted erroneously. Then the probability of being wrong somewhere when decoding a word of $C$ is

$$P_C = \frac{1}{M} \sum_{i=1}^{M} P_i.$$

We have the following result.

**Theorem 8.3.6 (Shannon's theorem).** *Suppose we are using codes with alphabet $\mathbb{F}_2$ over a transmission channel admitting an error probability $p$. For every real number $\varepsilon > 0$ and for every positive real number $R < 1 + p \log p + (1-p) \log(1-p)$ there is a code $C$ with information rate at least equal to $R$ and for which the error probability $P_C$ is smaller than $\varepsilon$.*

A more general version of this theorem may be given for codes over an arbitrary alphabet. For its proof, see for instance [36], pp. 27–29.

## 8.4 Bounds on the invariants

The Hamming $(7, 4)$ code is a *good* code, in the sense that it allows not only to detect if in transmitting data an error has occurred, but also to correct it, if it occurred. This code, as we have seen, is of type $(7, 16, 3)_2$, that is to say, it has parameters $n = 7$, $M = 16$ and $d = 3$.

There is a natural question: is it possible to construct a *better code*? To make this question meaningful, we must ask first when a code may be considered *better* than another.

It is clear that we would consider a code better than the Hamming $(7,4)$ code if it were of type $(n, M, d)_2$ with, for instance, $n < 7$, $M \geq 16$ and $d \geq 3$, or $n \leq 7$, $M > 16$ and $d \geq 3$, or $n \leq 7$, $M \geq 16$ and $d > 3$. In the first case, the rest of the performance being equal, the code would be shorter because $n < 7$, and so less expensive; in the second case, the code would have more words because $M > 16$, in the third case it might detect and even correct more errors, because $d > 3$. The fact is that the parameters of a code are not at all independent of each other, as we shall see shortly. In fact, there are several relations between $n$, $M$, and $d$; we shall describe the main ones. As a consequence of these relations, it is possible to prove that there is no better code, in the sense just discussed, than the Hamming $(7,4)$ code.

The relations between the parameters of a code are of two kinds: upper bounds for $M$ and lower bounds for $M$, in terms of $n$ and $d$. For instance, it is clear that $M \leq q^n$, because $C$, which is a set with $M$ elements, may be regarded as a subset of $\mathbb{F}_q^n$, which has $q^n$ elements. The bound $M \leq q^n$ is clearly very rough. A better one is given by the following theorem.

**Theorem 8.4.1 (Singleton bound).** *Let $n$, $M$, $d$ and $q$ be the parameters of a code $C$. Then:*

$$\boxed{M \leq q^{n-d+1}.}$$

PROOF. Write all words of the code $C$ in a table as follows:

$$
\begin{array}{ccccccc}
a_{1,1} & a_{1,2} & \cdots & a_{1,d-1} & a_{1,d} & \cdots & a_{1,n} \\
a_{2,1} & a_{2,2} & \cdots & a_{2,d-1} & a_{2,d} & \cdots & a_{2,n} \\
a_{3,1} & a_{3,2} & \cdots & a_{3,d-1} & a_{3,d} & \cdots & a_{3,n} \\
\vdots & \vdots & \ddots & \vdots & \vdots & \ddots & \vdots \\
a_{M,1} & a_{M,2} & \cdots & a_{M,d-1} & a_{M,d} & \cdots & a_{M,n}
\end{array}
\tag{8.5}
$$

where every line $(a_{1,1}, \ldots, a_{1,n})$ is a word of the code $C$. Now consider the strings consisting of the last $n - d + 1$ coordinates of every word of $C$, that is to say, the right part of Table (8.5). They make up a new code $C'$ having length $n - d + 1$. Two words of $C'$ are never equal, or else the corresponding words of $C$ would differ in at most $d - 1$ coordinates, which is impossible by the hypothesis that $C$ has minimum distance $d$. So the size of $C'$ is again $M$. On the other hand, since $C$ has length $n - d + 1$, we have $M \leq q^{n-d+1}$.    □

Notice that equality in Singleton bound holds if and only if, however $n - d + 1$ coordinates are chosen, the codewords have on them all possible values (see Exercise A8.7).

A more refined upper bound for $M$ is obtained by a *geometric* reasoning due to Hamming, which shall now be described.

Assume we have a code $C$ of type $(n, M, d)_q$. The code is a subset of $\mathbb{F}_q^n$. Consider the latter as an $n$-dimensional vector space. Now, for every point $x \in C$ consider the set of points $S_q(x, r)$ of $\mathbb{F}_q^n$ having Hamming distance from $x$ smaller than or equal to $r$, that is,

$$S_q(x, r) = \{y \in \mathbb{F}_q^n : d(x, y) \leq r\}.$$

Carrying on the geometric analogy, it is natural to call $S_q(x, r)$ a *sphere* of centre $x$ and *radius* $r$. We leave the reader the easy task (see Exercise A8.8) of proving that the *volume* of $S_q(x, r)$, that is to say, the number of elements in $S_q(x, r)$, equals

$$V_q(n, r) = \sum_{i=0}^{r} \binom{n}{i} (q - 1)^i. \tag{8.6}$$

Using this formula the following estimate is found:

**Theorem 8.4.2 (Hamming bound).** *Let $C$ be a code of type $(n, M, d)_q$ that corrects $k$ errors. Then*

$$\boxed{M \leq \frac{q^n}{\displaystyle\sum_{i=0}^{k} \binom{n}{i} (q - 1)^i}.} \tag{8.7}$$

*In particular, Equation (8.7) holds for $k = \lfloor (d - 1)/2 \rfloor$.*

PROOF. By Theorem 8.3.4, we have $d \geq 2k + 1$, so for every pair of points $x, y \in C$, the two spheres $S_q(x, k)$ and $S_q(y, k)$ are disjoint, that is, have no points in common. On the other hand, $\mathbb{F}_q^n$ includes the union $S$ of all spheres $S_q(x, k)$ when $x$ ranges in $C$, and as there are $M$ such spheres and they are pairwise disjoint, $S$ contains $M \cdot V_q(n, k)$ elements, and so

$$M \cdot V_q(n, k) \leq q^n. \qquad \square$$

The Hamming bound is, most of the times, better than Singleton's. For instance, the following result holds (see Exercises A8.10–A8.13; see also Exercise A8.9):

**Proposition 8.4.3.** *The codes of type $(n, M, d)_2$ for which the Singleton bound is better than the Hamming bound are only those with*

*(i) $d = 2$ and arbitrary $n$;*
*(ii) $d = 4$ and $n = 4, 5, 6$;*
*(iii) $d = 6$ and $n = 6, 7$;*
*(iv) $d$ even, $d \geq 8$ and $n = d$.*

The codes $C$ for which in (8.7) the equality holds are called *perfect*. They are characterised by the fact that they correct $k$ errors and are such that the union of the spheres $S_q(x, k)$, with $x \in C$, equals the set of *all possible words*. In other words, perfect codes use very *wisely* the words that may be constructed with a $q$-symbol alphabet. In particular, a perfect code cannot ever correct more than $k$ errors. Finally, it is clear that a perfect code cannot be improved, in the sense discussed at the beginning of this section.

Notice that the Hamming $(7, 4)$ code is a perfect code (see Exercise A8.14).

Another upper bound for $M$ can be obtained by computing the maximum possible value of the distance between two distinct codewords.

**Theorem 8.4.4 (Plotkin bound).** *If $C$ is a code of type $(n, M, d)_q$, and if $qd - n(q - 1) > 0$, we have*

$$\boxed{M \leq \frac{qd}{qd - n(q - 1)}.}$$

PROOF. Let $\Sigma$ be the sum of the distances of all ordered pairs of distinct words of $C$. There are $M(M - 1)$ such ordered pairs (see Exercise A8.15) and, for each of them, the distance is at least $d$. So we have

$$\Sigma \geq M(M - 1)d. \tag{8.8}$$

Compute now $\Sigma$ in another way. Number from 0 to $q - 1$ the symbols in the alphabet of $C$. List the words of $C$ as in Table (8.5), which may be considered as a $M \times n$ matrix with values in $\mathbb{F}_q$. Consider the $i$th column in the list. Suppose that, for all $j = 0, \ldots, q - 1$, the symbol $j$ of the alphabet appears $m_{i,j}$ times in this column. As the column is a vector of length $M$, we have $\sum_{j=0}^{q-1} m_{i,j} = M$. Then the contribution of this column to the sum $\Sigma$ is

$$\sum_{j=0}^{q-1} m_{i,j}(M - m_{i,j}) = M^2 - \sum_{j=0}^{q-1} m_{i,j}^2.$$

By Cauchy–Schwarz inequality (see Exercise A8.16), we have

$$q \sum_{j=0}^{q-1} m_{i,j}^2 = \sum_{j=0}^{q-1} 1 \cdot \sum_{j=0}^{q-1} m_{i,j}^2 \geq \left( \sum_{j=0}^{q-1} 1 \cdot m_j \right)^2 = \left( \sum_{j=0}^{q-1} m_{i,j} \right)^2$$

and so

$$\sum_{j=0}^{q-1} m_{i,j}(M - m_{i,j}) \leq M^2 - \frac{(\sum_{j=0}^{q-1} m_{i,j})^2}{q} = M^2 \frac{q - 1}{q}.$$

In conclusion, summing over all columns, we have

$$\Sigma = \sum_{i=1}^{n}\sum_{j=0}^{q-1} m_{i,j}(M - m_{i,j}) \leq \sum_{i=1}^{n} M^2 \frac{q-1}{q} = M^2 n \frac{q-1}{q}.$$

Keeping in mind (8.8), the claim follows.                                   □

Next we estimate $M$ *from below*. The reasoning that led us to proving the Hamming bound also suggests how to obtain a remarkable *optimal* lower bound for the parameter $M$.

Indeed, define $A_q(n,d)$ as the greatest $M$ for which a code of type $(n, M, d)_q$ exists. We want to give an estimate for $A_q(n,d)$.

**Theorem 8.4.5 (Gilbert–Varshamov bound).** *We have*

$$A_q(n,d) \geq \frac{q^n}{\displaystyle\sum_{i=0}^{d-1} \binom{n}{i}(q-1)^i}.$$

PROOF. Suppose we want to explicitly construct a code $C$ of type $(n, M, d)_q$. In order to do so, we begin by choosing any word $x_1 \in \mathbb{F}_q^n$ and *putting it* in $C$. Next, we choose a second word $x_2 \in \mathbb{F}_q^n$ having distance at least $d$ from $x_1$ and *put it* in $C$. We proceed like this, recursively: after having chosen the first $h$ words $x_1, \ldots, x_h$ of $C$, all having pairwise distance at least $d$, we choose the $(h+1)$th word $x_{h+1}$ having distance at least $d$ from all words $x_1, \ldots, x_h$, if this is possible. Indeed, we shall reach a point where we shall have to stop because *there will be no more space* for choosing one more word. So we shall have constructed a code $C$ enjoying the following property: the union of all spheres $S_q(x, d-1)$ having centre in the points $x \in C$ coincides with the whole $\mathbb{F}_q^n$. Otherwise, clearly, we might choose one more word. So for this code we have

$$M \cdot V_q(n, d-1) \geq q^n;$$

hence immediately the bound for $A_q(n,d)$ follows.                          □

In the applications, we are often more interested in *asymptotic* estimates of codes' efficiency than in the estimate of the parameters of a single code. For instance, we want to answer the following question: is it possible to construct a sequence of codes, each more efficient than the previous one?

A way of measuring the efficiency of a code $C$ of type $(n, M, d)_q$ is, as mentioned, the information rate $R = (\log_q M)/n$.

Another measure could be the number $\delta = d/n$, which could be called *separation ratio* or *normalised distance* of the code. The larger $\delta$, the abler the code to correct errors.

Consider now sequences of codes of increasing length and suppose we want an asymptotic description of the behaviour of their information rate as a function of the separation ratio, to answer questions such as: by increasing the separation ratio is it possible, and to what extent, to increase the information rate?

**Remark 8.4.6.** Recall that, given a sequence $\{x_n\}_{n \in \mathbb{N}}$ of real numbers, we may define two new sequences $\{a_n\}_{n \in \mathbb{N}}$ and $\{b_n\}_{n \in \mathbb{N}}$ as follows:

$$a_n = \inf\{x_m\}_{m \geq n}, \quad b_n = \sup\{x_m\}_{m \geq n}.$$

Call *limit inferior* [resp., *limit superior*] of $\{x_n\}_{n \in \mathbb{N}}$, and denote by $\liminf_{n \to \infty} x_n$ [resp., $\limsup_{n \to \infty} x_n$], the number $\sup\{a_n\}_{n \in \mathbb{N}}$ [resp., $\inf\{b_n\}_{n \in \mathbb{N}}$].

Clearly, $\limsup_{n \to \infty} x_n = +\infty$ if and only if the sequence $\{x_n\}_{n \in \mathbb{N}}$ has no upper bound (see Exercise A8.17). Moreover, $\limsup_{n \to \infty} x_n = -\infty$ if and only if $\lim_{n \to \infty} x_n = -\infty$ (see Exercise A8.18).

If, on the other hand, the sequence has an upper bound and has not as its limit $-\infty$, then the limit superior exists and is finite, and it can be shown (see Exercise A8.19) that in this case the limit superior $\xi$ is characterised by the following property: for all real number $\epsilon > 0$ there is an integer $M \in \mathbb{N}$ such that, for all $n > M$, we have $x_n < \xi + \epsilon$. Analogous properties hold for the limit inferior (see Exercise A8.19). A sequence converges if and only if the limits inferior and superior coincide.

It is useful to define the so-called *entropy* function $H_q(t)$ on the interval $[0, (q-1)/q]$, such that $H_q(0) = 0$ and

$$H_q(t) := t \log_q(q-1) - t \log_q t - (1-t) \log_q(1-t)$$

for $0 < t \leq (q-1)/q$.

We may translate Plotkin and Gilbert–Varshamov bounds into bounds for the limit

$$\alpha(\delta) := \limsup_{n \to \infty} \frac{\log_q A_q(n, d)}{n} = \limsup_{n \to \infty} \frac{\log_q A_q(n, n\delta)}{n},$$

which is an asymptotic measure of the information rate.

We have the following lemma; for its proof see Exercise A8.21:

**Lemma 8.4.7.** *For $0 \leq \delta \leq (q-1)/q$ we have*

$$\lim_{n \to \infty} \frac{\log_q V_q(n, [\delta n])}{n} = H_q(\delta). \tag{8.9}$$

Hence immediately follows (see Exercise A8.22):

**Theorem 8.4.8 (Asmyptotic Gilbert–Varshamov bound).** *We have*

$$\boxed{\alpha(\delta) \geq 1 - H_q(\delta).}$$

In an analogous way (see Exercise A8.23) it is possible to give the asymptotic version of the Plotkin bound:

**Theorem 8.4.9 (Asmyptotic Plotkin bound).** *We have*

$$
\begin{cases}
\alpha(\delta) \leq 1 - \dfrac{q\delta}{q-1} & \text{for } 0 \leq \delta \leq \dfrac{q-1}{q}, \\[2ex]
\alpha(\delta) = 0 & \text{for } \dfrac{q-1}{q} \leq \delta \leq 1.
\end{cases}
$$

These two estimates leave as possible values of $\alpha(\delta)$ the region denoted by $A$ in Figure 8.1.

## 8.5 Linear codes

As we have already seen for cryptography, it is very useful, when considering codes of length $n$, to use subsets of $\mathbb{F}_q^n$ having much *algebraic structure*. This makes it greatly easier to perform several operations such as:

- computing code parameters;
- coding and transmitting messages;
- decoding messages;
- detecting and correcting errors;
- computing the probability of a correct decoding.

We shall be able to discuss only some of these points. We hope this will suffice to show, once more, the power of algebraic methods and also to rouse the readers' curiosity, leading them to further exploration of the subject.

The most natural algebraic structure to be used on $\mathbb{F}_q^n$ is that of a *vector space* over the field $\mathbb{F}_q$. The elements of $\mathbb{F}_q^n$ are called *numerical vectors* of length $n$ over the field $\mathbb{F}_q$, whose elements are called *scalars*. When we want

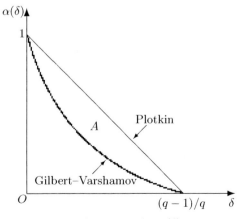

**Fig. 8.1.** Estimates for $\alpha(\delta)$

to emphasise the vector space structure of $\mathbb{F}_q^n$, we shall often denote its elements, that is to say, the numerical vectors, by boldface symbols, such as $\mathbf{x} = (x_1, \ldots, x_n)$.

A $(n, M, d)_q$–code $C$ is said to be *linear* if $C$ is a vector subspace of $\mathbb{F}_q^n$, that is, if and only if the sum of two words is still a word and the product of a word by a scalar is still a word of the code. If $C$ has dimension $k$, we say that $C$ is a $[n, k]_q$-code. Notice that in this case $M = q^k$, as a vector space of dimension $k$ over $\mathbb{F}_q$ is isomorphic to $\mathbb{F}_q^k$ and so has size $q^k$. As to $d$, we immediately see that computing it is far easier for linear codes than in the general case.

First of all, from Singleton bound it follows that:

**Proposition 8.5.1.** *For a $[n, k]_q$-code:*

$$\boxed{d \leq n - k + 1.} \tag{8.10}$$

**Definition 8.5.2.** *A linear code for which in Formula (8.10) the equality holds is said to be a* maximum-distance separable *code.*

These codes are characterised by the fact that, for any choice of $k$ coordinates, the codewords have on them all possible $q^k$ values (see Exercise A8.7).

Now, for every vector $\mathbf{x} \in \mathbb{F}_q^n$ define the *weight* $w(\mathbf{x})$ of $\mathbf{x}$ as the number of non-zero coordinates of $\mathbf{x}$. Notice that

$$d(\mathbf{x}, \mathbf{y}) = w(\mathbf{x} - \mathbf{y}) \tag{8.11}$$

(see Exercise A8.24). Hence the reader will easily deduce (see Exercise A8.25) the following result:

**Proposition 8.5.3.** *If $C$ is a linear code, its minimum distance $d$ equals the minimum weight of its non-zero words.*

Basically, for a generic code, finding the minimum distance implies computing *the distance of all pairs of codewords* (see Exercise A8.38), while, for linear codes, the problem is solved by computing the *weight of the single codewords*. The computational benefit is obvious.

Another benefit is given by the possibility of describing rapidly the code and the consequent easier coding of the words, and so their fast transmission. One of the methods to do so is the following one.

Consider a linear $[n, k]_q$-code and fix a *basis* $\mathbf{x}_1, \ldots, \mathbf{x}_k$ of the code, seen as a vector space over $\mathbb{F}_q$. For all $i = 1, \ldots, k$ write

$$\mathbf{x}_i = (x_{i,1}, \ldots, x_{i,n})$$

and consider the $k \times n$ matrix $\mathbf{X}$ having as its rows the vectors $\mathbf{x}_i$, that is,

$$\mathbf{X} = \begin{pmatrix} x_{1,1} & x_{1,2} & \cdots & x_{1,n} \\ x_{2,1} & x_{2,2} & \cdots & x_{2,n} \\ \vdots & \vdots & \ddots & \vdots \\ x_{k,1} & x_{k,2} & \cdots & x_{k,n} \end{pmatrix}.$$

A matrix of this kind is called a *generating matrix* of the code. As the code-words are linear combinations of the matrix rows, coding works as follows.

Given a word $\mathbf{a} = (a_1, \ldots, a_k) \in \mathbb{F}_q^k$, it is *coded* mapping it to the word $\mathbf{x} = a_1\mathbf{x}_1 + \cdots + a_k\mathbf{x}_k$. On the other hand, this vector is just

$$\mathbf{x} = \mathbf{a} \cdot \mathbf{X}$$

with the usual *row-by-column* multiplication. That is, the $i$th coordinate of the product $\mathbf{a} \cdot \mathbf{X}$ is the *scalar product*

$$a_1 x_{1,i} + \cdots + a_k x_{k,i}$$

of $\mathbf{a}$ by the $i$th column of $\mathbf{X}$. In conclusion, coding is performed by simple algebraic operations of product and sum.

Notice that the generating matrix of a linear code is not unique, as a subspace has, in general, several bases, and to each basis corresponds a different generating matrix. In particular, it makes sense to choose generating matrices that are as simple as possible. For instance, a very simple form for a generating matrix is the so-called *standard form*, that is

$$\mathbf{X} = \begin{pmatrix} 1 & 0 & \cdots & 0 & x_{1,k+1} & \cdots & x_{1,n} \\ 0 & 1 & \cdots & 0 & x_{2,k+1} & \cdots & x_{2,n} \\ \vdots & \vdots & \ddots & \vdots & \vdots & \ddots & \vdots \\ 0 & 0 & \cdots & 1 & x_{k,k+1} & \cdots & x_{k,n} \end{pmatrix}.$$

In a more compact form, this matrix can be written as follows:

$$\mathbf{X} = (\mathbf{I}_k \,|\, \mathbf{A}),$$

where $\mathbf{I}_k$ is the *identity matrix* of order $k$, here consisting of the first $k$ columns of $\mathbf{X}$, and $\mathbf{A}$ is the remaining matrix. The convenience of such a matrix lies in the fact that, having codified the vector $\mathbf{a} = (a_1, \ldots, a_k)$, we get a vector whose first $k$ coordinates *coincide* with those of $\mathbf{a}$. So it is clear that the *redundancy* of the code is concentrated exactly in the last $n - k$ coordinates.

Using some linear algebra, it is easy to prove (see Exercise A8.26) the following result:

**Proposition 8.5.4.** *Given a $[n, k]_q$-code, it is possible, may be up to renaming some of its coordinates, to find a generating matrix in standard form.*

On the other hand, a $[n, k]_q$-code can be *assigned* by giving a $k \times n$ generating matrix for it. The only restriction is for the *rank* of the matrix to be equal to $k$, that is to say, for the matrix rows to give a basis for the code and, to this end, it is necessary and sufficient that the $k$ rows are *linearly independent*, that is, the only linear combination of these vectors giving the zero vector is the one in which the scalar coefficients are all equal to zero.

Finally, notice that if, instead of a basis of a code $C$, we took a *spanning set* $\mathbf{x}_1, \ldots, \mathbf{x}_h$ for it, with $h \geq k$, we would arrive, proceeding as above, to an $h \times n$ matrix having rank $k$, which may still be viewed as a *generating matrix* for $C$. Basically, the restriction about the rows of the generating matrix having to be independent is not strictly necessary, even if it is useful to *minimise* the algorithmic and data-recording encumbrance.

**Example 8.5.5.** Consider the linear code $C$ of type $[3, 2]_3$ having as its generating matrix

$$\mathbf{X} = \begin{pmatrix} 2 & 0 & 1 \\ 1 & 1 & 2 \end{pmatrix}. \tag{8.12}$$

Clearly, the matrix has rank 2. The words of $C$ are all the vectors of the form $a(2, 0, 1) + b(1, 1, 2)$, with $0 \leq a, b \leq 2$, so they are the nine vectors

$$(2, 0, 1), \ (1, 1, 2), \ (0, 0, 0), \ (1, 0, 2), \ (2, 2, 1),$$
$$(0, 1, 0), \ (1, 2, 2), \ (2, 1, 1), \ (0, 2, 0).$$

The minimum distance of the code is 1, as this is the minimum weight of the non-zero elements: for instance, that of $(0, 1, 0)$ and of $(0, 2, 0)$. Unfortunately, this is not a good code, as it does not correct errors.

A *dual* way of individuating a code is as follows. Consider a subspace $V$ of $\mathbb{F}_q^n$ and consider the set $V^\perp$ of vectors of $\mathbb{F}_q^n$ having zero scalar product with all vectors of $V$ (see Exercise A8.16). Formally:

$$V^\perp = \{\, \mathbf{p} \in \mathbb{F}_q^n \mid \mathbf{p} \times \mathbf{x} = 0, \ \text{for all } \mathbf{x} \in V \,\}.$$

The reader will not miss the analogy with the usual notion of perpendicularity from elementary geometry. However, the analogy stops with the notation. Not many usual properties of perpendicularity extend to the situation we are studying. For instance, it may well happen that a non-zero vector of $\mathbb{F}_q^n$ is orthogonal to iself, as the vector $(1, 1) \in \mathbb{F}_2^2$ does.

In any case, if $V$ has dimension $k$, then $V^\perp$ is itself a subspace and has dimension $n - k$ (see Exercise A8.27). In particular, if $C$ is a linear $[n, k]_q$-code, then $C^\perp$ is a linear $[n, n - k]_q$-code, called *dual code* of $C$. Moreover, $(C^\perp)^\perp = C$ (see again Exercise A8.27).

Consider a generating matrix $\mathbf{H}$ of $C^\perp$. It is called *parity check matrix* for $C$. It is a $(n - k) \times n$ matrix enjoying the following, easy to verify (see Exercise A8.28), property:

$$\mathbf{x} \in C \quad \text{if and only if} \quad \mathbf{x} \cdot \mathbf{H}^t = \mathbf{0}, \tag{8.13}$$

where $\mathbf{0}$ is the zero vector of length $n - k$. Recall that if $\mathbf{A}$ is an arbitrary matrix, $\mathbf{A}^t$ denotes the *transpose* of $\mathbf{A}$, that is to say, the matrix having as columns the rows of $\mathbf{A}$. Clearly, the rows of $\mathbf{A}^t$ coincide with the columns of $\mathbf{A}$.

Notice that, analogously to what happened for the generating matrix, the parity check matrix of a code is in general not unique.

**Example 8.5.6.** If $C$ has generating matrix in standard form

$$\mathbf{X} = (\mathbf{I}_k \mid \mathbf{A}),$$

a parity check matrix for $C$ is given by (see Exercise A8.29)

$$\mathbf{H} = (-\mathbf{A}^t \mid \mathbf{I}_{n-k}).$$

A $[n, k]_q$-code may also be assigned by a parity check matrix. In other words, if we give a $(n - k) \times n$ matrix $\mathbf{H}$ having rank $n - k$, we may define a $[n, k]_q$-code $C$ by the condition (8.13). The reader may verify that the definition is meaningful and that $C$ actually has dimension $k$ (see Exercise A8.30).

**Example 8.5.7.** With the above remarks in mind, we may now see that our old friend the Hamming $(7, 4)$ code is a $[7, 4]_2$-code. Keeping in mind the linear system (8.3), we see that a parity check matrix for this code is

$$\mathbf{H} = \begin{pmatrix} 1 & 0 & 1 & 1 & 1 & 0 & 0 \\ 1 & 1 & 0 & 1 & 0 & 1 & 0 \\ 1 & 1 & 1 & 0 & 0 & 0 & 1 \end{pmatrix},$$

which has rank 3.

Knowing a parity check matrix of a linear code is very useful to decode messages. The method we are about to explain is called *syndrome decoding* method.

Assume a $[n, k]_q$-code is given by a parity check matrix $\mathbf{H}$. Given any vector $\mathbf{x} \in \mathbb{F}_q^n$, we define its *syndrome* as the vector $\mathbf{x} \cdot \mathbf{H}^t$ di $\mathbb{F}_q^{n-k}$. The syndrome is the zero vector only for the words of $C$.

Let $\mathbf{e}$ be any vector of $\mathbb{F}_q^n$. By summing $\mathbf{e}$ to each codeword, we get a set containing $\mathbf{e}$, which we shall denote by $\mathbf{e} + C$ and call a *coset* of $C$. Notice that the cosets of $C$ are pairwise disjoint and each of them is bijective to $C$, so each contains $q^k$ elements and there are $q^{n-k}$ of them (see Exercise A8.33). Remark further that the vectors in the coset $\mathbf{e} + C$ are exactly those having the syndrome $\mathbf{e} \cdot \mathbf{H}^t$ (see Exercise A8.34).

From every coset of $C$ we choose a minimum weight element, which will be called the *leader* of the coset. Make a list of all leaders and corresponding syndromes, which are all possible syndromes of elements of $\mathbb{F}_q^n$. If we receive a message and we discern in it a word $\mathbf{x}$ having non-zero syndrome, we realise that an error has occurred. Looking at the leader table, we find the leader $\ell$ having the same syndrome as $\mathbf{x}$ and decode $\mathbf{x}$ by subtracting from it the leader. So we obtain an element $\mathbf{y} = \mathbf{x} - \ell$ having zero syndrome, and so an element of the code. Moreover, by the definition itself of a leader as an element having minimum weight in its coset, $\mathbf{y}$ is an element of $C$ having minimum Hamming distance from $\mathbf{x}$. So the decoding is carried out adhering to the maximum likelihood principle.

**Example 8.5.8.** Consider the $[4,2]_3$-code having as its parity check matrix

$$\mathbf{H} = \begin{pmatrix} 1 & 1 & 0 & 2 \\ 0 & 2 & 1 & 2 \end{pmatrix}.$$

The reader will easily verify (see Exercise B8.27) that the nine leaders and the corresponding syndromes are

| leaders | $(0,0,0,0)$ | $(1,0,0,0)$ | $(2,0,0,0)$ | $(0,1,0,0)$ | $(0,2,0,0)$ |
|---|---|---|---|---|---|
| syndromes | $(0,0)$ | $(1,0)$ | $(2,0)$ | $(1,2)$ | $(2,1)$ |

| leaders | $(0,0,1,0)$ | $(0,0,2,0)$ | $(0,0,0,1)$ | $(0,0,0,2)$ |
|---|---|---|---|---|
| syndromes | $(0,1)$ | $(0,2)$ | $(2,2)$ | $(1,1)$ |

If we receive $(2,1,2,1)$ we observe that its syndrome is $(2,0)$. The corresponding leader is $(2,0,0,0)$, so we decode $(2,1,2,1)$ as $(2,1,2,1) - (2,0,0,0) = (0,1,2,1)$. The reader should verify that in each coset of $C$ different from $C$ there is a single leader and that this leader has weight 1. This implies that, if there is a single transmission error, the syndrome decoding correctly removes it.

We conclude this section by showing how it is possible to compute the minimum distance of a code from its parity check matrix.

**Theorem 8.5.9.** *The minimum distance $d$ of a $[n,k]_q$–code having parity check matrix $\mathbf{H}$ is the minimum size of a linearly dependent set of columns of matrix $\mathbf{H}$.*

PROOF. Let $\mathbf{h}_1, \ldots, \mathbf{h}_n$ be the columns of $\mathbf{H}$ and let $\mathbf{x} = (x_1, \ldots, x_n)$ be a non-zero word of $C$. We have $x_1\mathbf{h}_1 + \cdots + x_n\mathbf{h}_n = \mathbf{0}$, that is $\mathbf{x} \in C$ determines a non-trivial linear dependency relation among the columns of $\mathbf{H}$ and vice versa. The weight of $\mathbf{x}$, on the other hand, determines the number of columns of $\mathbf{H}$ involved in this relation. From this and from Proposition 8.5.3 the claim immediately follows.                                                                            □

**Example 8.5.10.** Considering again Example 8.5.8, we may see that every pair of columns of the matrix $\mathbf{H}$ consists of independent columns. So $d = 3$ and the code, as already seen, corrects one error. Moreover, it is *maximum-distance separable*, that is, $d = n - k + 1$ (see page 420).

## 8.6 Cyclic codes

In this section we shall discuss a class of codes which are very interesting, due to their rich algebraic structure. They are the *cyclic codes*, that is to say, the codes that, whenever a word belongs to them, all its *cyclic permutations* belong to the code. Let us fix notation and definitions necessary to introduce these codes.

Recall the notion of a permutation of the set $I_n = \{1, \ldots, n\}$ consisting of the first $n$ natural numbers (see Exercise A1.11). Consider the permutation $\gamma$ such that

$$(\gamma(1), \gamma(2), \ldots, \gamma(n)) = (n, 1, 2, \ldots, n-1).$$

Following the notation introduced in Exercise A1.11, we shall say that $\gamma$ has symbol $(n, 1, 2, \ldots, n-1)$. Notice that, for all $j = 0, \ldots, n-1$, the symbol of $\gamma^j$ is

$$(n - j + 1, \ldots, n, 1, 2, \ldots, n - j),$$

while $\gamma^n$ is the identical permutation. So, $\gamma$ generates a cyclic subgroup of order $n$ of the group $\mathfrak{S}_n$ of permutations of $I_n$. By this reason, $\gamma$ is called a *cyclic permutation* on the elements of $I_n$.

Notice now that, if $f$ is a permutation of $\mathfrak{S}_n$ having symbol $(i_1, \ldots, i_n)$, we may define the mapping

$$\omega_f : (x_1, \ldots, x_n) \in \mathbb{F}_q^n \to (x_{i_1}, \ldots, x_{i_n}) \in \mathbb{F}_q^n,$$

which clearly is an isomorphism of $\mathbb{F}_q^n$ in itself (see Exercise A8.35). In particular,

$$\omega_\gamma : (x_1, \ldots, x_n) \in \mathbb{F}_q^n \to (x_n, x_1, x_2, \ldots, x_{n-1}) \in \mathbb{F}_q^n$$

generates a cyclic group $\Gamma(n, q)$ of order $n$ of isomorphisms of $\mathbb{F}_q^n$ in itself (see again Exercise A8.35). For the sake of notational simplicity, we shall denote the linear mapping $\omega_\gamma$ again by $\gamma$.

Now we may give the definition of a cyclic code.

**Definition 8.6.1.** *A linear code $C$ of type $[n, k]_q$ is said to be cyclic if $C$ is stable under the action of $\Gamma(n, q)$, that is to say, if for every word $x \in C$ we have $\gamma(x) \in C$.*

**Example 8.6.2.** Given an element $\mathbf{a} = (a_1, \ldots, a_n) \in \mathbb{F}_q^n$, the vector space generated by the elements $\gamma^i(\mathbf{a})$, for $i = 0, \ldots, n-1$, is an example of a cyclic code, called *cyclic code generated by the element* $\mathbf{a}$ (see Exercise A8.44).

We are now going to discuss an efficient way to describe cyclic codes using polynomials. It relies on an $\mathbb{F}_q$-vector space isomorphism between $\mathbb{F}_q^n$ and the quotient ring $\mathbb{F}_q[x]/(x^n - 1)$. With an abuse of notation, we shall denote a polynomial $p(x) \in \mathbb{F}_q[x]$ and its class in $\mathbb{F}_q[x]/(x^n - 1)$ with the same symbol. In particular, $x^n = 1$ in $\mathbb{F}_q[x]/(x^n - 1)$.

So we have the following proposition, whose easy proof is left as an exercise for the reader (see Exercise A8.45):

**Proposition 8.6.3.** *Let* $f : \mathbb{F}_q^n \longrightarrow \mathbb{F}_q[x]/(x^n - 1)$ *be the mapping defined as follows: for all* $\mathbf{a} = (a_0, \ldots, a_{n-1}) \in \mathbb{F}_q^n$, *we have* $f(\mathbf{a}) = a_0 + a_1 x + \cdots + a_{n-1} x^{n-1}$. *Then:*

- *f is a vector space isomorphism;*
- $f(\gamma(\mathbf{a})) = x \cdot f(\mathbf{a})$.

Proposition 8.6.3 allows us to identify $\mathbb{F}_q^n$ with $\mathbb{F}_q[x]/(x^n - 1)$ as vector spaces, so we may equally consider the element $(a_0, a_1, \ldots, a_{n-1})$ in $\mathbb{F}_q^n$ or the polynomial $a_0 + a_1 x + \cdots + a_{n-1} x^{n-1}$ in $\mathbb{F}_q[x]/(x^n - 1)$.

**Remark 8.6.4.** The proposition shows that cyclic codes may be identified with vector subspaces of $\mathbb{F}_q[x]/(x^n - 1)$ that are stable under multiplication by $x$, and so also under multiplication by any polynomial $g(x) \in \mathbb{F}[x]$. This is the same as saying that cyclic codes are exactly the ideals of the ring $\mathbb{F}_q[x]/(x^n-1)$, and correspond to the ideals of $\mathbb{F}_q[x]$ containing the polynomial $x^n - 1$ (see Exercise A3.10).

As $\mathbb{F}_q[x]$ is a principal ideal ring (see § 1.3.6), every ideal has a generator $a(x)$, which is the monic polynomial having least degree in the ideal. Moreover, the ideal contains $x^n - 1$ if and only if $a(x)$ divides $x^n - 1$. In conclusion, cyclic codes are in bijection with monic divisors $a(x)$ of the polynomial $x^n - 1$ in $\mathbb{F}_q[x]$. The polynomial $a(x)$ is said to be a *generator* of the corresponding code.

Assume $x^n - 1 = a(x) \cdot b(x)$, with monic $a(x), b(x)$. Let $h$ be the degree of $a(x)$, so $n - h = k$ is the degree of $b(x)$. Modulo $x^n - 1$, we have the relation $a(x) \cdot b(x) = 0$, so $a(x), xa(x), \ldots, x^k a(x)$ are linearly dependent in $\mathbb{F}_q[x]/(x^n - 1)$, while clearly $a(x), xa(x), \ldots, x^{k-1} a(x)$ are not. Thus, the dimension of the cyclic code corresponding to the monic divisor $a(x)$ of $x^n - 1$ is $k = n - h$, where $h$ is the degree of $a(x)$ and $a(x), xa(x), \ldots, x^{k-1} a(x)$ form a basis.

The following proposition sums up and completes what we have been saying and characterises the cyclic codes $C$ of type $[n, k]_q$.

**Proposition 8.6.5.** *Let* $a(x) = a_0 + a_1 x + \cdots + a_{n-k} x^{n-k}$, *with* $a_{n-k} = 1$, *be a polynomial in* $\mathbb{F}_q[x]$ *that divides* $x^n - 1$, *and let* $\mathbf{a} = (a_0, a_1, \ldots, a_{n-k}, 0, \ldots, 0)$ *be the corresponding word in* $\mathbb{F}_q^n$. *Then the* $k$ *words* $\mathbf{a}, \gamma(\mathbf{a}), \ldots, \gamma^{k-1}(\mathbf{a})$ *give a basis for a cyclic code* $C$ *of type* $[n, k]_q$. *Vice versa, all cyclic codes of length* $n$ *over* $\mathbb{F}_q$ *can be obtained in this way.*

Now it is clear why it is interesting to find the divisors of $x^n - 1$ in $\mathbb{F}_q[x]$, a problem we have already encountered, under different guises (see for example § 5.1.8 and § 6.3). Without too many details, assume that $\mathrm{GCD}(n, q) = 1$ and let $x^n - 1 = p_1(x) \cdots p_s(x)$ be the decomposition of $x^n - 1$ into irreducible, monic factors in $\mathbb{F}_q[x]$. These polynomials are all distinct by the hypothesis $\mathrm{GCD}(n, q) = 1$ (see Proposition 5.1.16 or Exercise A5.18). The divisors of $x^n - 1$ may be obtained by taking in all possible ways a subset of $\{p_1(x), \ldots, p_s(x)\}$

and multiplying its elements. Thus, these divisors are in bijection with the power set of $\{p_1(x), \ldots, p_s(x)\}$, so there are $2^s$ of them (see Exercise A1.22).

Let $a(x) = \sum_{i=0}^{n-k} a_i x^i$ be the monic polynomial generator of a code $C$ of length $n$. If $a(x)$ has degree $n - k$, we have seen that the words $a(x)$, $xa(x), \ldots, x^{k-1}a(x)$ form a base for $C$, that is, $C$ is a $[n, k]$-code. So the matrix

$$\mathbf{G} = \begin{pmatrix} a_0 & a_1 & \cdots & a_{n-k} & 0 & 0 & \cdots & 0 \\ 0 & a_0 & a_1 & \cdots & a_{n-k} & 0 & \cdots & 0 \\ 0 & 0 & a_0 & a_1 & \cdots & a_{n-k} & \cdots & 0 \\ 0 & 0 & 0 & \ddots & \ddots & \ddots & \ddots & 0 \\ 0 & 0 & 0 & 0 & a_0 & a_1 & \cdots & a_{n-k} \end{pmatrix}$$

is a *generating* matrix for $C$. As we have seen, we may use it to encode information as follows: if $\mathbf{c} = (c_0, c_1, \ldots, c_{k-1})$ is the word we have to transmit, we may encode it by the product of matrices $\mathbf{c} \cdot \mathbf{G}$, which corresponds to the polynomial

$$c(x) = (c_0 + c_1 x + c_2 x^2 + \cdots + c_{k-1} x^{k-1}) a(x)$$

modulo $x^n - 1$.

Let again $x^n - 1 = a(x) \cdot b(x)$, with $b(x) = \sum_{i=0}^{k} b_i x^i$ a monic polynomial. The polynomial $b(x)$ is called *check polynomial* of the code $C$. From the above we deduce that the code $C$ consists exactly of all polynomials $c(x)$ in $\mathbb{F}_q[x]/(x^n - 1)$ such that $c(x)b(x) = 0$ modulo $x^n - 1$ (see Exercise A8.46).

The reason for the name of $b(x)$ comes from the following remark. From $x^n - 1 = a(x) \cdot b(x)$, we may deduce that

$$\sum_{r+s=i} a_r b_s = 0, \qquad \text{for } i = 1, \ldots, n - 1.$$

So the matrix

$$\mathbf{H} = \begin{pmatrix} 0 & \cdots & 0 & 0 & b_k & b_{k-1} & \cdots & b_0 \\ 0 & \cdots & 0 & b_k & b_{k-1} & \cdots & b_0 & 0 \\ 0 & \cdots & b_k & b_{k-1} & \cdots & b_0 & 0 & 0 \\ 0 & \cdots & \cdots & \cdots & \cdots & \cdots & \cdots & 0 \\ b_k & b_{k-1} & \cdots & b_0 & 0 & 0 & \cdots & 0 \end{pmatrix}$$

is a parity check matrix for code $C$.

In conclusion, it is clear that the codes determined by $a(x)$ and $b(x)$ are dual each other.

**Example 8.6.6.** Determine all binary cyclic codes of length 7.

We have to factor the polynomial $x^7 - 1$ over $\mathbb{Z}_2$. We find

$$x^7 - 1 = (x - 1)(x^3 + x + 1)(x^3 + x^2 + 1).$$

The third degree polynomials are irreducible because they have no roots in $\mathbb{Z}_2$. So there are $2^3 = 8$ cyclic codes of length 7 over $\mathbb{Z}_2$, and the generating polynomials are

$$1,$$
$$x + 1,$$
$$x^3 + x + 1,$$
$$x^3 + x^2 + 1,$$
$$(x + 1)(x^3 + x + 1) = x^4 + x^3 + x^2 + 1,$$
$$(x + 1)(x^3 + x^2 + 1) = x^4 + x^2 + x + 1,$$
$$(x^3 + x + 1)(x^3 + x^2 + 1) = x^6 + x^5 + x^4 + x^3 + x^2 + x + 1,$$
$$x^7 - 1.$$

The corresponding codes are:

| Generating polynomial | Corresponding code |
|---|---|
| (1)  1 | $\mathbb{F}_2^7$, |
| (2)  $x + 1$ | called *parity check code*, |
| (3)  $x^3 + x + 1$ | Hamming code, |
| (4)  $x^3 + x^2 + 1$ | Hamming code, |
| (5)  $x^4 + x^3 + x^2 + 1$ | dual code of Hamming code, |
| (6)  $x^4 + x^2 + x + 1$ | dual code of Hamming code, |
| (7)  $x^6 + x^5 + x^4 + x^3 + x^2 + x + 1$ | called *repetition code*, |
| (8)  $x^7 - 1$ | zero code. |

In case (2), the parity check matrix is the vector $(1, 1, 1, 1, 1, 1, 1)$. So $x = (x_1, \ldots, x_7)$ belongs to the code if and only if $x_1 + \cdots + x_7 = 0$ in $\mathbb{Z}_2$, that is, if and only if $x_1 + \cdots + x_7$ is even; whence the name of the code.

Dually, in case (7), the generating matrix is $(1, 1, 1, 1, 1, 1, 1)$. So the code consists of the words $x = (x_1, \ldots, x_7) \in \mathbb{F}_2^7$ such that $x_1 = x_2 = x_3 = x_4 = x_5 = x_6 = x_7$. Hence the code consists of the two 7-tuples $(0, 0, 0, 0, 0, 0, 0)$ and $(1, 1, 1, 1, 1, 1, 1)$. This justifies the name of this code (compare it with the code used by the Australian coin tosser in § 8.2).

In case (4), the parity check matrix is

$$\mathbf{H} = \begin{pmatrix} 0 & 0 & 1 & 1 & 1 & 0 & 1 \\ 0 & 1 & 1 & 1 & 0 & 1 & 0 \\ 1 & 1 & 1 & 0 & 1 & 0 & 0 \end{pmatrix}.$$

It may be immediately verified that, by changing names to the coordinates, it coincides with the parity check matrix of the Hamming code (see Example 8.5.7). An analogous reasoning holds for case (3) (see Exercise A8.47).

An interesting class of codes is that of the so-called BCH codes, from the initials of their inventors R. C. Bose, D. K. Ray–Chaudhuri and A. Hocquenghem (1959–60).

**Definition 8.6.7.** *A cyclic code of length $n$ over $\mathbb{F}_q$ is called* BCH *code with assigned distance $\delta$ if, denoting by $a(x)$ its generator and by $\beta$ an $n$th primitive root of unity, we have*

$$a(x) = l.c.m. \text{ of the minimal polynomials of } \beta^l, \beta^{l+1}, \dots, \beta^{l+\delta-2} \text{ for some } l.$$

In what follows we shall take $n = q^m - 1$, so $\beta$ is a generator of $\mathbb{F}_{q^m}$ (see § 5.1.4 and Corollary 5.1.31). Then $\beta$ and its powers may be regarded as numerical vectors of length $m$ over $\mathbb{F}_q$. Moreover, we shall assume $l + \delta - 2 < n$.

Why are these codes called *with assigned distance*? The answer comes from the following result.

**Proposition 8.6.8.** *The minimum distance of a BCH code with assigned distance $\delta$ over $\mathbb{F}_q$, with $n = q^m - 1$, is greater or equal than $\delta$.*

PROOF. Consider the matrix

$$\mathbf{H} := \begin{pmatrix} 1 & \beta^l & \beta^{2l} & \dots & \beta^{(n-1)l} \\ 1 & \beta^{l+1} & \beta^{2(l+1)} & \dots & \beta^{(n-1)(l+1)} \\ \vdots & \vdots & \vdots & \dots & \vdots \\ 1 & \beta^{l+\delta-2} & \beta^{2(l+\delta-2)} & \dots & \beta^{(n-1)(l+\delta-2)} \end{pmatrix}.$$

Considering every entry of the matrix as a column vector with $m$ components over $\mathbb{F}_q$, the matrix $\mathbf{H}$ can be thought of as a $m(\delta - 1) \times n$ matrix. A word $\mathbf{c}$ belongs to the BCH code if and only if $\mathbf{c} \cdot \mathbf{H}^t = \mathbf{0}$. So $\mathbf{H}$ is a parity check matrix of the code.

The first $\delta - 1$ columns of $\mathbf{H}$ are linearly independent, as the corresponding determinant is the Vandermonde determinant $V(\beta^l, \beta^{l+1}, \dots, \beta^{l+\delta-2})$ (see Exercise A4.64). On the other hand, every other $(\delta - 1)$-tuple of columns is linearly independent because the corresponding determinant is the product of a power of $\beta$ by an analogous Vandermonde determinant. The claim follows from Theorem 8.5.9.     □

We shall not dwell further on these codes. Suffice it to remark that the reasoning relying on Vandermonde determinants that led us to conclude that every BCH code of length $n = q^m - 1$ and assigned distance $\delta$ over $\mathbb{F}_q$ has minimum distance at least $\delta$, also holds for codes with parity check matrices of the form

$$\mathbf{H}' := \begin{pmatrix} h_0 & h_1 & \dots & h_{n-1} \\ h_0\beta_0 & h_1\beta_1 & \dots & h_{n-1}\beta_{n-1} \\ \vdots & \vdots & \dots & \vdots \\ h_0\beta_0^{\delta-1} & h_1\beta_1^{\delta-1} & \dots & h_{n-1}\beta_{n-1}^{\delta-1} \end{pmatrix},$$

where the $h_i$s and the $\beta_j$s are distinct elements of $\mathbb{F}_{q^m}^*$. In fact, the codes we are going to discuss in next section, called *Goppa codes*, are exactly of this kind.

## 8.7 Goppa codes

The world of linear codes is very heterogeneous and interesting, and it would be both pleasant and useful to explore it, armed with the algebraic equipment we have built up so far. The reader could appreciate how many algebraic notions expounded so far are put to use in this setting. Moreover, there are very simple linear codes, used since the 1950's or the 1960's, which are unbelievably efficient, just like the Hamming $(7, 4)$ code, and still used in many applications. However, we have already covered together a very long path and, for the sake of brevity, we cannot explore more together: we encourage the most curious readers to browse the references we gave so far. However, we want to give one last example, *Goppa codes*. The reason why we want to discuss them before bringing this chapter to an end is that they pave the way for the use of ideas and techniques from algebraic geometry in coding theory, as we shall mention at the end of this section. As we have already seen, this is what has already happened in the most recent advances in cryptography.

In order to define a Goppa code, we need some preliminary remarks. Consider the polynomial ring over a field $\mathbb{K}$ and fix an element $\gamma$ of the field and a polynomial $f(x) \in \mathbb{K}[x]$ of degree $t > 0$ such that $f(\gamma) \neq 0$, and so such that $\mathrm{GCD}(x - \gamma, f(x)) = 1$. Then there exists a unique polynomial $g(x) \in \mathbb{K}[x]$ of degree smaller than $t$, given by

$$g(x) = \frac{f(\gamma) - f(x)}{f(\gamma) \cdot (x - \gamma)}, \tag{8.14}$$

such that

$$(x - \gamma) \cdot g(x) \equiv 1 \pmod{f(x)}, \tag{8.15}$$

that is, such that $(x - \gamma) \cdot g(x) - 1$ is divisible by $f(x)$ (see Exercise A1.54). As usual, with a slight abuse of notation, we shall identify a polynomial with its class in $\mathbb{K}[x]/(f(x))$. We may say that $g(x)$ is the multiplicative inverse of $x - \gamma$ in $\mathbb{K}[x]/(f(x))$. So $1/(x - \gamma)$ is not a polynomial, but we consider it as such *modulo* $f(x)$, that is, in the ring $\mathbb{K}[x]/(f(x))$. Thus from now on we shall consider the symbol $1/(x - \gamma)$ as meaningful, when we shall work *modulo a polynomial that has not $\gamma$ as a root*.

**Example 8.7.1.** If $f(x) = x$ and if $\gamma \neq 0$ then the polynomial $g(x)$ of (8.14) is given by $-1/\gamma$, or

$$\frac{1}{x - \gamma} \equiv -\frac{1}{\gamma} \pmod{x}.$$

This is obvious, as $x = 0 \pmod{x}$.

This said, we may define Goppa codes:

**Definition 8.7.2.** Let $\mathbb{F}_q$ be a finite field, let $m$ be a positive integer and $f(x)$ a monic polynomial of degree $t > 0$ over $\mathbb{F}_{q^m}$. Let further $L = \{\gamma_1, \ldots, \gamma_n\}$ be a set of $n$ distinct elements of $\mathbb{F}_{q^m}$ such that $f(\gamma_h) \neq 0$, for $h = 1, \ldots, n$. The Goppa code $C(L, f(x))_{q,m}$ is the subset of $n$-tuples $(c_1, \ldots, c_n)$ of elements of $\mathbb{F}_q$ such that

$$\sum_{i=1}^{n} \frac{c_i}{x - \gamma_i} \equiv 0 \pmod{f(x)}. \tag{8.16}$$

Notice that a Goppa code is linear (see Exercise A8.48).

**Example 8.7.3.** If $f(x) = x$ and $L = \mathbb{F}_{q^m} \setminus \{0\}$, that is $n = q^m - 1$, then by Example 8.7.1, $C(L, x)_{q,m}$ is the $[q^m - 1, q^m - 2]_q$-code with parity check matrix the vector having as entries the non-zero elements of $\mathbb{F}_q$. By applying Theorem 8.5.9 we see that it has minimum distance $d = 2$.

**Example 8.7.4.** Consider a finite field $\mathbb{F}_q$ and let $\beta$ be an $n$th primitive root of unity, with $n = q^m - 1$, that is, $\beta$ is a generator of $\mathbb{F}_{q^m}$. Consider the polynomial $f(x) = x^{\delta - 1}$ and let $L = \mathbb{F}_{q^m} \setminus \{0\} = \{\beta^{-i}, 0 \leq i \leq n - 1\}$. The Goppa code $C(L, x^{\delta - 1})_{q,m}$ coincides with the BCH code corresponding to the case $l = 1$ considered in the previous section.

To verify this, proceed as follows. First of all, notice the polynomial identity (see Exercise A8.49):

$$x^n - 1 = (x - \beta^{-i}) \cdot \sum_{k=0}^{n-1} (\beta^{-i})^{n-k-1} x^k = (x - \beta^{-i}) \cdot \sum_{k=0}^{n-1} \beta^{i(k+1)} x^k, \qquad (8.17)$$

Next, keep in mind (see the proof of Proposition 8.6.8) that the word $(c_0, \ldots, c_{n-1})$ belongs to the BCH code corresponding to the case $l = 1$ if and only if

$$c_0 + c_1 \cdot \beta^j + c_2 \cdot \beta^{2j} + \cdots + c_{n-1} \cdot \beta^{(n-1)j} = 0, \quad \text{for all } 1 \leq j < \delta.$$

Notice that by (8.17) we have

$$\frac{x^n - 1}{x - \beta^{-i}} = \sum_{k=0}^{n-1} \beta^{i(k+1)} x^k;$$

hence,

$$(x^n - 1) \cdot \sum_{k=0}^{n-1} \frac{c_i}{x - \beta^{-i}} = \sum_{k=0}^{n-1} x^k \sum_{k=0}^{n-1} c_i \beta^{i(k+1)}, \qquad (8.18)$$

so it follows that the word $(c_0, \ldots, c_{n-1})$ belongs to the BCH code if and only if the polynomial appearing in the right-hand side of (8.18) is divisible by $x^{\delta - 1}$. In conclusion, $(c_0, \ldots, c_{n-1})$ belongs to the BCH code if and only if $\sum_{k=0}^{n-1} c_i / (x - \beta^{-i})$ is zero modulo $x^{\delta - 1}$, that is, if and only if the word $(c_0, \ldots, c_{n-1})$ belongs to the code $C(L, x^{\delta - 1})_{q,m}$.

Let us now determine a parity check matrix for a Goppa code $C(L, f(x))_{q,m}$.

Keeping in mind Equation (8.16), which determines the codewords, we might be tempted to say that the matrix

$$\left( \frac{1}{x - \gamma_1} \quad \frac{1}{x - \gamma_2} \quad \cdots \quad \frac{1}{x - \gamma_n} \right) \qquad (8.19)$$

is the parity check matrix of $C(L, f(x))_{q,m}$, were it not the case that this is *not a matrix of constants*, but of polynomials and what's more not even polynomials over $\mathbb{F}_q$ but over the larger field $\mathbb{F}_{q^m}$ or, rather, over a quotient ring of $\mathbb{F}_{q^m}[x]$. It is only in special cases, as the one discussed in Example

8.7.3, that it is a matrix of constants. But from (8.19) we may get an actual parity check matrix.

Recall that we are taking everything modulo $f(x)$ and that, for $i = 1, \ldots, n$, we have

$$\frac{1}{x - \gamma_i} = \frac{f(\gamma_i) - f(x)}{f(\gamma_i) \cdot (x - \gamma_i)}.$$

Setting

$$f(x) = \sum_{i=0}^{t} f_i x^i,$$

the following polynomial identity is easily verified (see Exercise A8.50):

$$f(x) - f(y) = (x - y) \cdot \sum_{0 \le h + k \le t - 1} f_{h+k+1} x^h y^k; \qquad (8.20)$$

it may be rewritten as

$$\frac{f(x) - f(y)}{x - y} = \sum_{0 \le h + k \le t - 1} f_{h+k+1} x^h y^k.$$

Set now $g_i = 1/f(\gamma_i)$, $i = 1, \ldots, n$, hence we may write

$$\frac{1}{x - \gamma_i} = -g_i \cdot \sum_{0 \le h + k \le t - 1} f_{h+k+1} x^h \gamma^k.$$

So the necessary and sufficient condition for a word $\mathbf{c} = (c_1, \ldots, c_n)$ to belong to the code $C(L, f(x))_{q,m}$, that is, the *orthogonality* of $\mathbf{c}$ to the vector (8.19), is translated into the orthogonality condition of $\mathbf{c}$ to the matrix

$$\mathbf{K} = \begin{pmatrix} g_1 f_t & \cdots & g_n f_t \\ g_1 (f_{t-1} + f_t \gamma_1) & \cdots & g_n (f_{t-1} + f_t \gamma_n) \\ \vdots & \cdots & \vdots \\ g_1 (f_1 + f_2 \gamma_1 + \cdots + f_t \gamma_1^{t-1}) & \cdots & g_n (f_1 + f_2 \gamma_n + \cdots + f_t \gamma_n^{t-1}) \end{pmatrix},$$

that is,

$$\mathbf{c} \cdot \mathbf{K}^t = \mathbf{0},$$

so $\mathbf{K}$ is a parity check matrix for the code $C(L, f(x))_{q,m}$. On the other hand, keeping in mind Exercise A8.32, we deduce that another parity check matrix of $C(L, f(x))_{q,m}$, having a simpler form, is given by

$$\mathbf{H} = \begin{pmatrix} g_1 & \cdots & g_n \\ g_1 \gamma_1 & \cdots & g_n \gamma_n \\ \vdots & \ddots & \vdots \\ g_1 \gamma_1^{t-1} & \cdots & g_n \gamma_n^{t-1} \end{pmatrix}. \qquad (8.21)$$

Beware! We are somewhat cheating, here. In fact, stricly speaking $\mathbf{H}$ is not yet a parity check matrix, as its elements *are not* constants, that is, elements of $\mathbb{F}_q$. Indeed, as $f(x)$ is a polynomial with coefficients in $\mathbb{F}_{q^m}$, the elements of $\mathbf{H}$, for instance $g_1, \ldots, g_n$, belong to $\mathbb{F}_{q^m}$ rather than to $\mathbb{F}_q$. But this is easily fixed, in a way analogous to what has been done for BCH codes in the previous section. Indeed, it suffices to keep in mind that $\mathbb{F}_{q^m}$ is a vector space of dimension $m$ over $\mathbb{F}_q$, so its elements can been regarded as column vectors of length $m$ over $\mathbb{F}_q$. Thus, we are considering the elements of $\mathbf{H}$ as *columns* of $m$ elements of $\mathbb{F}_q$. So $\mathbf{H}$, as a matrix over $\mathbb{F}_q$, becomes a $mt \times n$ matrix, rather than a $t \times n$ one. Of course, the rows of this matrix are not necessarily independent: we might have to remove some to make $\mathbf{H}$ an actual parity check matrix in the sense of the previous section (see Exercise A8.31).

A first conclusion to be drawn from this analysis is the following proposition:

**Proposition 8.7.5.** *A Goppa code $C(L, f(x))_{q,m}$ with $L$ of size $n$ and $f(x)$ of degree $t$ is a $[n, k]_q$-code, with $k \geq n - mt$.*

What can be said about the minimum distance of a Goppa code? This question can be given only a partial answer. Indeed, we have the following proposition, whose proof can be left as an exercise to the reader (see Exercise A8.51), who might want to argue in a similar as for BCH codes, using Theorem 8.5.9 and the properties of Vandermonde determinants. Here we provide instead a different proof, which will enable us to give an interesting extension.

**Proposition 8.7.6.** *A Goppa code $C(L, f(x))_{q,m}$ with $f(x)$ of degree $t$ has minimum distance $d \geq t + 1$.*

PROOF. Identify $\mathbf{c} = (c_1, \ldots, c_n) \in \mathbb{F}_q^n$ with the *rational function*

$$\phi(x) = \sum_{i=0}^{n} \frac{c_i}{x - \gamma_i},$$

which is an element of the field of fractions of $\mathbb{F}_p[x]$.

We may write $\phi(x) = n(x)/d(x)$, where $d(x) = (x - \gamma_1)^{\epsilon_1} \cdots (x - \gamma_n)^{\epsilon_n}$, where $\epsilon_i = 0$ if $c_i = 0$, while $\epsilon_i - 1$ if $c_i \neq 0$, for all $i = 1, \ldots, n$. So the degree of $d(x)$ equals the weight $w(\mathbf{c})$. Moreover, $n(x)$ is a polynomial of degree smaller than that of $d(x)$ and $\mathrm{GCD}(n(x), d(x)) = 1$.

On the other hand, $\mathbf{c}$ belongs to the code if and only if $\phi(x) \equiv 0 \pmod{f(x)}$, that is, if and only if $f(x)$ divides $n(x)$. In this case, $\partial(n(x)) \geq t$, so $\partial(d(x)) \geq t + 1$. But as $\partial(d(x)) = w(\mathbf{c})$, the claim follows from Proposition 8.5.3.    □

The previous proof suggests that, under particular conditions, it is possible to improve the estimate on the minimum distance. For instance, we have:

**Proposition 8.7.7.** *A Goppa code $C(L, f(x))_{2,m}$ with $f(x)$ of degree $t$ without multiple roots has minimum distance $d \geq 2t + 1$, so it corrects $t$ errors.*

PROOF. Consider a word $\mathbf{c} = (c_1, \ldots, c_n)$ of the Goppa code $C(L, f(x))_{2,m}$. Setting

$$g(x) = (x - \gamma_1)^{c_1} \cdot (x - \gamma_2)^{c_2} \cdot \cdots \cdot (x - \gamma_n)^{c_n},$$

we have

$$\phi(x) = \sum_{i=1}^{n} \frac{c_i}{x - \gamma_i} = \frac{g'(x)}{g(x)}, \tag{8.22}$$

where $g'(x)$ is the derivative of $g(x)$ (see Exercise A8.52). Moreover, $\mathrm{GCD}(g(x), g'(x)) = 1$ as $g(x)$ has no multiple roots.

Notice that, as we are working in characteristic 2, in $g'(x)$ only powers of $x$ with even exponent appear, that is, $g'(x)$ is a perfect square.

On the other hand, as $f(x)$ divides $\phi(x)$, then $f(x)$ divides $g'(x)$. But in this case, as $g'(x)$ is a perfect square and $f(x)$ as no multiple roots, $g'(x)$ is divisible by $f(x)^2$ too. So reasoning as in the proof of Proposition 8.7.6, we conclude that the degree of the denominator of $\phi(x)$, that is, the weight of $\mathbf{c}$, is at least $2t + 1$.     $\square$

The previous propositions make it plain that Goppa codes have *good separation characteristics* and so *good error-correcting capabilities*.

Now, to conclude, we shall see that Goppa codes are efficient in other ways too. Indeed, the following result holds:

**Theorem 8.7.8.** *There is a sequence of Goppa codes in which the information rate tends to the Gilbert–Varshamov bound.*

PROOF. Choose the parameters $n = q^m$, $t$ and $d$, and try to determine a Goppa code $C(L, f(x))_{q,m}$ having these parameters. As $n = q^m$, the choice of $L$ is forced: we have to take $L = \mathbb{F}_{q^m}$. Try next to find a monic polynomial $f(x) \in \mathbb{F}_{q^m}[x]$, irreducible over $\mathbb{F}_{q^m}$, of degree $t$. In this case $f(x)$ has no roots in $\mathbb{F}_{q^m}$, so we have $f(c) \neq 0$ for all $c \in L = \mathbb{F}_{q^m}$.

Finally, we want the code $C(L, f(x))_{q,m}$ to have minimum distance at least equal to $d$. For this to happen, we must exclude all polynomials $f(x)$ of degree $t$ such that there is a word $\mathbf{c} = (c_1, \ldots, c_n) \in C(L, f(x))_{q,m}$ with $w(\mathbf{c}) = w < d$. Let us give an estimate for the number $\nu(t, d)$ of such polynomials.

In order to do so, let $\mathbf{c} = (c_1, \ldots, c_n)$ be an element of $\mathbb{F}_q^n$ of weight $w(\mathbf{c}) = w < d$. Notice that the number of such elements is

$$\mu_w = \binom{n}{w}(q - 1)^w$$

(see the solution of Exercise A8.8). As we have seen, $w(\mathbf{c}) = w$ means that the numerator $n(x)$ of $\phi(x) = \sum_{i=0}^{n} \gamma_i/(x - c_i)$ has degree at most $w - 1$. Recall that $n(x)$ must be divisible by $f(x)$ for $\mathbf{c}$ to be in $C(L, f(x))_{q,m}$. Thus, $f(x)$ can be chosen in at most $\lfloor (w - 1)/t \rfloor$ ways for $\mathbf{c}$ to be in the code. In conclusion, we have the following estimate for $\nu(t, d)$:

$$\nu(t,d) \le \sum_{w=1}^{d-1} \left\lfloor \frac{w-1}{t} \right\rfloor \mu_w = \sum_{w=1}^{d-1} \left\lfloor \frac{w-1}{t} \right\rfloor \binom{n}{w} (q-1)^w.$$

Keeping in mind formula (8.6) on page 415, we find that

$$\nu(t,d) \le \frac{d}{t} V_q(n, d-1).$$

We use now the estimate for the number $n_{t,q^m}$ of monic irreducible polynomials of degree $t$ found in Exercise A5.33, that is,

$$n_{t,q^m} > \frac{q^{mt}}{t} \left(1 - q^{-mt/2+m}\right).$$

So, in conclusion, a sufficient condition for the existence of our code $C(L, f(x))_{q,m}$, that is, of the polynomial $f(x)$, is

$$d\, V_q(n, d-1) \le q^{mt} \left(1 - q^{-mt/2+m}\right).$$

Take the logarithms in base $q$ of both sides of the previous inequality, divide by $n$ and take the limit for $n \to \infty$, so that the separation ratio $d/n$ tends to a number $\delta$. This may be done because we may choose $n$ and $d$ arbitrarily. Then, keeping in mind formula (8.9) on page 418, we find that for our Goppa codes to exist it suffices that

$$H_q(\delta) + \mathcal{O}(1) \le \frac{mt}{n} + \mathcal{O}(1). \tag{8.23}$$

Having fixed $\delta$, it is clear that this relation can be satisfied by some $t$, so there are infinitely many codes verifying the properties necessary for $n \to \infty$. By Proposition 8.7.5, a code obtained in this way has dimension at least $n - mt$. The most *unlucky* situation is when this dimension is the lowest possible one, that is $n - mt$. In this case, the information rate is $R = 1 - mt/n$. To make this ratio as large as possible, we shall have to take $t$ minimal under the condition (8.23). So, for $n \to \infty$, we have that $mt/n$ tends to $H_q(\delta)$ and so the information rate tends to $1 - H_q(\delta)$. Keeping in mind Theorem 8.4.8, this concludes the proof. □

Notice that, by Equation (8.23), which gives a sufficient condition to construct an infinite sequence of Goppa codes with a given separation ratio $\delta$, these codes have, at least asymptotically, an information rate that cannot exceed the Gilbert–Varshamov bound $1 - H_p(\delta)$, but tends to it *from below*.

To construct sequences of Goppa codes with a given separation ratio $\delta$ for which the information rate tends to a number greater than the Gilbert–Varshamov bound $1 - H_p(\delta)$, one may use analogous but more complex constructions. Here algebraic curves make their appearance, as in cryptography. The starting point is given by the following remark.

The *classical* Goppa codes considered so far use rational functions over a finite field $\mathbb{F}$. Considered *geometrically*, $\mathbb{F}$ may be regarded as a *line* of the numerical affine plane $\mathbb{F}^2$, for instance the $x$-axis having equation $y = 0$ (see the remarks in § 7.9.2).

Now we may take a step forward: we may use *algebraic curves* in $\mathbb{F}^2$, defined by equations of the form $f(x, y) = 0$ with $f(x, y) \in \mathbb{F}[x, y]$. If we consider such a curve $\Gamma$ rather than the line $y = 0$, by using suitable rational functions defined on it we may construct natural extensions of the classical Goppa codes. The only restriction for everything to work well is that $\Gamma$ has many *rational points*, that is, that there are many points $(a, b) \in \mathbb{F}^2$ verifying the equation $f(x, y) = 0$. This condition is obviously verified for the equation $y = 0$ defining $\mathbb{F}$. Now, by using these *generalised*, or as they are usually called, *geometric Goppa codes*, for suitable algebraic curves with many rational points, one may improve Theorem 8.7.8. One proves indeed the following result, due to three Russian mathematicians, M. A. Tsfaman, S. G. Vladut and T. Zink (see [60]), which will be only stated here:

**Theorem 8.7.9.** *There is a sequence of geometric Goppa codes over $\mathbb{F}_{p^2}$, for $p \geq 7$, such that the information rate tends to a number greater than the Gilbert–Varshamov bound.*

# Appendix to Chapter 8

## A8 Theoretical exercises

**A8.1.**\* Prove that the error probability of a repetition binary code repeated $n$ times is

$$P_n = \sum_{0 \leq i < \lfloor n/2 \rfloor} \binom{n}{i} p^{n-i}(1-p)^i,$$

where $p$ is the probability of an error along the transmission channel. Verify that $0 < P_n < 1$ and prove that $P_n$ tends to 0 as $n$ approaches infinity.

**A8.2.** Prove that a word of the Hamming $(7, 4)$ code is a solution of the linear system (8.3) on page 410, and vice versa (see formula (8.2)).

**A8.3.** Prove that two distinct words of the Hamming $(7, 4)$ code differ in at least three coordinates.

**A8.4.**\* Prove the three properties of the Hamming distance $d(x, x')$ between two words $x, x'$ of a code (see Definition 8.3.2 on page 412).

**A8.5.**\* Prove Theorem 8.3.4 on page 412.

**A8.6.** Prove formula (8.4) in Remark 8.3.5 on page 413.

**A8.7.** Prove that for a code of type $(n, M, d)_q$ equality holds in Singleton bound if and only if, for every choice of $n - d + 1$ coordinates, the codewords assume on them all possible values.

**A8.8.**\* Prove formula (8.6) on page 415.

**A8.9.** Let $C$ be a code of type $(n, M, d)_q$. Prove that $n \geq d$ and that if $q = 2$ and $d = n$ then $M = 2$.

In the following Exercises A8.10–A8.13 we sketch the proof of Proposition 8.4.3.

**A8.10.** Verify that, for a binary code, Singleton bound is better than Hamming bound if and only if $\sum_{i=1}^{k} \binom{n}{i} < 2^{d-1}$, where $k = \lfloor (d-1)/2 \rfloor$.

**A8.11.**\* Verify that, if $d$ is odd, we have $\sum_{i=1}^{k} \binom{n}{i} = 2^{d-1}$.

**A8.12.** Verify that, if $d$ is even, the inequality $\sum_{i=1}^{k} \binom{n}{i} < 2^{d-1}$ holds: (i) if $d = 2$, for all $n$; (ii) if $d = 4$, only for $n = 4, 5, 6$; (iii) if $d = 6$, only for $n = 6, 7$; (iv) if $d = 8, 10$, only for $n = d$.

**A8.13.**\* Complete the proof of part (iv) of Proposition 8.4.3.

**A8.14.** Prove that the Hamming $(7, 4)$ code is a perfect code.

**A8.15.** Prove that the ordered pairs of distinct elements of a set consisting of $M$ elements are exactly $M(M - 1)$. Hint: recall Exercise A1.17.

**A8.16.**\* Prove Cauchy–Schwarz inequality: given $\mathbf{x} = (x_1, \ldots, x_n)$ and $\mathbf{y} = (y_1, \ldots, y_n)$ in $\mathbb{R}^n$, we have

$$(\mathbf{x} \times \mathbf{x}) \cdot (\mathbf{y} \times \mathbf{y}) \geq (\mathbf{x} \times \mathbf{y})^2, \qquad (8.24)$$

where $\mathbf{x} \times \mathbf{y}$ denotes the usual *euclidean scalar product*

$$\mathbf{x} \times \mathbf{y} = \sum_{i=1}^{n} x_1 y_1 + \cdots + x_n y_n.$$

Moreover, in (8.24) equality holds if and only if $\mathbf{x}$ and $\mathbf{y}$ are linearly dependent.

**A8.17.** Prove that the limit superior [respectively, the limit inferior] of a sequence is $+\infty$ [resp., $-\infty$] if and only if the sequence is not bounded above [resp., below].

**A8.18.** Prove that the limit superior [respectively, the limit inferior] of a sequence is $-\infty$ [resp., $+\infty$] if and only if the sequence has limit $-\infty$ [resp., $+\infty$].

**A8.19.**\* Prove that, if $\{x_n\}_{n \in \mathbb{N}}$ is bounded above [resp., below] and has not infinite limit, then its limit superior $\xi$ [resp., inferior $\eta$] exists and is finite and is characterised by the property that for all real number $\epsilon > 0$ there is an $M \in \mathbb{N}$ such that for all $n > M$ we have $x_n \leq \xi + \epsilon$ [resp., $\eta - \epsilon \leq x_n$]. Deduce that a sequence has limit if and only if the limits inferior and superior coincide.

**A8.20.**\* Prove *Stirling formula*: $\log n! = n \log n - n + \mathcal{O}(\log n)$.

**A8.21.**\* Prove Lemma 8.4.7.

**A8.22.**\* Prove the asymptotic Gilbert–Varshamov bound (see Theorem 8.4.8).

**A8.23.**\* Prove the asymptotic Plotkin bound of Theorem 8.4.9.

**A8.24.** Prove formula (8.11) on page 420.

**A8.25.** Prove Proposition 8.5.3 on page 420.

**A8.26.*** Prove Proposition 8.5.4 on page 421.

**A8.27.** Prove that if $V$ is a vector subspace of $\mathbb{F}_q^n$, then $V^\perp$ is too. Moreover, if $V$ has dimension $k$, then $V^\perp$ has dimension $n - k$. Finally, prve that $(V^\perp)^\perp = V$.

**A8.28.** Prove formula (8.13) on page 422.

**A8.29.** Prove the claim in Example 8.5.6 on page 423.

**A8.30.** Prove that an $(n-k) \times n$ matrix $\mathbf{H}$ of rank $n-k$ defines, by formula (8.13), a code $C$ of type $[n, k]_q$.

**A8.31.** Prove that it is not necessary for the rows of a parity check matrix to be linearly independent.

**A8.32.** Prove that, acting on the rows of a generating matrix [resp., of a parity check matrix] of a code $C$ by *elementary operations*, which consist in: (1) exchanging two rows; (2) multiplying a row by a non-zero scalar; (3) add to a row a multiple of another row; we get again a generating matrix [resp., a parity check matrix] for the same code.

**A8.33.** Prove that a $[n, k]_q$-code $C$ has $q^{n-k}$ cosets (see page 423), and each coset $\mathbf{e} + C$ of $C$ contains exactly $q^k$ elements.

**A8.34.** Consider a $[n, k]_q$-code $C$ with parity check matrix $\mathbf{H}$. Prove that the vectors in the coset $\mathbf{e} + C$ are all the vectors having the same syndrome $\mathbf{e} \cdot \mathbf{H}^t$. Show by an example that the syndrome may change if the parity check matrix $\mathbf{H}$ is modified.

**A8.35.** Prove that, if $f$ is a permutation in $\mathfrak{S}_n$, the mapping $\omega_f$ defined on page 425 is linear and bijective. Let $\mathrm{GL}(n, q)$ be the group of isomorphisms of $\mathbb{F}_q^n$ in itself. Prove that the mapping $f \in \mathfrak{S}_n \to \omega_f \in \mathrm{GL}(n, q)$ is an injective group homomorphism.

**A8.36.** Two codes $C$, $C'$ of length $n$ over the same alphabet $\mathbb{F}_q$ are said to be *equivalent* if there is a permutation $p$ of $\mathbb{F}_q$ and a permutation $f \in \mathfrak{S}_n$ having symbol $(i_1, \ldots, i_n)$ such that the mapping

$$\omega_{p,f} : (x_1, \ldots, x_n) \in \mathbb{F}_q^n \to (p(x_{i_1}), \ldots, p(x_{i_n})) \in \mathbb{F}_q^n$$

tranforms $C$ into $C'$. If, in the above situation, $p$ is the identity permutation, $C$ and $C'$ are said to be *equivalent by positional permutation*; if instead $f$ is the identity permutation $C$ and $C'$ are said to be *equivalent by symbol permutation*. Prove that two equivalent codes have the same size.

**A8.37.** Two codes $C$ and $C'$ over the same alphabet $\mathbb{F}_q$ are said to be *distance equivalent* if there is a bijective mapping $f : C \to C'$ that is an *isometry* for the Hamming distance, that is to say, for all $x, x' \in C$ one has $d(x, x') = d(f(x), f(x'))$. Prove that two equivalent codes are distance equivalent.

**A8.38.** Let $C$ be a code of size $M$. Order the elements of $C$ into an $M$-tuple $(x_1, \ldots, x_M)$. Let $\mathbf{D}(C)$ be the $M \times M$ matrix whose entry in position $(i, j)$ is $d(x_i, x_j)$. Verify that $\mathbf{D}(C)$ is a symmetric matrix whose entries on the principal diagonal are zero. Prove next that two codes $C$, $C'$ are distance equivalent if and only if there are orderings of the elements such that the matrices $\mathbf{D}(C)$ and $\mathbf{D}(C')$ are equal.

**A8.39.** Let $C$ be a code of size $M$. Order the elements of $C$ into an $M$-tuple $(x_1, \ldots, x_M)$. Let $\mathbf{d}(C)$ be the vector of length $M$ whose $i$th component is $w(x_i)$. Describe the connection between $\mathbf{d}(C)$ and the matrix $\mathbf{D}(C)$. Prove that two linear codes $C, C'$ are distance equivalent if and only if there are orderings of the elements such that the vectors $\mathbf{d}(C)$ and $\mathbf{d}(C')$ are equal.

**A8.40.*** Give an example of codes that are distance equivalent but not equivalent.

**A8.41.** Two codes $C, C'$ of length $n$ over the same alphabet $\mathbb{F}_q$ are said to be *linearly equivalent* if there is a vector space isomorphism $f$ of $\mathbb{F}_q^n$ such that $f(C) = C'$. Show that two codes that are equivalent by positional permutation are linearly equivalent.

**A8.42.** Give an example of codes that are equivalent but not linearly equivalent.

**A8.43.** Give an example of codes that are linearly equivalent but not distance equivalent.

**A8.44.** Verify the claim made in Example 8.6.2.

**A8.45.** Prove Proposition 8.6.3.

**A8.46.** Let $b(x)$ be the check polynomial of a cyclic code $C$ of type $[n, k]_q$. Prove that $C$ consists of all polynomials $c(x)$ in $\mathbb{F}_q[x]/(x^n - 1)$ such that $c(x)b(x) = 0$ modulo $x^n - 1$.

**A8.47.** Verify that, by renaming coordinates, the parity check matrices corresponding to cases (3) and (4) of Example 8.6.6 coincide with the parity check matrix of the Hamming code of Example 8.5.7.

**A8.48.** Verify that Goppa codes (see Definition 8.7.2) are linear.

**A8.49.** Verify identity (8.17) on page 431.

**A8.50.** Verify identity (8.20) on page 432.

**A8.51.** Prove Proposition 8.7.6 on page 433, by showing that a Goppa code $C(L, f(x))_{q,m}$ with $f(x)$ of degree $t$ has a parity check matrix in which every $t$-tuple of columns is linearly independent.

**A8.52.** Prove formula (8.22) on page 434.

# B8 Computational exercises

**B8.1.** Is $\mathbf{c} = (0, 1, 1, 1, 1, 1, 1)$ a word of the Hamming $(7, 4)$ code? If it is not, in which position did the error occur, assuming that no more than one transmission error has occurred?

(a) $\mathbf{c}$ is a word of the Hamming code.
(b) The first coordinate of $\mathbf{c}$ is wrong.
(c) The third coordinate of $\mathbf{c}$ is wrong.
(d) The seventh coordinate of $\mathbf{c}$ is wrong.

**B8.2.** Is $c = (0, 1, 1, 0, 1, 1, 0)$ a word of the Hamming $(7, 4)$ code? If it is not, in which position did the error occur, assuming that no more than one transmission error has occurred?

(a) $c$ is a word of the Hamming code.
(b) The first coordinate of $c$ is wrong.
(c) The third coordinate of $c$ is wrong.
(d) The seventh coordinate of $c$ is wrong.

**B8.3.** Is $c = (0, 1, 0, 0, 0, 1, 0)$ a word of the Hamming $(7, 4)$ code? If it is not, in which position did the error occur, assuming that no more than one transmission error has occurred?

(a) $c$ is a word of the Hamming code.
(b) The first coordinate of $c$ is wrong.
(c) The third coordinate of $c$ is wrong.
(d) The seventh coordinate of $c$ is wrong.

**B8.4.** Compute the Hamming distance of $(0, 0, 0, 0, 0, 0, 0)$ from $(0, 1, 1, 0, 1, 1, 0)$.

(a) 2.
(b) 3.
(c) 4.
(d) 5.

**B8.5.** Compute the Hamming distance of $(1, 0, 1, 0, 0, 1, 0)$ from $(0, 0, 1, 1, 0, 1, 1)$.

(a) 3.
(b) 4.
(c) 5.
(d) 7.

**B8.6.** Is $\{(0, 0, 1), (0, 0, 0), (0, 1, 0), (0, 1, 1)\}$ a vector subspace of $\mathbb{F}_2^3$? If so, find its dimension.

(a) No, it is not a subspace.
(b) Yes, it is a subspace of dimension 1.
(c) Yes, it is a subspace of dimension 2.
(d) Yes, it is a subspace of dimension 3.

**B8.7.** Is $\{(1, 0, 1, 0, 1, 0), (0, 0, 0, 0, 0, 0), (0, 1, 0, 0, 0, 1), (1, 1, 1, 1, 1, 1)\}$ a vector subspace of $\mathbb{F}_2^6$? If so, find its dimension.

(a) No, it is not a subspace.
(b) Yes, it is a subspace of dimension 1.
(c) Yes, it is a subspace of dimension 2.
(d) Yes, it is a subspace of dimension 3.

**B8.8.** Compute the dimension and the minimum distance of the linear code $\{(0, 0, 0, 0, 0), (1, 1, 1, 1, 0), (1, 0, 0, 0, 1), (0, 1, 1, 1, 1)\} \subset \mathbb{F}_2^5$.

(a) The dimension is 2 and the minimum distance is 2.
(b) The dimension is 2 and the minimum distance is 3.

(c) The dimension is 3 and the minimum distance is 2.
(d) The dimension is 3 and the minimum distance is 3.

**B8.9.** Is the linear code of the previous exercise maximum-distance separable? Does it detect or correct errors?

(a) It is not maximum-distance separable and detects and corrects one error.
(b) It is not maximum-distance separable and detects and corrects two errors.
(c) It is not maximum-distance separable and detects, but does not correct, one error.
(d) It is maximum-distance separable and detects and corrects one error.

**B8.10.** Determine the dimension of the vector subspace of $\mathbb{F}_2^4$ generated by the set $\{(1, 1, 0, 0), (1, 0, 1, 0), (0, 0, 0, 0), (1, 0, 0, 1), (0, 1, 0, 1)\}$.

(a) 1.
(b) 2.
(c) 3.
(d) 4.

**B8.11.** Determine the dimensione of the vector subspace of $\mathbb{F}_3^4$ generated by the set $\{(1, 0, 1, 1), (0, 1, 1, 2), (1, 1, 2, 0)\}$.

(a) 1.
(b) 2.
(c) 3.
(d) 4.

**B8.12.** Compute the minimum distance of the linear code given by the vector subspace of the previous exercise, determine if it is maximum-distance separable and if it detects or corrects errors.

(a) The minimum distance is 2, the code is not maximum-distance separable and does neither detect nor correct errors.
(b) The minimum distance is 2, the code is maximum-distance separable and detects and corrects one error.
(c) The minimum distance is 3, the code is not maximum-distance separable and detects, but does not correct, one error.
(d) The minimum distance is 3, the code is maximum-distance separable, detects two errors, and corrects one error.

**B8.13.** Compute $V_q(n, r)$ for $q = 2$.

**B8.14.** Compute $V_7(9, 4)$.

**B8.15.** Consider a binary code defined by the parity check matrix

$$\begin{pmatrix} 1 & 0 & 0 & 1 & 0 & 0 & 0 \\ 0 & 0 & 1 & 0 & 1 & 0 & 1 \\ 0 & 0 & 1 & 0 & 0 & 1 & 1 \\ 0 & 1 & 0 & 1 & 0 & 0 & 1 \end{pmatrix}.$$

Which is the minimum distance of this code?

(a) 1.
(b) 2.
(c) 3.
(d) 4.

**B8.16.** Is the code of the previous exercise perfect?

**B8.17.** Consider a code over $\mathbb{F}_7$ defined by the parity check matrix

$$\begin{pmatrix} 2 & 1 & 0 & 6 & 0 & 0 & 1 \\ 0 & 1 & 3 & 0 & 2 & 3 & 4 \\ 4 & 0 & 4 & 6 & 5 & 3 & 0 \end{pmatrix}.$$

Which is the minimum distance of this code?

(a) 1.
(b) 2.
(c) 3.
(d) 4.

**B8.18.** Consider the binary code defined by the parity check matrix

$$\begin{pmatrix} 1 & 0 & 0 & 0 & 1 & 0 & 1 & 1 \\ 0 & 1 & 0 & 0 & 0 & 1 & 1 & 1 \\ 0 & 0 & 1 & 0 & 1 & 1 & 0 & 1 \\ 0 & 0 & 0 & 1 & 1 & 1 & 1 & 0 \end{pmatrix}.$$

Which is the minimum distance of this code?

(a) 1.
(b) 2.
(c) 3.
(d) 4.

**B8.19.** Determine the dual code of the code defined in the previous exercise.

**B8.20.** Determine all cosets and syndromes of the code of Exercise B8.17.

**B8.21.** Determine the dual of the code defined in Exercise B8.18.

**B8.22.** Determine all cosets and syndromes of the code of Exercise B8.18.

**B8.23.** Determine a generating matrix in standard form of the code defined in Exercise B8.17.

**B8.24.** Determine a generating matrix in standard form of the code defined in Exercise B8.18.

**B8.25.** Compute the information rate of the code defined in Exercise B8.12 and of its dual.

**B8.26.** Compute the separation ratio of the code defined in Exercise B8.12 and of its dual.

**B8.27.** Verify all calculations and claims in Example 8.5.8 on page 424.

**B8.28.** Prove that the cyclic codes of Example 8.6.6 of page 427, for cases (3) and (4) are equivalent by positional permutation to the Hamming code.

**B8.29.** Determine all binary cyclic codes of length smaller or equal than 6.

**B8.30.** Determine the cyclic codes over $\mathbb{Z}_3$ of length 5.

**B8.31.** Determine the generating polynomial and the dimension of the smallest cyclic code over $\mathbb{Z}_3$ that contains the word $(1, 2, 0, 0, 2)$.

**B8.32.** Determine the generating polynomial and the dimension of the smallest cyclic code over $\mathbb{Z}_5$ that contains the word $(4, 2, 0, 3, 1, 0, 2)$.

**B8.33.** Determine the BCH code with assigned distance $\delta = 3$ over $\mathbb{F}_2$ with parameters $l = 1$ and $n = 3$.

**B8.34.** Determine the Goppa code over $\mathbb{F}_2$ with $f(x) = x^2 + x + 1$ and $L = \mathbb{F}_2$.

**B8.35.** Determine the Goppa code over $\mathbb{F}_7$ with $f(x) = x^2 + 1$ and $L = \mathbb{F}_7$.

# C8 Programming exercises

**C8.1.** Write a program that takes in input an arbitrary sequence of binary digits and outputs the $N$-time repetition of each digit, where $N$ is a fixed digit.

**C8.2.** Write a program that implements the Hamming $(7, 4)$ code, that is, given in input 4-tuples of binary digits, it outputs 7-tuples according to formula (8.2) on page 410.

**C8.3.** Write a program that checks if a 7-tuple of binary digits is a word of the Hamming code. If it is not, it corrects the 7-tuple by applying the maximum likelihood principle, assuming a single transmission error has occurred.

**C8.4.** Write a program that measures the distance between two words of the Hamming code.

**C8.5.** Write a program that carries out the syndrome decoding.

**C8.6.** Write a program that determines all binary cyclic codes of length $n$, for $n \leq 10$.

**C8.7.** Write a program that determines all Goppa codes over $\mathbb{Z}_p$, with $p$ a prime smaller than 20, and $L = \mathbb{Z}_p$, with $f(x)$ a polynomial of degree at most 4.

# 9

# The future is already here: quantum cryptography

In this last chapter we shall have a look at the new frontier of cryptography, the *quantum cryptography*, which relies on ideas originating in *quantum mechanics*. On the one hand, quantum cryptography envisions the creation of a computer completely different and unprecedented with respect to classical computers, the so-called *quantum computer*. At present, it is only conceived as a theoretical possibility; but if it were actually developed, it would be able to perform in polynomial time some of the computations that a classical computer performs in exponential time. As we know, this would make all present cryptosystems, such as $RSA$, vulnerable, seriously jeopardising civil, military, financial security systems. The result could lead to the collapse of our very civilisation, which largely relies on such systems. On the other hand, the same ideas on which the notion of a quantum computer is based lead to new, completely unbreakable *quantum cryptosystems*, impervious even to an hypothetic quantum computer, and having further the surprising feature of discovering if an eavesdropper has attempted, even unsuccesfully, to intrude a private communication.

As we said, quantum cryptography is based on fundamental ideas of quantum mechanics. This is a relatively recent branch of physics, developed during the last century, which builds on the observation that the mostly *deterministic* laws of physics, which explain macroscopic phenomena cannot be succesfully applied to microscopic ones. Physicists had to devise completely new ways of looking at phenomena regarding the *extremely small scale*, sometimes apparently weird ones, opposite to everyday intuition and often controversial. Moreover, very elegant mathematical models have been constructed in order to deal with these phenomena. All this is crucial in quantum cryptography, but we shall be able to mention it only fleetingly. The interested reader can delve further into this subject by browsing the works listed in bibliography. This chapter, even more than the previous one, is meant to give general information rather than going into details, about a non-classical part of cryptography which is certain to have a major role in the future. A future that has already begun because, as we shall see, the theory, if not the practice, of quantum computers is already quite developed, and there are already working quantum cryptosystems, although only on a small scale.

## 9.1 A first foray into the quantum world: Young's experiment

Recall that, in classical mechanics, the position of a point or of a particle is described by the vector $\mathbf{x}$ of its coordinates in a given spatial reference system. The vector $\mathbf{x} = \mathbf{x}(t)$ is a function of time $t$ and satisfies a system of differential equations, the *equations of motion*. The classical physical problem consists in determining the motion of a point, that is to say, in computing the function $\mathbf{x}(t)$ when its value $\mathbf{x}_0$ and possibly some of its derivatives at *initial time* $t = t_0$ are known. The answer is found by solving the system of differential equations describing the motion of the particle with the given initial conditions. As is known from mathematical analysis, under suitable conditions, usually verified for ordinary physical systems, the equations of motion have exactly one solution verifying the initial conditions. This means that the whole motion of the particle, and in particular its position and its velocity in every instant, are completely determined by the information we have about the particle in a given instant. This is the reason why classical mechanics is said to be *deterministic*, because everything within its scope is uniquely determined by initial conditions.

As we shall see, this model, although valid for macroscopic systems, fails for *very small* systems, such as elementary particles, which are ruled by *probabilistic*, rather than deterministic, laws. This is the scope of quantum mechanics.

We begin our brief travel through the world of quantum mechanics from afar, that is, from an experiment carried out by an English scientist who lived in Cambridge at the end of 18th century, Thomas Young. Recall that at that time there was a heated debate among physicists about whether light had the nature of a particle or of a wave. The scientists supporting the former theory claimed that light consisted of particles, called *photons*, which, travelling through space, hit objects and, to put it roughly, light them up. The supporters of the wave theory contended instead that light is, like sound, carried by waves that somehow propagate in space. Modern quantum physics settled the question by proving both theories right: it is true that light consists of single particles, photons, but they also have a wavelike behaviour. Our perceiving light as a undulatory or as a particle phenomenon depends on the circumstances. This seeming paradox is part of an unavoidable ambivalence, called *wave–particle duality*, which is a particular instance of a general principle, known as *Heisenberg uncertainty principle* which is, in turn, the cornerstone of quantum mechanics, marking its radical difference from the determinism of classical mechanics. Heisenberg uncertainty principle, put forward in the 1920's, states, in a qualitative form, that *there are pairs of observable properties of a microscopic physical system, such as position and velocity, or energy and time, that cannot be both determined or measured exactly at the same time.* In other words, measuring one of the two conjugated properties irreparably modifies the other. And this modification, and the consequent impossibility of simultaneously measuring two conjugated properties, does not depend on the limits of our ways of measuring, but is an objective impossibility that has a mathematical proof, and we shall come back to it later (see Section 9.4).

Let us turn back our attention to Young, who was far from dreaming up the uncertainty principle. With his experiment he succeeded instead in giving a convincing evidence to the wavelike nature of light. As apparently happens to English physicists – remember Newton's apple – he got his idea while relaxing and enjoying nature. Unlike Newton, Young was not under an apple tree, but on the shore of a lake. He saw two swans swimming on the surface of the lake, parallel to each other. Young noticed that the two swans left behind them two semicircles of waves, which interfered creating on the calm surface of the lake a peculiar pattern. This was due to the fact that when two wave crests met, a crest higher than each of the two appeared; when two wave troughs met, a trough lower than each of the two starting troughs formed, and when a crest and a trough met they cancelled out. All this has nothing special; it is a scene we have likely witnessed several times. What made the episode so interesting for Young was that he recalled in that moment having seen *exactly the same pattern* formed by the waves left by the swans, during on optical experiment. The experiment was as follows (see figure 9.1).

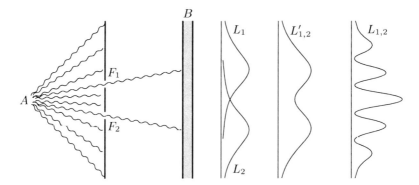

**Fig. 9.1.** Young's experiment

Assume we have a light source $A$ in front of a plate with two narrow slits $F_1$ and $F_2$; behind the plate there is a screen $B$. The distribution of light on the screen, that is, from a particle viewpoint, the distribution of photons absorbed by the screen $B$, is described by a surface whose curvilinear section is given by curves $L_1$, $L_2$ and $L_{1,2}$ is shown in figure 9.1. The curve $L_1$ describes the distribution in the case in which the slit $F_1$ is open and $F_2$ is closed, while the curve $L_2$ describes the distribution in the case in which the slit $F_2$ is open and $F_1$ is closed. According to classical mechanics and considering light as a particle phenomenon, that is, having in mind the propagation of photons, Young would have expected that, when both slits are open, the distribution of photons would have followed the the curve $L'_{1,2}$, that is, the *sum* of $L_1$ and $L_2$. The experiment showed instead that this is not true: the distribution was described by the curve $L_{1,2}$ and the pattern on the screen was exactly that formed by the waves left by the swans on the lake!

This, Young thought reasonably enough, was the *final proof* of the wavelike nature of light: the light rays passing through the slits $F_1$ and $F_2$ behave exactly like the waves left by the swans on the lake and this is why the image on the screen is analogous to the pattern formed by the waves.

Quantum physics seems not to be involved so far: in fact, we are talking about 1799. Nevertheless, there is a *modern* way of repeating the experiment using the most recent technology, to obtain truly surprising results. Modern technology, indeed, allows us to emit from the light source $A$ *single photons*, one each second, say, at a fixed speed but with randomly variable directions within a sector $D$, as shown in figure 9.1. Now each photon travels alone toward the plate, some passing through slit $F_1$, some other through slit $F_2$, some more not passing through any of the two slits, everything happening in a completely random way. Our eye cannot of course *see* a single photon, but there are photon detectors which can be put on the screen $B$. Let the experiment proceed for some hours, until as many photons passed as with Young's original experiment, which used a substantial light source, such as a candle or a lightbulb. Which image do we expect on the screen? A moment's thought would lead us to reason like this: the image we saw in Young's original experiment was caused by undulatory *interaction* of several photons among them, while here, where photons travel *independently* of one another, there is no reason why they should interfere, so we do *not* expect to see on the screen the same image as before. If we trust classical mechanics, we rather expect to see on the screen two bright zones, which are the projections on $B$ of the slits $F_1$ and $F_2$. On the contrary – lo and behold! – the result of this second experiment *is absolutely analogous* to that of Young's original experiment. This, as we have seen, is completely unexplainable within classical mechanics: indeed, it is a *quantum phenomenon*, whose deep explanation is at the heart of this modern branch of physics.

The explanation of this phenomenon given by quantum mechanics lies in the so-called *superposition of states*. Alternative interpretations, based for instance on the idea of *parallel uninverses* or *multiverses*, or of *hidden variables*, have not yet been experimentally verified. We shall now give an intuitive exposition of it. The description of the mathematical machinery necessary for modelling it shall be given later (see Section 9.4).

Something is certain: each photon starts from $A$ and, if it passes through one of the two slits $F_1$ or $F_2$, then it reaches the screen $B$. What happens in the interval between the photon's departure from $A$ and its arrival on $B$ is for us a mystery which appears not to be ruled by deterministic laws, but by probabilistic ones. Basically, the photon, if it passes through one of the slits, has the same probability of passing through $F_1$ or $F_2$ and, strange as it may seem, we may adopt the viewpoint that it passes through both $F_1$ and $F_2$, so interacting with itself and determining the undulatory effect which emerges in Young's experiment. Each of the two possibilities, the passage through $F_1$ or $F_2$, is called a *state* of the photon, and as we are supposing that in the intermediate phase of the travel from $A$ to $B$, when we do not carry out any observation, the two states somehow take place simultaneously, this explains why this phenomenon is called *superposition of states*. This viewpoint, unusual as it may seem, can be formalised mathematically – in a while we shall briefly outline how – and leads to an explanation of the result of Young's experiment. In conclusion the superposition of states is a way of describing an object during a period of ambiguity when no observations or measures are performed.

Clearly, when we perform an observation or a measure to find out the actual state of the photon, the ambiguity stops and consequently the super-position of states ceases. But this, in accordance with the uncertainty prin-ciple, irreversibly *modifies* the system itself. So we expect that, if we modify Young's experiment in such a way that it is possible to ascertain, for each photon, the slit it passes through, this very act modifies the result of the ex-periment. And, however contrary to intuition, this is what happens. In this second case, indeed, the distribution of the photons absorbed by the screen $B$ is no longer ruled by the curve $L_{1,2}$, but by the curve $L'_{1,2}$: it is as if, by observing and measuring the trajectory of the photons, we treated them as particles having motions ruled by classical mechanics, obtaining exactly the result classical mechanics would have predicted!

## 9.2 Quantum computers

We are now going to discuss how Young's experiment and its odd explanation may lead to the invention of a *quantum computer* able to perform computa-tions in such a way to reduce the time necessary for some algorithms from exponential to polynomial.

The idea is due to an English physicist, David Deutsch, who introduced this notion in 1984 (see [17], [18]). Deutsch first remarked that usual computers operated essentialy by using the laws of classical physics, while it would have been desirable to have computers operating in accordance with the laws of quantum physics, as they can lead to a quicker carrying out of operations. Let us see why.

If a classical computer is to examine a problem requiring a number of attempts, it has to proceed *serially*. Consider, for instance, the factorisation of a number $n$. Basically, the computer may proceed by using the sieve of Eratosthenes, so it divides the number first by 2, then by 3, and so on, up to $[\sqrt{n}]$, if necessary. And this seriality is responsible for the exponential time needed by the computation. On the contrary, if we had at our disposal a quantum computer, we might envision using the superposition of states to avoid specifying serially the numbers from 2 to $[\sqrt{n}]$. Basically, just as the photon in Young's experiment, if not observed, is in a superposition of states when passing from $A$ to $B$, and thus we may regard it as simultaneously passing through both $F_1$ and $F_2$, in the same way we may imagine the existence of a computer in which, by exploiting the same principle, while a computation is carried out and it is not interfered with by external factors, the input may assume simultaneously a whole set of numerical values, somewhat as if it were a variable rather than a number. So the quantum computer, which will work, rather than with simple *bits*, with *quantum bits* or *qubits*, might be able to factor the number $n$ proceeding not serially, but performing *a single division*. Of course, this reduces the computing time for the sieve of Eratosthenes from exponential to polynomial.

This, which might look like pure fantasy, is actually not such, at least from a theoretical viewpoint. Indeed, it suffices to use the properties of elementary particles and their quantum mechanics to represent numbers and operate on them. The basic idea is as follows. Many elementary particles have an observable property, or simply an *observable*, called *spin*, which is, broadly speaking, analogous to the angular momentum of a macroscopic ball rotating around its axis. It is important to emphasise the words *broadly speaking*, warning that this analogy is not to be taken too literally. In any case, some particles called *fermions* have half-integer spin and some of these, like *electrons*, have spin with absolute value $1/2$. The spin, that is, this *intrinsic angular momentum*, if computed with respect to a particular directed axis, which we shall call $z$-axis, may assume the values $S_z = 1/2$ and $S_z = -1/2$, corresponding intuitively to a *clockwise* or a *counterclockwise rotation*. So consider such a particle $p$ and decide that it represents 0 if it has clockwise spin and 1 if it has counterclockwise spin. Take $h$ particles of this kind, $p_1, \ldots, p_h$, and put them in an equal number of boxes numbered from 1 to $h$, one each, not communicating with each other, so as not to interfere. Thus, proceeding as usual we may represent all numbers with $h$ binary digit $(m_1 \cdots m_h)_2$, that is to say, all numbers from 0 to $2^h - 1$. Actually, in order to represent different numbers we have to be able to give the particles different spin components $S_z$ along the $z$-axis. This can be done by subjecting the particle to an energy pulse: if the pulse is high enough, the particle changes its spin, else it keeps the spin it had. But we want to use the superposition of states, to let the particle have simultaneously, in a quantum sense, different components $S_z$ of the spin. To obtain this, we enclose each particle in its box, so as not to observe it, and subject each particle to an energy pulse of random strength. So each particle enters a superposition of states and represents *simultaneously* 0 or 1. This is a qubit. Now, a number consisting of $h$ qubits is a *quantum* number which may be any number from 0 to $2^h - 1$. If we learn how to operate on these numbers, we have solved, at least theoretically, the problem of constructing a quantum computer.

This is Deutsch's contribution, which left open an important theoretical problem and innumerable, huge practical ones. The theoretical problem amounted to conceiving algorithms that actually worked, at least in principle, on such a thing as a quantum computer. This problem has been solved in 1994 by Peter Shor, who found polynomial algorithms for a quantum computer to factor large integers and to determine discrete logarithms (see [54]).

The union of Deutsch's and Shor's results would seem to be disastrous for cryptography: are all cryptographic systems that, such as $RSA$, rely on the difficulty of factoring large numbers or on the Diffie–Hellman hypothesis (see Section 7.7, Section 7.7.6), unreliable? Can they be easily eluded by using a quantum computer? The answer is "yes", but, at present, only in theory. Indeed, the practical issues related with the actual construction of quantum computers seem currently difficult to overcome. In particular, for a quantum computer to be able to operate, it must be completely isolated from the exterior, an isolation impossible to attain with today knowledge. This problem, the so-called *quantum decoherence*, like other problems related to quantum computing, is among those most actively investigated by physicists and might one day be solved. However, it is not easy to estimate when this will happen. Certainly, the day a quantum computer will become a reality, all cryptographic systems we have been discussing so far will be easily circumvented, and this shall put in serious danger our very safety, which relies on them. However, as we shall briefly see, the same ideas that lead to quantum computers also lead to unassailable techniques of quantum cryptography. We shall discuss this subject in the next sections.

## 9.3 Vernam's cipher

Before discussing quantum cryptography, let us get momentarily back to the classical case, and give the reader some good news. There are ciphers that are theoretically completely unassailable, and we shall deal with them in this section. So, even if one day quantum computers will be made a reality, we will be able to resort to these ciphers, and trust them with our safety. Why then, the reader will ask, did not we mention them till now, taking pains to discuss classical cryptography, $RSA$, elliptic curves and so on? The answer will be given in due course: we shall explain first how these ciphers work, and then we shall answer the above natural and crucial question. Finally we shall see, in the next sections, how quantum cryptography may yield a method for reasonably and effectively using these ciphers.

The cipher we are talking about is the so-called *Vernam's cipher*, which bears the name of Gilbert S. Vernam, an employee of the American Telephone & Telegraph Company who, together with Joseph O. Mauborgne, a Major General in the United States Army, proposed it during World War I. In his fundamental text [53], already cited in the chapter about codes, C. E. Shannon proved that Vernam's ciphers are secure against cryptanalysis and, further, that every cipher that is secure against cryptanalysis is a Vernam's cipher. In other words, there exists a unique perfectly safe cryptographic system, and it is the Vernam's cipher.

Let us describe it. This cipher is also called *one time pad*, because the enciphering key used to be written on the sheets of a notepad to be made known to the users (the sender and receiver) and should not be used more than *one time*.

The system is very easy. The sender and the receiver have the same key, which has to satisfy the following properties:

(i) it must be the same length as the message to be sent;
(ii) it must be a completely random sequence of characters;
(iii) it must never be used more than once.

Under these hypotheses it is not necessary to use an intricate enciphering function: we may use a simple one, say, addition or subtraction. For instance, if we choose the binary alphabet $\{0, 1\}$ consisting of the two digits 0 and 1, we may define a Vernam's cipher as follows:

**Definition 9.3.1.** *Let $m = m_1 m_2 \ldots m_r$ be a binary message to be sent. Let $K = k_1 k_2 \ldots k_r$ be the key, which is a binary string of the same length of the message, consisting of random digits. The enciphering is carried out by substituting the message $m$ with the message $c = c_1 c_2 \ldots c_r$, where*

$$c_i = m_i + k_i, \qquad i = 1, \ldots r.$$

*The key must not be used again. This cipher is called a Vernam's cipher.*

Clearly, we may do the same if we use any other alphabet, as shown in the following example.

**Example 9.3.2.** Suppose the message to be sent is the word $HOME$. We choose as a key the sequence of random letters $DXTG$, having the same length as the message. Associating, as usual, with each letter a number in $\{0, 1, \ldots, 25\}$ (see Table 7.3 on page 324) we may encipher the message by summing the word $DXTG$ to the word $HOME$, where the sum is taken modulo 26:

$$
\begin{aligned}
H + D &\longrightarrow 7 + 3 = 10 & &\longrightarrow K, \\
O + X &\longrightarrow 14 + 23 = 37 \equiv 11 \ (\mathrm{mod}\,26) & &\longrightarrow L, \\
M + T &\longrightarrow 12 + 19 = 31 \equiv 5 \ (\mathrm{mod}\,26) & &\longrightarrow F, \\
E + G &\longrightarrow 4 + 6 = 10 & &\longrightarrow K.
\end{aligned}
$$

So the enciphered message is $KLFK$. To reconstruct the original message, the receiver will only have to subtract the sequence $DXTG$ from the message he received.

Let us understand why Vernam's ciphers are unbreakable. The reason lies in the randomness of the character constituting the key. Indeed, if a person who does not possess the key wanted to decipher the message, he could in principle try out each possible key. This operation has an enormous computational complexity, as the number of keys increases exponentially as a function of the length $r$ of the message (see Exercise A9.1). However, this is not the reason why the cipher is unbreakable. In fact, we saw that, with a possible future appearance of quantum computers, computing limitations like this might not be relevant anymore. The true reason for this cipher's unbreakability is the fact that, by the arbitrariness of the key, one would obtain, in the course of the analysis, all possible plaintexts of length $r$. Moreover, by the randomness of the key, all these plaintexts would be equally probable. Now, most of them would certainly be meaningless, but all meaningful ones of length $r$ would be possible. In other words, if a 4-letter words is sent, as in Example 9.3.2, the cryptanalysis we just outlined – the only possible one – gives as a result that the plaintexts $HOME$, $AWAY$, $NICE$, $UGLY$, $BOOK$, and so on, are all equally probable!

**Example 9.3.3.** If we try to carry out the cryptanalysis of the message $KLFK$ sent in Example 9.3.2, we have to try out all 4-letter keys. Among them, we shall meet $IDFW$, which leads to the following deciphering:

$$
\begin{aligned}
K - I &\longrightarrow 10 - 8 = 2 & &\longrightarrow C, \\
L - D &\longrightarrow 11 - 3 = 8 & &\longrightarrow I, \\
F - F &\longrightarrow 5 - 5 = 0 & &\longrightarrow A, \\
K - W &\longrightarrow 10 - 22 = -12 \equiv 14 \,(\mathrm{mod}26) &&\longrightarrow O.
\end{aligned}
$$

So we obtain the word $CIAO$.

However, it is crucial that the key is never used twice. Indeed, if the sender made the fatal mistake of using again the same key, the cipher would become open to attempts of cryptanalysis. To see why, rather than expound it theoretically, let us consider an example.

**Example 9.3.4.** Consider again Example 9.3.2. Suppose the sender uses again the key $DXTG$ to encipher the word $TOWN$. As is readily seen, the result is the word $WLPT$. Suppose further the cryptanalyst knows that the letter $O$ appears in both words in the same place. By comparing the words $WLPT$ and $KLFK$, he deduces that the letter $O$ corresponds to the letter $L$, and so that the key has $X$ in second position. It is not a lot of information, but it is better than nothing. If the sender keeps using the same key, the cryptanalyst, in an analogous way, sooner or later will discover it.

This example shows how frequency analysis may help cryptanalysts when the sender makes the ill-omened mistake of using several times the same key.

As already mentioned, there must be some weak point in this cipher, in itself unbreakable. Otherwise, it would be used universally, yielding security and satisfaction. Actually, there are weak points, and quite serious ones.

The first, but not the main one, lies in how to generate keys. They must be long enough to allow exchanging complex messages, randomly generated, and, as they must not be used twice, lots of them are needed, to be able to communicate frequently. Now, generating random numbers is not at all a trivial problem in computer science, and even less so generating long strings of random numbers. But, as we have said, this is not the main problem of this cipher, so we do not dwell further on this issue, important as it may be.

The main problem lies in the fact that, in order to be able to communicate securely using a Vernam's cipher, it is necessary to send in advance the key, through a channel that must be absolutely secure. In other words, before being able to communicate secretly, we must communicate secretly the key. But the key has the same length as the message to be sent, so we have a vicious circle.

In conclusion, the only ciphers that are theoretically secure, Vernam's ciphers, are very difficult to use in practice. In fact, they have been used very rarely: for instance for communicating between the White House and the Kremlin, where the keys were transferred by hand, in the presence of witnesses, in conditions of maximum security. Of course, in general it is not possible to operate this way, so other methods are used, as $RSA$, which, even if not theoretically completely secure, ensure a reasonable level of security.

Nonetheless, if we were able to produce random, sufficiently long keys and we could send them in a secure way, that is to say, with the certainty that no third party eavesdropped and intercepted the key, we migh use, securely and without qualms, Vernam's ciphers. This is made possible by quantum cryptography, as we shall see briefly.

## 9.4 A short glossary of quantum mechanics

In this section we collect, without any claim of completeness, some basics of the mathematical machinery necessary to discuss quantum mechanics. We shall need all of this when, in next section, we shall talk about quantum cryptography.

First of all, the mathematical setting is a real vector space $\mathcal{H}$ endowed with a scalar product

$$\langle \ , \ \rangle : \mathcal{H} \times \mathcal{H} \to \mathbb{R}$$

with the property of being positive-definite, that is to say, such that $\langle u, u \rangle > 0$ for every non-zero vector $u \in \mathcal{H}$. As usual, we also define the *norm* of a vector $u \in \mathcal{H}$ as

$$||u|| = \langle u, u \rangle$$

and the *length* of $u$ as

$$|u| = \sqrt{||u||} = \sqrt{\langle u, u \rangle}.$$

The vectors of length 1 are called *unit vectors* or *versors*. Two vectors $u, v$ are said to be *orthogonal* if $\langle u, v \rangle = 0$.

Sometimes, in addition to $\mathcal{H}$, we also consider the *complexification* of $\mathcal{H}$, which with a slight abuse of notation we shall sometimes again denote by $\mathcal{H}$ rather than, as usual, by $\mathcal{H}_{\mathbb{C}}$. This is a complex vector space of the same dimension as $\mathcal{H}$, containing $\mathcal{H}$ as a real subspace: we have $\mathcal{H}_{\mathbb{C}} = \mathcal{H} \oplus i\mathcal{H}$. Moreover, $\mathcal{H}_{\mathbb{C}}$ is endowed with a positive-definite hermitian scalar product extending the scalar product on $\mathcal{H}$ and denoted again by $\langle\ ,\ \rangle$. It is uniquely determined by these conditions and is called *complexification* of the scalar product in $\mathcal{H}$ (see [13]). In this chapter we shall mostly need the real case. Moreover, in what follows, it will suffice to fix our attention to the case in which $\mathcal{H}$ has dimension 2.

The vector space $\mathcal{H}$ is called the *state space* and each of its non-zero vectors is said to be a *state vector*. Usually a state vector is assigned up to a multiplicative constant $t$, that is, we consider two non-zero vectors $u$ and $tu$, with $t \neq 0$, as equivalent. So, rather than in $\mathcal{H}$, we might work in the associate projective space $\mathbb{P}(\mathcal{H})$, which is a projective line. Alternatively, and we shall take this approach, we may represent an equivalence class of state vectors by a versor. This leaves an indeterminacy we may live with: namely, two versors $u, v$ are equivalent if and only if $u = tv$, with $|t| = 1$.

To begin browsing our glossary, let us see a physical situation where this mathematical model may be applied. Recall that the smallest unit or *quantum* of light is the photon. As we saw, in quantum physics it may be regarded as a particle, but it has a wavelike behaviour too. In a very simplified way, we may describe this wavelike behaviour as follows: the photon may be regarded as a tiny electromagnetic field that propagates describing a sinusoid along an axis given by a line $d$, as shown in Figure 9.2. Of course, the sinusoid lies in a plane $\pi$ containing the line $d$. The direction of the lines of $\pi$ that are orthogonal to $d$ is the direction along which the photon oscillates, and is called *polarisation* of the photon.

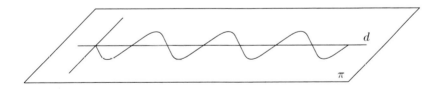

**Fig. 9.2.** Polarisation of the photon

So, the way of representing mathematically this undulatory behaviour of the photon is to consider it not as a point, but rather as a vector whose direction is the polarisation of the photon. This explains the mathematical setting we introduced: namely, we may consider our vector space $\mathcal{H}$ as the set of polarisation states of photons. The reason why we may assume $\mathcal{H}$ of dimension 2 is clear: indeed, if the photons, leaving from a light source, propagate along direction $d$ in the three-dimensional space, then the polarisations of the photons emitted by the light source are space vectors lying in the plane orthogonal to $d$.

It is well known that in $\mathcal{H}$ there are *orthonormal bases*, that is, ordered pairs $(u, v)$ of mutually orthogonal versors. Such pairs of vectors form a basis of $\mathcal{H}$ and allow us to identify $\mathcal{H}$ with $\mathbb{R}^2$ [with $\mathbb{C}^2$ in the complex case], by identifying the vector $xu + yv$ with the ordered pair $(x, y)$. Under this identification, the scalar product $\langle \ , \ \rangle$ is identified with the euclidean scalar product $(x_1, y_1) \times (x_2, y_2) = x_1 x_2 + y_1 y_2$ [with the *standard hermitian product* $(x_1, y_1) \times (x_2, y_2) = x_1 \overline{x_2} + y_1 \overline{y_2}$ in the complex case] (see Exercises A9.2–A9.3).

Of course, there are infinitely many orthonormal bases of $\mathcal{H}$. However, suppose we have chosen one, by which we identify $\mathcal{H}$ with $\mathbb{R}^2$ as we just explained: in other words, we *introduced a coordinate system* on $\mathcal{H}$. This orthonormal reference system is usually denoted by $(\uparrow, \rightarrow)$, and its states correspond to *vertically* and *horizontally* polarised photons. We shall call this reference frame *rectilinear*. Another natural orthonormal reference frame is the *diagonal* reference $(\nearrow, \searrow)$, where $\nearrow$ and $\searrow$ correspond to the states of the photons polarised at 45 and $-45$ degrees with respect to the rectilinear reference, respectively. The connection between the two references is given by the following relations:

$$\begin{pmatrix} \nearrow \\ \searrow \end{pmatrix} = \frac{1}{\sqrt{2}} \begin{pmatrix} 1 & 1 \\ 1 & -1 \end{pmatrix} \begin{pmatrix} \uparrow \\ \rightarrow \end{pmatrix}, \qquad \begin{pmatrix} \uparrow \\ \rightarrow \end{pmatrix} = \frac{1}{\sqrt{2}} \begin{pmatrix} 1 & 1 \\ 1 & -1 \end{pmatrix} \begin{pmatrix} \nearrow \\ \searrow \end{pmatrix}.$$

It is necessary now to formalise mathematically the notion of observation or measure of a photon. In our mathematical setting consisting of the vector space $\mathcal{H}$, we shall call *observable* any linear mapping

$$A : \mathcal{H} \to \mathcal{H}$$

that is *symmetric* [*hermitian* in the complex case], namely, such that for every pair of vectors $u, v \in \mathcal{H}$ we have

$$\langle A(u), v \rangle = \langle u, A(v) \rangle.$$

As is well known, given a reference frame, and so having identified $\mathcal{H}$ with $\mathbb{R}^2$ [with $\mathbb{C}^2$ in the complex case], every mapping $A$ as above is identified with a square $2 \times 2$ matrix over $\mathbb{R}$ [over $\mathbb{C}$, resp.]. If the reference is orthonormal, the matrix is symmetric [hermitian in the complex case].

It is further well known that a linear mapping $A : \mathcal{H} \to \mathcal{H}$ that is symmetric [hermitian, resp.] is *orthogonally diagonalisable*, that is to say, there is an orthonormal basis $(v_1, v_2)$ consisting of eigenvectors of $A$. This is the content of the so-called spectral theorem (see [13], pag. 428). In such a basis, the matrix of $A$ becomes diagonal

$$\begin{pmatrix} \lambda_1 & 0 \\ 0 & \lambda_2 \end{pmatrix}, \tag{9.1}$$

where $\lambda_1, \lambda_2$ are the eigenvalues of $A$. They are the possible outcomes of a *measurement* of the observable $A$. So the observable $A$ has measure $\lambda_i$ only if applied to the eigenvector $v_i$, $i = 1, 2$, or to a multiple of it.

In general, give a system state $v \in \mathcal{H}$, which may be assumed to be a versor, we have

$$v = \langle v, v_1 \rangle v_1 + \langle v, v_2 \rangle v_2 \qquad (9.2)$$

and so

$$A(v) = \lambda_1 \langle v, v_1 \rangle v_1 + \lambda_2 \langle v, v_2 \rangle v_2. \qquad (9.3)$$

(see Exercise A9.2). By Cauchy–Schwarz inequality (see Exercise A9.5 and Exercise A9.6), we have $0 < |\langle v, v_i \rangle| \leq 1$ and $|\langle v, v_i \rangle| = 1$ if and only if $v$ is proportional to $v_i$, $i = 1, 2$. As (see Exercise A9.4)

$$1 = ||v|| = |\langle v, v_1 \rangle|^2 + |\langle v, v_2 \rangle|^2,$$

the larger $|\langle v, v_1 \rangle|$, the smaller $|\langle v, v_2 \rangle|$, and so the closer $v$ is to $v_1$. We get an analogous result if we exchange $v_1$ and $v_2$.

So it is natural to define $|\langle v, v_i \rangle|^2$ as the *probability* $P(A, v, \lambda_i)$ for $A$ to have measure $\lambda_i$ in state $v$, $i = 1, 2$. In particular, if this value equals one, that is if the probability is 1, then the measure of $A$ in state $v$ is exactly the eigenvalue $\lambda_i$. Moreover, as can be expected,

$$P(A, v, \lambda_1) + P(A, v, \lambda_2) = |\langle v, v_1 \rangle|^2 + |\langle v, v_2 \rangle|^2 = 1.$$

Notice further that, by (9.2) and (9.3), we have

$$\langle A(v), v \rangle = \lambda_1 P(A, v, \lambda_1) + \lambda_2 P(A, v, \lambda_2).$$

This quantity is denoted by the symbol $E(A)(v)$ and is called *mean* or *expected value* of the measure of the observable $A$ in state $v$. We may also consider the non-negative quantity $\Delta A(v)$ such that

$$\Delta A(v)^2 = E\left((A - E(A)(v) \cdot I)^2\right)(v) = \langle (A - E(A)(v) \cdot I)^2(v), v \rangle =$$
$$= \langle A - E(A)(v) \cdot v, A - E(A)(v) \cdot v \rangle.$$

Notice that, if $v$ is a versor,

$$\Delta A(v)^2 = \langle A(v), A(v) \rangle - E(A)(v)^2 = \langle A^2(v), v \rangle - E(A)(v)^2 =$$
$$= E(A^2)(v) - E(A)(v)^2.$$

The number $\Delta A(v)^2$ [$\Delta A(v)$, resp.] is called the *variance* [*standard deviation*, resp.] of the measure of the observable $A$ in state $v$. These numbers quantify how the measure of the observable $A$ in state $v$ deviates from its expected value. This terminology is consistent with the usual one in probability theory (see [29]).

Let us return to our glossary and have a look at how this mathematical language applies to specific physical questions.

So, consider again light and photons which, as we have seen, are represented as state vectors keeping trace of their polarisation. But the light we see usually consists of a huge number of photons with widely different polarisations, so usually we do not perceive at all the polarisation phenomenon. Unless, as we know from experience, we wear polarising glasses. Let us explain their effect. There are transparent materials having a unidirectional crystalline structure. If light passes through these materials, the photons that are polarised in the same direction as the crystals making up the material get through undisturbed, those having orthogonal polarisation are absorbed, that is, their passage is stopped, and the remaining ones have a probability of getting through but, if they pass, they emerge all with the same polarisation. In conclusion, if such a filter is located between the light source and our eye, we only perceive light that is *polarised* in the direction allowed by the filter. It is clear that, by rotating the filter, we may let through light with different polarisations. For instance, if a filter only lets through vertically polarised light, rotating it by 45 degrees it will let through diagonally polarised light, rotating it by 90 degrees it will let through horizontally polarised light and so on.

Let us see now the mathematical formulation of this phenomenon. In order to do so, let us fix our attention on the observable $A$, which in the rectilinear reference frame has matrix

$$\begin{pmatrix} 1 & 0 \\ 0 & 0 \end{pmatrix}.$$

Which effect do we obtain by applying $A$ to an arbitrary state given by a versor $v$? Clearly, if $v = \uparrow$ then $A(v) = v = \uparrow$, if instead $v = \rightarrow$ then $A(v) = 0$. In general, if

$$v = (x, y) = x \cdot \uparrow + y \cdot \rightarrow,$$

we have

$$A(v) = \langle v, \uparrow \rangle \cdot \uparrow = x \cdot \uparrow .$$

So the effect of $A$ is that of a vertical polarising filter, which we shall call filter of type $\uparrow$. The photons with vertical polarisation get through undisturbed, those with horizontal polarisation do not get through, and those with polarisation $v = (x, y)$ have, according to our mathematical description, a probability equal to $x^2$ of getting through, with vertical polarisation.

For instance, the photon $\nearrow = (\uparrow + \rightarrow)/\sqrt{2}$ has probability $1/2$ of passing, and the same happens for the photon $\searrow$. In other words, recalling a concept introduced in Section 9.1, the photons $\nearrow$ and $\searrow$ are in a situation of *superposition of states* with respect to the observable $A$. Having photon $\nearrow$ pass through a polariser of type $\uparrow$, we subject it to an observation: we *observe* whether the light goes out vertically polarised or does not go out. The effect of this observation is, in the words of P. A. M. Dirac, one of the fathers of quantum mechanics, of *forcing the photon entirely in the state of vertical or horizontal polarisation. It will have to jump abruptly from the condition of partially belonging to each of these states to exclusively belonging to one of them.*

*We cannot foresee which of the two states it will jump to: the phenomenon is governed by probabilistic laws* (see also [20]).

Of course, the observable

$$\begin{pmatrix} 0 & 0 \\ 0 & 1 \end{pmatrix}$$

acts as a horizontal polarising filter, that is, of type →, while the observables

$$\frac{1}{\sqrt{2}} \begin{pmatrix} 1 & 1 \\ 1 & 1 \end{pmatrix}, \quad \frac{1}{\sqrt{2}} \begin{pmatrix} 1 & -1 \\ -1 & 1 \end{pmatrix}$$

act as polarising filters of type ╱ and ╲, respectively.

In general, the observable (9.1), with $\lambda_1 \neq \lambda_2$ both non zero, acts as a polarising filter that lets through the photons having rectilinear polarisation, preserving the polarisation. We shall call this a polarising filter of *rectilinear type*, or of type +. The photons having polarisation ╱ and ╲ always have probability $1/2$ of passing with polarisation ↑ or →. Of course there are analogous filters of *diagonal type* or of type ×, which let through the photons having diagonal polarisation preserving their polarisation, while the photons having rectilinear polarisation subjected to them have probability $1/2$ of passing with polarisation ╱ or ╲ (see Exercise A9.11).

We conclude this section giving a sketch of the mathematical formulation of Heisenberg uncertainty principle.

First of all, some remarks about the product of two observables are in order. Given two observables $A$ and $B$, we define their *commutator* $[B, A]$ as the observable such that

$$[B, A](v) := B(A(v)) - A(B(v))$$

for all $v \in \mathcal{H}_{\mathbb{C}}$. It is useful to remark that, for each pair of observables $A, B$, we have

$$\langle [B, A](v), v \rangle = \langle A(v), B(v) \rangle - \langle B(v), A(v) \rangle =$$
$$= \langle A(v), B(v) \rangle - \overline{\langle A(v), B(v) \rangle} = 2\,\mathrm{Im}(\langle A(v), B(v) \rangle).$$

$$(9.4)$$

Clearly, we have $[B, A] = 0$ if and only if $A$ and $B$ *commute*, that is to say, if and only if $A \cdot B = B \cdot A$, where the product is the composition of mappings. Notice further that if $A$ and $B$ are observables, it is not always the case that $A \cdot B$ is an observable, but it is if $A$ and $B$ commute (see Exercises A9.8 and A9.9). Moreover, $A$ and $B$ have a common orthonormal basis consisting of eigenvectors if and only if $A$ and $B$ commute (see Exercise A9.10). Physically, this means that two observables commute if and only if *they may be measured simultaneously*, that is, if and only if there are states $v$ on which, for both observables, it is possible to compute the measure with probability 1. In this case, the observables are said to be *compatible*. If instead $[B, A] \neq 0$, the observables are *not simultaneously measurable* and are said to be *incompatible*.

After these premises, we may state Heisenberg uncertainty principle in the form given by Dirac:

**Theorem 9.4.1 (Heisenberg uncertainty principle).** *Let $A$ and $B$ be observables and let $v$ be a state determined by a versor in $\mathcal{H}_{\mathbb{C}}$. We have*

$$\Delta A(v)^2 \cdot \Delta B(v)^2 \geq \frac{1}{4}|\langle [B,A](v), v \rangle|^2. \qquad (9.5)$$

PROOF. Notice that it suffices to prove (9.5) for the observables $A' = A - E(A)(v) \cdot I$, and $B' = B - E(B)(v) \cdot I$, where $I$ is identity. Indeed, we have $[B,A] = [B', A']$, $\Delta A(v) = \Delta A'(v)$ and $\Delta B(v) = \Delta B'(v)$. Notice further that $E(A')(v) = E(B')(v) = 0$ and so $\Delta A'(v)^2 = \langle A'(v), A'(v) \rangle$ and $\Delta B'(v)^2 = \langle B'(v), B'(v) \rangle$.

From (9.4) it follows that

$$|\langle [B', A'](v), v \rangle| = 2|\operatorname{Im}(\langle A'(v), B'(v) \rangle)| \leq 2|\langle A'(v), B'(v) \rangle|.$$

We may conclude by applying Cauchy–Schwarz inequality. $\qquad \square$

Notice that here it is actually important to consider everything in $\mathcal{H}_{\mathbb{C}}$ rather than in $\mathcal{H}$. Indeed, if $A$ and $B$ are observables over $\mathcal{H}$, by (9.4) we have $\langle [B,A](v), v \rangle = 0$ for all $v \in \mathcal{H}$, and Heisenberg uncertainty principle becomes trivial. Heisenberg principle becomes relevant for pairs of observables $A$, $B$ such that $\langle [B,A](v), v \rangle = 2\operatorname{Im}(\langle A(v), B(v) \rangle)$ is non-zero for every versor $v$. In this case, $(\langle [B,A](v), v \rangle^2)/4$ is a continuous function that never becomes zero on the compact set $U = \{v \in \mathcal{H}_{\mathbb{C}} : |v| = 1\}$. So it has a minimum $M > 0$ and Equation (9.5) implies that

$$\Delta A(v)^2 \cdot \Delta B(v)^2 \geq M.$$

This relation says that the smaller we make the standard deviation $\Delta A(v)$, the larger $\Delta B(v)$ is forced to become, so: *whenever we try to improve the measurement of one of the two observables, making it arbitrarily precise, we are compelled to increase the inaccuracy with which we measure the other observable.*

## 9.5 Quantum cryptography

The principles of quantum mechanics we have very quickly recalled may be used, as already mentioned, to solve the problem of creating and transmitting random keys, so making it possible to use Vernam's ciphers. The transmission will be, as we shall see, completely secure, that is to say, both the sender and the receiver are able to detect possible third parties trying to eavesdrop in order to get, even partially, information about the key being transmitted. As we shall see, the *uncertainty* of quantum world provides *certainty* about communication security.

The first ideas about quantum cryptography can be found in an important contribution by S. Wiesner (see [62]), which already circulated as a manuscript around 1970, but was published only in 1983, having been rejected by several journals that did not realise its groundbreaking importance. These ideas were then developed by C. H. Bennet and G. Brassard around 1980. They worked out a protocol, that we shall describe here, for a *quantum distribution of keys*. This protocol is known today as the *BB*84 protocol, as the paper [7] in which it was described appeared in 1984 (see also the expository article [9], which contains interesting remarks about technical aspects, which we shall overlook here, concerning the setting up of the machinery for the actual implementation of the protocol).

Here is the gist of the idea of Bennet and Brassard. To discuss it, consider again what we said at the end of section 9.1. In that situation we may suppose that $A$, the photon source, is the sender of a message, call her *Alice*, and the screen $B$ is the receiver of the message, call him *Bob*. As we saw, the message, in the absence of intrusions, consists in an image which is described by the curve $L_{1,2}$. But if there is an eavesdropper, calk her *Eve*, who tries to measure the photons while they pass through the slits $F_1$ and $F_2$, this attempt would modify the image, which would now be described by the curve $L'_{1,2}$ rather than by the curve $L_{1,2}$. So Bob, conversing, say by telephone, with Alice and describing, if not the whole image at least some of its features, would certainly notice the intrusion and could communicate it to Alice.

The idea of quantum cryptography is not very different. Transmitting informations between Alice and Bob takes place through two channels: the first part through a *quantum channel*, for instance an optical fibre transmitting polarised photons, the second one, which consists itself of two parts, through an ordinary, not necessarily secure, channel.

In describing the $BB84$ protocol we shall use the mathematical machinery we introduced in the previous section. In particular, the polarised photons will be vectors in $\mathcal{H}$. We shall assume that a rectilinear orthonormal reference frame $(\uparrow, \rightarrow)$ has been introduced. In addition to photons having state $\uparrow$ and $\rightarrow$, we shall only use those with states $\nearrow$ and $\searrow$. We shall also use filters of type $+$ and $\times$.

**Step 1:** *communicating by transmitting photons through a quantum channel.*

**A**) Alice has at her disposal, for the transmission, four polarised filters which she may use to send through the transmission channel photons having one of the four polarisations $\uparrow$, $\rightarrow$, $\nearrow$ and $\searrow$. Alice and Bob, who receives, fixed beforehand a binary numerical value to be attributed to a polarised photon, as follows:

| Polarised photons | $\uparrow$ | $\rightarrow$ | $\nearrow$ | $\searrow$ |
|---|---|---|---|---|
| Corresponding binary digits | 1 | 0 | 1 | 0 |

Alice chooses a number $r$ far greater than the length of the key she wants to send Bob. Moreover, Alice randomly chooses a string, that is an ordered $r$-tuple of polarisers, and sends Bob the corresponding string of polarised photons, keeping a record of it. So Alice has, in fact, sent Bob a sequence of $r$ binary digits, which represents the number that is Alice's actual message to Bob.

For instance, we might have the following situation:

| Polarisation of photons sent by Alice | ↑ | ↑ | ↗ | ↘ | ↑ | ↘ | ↑ | → |
|---|---|---|---|---|---|---|---|---|
| Message sent by Alice | 1 | 1 | 1 | 0 | 1 | 0 | 1 | 0 |

Of course, if Bob just received the polarised photons and read the number sent by Alice and used it as a key, then no security would be guaranteed, because Eve, the possible eavesdropper, could, just like Bob, read the same number spying on the channel. On the other hand, so far we have not used any quantum phenomenon. Let us see now how Bob may proceed to ensure the secrecy of the key.

**B)** Bob has at his disposal four filters, two of rectilinear type +, that is, ↑ and →, and two of diagonal type ×, that is, ↗ and ↘, which operate as explained in section 9.4. Bob does not know the polarisations of the photons sent him by Alice. So he lets the received photons pass randomly through his filters and *measures* their polarisation. He attributes next to each polarised photon its numerical equivalent and determines a received message.

For instance, we might have:

| Message sent by Alice | 1 | 1 | 1 | 0 | 1 | 0 | 1 | 0 |
|---|---|---|---|---|---|---|---|---|
| Polarisation of the photons sent by Alice | ↑ | ↑ | ↗ | ↘ | ↑ | ↘ | ↑ | → |
| Filters used by Bob | ↑ | ↗ | ↘ | → | ↑ | ↘ | ↗ | ↗ |
| Polarisation of the photons measured by Bob | ↑ | ↘ | ∅ | ↑ | ↑ | ↘ | ↗ | ↘ |
| Message received by Bob | 1 | 0 | 1 | 1 | 1 | 0 | 1 | 0 |

Notice that, for the third photon sent by Alice during the transmission, Bob used a filter that absorbs it and does not let it go through, so in this position no light arrived. We denoted this situation by the symbol ∅. Nevertheless, Bob deduces that the message sent by Alice is 1, as the only polarisation incompatible with the filter he used corresponds to 1.

Notice that, in doing so, Bob might or might not make some *error*. Indeed, if he uses the filter + [the filter ×, resp.] on a photon having rectilinear [diagonal, resp.] polarisation, no error occurs. But if he uses the filter + [the filter ×, resp.] on a photon having diagonal [rectilinear, resp.] polarisation, this modifies the reception of the corresponding photon. As both the polarisation of the photons sent by Alice and the filters used by Bob are randomly chosen, we may expect such errors to occur with probability $1/2$. But these errors do not necessarily imply an error in the final received message. Indeed, for each error caused by using the wrong filter on one of Alice's photons, there is an even chance that the corresponding digit is the right one, although the polarisation of the photon is wrong. For instance, as the previous table shows, if Alice sends a photon with polarisation ↑, as in the steps 2 and 7 of the transmission, and if Bob erroneously uses the filter ×, it may happen with probability $1/2$ that either Bob receives a photon with polarisation ↘, which is what happens in step 2 of the transmission, or he receives a photon with polarisation $\nearrow$, as happens in step 7 of the transmission. In the first case, the corresponding digit will be wrong, in the second one it will be correct. So the probability $P$ of Bob receiving correctly the message is

$$P = \frac{1}{2} \cdot 1 + \frac{1}{2} \cdot \frac{1}{2} = \frac{3}{4} = 75\%.$$

The first summand corresponds to the case, having probability $1/2$, in which Bob chooses the correct filter, while the second one corresponds to the case, having again probability $1/2$, in which the filter is not the correct one. So the probability of an *error* by Bob is $25\%$. This is also the probability of anybody else committing an error, including Eve, if she taps the transmission channel without authorisation and tries to read the message.

The table above demonstrates what we have been saying: Bob's errors in choosing filters occur in steps 2, 4, 7 and 8 of the transmission, that is, in one half of the cases, while there is an error in the received message only in steps 2 and 4 of the transmission, that is, in one quarter of the cases.

Now, the message has been transmitted and, in order to do so, Alice and Bob have used in an essential way quantum properties. However, due to the presence of errors, they cannot stop here. These errors, as we are going to see, can be removed, and their existence is in fact crucial to discover potential attempts of eavesdropping by Eve. So we need further steps in the protocol.

**Step 2:** *communicating through an unprotected channel to remove errors and extract the* raw key.

In this step Alice and Bob, before worrying about potential attempts of eavesdropping by Eve, act to detect and remove the errors made by Bob in receiving the message, that is to say, in choosing filters. If there were no attempts of eavesdropping, the resulting message, with the errors removed, could be used as the key for a Vernam's cipher. But as Alice and Bob cannot be certain that Eve did not interfere, the key they obtained, called a *raw key*, could be both insecure and erroneous. So a further step is needed in order to verify the possible presence of interferences.

Removing the errors committed by Bob is easy. Indeed, now both Alice and Bob know each a part of the information: Alice knows the string of photon polarisations sent to Bob, while Bob knows which filters he chose for each of them. Then by using a possibly insecure channel, like email or telephone, Bob lets Alice know which filters ($+$ or $\times$) were used and Alice tells him which ones were the right ones. After this exchange of information, Alice and Bob delete in both the sent and the received message all digits corresponding to the positions in which Bob used a wrong filter. In the remaining positions Bob did not commit errors, so *if there are no intrusions in the quantum channel,* the remaining part of the message, the raw key, is the same for Alice and Bob. In the example we are studying, which continues the one of the previous tables, Bob communicates to Alice the filter sequence

$$+ \quad \times \quad \times \quad + \quad + \quad \times \quad \times \quad \times$$

and Alice lets him know that Bob only used the right filter in steps 1, 3, 5 and 6 of the transmission. So, deleting from the sent and received messages the digits in positions 2, 4, 7 and 8, which correspond to the wrong filters used by Bob, they deduce the raw key

$$1 \quad 1 \quad 1 \quad 0,$$

which is kept secret. Notice that in exchanging information about the errors committed by Bob, he and Alice do not disclose to Eve, were she to illicitly listen in on the insecure channel, any information about the raw key. Indeed, if Eve did not tamper with the quantum channel, her knowing that in steps 1, 3, 5 and 6 of the transmission Bob used the right filter on the photon sent by Alice does not give her any information about the photons actually sent, so the corresponding digits, which make up the raw key, remain secret.

**Step 3:** *communicating on an unprotected channel to verify Eve's presence.*

Now Alice and Bob must ascertain whether Eve tried to eavesdrop on the quantum channel or not. Recall that if Eve *only* eavesdropped on the unprotected channel over which they communicated in step 2, this has no importance at all. If no intrusion occurred, Alice and Bob may use the raw key, or a part of it, as the key for a Vernam's cipher. If on the other hand an intrusion occurred, Alice and Bob have to throw everything away and start afresh, in the hope that Eve grows tired of eavesdropping, or they should use a more secure quantum channel.

To understand how to proceed in this step, some remarks are in order. Notice first that detecting whether an attempt of eavesdropping occurred is the reason why Alice and Bob use both rectilinearly and diagonally polarised photons. Indeed, if they only used rectilinearly polarised photons, they might not find out an intrusion by Eve, as she could intercept Alice transmission with a 100% accuracy, using for instance a vertical filter, and then mimic Alice by sending again the same data to Bob. Such a strategy on the part of Eve is called *opaque eavesdropping.*

By contrast, as we are going to see, the use by Alice and Bob of two different kinds of polarisation for the photons makes any eavesdropping by Eve plain. Indeed, if Eve listens in on the quantum channel and wants to read Alice's message, she must first measure the polarisation of the transmitted photons. As she is now in the same conditions as Bob, she has no choice but to proceed like him, which forces her, as we know, to committ some error. So, if she later resends Bob the message she read, the message she sends will be different from the one Alice sent, that is to say, it will *contain some errors*, so Alice and Bob, comparing their raw keys, have a good chance of detecting any eavesdropping.

To give a mathematical formulation to what we have been saying, notice that an attempt of eavesdropping by Eve has an immediate effect on the probability of the presence of errors in the message received by Bob, which has been computed in step 1 to be equal to $P = 1/4$.

Assume Eve interferes with probability $s$, with $0 \leq s \leq 1$, that is, $s$ is the proportion of photons Eve tampers with when eavesdropping on Alice and Bob. If $s = 0$, this means that Eve never interferes, while $s = 1$ means that Eve measures every photon Alice sends to Bob. Intermediate values of $s$ mean that Eve sometimes listens on, sometimes does not.

Due to the randomness hypotheses we made, Bob's and Eve's choices when using filters are independent of each other and of the polarisations chosen by Alice when sending photons, so an error in the message received by Bob may occur only if:

- Bob is wrong, which as we know happens with probability $1/4$, and Eve does not interfere, which has a probability equal to $1 - s$;
- Bob is correct, which happens with probability $3/4$, but Eve interferes, with probability $s$, and causes an error in the choice of the filter, which in turn causes an error in the received message with probability $1/2$.

In conclusion, the new error probability $P'$ when Bob receives the message is given by the formula

$$P' = \frac{1}{4}(1 - s) + \frac{3}{4} \cdot \frac{1}{2} s = \frac{1}{4} + \frac{s}{8}.$$

For instance, if $s = 1$, that is, if Eve interferes each time, then Bob's error probability of $1/4$ becomes $3/8$, a substantial and appreciable increase.

Let us understand how this reflects on the probability that there are discrepancies between Alice's and Bob's raw key: recall that, when Eve interferes, these keys could be different. So there is a discrepancy only if Eve interferes, and this happens with probability $s$. In this case Eve commits a mistake in choosing the filter with probability $1/2$, and an error occurs in the message with probability, again, $1/2$. In conclusion, the probability $P_K$ of discrepancy between the two keys is

$$P_K = \frac{1}{2} \cdot \frac{1}{2} s = \frac{s}{4}.$$

This argument leads us to understand how Alice and Bob may act to detect Eve's potential presence. As every discrepancy between Alice's and Bob's raw keys is due to Eve's presence, they will be certain of Eve's eavesdropping even if they find a *single* discrepancy. So they fix, by communicating through the insecure channel, a random $m$-digit subset of their raw keys, which might now be different, and compare these digits, again communicating through the insecure channel. If Eve interfered, they will find about $ms/4$ differences between the two keys. So, if $m$ is very large, it makes sense to foresee that Alice and Bob will sooner or later find a discrepancy, thus becoming certain of Eve's interference.

For instance, if they estimate the probability of Eve eavesdropping around $s \leq 1/1000$, by comparing $20,000$ digits of the raw key they may expect to find about 5 discrepancies.

If, on the other hand, Alice and Bob found no discrepancies in any of the $m$ digits they have chosen, the probability of Eve having interfered without her interference being detected is

$$\left(1 - \frac{s}{4}\right)^m .$$

If $m$ is very large, this probability is very low. For instance, if $s = 1$ and $m = 150$, we have

$$\left(\frac{3}{4}\right)^{150} < 2^{-50},$$

a very small number.

In conclusion, if after this check Alice and Bob are reasonably, even if not completely, confident that Eve did not interfere, then they may use as the key for a Vernam's cipher what remains after deleting from the raw key the $m$ digits used up during the check. If, on the other hand, they find a discrepancy between the two raw keys, so they know that Eve interfered, they must start over. This is not pleasant, but it is better than laying themselves open to Eve's spying.

Before concluding, there are some final remarks about the $BB84$ protocol. What we have been describing here gives only its basic traits, leaving out the technical issues emerging when actually implementing it, which in turn leads to further interesting theoretical problems. Without giving details, it is useful to sketch the most important questions of this kind, referring the reader to the article [9] for further information.

The main problem lies perhaps in the fact that the model we outlined does not consider the potential errors that might occur during the transmission of the photons through the quantum channel, that is errors due to *noise* in the channel. This is, in other word, a *noiseless* model. To remedy the presence of noise, it is possible to use the coding theory techniques discussed in Chapter 8. It is not possible, however, to completely ignore errors due to noise, and their presence seriously jeopardises step 3 of our protocol, as it may induce Alice and Bob to believe that Eve interfered even if she did not, so it might block the communication between them. This problem may be obviated by estimating the measure in which noise affects the error probability in the raw key, and then proceeding as in step 3 to determine whether Eve really interfered or not. The procedure for determining a common key without errors due to noise, after Alice and Bob ascertained that Eve did not eavesdrop, is more complex. We do not describe it, referring instead to [8].

Another important technical problem is as follows: in the protocol outlined above it is vital for the transmission to send *one photon at a time*. This is a very complex operation, which is virtually impossible when transmitting in a vacuum or, even worse, through air. The way to manage it is transmitting through an optical fibre, and this presently limits the practical realisation of this system. Anyway, imagine sending, rather than a single photon at a time, beams of polarised light consisting of many photons. This immediately lay Alice and Bob open to Eve's eavesdropping. Indeed, she could subtract one of the photons from each photon packet, measure it, sending the remaining ones the Bob with no alteration. Clearly, this thwarts the search for Eve's interferences in step 3 of the protocol. Of course, the reader may think, when Bob detects a missing photon among those sent by Alice he realises from this that Eve stepped in. This is true, but only if the number of photons sent by Alice in each ray is low. Otherwise, it is very difficult on the part of Bob to estimate the absence of a single photon from each packet and, even if he were able to do so, he might ascribe the fact to a natural loss of photons due to the transmission medium.

In any case, quantum cryptography, unlike quantum computer which is presently only hypotethical, actually exists. C. Bennet and J. Smolin, in 1988, have brought about a first system to send keys using the $BB84$ protocol. At the time, they could only send messages over a distance of a few centimetres. Presently, by using optical fibres, it is possible to send messages over quite long distances. For instance, in 1995, researchers of the University of Geneva were able to send numerical keys with the $BB84$ protocol over a distance of 23 kilometres. Research is presently carried out about the possibility of sending messages through air or via satellite. So, as we have often seen in history so far, cryptologists are once more at an advantage over cryptanalysts. And this time the gap appears to be unbridgeable!

# Appendix to Chapter 9

## A9 Theoretical exercises

**A9.1.** Prove that the number of random $r$-letter keys out of an alphabet of length $k$ is $k^r$.

**A9.2.** Let $V$ be a real [complex, resp.] vector space of dimension $n$ endowed with a scalar [hermitian, resp.] product. Let $v_1, \ldots, v_n$ be an *orthonormal* system of vectors, that is to say, $\langle v_i, v_j \rangle = 0$ if $1 \leq i < j \leq n$ while $\langle v_i, v_i \rangle = 1$, for all $i = 1, \ldots, n$. Prove that $v_1, \ldots, v_n$ is a basis of $V$ and that for every vector $v \in V$ we have $v = \langle v, v_1 \rangle v_1 + \cdots + \langle v, v_n \rangle v_n$.

**A9.3.** In the same hypotheses as the previous exercise, prove that if $v = a_1 v_1 + \cdots + a_n v_n$ e $w = b_1 v_1 + \cdots + b_n v_n$, then we have

$$\langle v, w \rangle = a_1 b_1 + \cdots + a_n b_n, \quad [\langle v, w \rangle = a_1 \overline{b_1} + \cdots + a_n \overline{b_n}, \text{ resp.}].$$

**A9.4.** In the same hypotheses as the previous exercise, prove that

$$||v|| = \langle v, v \rangle = |\langle v, v_1 \rangle|^2 + \cdots + |\langle v, v_n \rangle|^2$$

and consequently that

$$|v| = \sqrt{||v||} = \sqrt{|\langle v, v_1 \rangle|^2 + \cdots + |\langle v, v_n \rangle|^2}.$$

**A9.5.**\* Extend Cauchy–Schwarz inequality to an arbitrary real vector space $V$ of dimension $n$ endowed with a *positive-definite* scalar product $V \times V \to \mathbb{R}$, that is, such that $\mathbf{v} \times \mathbf{v} \geq 0$ with equality holding only if $\mathbf{v} = 0$.

**A9.6.**\* Extend Cauchy–Schwarz inequality to an arbitrary complex vector space $V$ of dimension $n$ endowed with a *positive-definite* hermitian scalar product (see [13], Ch. 18) $V \times V \to \mathbb{C}$, that is, such that $\mathbf{v} \times \mathbf{v} \geq 0$ with equality holding only if $\mathbf{v} = 0$.

**A9.7.**\* Extend the constructions given in section 9.4 to an arbitrary complex vector space of dimension $n$ endowed with a positive-definite hermitian scalar product.

**A9.8.** Prove that if two observables $A$, $B$ in a real or complex vector space of dimension $n$ commute, then $A \cdot B$ and $B \cdot A$ are observables too.

**A9.9.** Give an example of two noncommuting observables $A$, $B$ whose product is not an observable.

**A9.10.**\* Prove that if $A$ ad $B$ are observables in a real or complex vector space of dimension $n$, they have a common orthonormal basis consisting of eigenvectors if and only they commute.

**A9.11.** Write the equation of an observable acting as a filter of type $\times$.

# B9 Computational exercises

**B9.1.** Mimicking Example 9.3.2, send the message *NOON* using a Vernam's cipher with key *UCDP*.

**B9.2.** Consider the ciphertext obtained in the previous exercise and find the key with which it would be eciphered as *DAWN*.

**B9.3.** Assume that in exercise B9.1 the sender uses again the key *UCDP* to send the message *MOON* and that a cryptanalyst knows that the letter *O* appears twice in both messages. What may the cryptanalyst deduce about the key?

**B9.4.** Mimicking Example 9.3.2, send the message *TOMORROW AT NOON* using a Vernam's cipher with key *PDTKCDPTMFTGTQ*.

**B9.5.** Consider the ciphertext obtained in the previous exercise and find the key with which it would be deciphered as *EAT TWO PORK PIES*.

**B9.6.** Verify that the product of the two observables

$$A = \begin{pmatrix} 1 & 1 \\ 1 & 0 \end{pmatrix}, \quad B = \begin{pmatrix} 0 & 1 \\ 1 & 1 \end{pmatrix}$$

over $\mathbb{C}^2$ is not an observable. Are the two observables compatible?

**B9.7.** Write the commutator $[B, A]$ of the observables of the previous exercise.

**B9.8.** Determine the eigenvalues and the eigenvectors of the observables defined in Exercise B9.6.

# C9 Programming exercises

**C9.1.** Write a program that produces a Vernam's cipher.

**C9.2.** Write a program that verifies whether the product of two observables is an observable.

**C9.3.** Write a program that verifies whether two observables are compatible.

**C9.4.** Write a program that simulates the transmission of a key using the $BB84$ protocol.

# Solution to selected exercises

## Exercises of Chapter 1

**A1.1.** The basis of induction, that is, property 1 of (CI), is the same as for mathematical induction. Property 2 of (CI) is weaker than 2 of mathematical induction, so if (CI) is true, mathematical induction is true a fortiori.

**A1.2.** Assume by contradiction that there is a non-empty subset $T$ of $\mathbb{N}$ that has no least element. Let $A$ be the complement of $T$ in $\mathbb{N}$. Then $0 \in A$, or else $0$ would be the least element in $T$. Moreover, if $n \in A$, then $n + 1 \in A$ as well, or else $n + 1$ would be the least element in $T$. Then, by mathematical induction, $A = \mathbb{N}$, against the hypothesis that $T$ is non-empty.

**A1.3.** Let $A$ be a non-empty subset of $\mathbb{N}$ that satisfies the two properties of (CI), where we assume, for the sake of simplicity, $n_0 = 0$. Assume by contradiction that $A$ is not equal to $\mathbb{N}$. Then the complement $U$ of $A$ is not empty, so it has a least element $m \in U$ by the well-ordering principle, with $m \neq 0$ because $0 \in A$. As $m$ is the least element of $U$, for all $k$ such that $0 \leq k < m$ we have $k \in A$, so from property 2 it follows that $m \in A$, yielding a contradiction.

**A1.5.** The correct answer is (d). It is necessary to define mathematically what we mean by a "small city", or else neither the basis of the induction nor the inductive step make sense.

**A1.6.** The correct answer is (c); indeed, the inductive step does not hold for $n = 49,999$.

**A1.8.** Recall that in the proof of Proposition 1.3.1 we have assumed $b > 0$. Assume further that $a \geq 0$. Consider the subset $S = \{\, n \mid (n + 1)b > a \,\}$ of $\mathbb{N}$. Clearly $S$ is not empty, because for instance $a \in S$. So, by the well-ordering principle, $S$ has a least element, which will be called $q$. Then $(q + 1)b > a$, because $q \in S$, and $qb \leq a$, or else we would have $q - 1 \in S$, contradicting the fact that $q$ is the least element of $S$.

**A1.9.** Let $A$ be the set of the $n$s for which we may compute $a_n$. Then $A$ satisfies the two properties of mathematical induction, so for all $n > n_0$ we have $n \in A$, while $n \in A$ for $n = 1, 2, \ldots, n_0$ by hypothesis.

**A1.12.** The claim is true for $n = 1$. Suppose the number of elements of $\mathfrak{S}_{n-1}$ is $(n-1)!$. Define the mapping $f : \mathfrak{S}_n \to \mathfrak{S}_{n-1}$ that maps the permutation having symbol $(i_1, \ldots, i_n)$ to the permutation having the symbol obtained deleting $n$ from $(i_1, \ldots, i_n)$. Prove that $f$ is surjective and that the preimage of every element of $\mathfrak{S}_{n-1}$ consists of $n$ elements of $\mathfrak{S}_n$. Deduce that the number of elements of $\mathfrak{S}_n$ is $n!$.

**A1.13.** The correct answer is (c). Indeed, a function between sets having the same (finite) size is injective if and only if it is bijective, and this happens if and only if it is surjective. So we prove by induction on $n$ that there are $n!$ bijective functions from $A$ to $B$. The basis of the induction, for $n = 1$, is trivial. Assume that $A$ and $B$ have $n+1$ elements. Fix an element $a \in A$. Then there are $n+1$ possible choices for the image of $A$ under a function $f$. Now, if $f$ is bijective, then $f : A \setminus \{a\} \to B \setminus \{f(a)\}$ is bijective too, and by the inductive hypothesis there are $n!$ such bijective functions. So there are $n! \cdot (n+1) = (n+1)!$ bijective functions from $A$ to $B$.

**A1.14.** The reader might want to prove Formula (1.51) by induction on $n$. We give here a direct proof. Start with the identity

$$\frac{n+1}{k(n-k+1)} = \frac{1}{k} + \frac{1}{n-k+1}.$$

Multiplying both sides by $\dfrac{n!}{(k-1)!(n-k)!}$ we obtain

$$\frac{(n+1)!}{k!(n-k+1)!} = \frac{n!}{k!(n-k)!} + \frac{n!}{(k-1)!(n-k+1)!},$$

that is, Formula (1.51), as wanted.

**A1.15.** The claim is true if $n = 1$. Notice next that the subsets of size $m$ of $I_n$ that include $n$ are as many as the subsets of size $m-1$ of $I_{n-1}$, while the subsets of size $m$ of $I_n$ that do not include $n$ are as many as the subsets of size $m$ of $I_{n-1}$. Apply now the inductive hypothesis and Formula (1.51).

**A1.19.** The correct answer is (c). We have assumed $n \le m$, else, if $n > m$ there would be no injective functions $A \hookrightarrow B$.

**A1.20.** The correct answer is (a), as may be proved by induction, or directly by noticing that for every element $a$ of $A$ there are $m$ possible choices for its image.

**A1.21.** It suffices to remark that every ordered $m$–tuple of elements of $A$ determines a mapping from $I_m$ to $A$; apply then the previous exercise.

**A1.22.** The correct answer is (b). Indeed, given a subset $Y$ of $X$, consider the function $f_Y : X \to I_2$, called *characteristic function of $Y$*, that takes the value 1 on all elements of $Y$ and nowhere else. Prove that the mapping associating with $Y \in \mathcal{P}(X)$ its characteristic function is a bijection of $\mathcal{P}(X)$ in the set of mappings from $X$ to $I_n$. Apply then Exercise A1.20.

**A1.24.** The formula for $s(n, h)$ is true for $n = 1$. As to $s(n, h)$ for $n > 1$, it is the sum of the number of monomials in which $x_1$ appears to the degree $i$, for all $i = 0, \ldots, h$, and this number is $s(n-1, h-i)$. Conclude by applying the inductive hypothesis and (1.52).

**A1.26.** In the hypotheses of the exercise we have

$$y_n + x_n = b_{k-1}y_{n-1} + \cdots + b_0 y_{n-k} + d_n + b_{k-1}x_{n-1} + \cdots + b_0 x_{n-k} =$$
$$= b_{k-1}(y_{n-1} + x_{n-1}) + \cdots + b_0(y_{n-k} + x_{n-k}) + d_n,$$

that is, $\{y_n + x_n\}$ is a solution of (1.4). Moreover, if $z_n$ is another solution of (1.4), then

$$y_n - z_n = b_{k-1}y_{n-1} + \cdots + b_0 y_{n-k} + d_n +$$
$$- b_{k-1}z_{n-1} - \cdots - b_0 z_{n-k} - d_n =$$
$$= b_{k-1}(y_{n-1} - z_{n-1}) + \cdots + b_0(y_{n-k} - z_{n-k}),$$

that is, $\{y_n - z_n\}$ is a solution of (1.53).

**A1.27.** We have

$$a_1 = ba_0 + c, \qquad\qquad a_2 = ba_1 + c$$

which may be interpreted as a linear system in $b$ and $c$. If $a_0 \neq a_1$, this system uniquely determines $b$ and $c$ and so the whole sequence starting from $a_0$. If $a_0 = a_1$, also $a_2 = a_1 = a_0$, else the system would not be compatible, which is not possible. Prove next by induction that the sequence $\{a_n\}$ is constant.

**A1.28.** The basis of the induction is obvious by the definition of $A$ and of Fibonacci numbers $f_n$. Suppose Proposition 1.2.3 is true for $n-1$ and prove it for $n$. We have

$$A^n = A^{n-1} \cdot A = \quad \text{(by the inductive hypothesis)}$$
$$= \begin{pmatrix} f_{n-2} & f_{n-1} \\ f_{n-1} & f_n \end{pmatrix} \cdot \begin{pmatrix} 0 & 1 \\ 1 & 1 \end{pmatrix} = \begin{pmatrix} f_{n-1} & f_{n-2} + f_{n-1} \\ f_n & f_{n-1} + f_n \end{pmatrix} = \begin{pmatrix} f_{n-1} & f_n \\ f_n & f_{n+1} \end{pmatrix},$$

where the last equality follows from Definition (1.5) of Fibonacci sequence.

**A1.30.** Fix $k$ and proceed by induction on $n$. For $n = 1$, Formula (1.54) becomes $f_{k+1} = f_k + f_{k-1}$, which is true. So, assume the truth of the formula for all $0 \leq m < n$ and prove it for $n$. By the inductive hypothesis we have

$$f_{n-1+k} = f_k f_n + f_{k-1} f_{n-1} \qquad \text{and} \qquad f_{n-2+k} = f_k f_{n-1} + f_{k-1} f_{n-2};$$

hence, summing, we get

$$f_{n+k} = f_{n-1+k} + f_{n-2+k} = f_k(f_n + f_{n-1}) + f_{k-1}(f_{n-1} + f_{n-2}) =$$
$$- f_k f_{n+1} + f_{k-1} f_n \cdot$$

We want to prove now that $f_{kn}$ is a multiple of $f_n$. Proceed by induction on $k$. For $k = 1$ this is obvious. Assume that $f_{mn}$ is a multiple of $f_n$ for all $m \leq k$ and prove it for $k + 1$. The previous relation implies

$$f_{(k+1)n} = f_{kn+n} = f_{kn} f_{n+1} + f_{kn-1} f_n,$$

so, by the inductive hypothesis, both $f_n$ and $f_{kn}$ are multiples of $f_n$, so $f_{(k+1)n}$ is too.

**A1.31.** Proceed by induction on $n$, the result being trivial for $n = 1$. Assume the result to be true for every integer smaller than $n$ and prove it for $n$. If $n$ is a Fibonacci number, the result is true. If $f_k < n < f_{k+1}$, then $0 < n - f_k < f_{k+1} - f_k = f_{k-1}$. By induction, $n - f_k$ is a sum of distinct Fibonacci numbers, $n - f_k = f_{k_1} + f_{k_2} + \cdots + f_{k_r}$. So, $n = f_k + f_{k_1} + f_{k_2} + \cdots + f_{k_r}$ is a sum of distinct Fibonacci numbers.

**A1.33.** For $n = 1$, Formula (1.55) is true. Define $\lambda = (1 + \sqrt{5})/2$, assume Formula (1.55) to be true for all $m < n$ and prove it for $n$. We have $f_n = f_{n-1} + f_{n-2} \geq \lambda^{n-3} + \lambda^{n-4} = \lambda^{n-4}(\lambda + 1) = \lambda^{n-4}\lambda^2 = \lambda^{n-2}$.

**A1.34.** The result is obtained by summing the following relations:

$$f_1 = f_2,$$
$$f_3 = f_4 - f_2,$$
$$f_5 = f_6 - f_4,$$
$$\vdots$$
$$f_{2n-1} = f_{2n} - f_{2(n-1)}.$$

**A1.36.** We know that there are $q', r'$ such that $a = q'b + r'$, with $0 \leq r' < |b|$. If $r' \leq |b|/2$, set $q = q'$, $r = r'$. If not, set $q = q' + |b|/b$ and $r = r' - |b|$ and verify that $a = qb + r$. As we are assuming $r' > |b|/2$, we have $0 > r > |b|/2 - |b| = -|b|/2$. The task of verifying the uniqueness is left to the reader.

**A1.37.** Let $d = \mathrm{GCD}(b, c)$. Then $d \mid b$ and $d \mid c$, so $d \mid a$. But then $d$, which divides both $a$ and $b$, must be invertible.

**A1.38.** Using Bézout's identity, verify that $b$ divides 1 and is so invertible.

**A1.39.** The subset $S$ is not empty (why?), so it has a least element $d$ by the well-ordering principle.

**A1.41.** Let $(\bar{x}, \bar{y})$ be an integer solution of (1.18). Let $(x_0, y_0)$ be a solution of $ax + by = 0$, that is, a pair such that $ax_0 + by_0 = 0$. Then $(\bar{x} + x_0, \bar{y} + y_0)$ is a solution of (1.18). Indeed,

$$a(\bar{x} + x_0) + b(\bar{y} + y_0) = a\bar{x} + b\bar{y} + ax_0 + by_0 = c + 0 = c.$$

Vice versa, let $(\bar{x}, \bar{y})$ and $(x', y')$ be two solutions of (1.18). Then

$$a(\bar{x} - x') + b(\bar{y} - y') = a\bar{x} + b\bar{y} - ax' - by' = c - c = 0,$$

so $(x', y')$ differs from $(\bar{x}, \bar{y})$ by a solution of the associate homogeneous equation $ax + by = 0$.

**A1.42.** Notice that $q$ represents the largest integer such that $q \cdot 365$ is not greater than $a$, so $q$ is the integer number 338. To determine $r$, it suffices to observe that $r = a - bq$, so

$$r = 123456 - 338 \cdot 365 = 86.$$

**A1.44.** If both $d$ and $d'$ are greatest common divisors of $a$ and $b$, we have $d \mid d'$ and $d' \mid d$. So $d$ and $d'$ are associate. On the other hand, if $d = \mathrm{GCD}(a, b)$ and if $d'$ is associate to $d$, then $d' \mid a$ and $d' \mid b$ and moreover $d' \mid d$, which implies that $d' = \mathrm{GCD}(a, b)$.

**A1.46.** Notice that $(x) = (y)$ if and only if $x \mid y$ and $y \mid x$.

**A1.51.** It suffices to prove that it has no zero-divisors. Let $p(x) = \sum_{i=0}^{n} a_i x^i$ and $q(x) = \sum_{j=0}^{m} b_j x^j$ be two *non-zero* polynomials (so they have at least one non-zero coefficient). Assume that $\partial p(x) = n$ and $\partial q(x) = m$; this means that $a_n \neq 0$ and $b_m \neq 0$. From the definition of product polynomial $p(x)q(x)$, it follows that the coefficient of $x^{m+n}$ is $a_n b_m$, which is different from zero because $a_n$ and $b_m$ are different from zero and are elements of an integral domain, in which there are no zero-divisors. So $p(x)q(x)$ cannot be the zero polynomial.

**A1.54.** By the factor theorem, $g(x) = (f(c) - f(x))/(f(c) \cdot (x - c))$ is a polynomial, as $f(c) - f(x)$ is divisible by $x - c$. Verify that $g(x)$ has degree less than $t$ and that $(x-c) \cdot g(x) - 1$ is divisible by $f(x)$. This proves the existence. As for the uniqueness, notice that, if $h(x)$ is another polynomial of degree less than $t$ such that $(x-c) \cdot h(x)$ is divisible by $f(x)$, then $(x - c) \cdot (g(x) - h(x))$ is divisible by $f(x)$ too. As $f(x)$ is relatively prime with $x - c$, $f(x)$ should divide $g(x) - h(x)$. The polynomials $g(x)$ and $h(x)$ having both degree less than $t$, the same holds for $g(x) - h(x)$, so the only possibility is $g(x) - h(x) = 0$. This proves the uniqueness.

**A1.55.** Proceed by induction on $n$. If $n = 0$, then $f(x)$ is a non-zero constant, which has no roots, so the thesis is trivial and the basis of induction is proved. Suppose then the thesis is true for every polynomial of degree less than $n$. Let $f(x)$ be a polynomial of degree $n$. If $f(x)$ has no roots, the thesis is trivial. If $f(x)$ has a root $\alpha$, then by the factor theorem we have $f(x) = (x - \alpha)q(x)$, where $q(x)$ has degree $n - 1$. Moreover, the set of roots of $f(x)$ consists exactly of $\alpha$ and of the roots of $q(x)$, which by the inductive hypothesis are at most $n - 1$, so $f(x)$ has at most $n$ roots.

**A1.56.** Proceed by induction on $n$. The basis of induction is obvious, because a linear polynomial has exactly a root. Let $n > 1$. If we denote by $\alpha_1 \in \mathbb{C}$ a root of $f(x)$ (which is sure to exist, by the Fundamental theorem of algebra) the factor theorem implies that

$$f(x) = (x - \alpha_1)q(x)$$

where $q(x)$ still has coefficients in $\mathbb{C}$ and $\partial q(x) = n - 1$. So, by the inductive hypothesis, $q(x)$ has exactly $n - 1$ roots $\alpha_2, \ldots, \alpha_n$ and by the factor theorem we have

$$q(x) = a(x - \alpha_2) \cdots (x - \alpha_n),$$

so $f(x)$ has exactly $n$ solutions and

$$f(x) = a(x - \alpha_1)(x - \alpha_2) \cdots (x - \alpha_n).$$

**A1.57.** Let $f(x) = \sum_{k=0}^{n} a_k x^k$. Then

$$0 = f(\alpha) = \sum_{k=0}^{n} a_k \alpha^k;$$

hence, by conjugating both sides, and observing that real numbers are self-conjugate,

$$0 = \bar{0} = \overline{f(\alpha)} = \overline{\sum_{k=0,n} a_k \alpha^k} = \sum_{k=0}^{n} a_k \bar{\alpha}^k = f(\bar{\alpha}),$$

that is, $\bar{\alpha}$ is also a root of $f(x)$.

**A1.59.** Keeping in mind the linearity of the derivative, it suffices to observe that Leibniz's law holds for monomials.

**A1.64.** Assume by contradiction that there is a $d > 1$ that divides $f_n$ and $f_{n+1}$. Then it also divides $f_{n-1} = f_{n+1} - f_n$. Going backwards, we shall find that $d$ divide $f_2 = 1$, which is impossible.

**A1.65.** Prove first that, if $m = nq + r$, then $\mathrm{GCD}(f_m, f_n) = \mathrm{GCD}(f_n, f_r)$. We have the following chain of equalities:

$$\mathrm{GCD}(f_m, f_n) = \mathrm{GCD}(f_{nq+r}, f_n) = \mathrm{GCD}(f_r f_{nq-1} + f_{r+1} f_{nq}, f_n),$$

where the last equality follows from (1.54). Now, $f_{nq}$ is a multiple of $f_n$, so

$$\mathrm{GCD}(f_r f_{nq-1} + f_{r+1} f_{nq}, f_n) = \mathrm{GCD}(f_r f_{nq-1}, f_n).$$

If we prove that $\mathrm{GCD}(f_{nq-1}, f_n) = 1$, we may conclude that $\mathrm{GCD}(f_r f_{nq-1}, f_n) = \mathrm{GCD}(f_r, f_n)$, which was what had to be proved. Let $\mathrm{GCD}(f_{nq-1}, f_n) = d$: then $d \mid f_n$ (so, also, $f_{qn}$) and $f_{qn-1}$. As it divides two consecutive Fibonacci numbers, it must be $d = 1$.

Having this result, the fact that $\mathrm{GCD}(f_n, f_m) = f_d$, with $d = \mathrm{GCD}(n, m)$, is immediate. Indeed, applying the Euclidean algorithm starting with $m$ and $n$, and denoting by $r_t$ the last non-zero remainder (which is so $\mathrm{GCD}(m, n)$), we find

$$\mathrm{GCD}(f_m, f_n) = \mathrm{GCD}(f_{r_1}, f_n) = \ldots = \mathrm{GCD}(f_{r_{t-1}}, f_{r_t}) = f_{r_t},$$

the last equality holding because, as $r_t$ divides $r_{t-1}$, then (by the above) $f_{r_t}$ divides $f_{r_{t-1}}$.

**A1.66.** It has already been proved in Exercise A1.30 that, if $n \mid m$, then $f_n \mid f_m$. We have now to prove the converse, that is to say, $f_n \mid f_m$ implies that $n \mid m$. Let $f_n \mid f_m$. Then $\mathrm{GCD}(f_n, f_m) = f_n$. But, by Exercise A1.65, $\mathrm{GCD}(f_n, f_m) = f_d$, with $d = \mathrm{GCD}(n, m)$. So $d = n$, and if $\mathrm{GCD}(n, m) = n$, this means that $n \mid m$.

**A1.67.** The correct answer is (b). Indeed, for $n \geq 1$, we have the relation

$$\frac{f_{n+1}}{f_n} = 1 + \frac{f_{n-1}}{f_n},$$

so, setting $x = \lim_{n \to \infty} f_{n+1}/f_n$, we have

$$x = 1 + \frac{1}{x}.$$

**A1.68.** For $k = 0$, $r_n = 0 = f_0 = 0$, so the basis of induction is verified. Assume the inequality to be verified for every $m$ such that $0 \leq m < k$ and prove it for $k$. From

$$r_{n-k} = r_{n-k+1} q_{n-k+2} + r_{n-k+2},$$

as the inductive hypothesis is $r_{n-k+1} \geq f_{k-1}$ and $r_{n-k+2} \geq f_{k-2}$, and as we have further $q_{n-k+2} \geq 1$, we get

$$r_{n-k} \geq f_{k-1} + f_{k-2} = f_k.$$

**A1.69.** If $a$ and $b$ ($a \geq b$) are two integers such that $D(a, b) \geq n$, this means that the last non-zero remainder is $\geq r_{n-1}$, so, using the result of the previous exercise, we have $r_{n-k} \geq f_k$. In particular, $b = r_0 \geq f_n$. So, if $b < f_n$, then certainly $D(a, b) < n$ must be true.

**A1.74.** If $\alpha = [a_0; a_1, \ldots, a_n, \ldots]$, with $a_0$ positive integer, we have $1/\alpha = [1; a_0, a_1, \ldots, a_n, \ldots]$.

**A1.75.** Proceed by induction, observing that $C_k = [a_0; a_1, \ldots, a_{k-1}, a_k + 1/a_{k+1}]$, so

$$C_k = \frac{\left(a_k + \dfrac{1}{a_{k+1}}\right) p_{k-1} + p_{k-2}}{\left(a_k + \dfrac{1}{a_{k+1}}\right) q_{k-1} + q_{k-2}}.$$

**A1.76.** From formula (1.43) we get

$$C_k - C_{k-2} = \frac{(-1)^k (q_k - q_{k-2})}{q_k q_{k-1} q_{k-2}}.$$

Hence, conclude by using (1.41).

**A1.78.** The first two claims follow from formula (1.44). The third one follows from formula (1.43).

**A1.80.** Let $\sqrt{n} \in \mathbb{Q}$, so $\sqrt{n} = p/q$, with $p/q$ a reduced fraction. Then we have $p^2 = nq^2$. By applying Corollary 1.3.9 we see that $p \mid n$. Deduce that $q \mid p$. By applying Exercise A1.38 deduce that $q = \pm 1$ and consequently that $n$ is a square.

**A1.84.** Keep in mind that $a/b = C_n$ and use formula (2.11).

**A1.86.** Prove that

$$|\alpha - C_{n+1}| + |\alpha - C_n| = |C_{n+1} - C_n| = \frac{1}{q_n q_{n+1}}$$

and conclude from here.

**A1.89.** We may write $\alpha = (a + \sqrt{b})/c$ with integers $a, b, c$, $b > 0$ and $c \neq 0$. So $\alpha = (a|c| + \sqrt{bc^2})/(c|c|)$. Set $p = a|c|, q = c|c|, d = bc^2$ and conclude.

**B1.1.** The correct answer is (d); indeed, $n < 2^n$ for all $n \in \mathbb{N}$. The basis of induction is true because $0 < 2^0 = 1$. Suppose $n < 2^n$ for a natural number $n$. Then $n + 1 < 2^n + 1 \leq 2^n + 2^n = 2^{n+1}$, where the inequality $\leq$ follows from the fact that $1 \leq 2^n$ for all $n \in \mathbb{N}$.

**B1.2.** The correct answer is (a); indeed $S(n)$ may be obtained from $S(n-1)$ by adding to it the $n$th odd natural number, that is, $2n - 1$.

**B1.3.** The correct answer is (c). The basis of induction is obvious. Suppose the formula holds for $n - 1$. From Exercise B1.2 we know that $S(n) = S(n-1) + 2n - 1$, so from the inductive hypothesis we find that $S(n) = (n-1)^2 + 2n - 1 = n^2$.

**B1.6.** The correct answer is (b), that is,

$$\sum_{k=0}^{n} (4k + 1) = (2n + 1)(n + 1). \tag{1}$$

For $n = 0$ formula (1) becomes $1 = 1$, so the basis of induction is true. Suppose (1) is true for $n$ and prove it for $n + 1$:

$$\sum_{k=0}^{n+1}(4k+1) = 4(n+1) + 1 + \sum_{k=0}^{n}(4k+1) = \text{(by the induct. hyp.)}$$

$$= 4n + 5 + (2n+1)(n+1) = (2n+3)(n+2).$$

**B1.7.** The correct answer is (c), that is

$$\sum_{k=1}^{n} k^2 = \frac{n(n+1)(2n+1)}{6}. \tag{2}$$

The basis of induction is obvious, because for $n = 1$ the equation (2) becomes $1 = 1$. Suppose (2) is true for $n$ and prove it for $n + 1$:

$$\sum_{k=1}^{n+1} k^2 = (n+1)^2 + \sum_{k=1}^{n} k^2 = \text{(by the induct. hyp.)}$$

$$= (n+1)^2 + \frac{n(n+1)(2n+1)}{6} = \frac{(n+1)(n+2)(2n+3)}{6}.$$

**B1.8.** The correct answer is (a). Verify it by induction.

**B1.9.** The correct answer is (c). The basis of induction is trivial. Assume (c) to be true for $n - 1$ and prove it for $n$:

$$\sum_{k=1}^{n}(2k)^3 = (2n)^3 + \sum_{k=1}^{n-1}(2k)^3 = (2n)^3 + 2(n-1)^2 n^2 =$$

$$= 2n^2[(n-1)^2 + 4n] = 2n^2(n+1)^2.$$

**B1.10.** The correct answer is (a). The basis of induction is trivial. Assume the result to be true for $n - 1$. Then $\sum_{k=0}^{n} 2 \cdot 3^k = 2 \cdot 3^n + (3^n - 1) = 3^{n+1} - 1$.

**B1.11.** The correct answer is (b).

**B1.14.** The correct answer is (a).

**B1.15.** The correct answer is (b).

**B1.16.** The correct answer is (c), because $(103/100)^{24}$ is about 2, so there will be about $2 \cdot 5 = 10$ billion people.

**B1.17.** Define $\lambda_1 = (-1 + i\sqrt{3})/2$ and $\lambda_2 = (-1 - i\sqrt{3})/2$. With the usual method the eigenvectors relative to $\lambda_1$, $\lambda_2$ and 1, respectively, are found:

$$\left(2\lambda_1, 1, \frac{\lambda_2}{3}\right), \qquad \left(2\lambda_2, 1, \frac{\lambda_1}{3}\right), \qquad (6, 3, 1).$$

Define next

$$D = \begin{pmatrix} \lambda_1 & 0 & 0 \\ 0 & \lambda_2 & 0 \\ 0 & 0 & 1 \end{pmatrix}, \qquad C = \begin{pmatrix} 2\lambda_1 & 2\lambda_2 & 6 \\ 1 & 1 & 3 \\ \lambda_2/3 & \lambda_1/3 & 1 \end{pmatrix}.$$

The inverse of $C$ is

$$C^{-1} = \frac{1}{3} \begin{pmatrix} \lambda_2/2 & 1 & 3\lambda_1 \\ \lambda_1/2 & 1 & 3\lambda_2 \\ 1/6 & 1/3 & 1 \end{pmatrix}.$$

We have $A = C \cdot D \cdot C^{-1}$, so a closed formula is found by multiplying $A^n = C \cdot D^n \cdot C^{-1}$ for $X_0$.

**B1.18.** The correct answer is (a), as can be verified directly, without using the closed formula found in Exercise B1.17. Indeed, after one year there will be 120 newborn beetles and 60 one-year beetles (180 altogether). After two years there will be 60 one-year ones and 20 two-year ones (80 altogether), while after three years there will be 120 newborn ones and 20 two-year ones (140 altogether), exactly as in the starting year. So the same situation repeats every third year; hence the result follows.

**B1.19.** Two eigenvectors corresponding to 1 and $7/10$, respectively, are $(2,1)$ and $(-1,1)$. Set

$$C = \begin{pmatrix} 2 & -1 \\ 1 & 1 \end{pmatrix}, \qquad \text{quindi} \quad C^{-1} = \frac{1}{3} \begin{pmatrix} 1 & 1 \\ -1 & 2 \end{pmatrix}.$$

So we may compute

$$A^n = C \cdot \begin{pmatrix} 1 & 0 \\ 0 & (7/10)^n \end{pmatrix} \cdot C^{-1} = \frac{1}{3} \begin{pmatrix} 2 + (7/10)^n & 2 - 2(7/10)^n \\ 1 - (7/10)^n & 1 + 2(7/10)^n \end{pmatrix};$$

hence it is possible to compute $X_n$ multiplying by $X_0$.

**B1.20.** The correct answer is (d), because there will be 5 million inhabitants, as may also be directly verified, without applying the formula found in Exercise B1.19.

**B1.21.** The correct answer is (c). Let $S(n)$ be the number sought. We have $S(1) = 2$. Moreover, given $n$ lines in general position, forming $S(n)$ regions, one more line, in general position with respect to the other ones, intersects them in $n$ points, so meets $n+1$ regions, dividing each into two parts. The number of regions, with those added after the $(n+1)$th line increases by $n+1$. So we have $S(n+1) - S(n) = n+1$. From here it is easy to conclude.

**B1.22.** The correct answer is (d). Reasoning as in the previous exercise, prove that the right answer is $n^2 - n + 2$.

**B1.23.** By induction on the number $n$ of all lines. If $n = 1$, clearly two colours suffice. Suppose then we have proved that it is possible to colour with just two colours the regions formed by less than $n$ lines, and prove it in the case in which the $n$th line, $r$, is added. Divide the regions into two groups, depending on which side of $r$ they are on. Then it suffices to leave the colour of those on one side as it was, and to change the colour of those on the other side. We have to verify that this is a "good" colouring: indeed, if two bordering regions are on the same side with respect to $r$, they will have different colours (they had different colours before the appearance of $r$, and now, either they both keep their colours or the colours are changed in both, but in any case will be different). If the two regions are on different sides with respect to $r$, their colours are different because one of them has had its colour changed.

**B1.24.** The correct answer is (a), because every third month Mark gets a 1% interest, which so becomes 4,060401% yearly.

**B1.28.** The correct answer is (a). Indeed, the roots of the characteristic equation of the recurrence relation are 2 and $-1$. So the solution has the form $c_1 2^n + c_2(-1)^n$, where $c_1$ and $c_2$ are determined by the initial values $a_0 = 2 = c_1 + c_2$ and $a_1 = 7 = 2c_1 - c_2$.

**B1.29.** The correct answer is (a). Indeed, a solution is found to be of the form $-n - 7$.

**B1.32.** The correct answer is (a).

**B1.33.** The correct answer is (c).

**B1.34.** The correct answer is (a). Indeed, $491 = 2 \cdot 245 + 1$ and $245 = 245 \cdot 1 + 0$.

**B1.35.** The correct answer is (b), because the Euclidean algorithm ends after just two steps, as shown in the solution of Exercise B1.34.

**B1.37.** We have $34567 = 457 \cdot 76 - 165$, then $457 = (-165)(-3) - 38$, after which $-165 = (-38) \cdot 4 - 13$, then $-38 = (-13) \cdot 3 + 1$ and finally $-13 = 1 \cdot (-13) + 0$.

**B1.38.** The correct answer is (a).

**B1.39.** The correct answer is (c). Indeed $28762 = 18 \cdot 1515 + 1492$, then $1515 = 1492 + 23$, hence $1492 = 64 \cdot 23 + 20$, then $23 = 20 + 3$, $20 = 6 \cdot 3 + 2$, $3 = 2 + 1$ and finally $2 = 2 \cdot 1 + 0$.

**B1.40.** The correct answer is (c), as the solution of Exercise B1.39 shows.

**B1.42.** The correct answer is (b). There are integer solutions because $\text{GCD}(92, 18) = 4$ and 4 divides 180. Using the Euclidean algorithm to find the GCD it is possible to find Bézout's identity

$$4 = 92 \cdot (-3) + 28 \cdot 10;$$

hence

$$180 = 45 \cdot 4 = 92 \cdot (-135) + 28 \cdot 450$$

so a solution is $(\bar{x}, \bar{y}) = (-135, 450)$. To find *all* solutions, consider the associate homogeneous equation:

$$0 = 92x + 28y = 4(23x + 7y).$$

It admits as its solutions the pairs $(x_0, y_0) = (-7t, 23t)$, with $t$ ranging in $\mathbb{Z}$. So all the solutions of $92x + 28y = 180$ are the pairs

$$(x, y) = (-135 - 7t, 450 + 23t), \qquad \text{for all } t \in \mathbb{Z}.$$

**B1.43.** The correct answer is (a); indeed, there are no integer solutions because $\text{GCD}(482, 20) = 2 \nmid 35$.

**B1.47.** The correct answer is (c), as the leading coefficient, that is to say, the coefficient of the highest degree term, is $-3 \neq 1$.

**B1.48.** The correct answer is (b).

**B1.49.** The correct answer is (d), because $-x-1$ is only *a* greatest common divisor, but *the* greatest common divisor of the two polynomials is $x + 1$.

**B1.51.** The correct answer is (b).

**B1.52.** The correct answer is (d), because the greatest common divisor of these polynomials is $x + 1$ (see Exercise B1.49), and if $h(x)$, $k(x)$ are the polynomials appearing in Bézout's identity, then $f(x) = 4h(x)$ and $g(x) = 4k(x)$.

**B1.53.** The correct answer is $-25x^4 - 8x^3 + 9x^2$, hence (d).

**B1.55.** The greatest common divisor is $1 + i$.

**B1.56.** In base 10, the two factors of the multiplication are 29 and 13, while the result is 377, so the operation is correct.

**B1.59.** The correct answer is (a).

**B1.60.** The correct answer is (a).

**B1.61.** The correct answer is (c).

**B1.62.** The correct answer is (d), because the sum is 10110100, while the product is 1000100011011.

**B1.63.** The correct answer is (c).

**B1.64.** The correct answer is (b).

**B1.65.** The correct answer is (c).

**B1.66.** $40/99$.

**B1.67.** $2491/45$.

**B1.68.** $101/111$.

**B1.69.** $10001001/11000$.

**B1.70.** $40/66 = 10/15$.

**B1.71.** $10342/60 = 3521/30$.

**B1.72.** $2, \overline{3}$.

**B1.73.** $4, \overline{1254}$.

**B1.74.** $11, \overline{10}$.

**B1.75.** The correct answer is (d), because the continued fraction is $[1; 3, 1, 10, 2]$.

**B1.76.** The correct answer is (a).

**B1.77.** The correct answer is (d), because the continued fraction is $[1; 1, 1, 1, 13, 2, 2]$, which consists of 7 terms.

**B1.81.** The continued fraction of $\alpha$ may be written as follows:

$$\alpha = 4 + \cfrac{1}{1 + \cfrac{1}{3 + \cfrac{1}{\alpha}}}.$$

Developing fractions, we get the equation

$$\alpha = \frac{19\alpha + 5}{4\alpha + 1}$$

which is equivalent to the second degree equation $4\alpha^2 - 18\alpha - 5 = 0$, so

$$\alpha = \frac{9 + \sqrt{101}}{4}.$$

**B1.82.** The correct answer is (d), because the continued fraction is $[5; \overline{5, 10}]$.

# Exercises of Chapter 2

**A2.1.** By hypothesis we have $f(x) \leq k_1 g(x)$ for $x > c_1$ and $g(x) \leq k_2 h(x)$ for $x > c_2$. Then, having set $c_3 = \max\{c_1, c_2\}$, we find that

$$f(x) \leq k_1 g(x) \leq k_1 k_2 h(x) = k_3 h(x)$$

for $x > c_3$, where $k_3 = k_1 k_2$.

**A2.5.** For $n$ even, we have that $f(n)/g(n) = 2n+1$ and so $\lim_{m\to\infty} f(2m)/g(2m) = \infty$. On the other hand, for $n$ odd we have $f(n)/g(n) = 1/(2n+1)$, so $\lim_{m\to\infty} f(2m+1)/g(2m+1) = 0$. It can be found analogously that $\lim_{m\to\infty} g(2m)/f(2m) = 0$ and $\lim_{m\to\infty} g(2m+1)/f(2m+1) = \infty$. So we deduce that those limits do no exist.

**A2.8.** The correct answer is (c).

**A2.9.** The correct answer is (a). Indeed, by observing that all factors of $n!$ have length less than or equal to $L(n)$ and using the previous exercise, we conclude that

$$L(n!) \leq nL(n) \in \mathcal{O}(n \ln n) = \mathcal{O}(nk).$$

Notice that this estimate can be improved, as many factors of $n!$ have length less than the length of $n$.

**A2.10.** None of (a), (b), (c) is a reasonable estimate. Indeed, $\binom{n}{m}$ is the ratio of the two numbers $n(n-1)\cdots(n-m+1)$ and $m!$ having length $\mathcal{O}(mh)$ and $\mathcal{O}(mk)$, respectively. So the length of $\binom{n}{m}$ is $\mathcal{O}(mh - mk)$, so it is $\mathcal{O}(mh)$ as well.

**A2.15.** Let $a/b$ be the number being considered. We may assume $a > b$. Developing $a/b$ as a continued fraction is equivalent to searching $\text{GCD}(a, b)$, which requires $\mathcal{O}(\log^3 a)$ bit operations.

**A2.16.** We have to verify the relation

$$p(x) = (x - \alpha)(p_n x^{n-1} + p_1(\alpha)x^{n-2} + \cdots + p_{n-2}(\alpha)x + p_{n-1}(\alpha)) + p(\alpha),$$

which can be easily done keeping in mind Table (2.14) and the meaning of its second row.

**A2.18.** Every $n \times n$ matrix has $n^2$ entries. We have to carry out $n$ multiplications and $n$ additions for each of them: so, in all, $n^3$ multiplications and $n^3$ additions are to be carried out. So the algorithm has complexity $\mathcal{O}(n^3(\log^2 m + \log n))$.

**A2.20.** See [13], Cap. 24, §1.

**A2.21.** Apply first Exercise A2.20 to reduce the matrices in row echelon form and notice that the determinant of the matrix is the product of the *pivot elements* (see [13], Cap. 8).

**A2.22.** Hint: use Gaussian algorithm to compute the inverse (see [13], pag. 146).

**A2.23.** The correct answer is (c). Indeed, to compute $n!$ we have to carry out successively $n$ multiplications of two integers whose length can be estimated with that of $n$ (which is $k$) and with that of $n!$ (which is $nk$).

**A2.24.** A reasonable estimate is $\mathcal{O}(m^2 \log^2 n)$. Keep in mind what was said in the solution to Exercise A2.10.

**A2.25.** A reasonable estimate is $\mathcal{O}(n^2 \log^2 m)$. Indeed, the length of $m^n$ is $\mathcal{O}(n \log m)$ and to compute $m^n$ it is necessary to carry out $n$ multiplications of two integers whose length can be estimated with that of $m$ (which is $\mathcal{O}(\log m)$) and with that of $m^n$ (which is $\mathcal{O}(n \log m)$).

**B2.1.** The correct answer is (b). In fact, (a) is an estimate of the complexity of $f(n)$ too, but one which is worse than (b).

**B2.2.** The correct answer is (d); indeed, $\log n$ is negligible with respect to $n$

**B2.7.** The correct answer is (c).

**B2.13.** The correct answer is (b).

**B2.14.** The correct answer is (b).

**B2.17.** The correct answer is (c).

**B2.19.** The correct answer is (b).

**B2.21.** The correct answer is (c). Each squaring requires at most $\mathcal{O}(\log^2 n)$ bit operations, and it is necessary to carry out $n$ of them, so obtaining the estimate $\mathcal{O}(n \log^2 n)$. The computation time of additions is negligible with respect to this one.

**B2.22.** The correct answer is (c).

**B2.23.** The correct answer is (c).

**B2.24.** The correct answer is (a); indeed, $1176 = 159 \cdot 7 + 63$, $159 = 63 \cdot 3 - 30$, $63 = -30 \cdot (-2) + 3$ and $-30 = 3 \cdot (-10) + 0$.

# Exercises of Chapter 3

**A3.1.** For every integer $a$, $a - a = 0$ is a multiple of $n$ for any $n$, so $a \equiv a \pmod{n}$, that is to say, congruence modulo $n$ is a reflexive relation. We prove next that this relation is symmetric. If $a \equiv b \pmod{n}$, this means that $a - b = hn$, for some $h \in \mathbb{Z}$; hence follows that $b - a = -(a - b) = -(hn) = (-h)n$, which proves that $b \equiv a \pmod{n}$. Finally, if $a \equiv b \pmod{n}$ and $b \equiv c \pmod{n}$, this means that $a - b = hn$ and $b - c = kn$ with $h, k \in \mathbb{Z}$, so $a - c = (a - b) + (b - c) = hn + kn = (h + k)n$, which implies that $a \equiv c \pmod{n}$. So we have proved that congruence relation is transitive too, for all $n$, so it is an equivalence relation.

**A3.2.** Let $A = \{a, b, c\}$ be a set consisting of three distinct elements. Consider the relation $R$ on $A$ defined as follows: $R = \{(a, a), (b, b), (c, c), (a, b), (b, a), (b, c), (c, b)\}$. Clearly, $R$ is reflexive and symmetric, but is not transitive as $a R b$ and $b R c$, while it is not true that $a R c$, so $R$ is an example of $a$). Consider now $R' = \{(a, a), (b, b), (c, c), (a, b)\}$ over the same $A$. Clearly, $R'$ is reflexive and transitive but not symmetric, because $a R' b$ but it is not true that $b R' a$. Other examples of relations of type $b$) are order relations on sets with at least two distinct elements. Finally, consider the set $B$ of the students in a same secondary school form, and let the relation $R''$ be defined on $B$ as follows: $a R'' b$ if and only if $a$ and $b$ got the same mark in the first maths test this year. The relation $R''$ is clearly symmetric and transitive, but is not necessarily reflexive, as some student might have been absent the day of the test. Clearly the relation $R''$ is an equivalence relation if we do not take it on the whole set $B$, but only on the subset of $B$ consisting of the students who actually took the test.

**A3.3.** It is the relation *being a divisor of*, which we have denoted by $|$, that is, $a R b$ if and only if $a \mid b$. Clearly, $a = 1 \cdot a$ for all $a \in \mathbb{N}$, so $a \mid a$ and $|$ is a reflexive relation. Moreover, it is transitive, because if $a \mid b$ and $b \mid c$, then $b = ah$ and $c = bj$, so $c = a(jh)$, that is $a \mid c$. Suppose now $a \mid b$ and $b \mid a$ for two natural numbers $a$ and $b$. Then there exist $h, k \in \mathbb{N}$ such that $a = bh$ and $b = ak$, so $a = a(hk)$ and, cancelling out $a$, we find $hk = 1$, which in $\mathbb{N}$ implies that $h = k = 1$, that is $a = b$ and $|$ is antisymmetric. So we have shown that $|$ is an order relation in $\mathbb{N}$. In $\mathbb{Z}$, on the other hand, we cannot conclude from $hk = 1$ that $h = k = 1$, because we might also have $h = k = -1$, so $|$ is not antisymmetric (nor an order relation) in $\mathbb{Z}$.

**A3.11.** If $ac = bc$ with $c \neq 0$, then $(a - b)c = 0$. In an integral domain (and in particular in $\mathbb{Z}$) there are no zero-divisors, so $a - b = 0$, that is $a = b$.

**A3.14.** The correct answer is (b). Indeed, a finite commutative ring with unity is a field if and only if it is an integral domain (see Exercise A1.48).

**A3.15.** The correct answer is (a).

**A3.17.** Consider the coefficients of the two polynomials as numbers in $\{0, \dots, n\}$, so they have length bounded by the length of $n$. To sum the two polynomials we need at most $m + 1$ sums of such numbers and then reducing modulo $n$. But as the coefficients so obtained are bounded by $2n$, to reduce them modulo $n$ it suffices to carry out at most one subtraction. From this it is possible to deduce the claim.

**A3.18.** Consider the coefficients of the two polynomials as numbers in $\{0, \dots, n\}$, so they have length bounded by the length of $n$. By applying Proposition 2.5.9, verify that the product of the two polynomials has complexity $\mathcal{O}(m^2 \log^2 n)$. As the coefficients of the polynomial so obtained have to be reduced modulo $n$, from this it is possible to deduce the claim.

**A3.19.** It suffices to recall that a number is congruent modulo 11 to the sum of its decimal digits, taken with alternate signs, starting from the rightmost one. Then proceed as when casting out nines.

**A3.24.** If $G$ is cyclic with generator $x$, there is a surjective homomorphism $n \in \mathbb{Z} \to x^n \in G$. If $G$ is infinite, this is an isomorphism. If not, $G$ is a quotient of $\mathbb{Z}$.

**A3.25.** If $G$ is cyclic with generator $x$ and if $H$ is a non-trivial subgroup, let $n$ be the smallest positive integer such that $x^n \in H$. Using the division algorithm, prove that $x^n$ is a generator of $H$.

**A3.27.** Suppose $m$ divides $n$. Then $n = mk$ for some integer $k$, so $x^n = (x^m)^k = 1^k = 1$. Vice versa, suppose $x^n = 1$. We may write $n = qm + r$, with $0 \le r < m$. So $1 = x^n = (x^m)^q x^r = x^r$. Then, by definition of order we have $r = 0$, or else the order of $x$ would be $r < m$. So $m \mid n$.

**A3.28.** The powers $x, x^2, \ldots, x^m$ are all different by definition of order. Apply now the same reasoning as in the previous exercise to show that every other power is equal to one of these.

**A3.30.** Let $\mathrm{GCD}(m, n) = 1$. For all $k \in Z$, there is an $h$ such that $hm \equiv k$ (mod $n$). Then $(x^m)^h = x^k$, so $x^m$ is a generator. Vice versa, if $x^m$ is a generator, there is an integer $h$ such that $(x^m)^h = x$. Then $hm \equiv 1$ (mod $n$), which implies $\mathrm{GCD}(m, n) = 1$.

**A3.32.** Let $n = dm$. Then $\langle x^m \rangle$ has order $d$. Vice versa, if $H$ is a subgroup of order $d$ of $G$, we have $n = dm$ by Lagrange's theorem. We have seen that a generator of $H$ is given by $x^h$ with $h$ the smallest positive integer such that $x^h \in H$ (see Exercise A3.25). As $(x^h)^d = 1$, we have a relation of the form $dh = nk$. Dividing by $d$ we have $h = mk$. So $x^h = (x^m)^k \in \langle x^m \rangle$; hence, $H \subset \langle x^m \rangle$. As $\langle x^m \rangle$ has order $d$, we have $H = \langle x^m \rangle$ (and $k = 1$).

**A3.33.** Hint: the element $x^h$ has order $d$ if and only if it is a generator of the subgroup $\langle x^m \rangle$.

**A3.34.** Hint: the relation $y_1^m = y_2$ is equivalent to the congruence equation $mn_1 \equiv n_2$ (mod $n$).

**A3.36.** Define in $G$ the relation $R_H$ as follows: $x R_H y$ if and only if $xy^{-1} \in H$. Verify that this is an equivalence relation. Verify that the equivalence class of an element $x$ is the set denoted by $Hx$ and called *right coset* of $H$, consisting of all the elements of the form $tx$ with $t \in H$. Prove that $Hx$ has the same number of elements as $H$. Conclude that the order of $G$ is equal to the order of $H$ times the order of the quotient set of $G$ with respect to the relation $R_H$.

**A3.37.** We know that $\varphi(n)$ is the order of the finite group $U(\mathbb{Z}_n)$. Let $m$ be the period of an element $a$ of $U(\mathbb{Z}_n)$. Then $m$ divides $\varphi(n)$, so $a^{\varphi(n)} = e$ (see Exercise A3.27).

**A3.38.** In the situation described, it suffices to divide $m$ by $\varphi(n)$, that is, $m = q\varphi(n) + r$, so $a^m \equiv a^r$ (mod $n$), then compute $a^r$ (mod $n$).

**B3.1.** The correct answer is (c).

**B3.2.** The correct answer is (a).

**B3.3.** The correct answer is (c): the zero-divisors are classes [2], [3], [4].

**B3.4.** The correct answer is (a) as 19 is a prime number. The reader might want to verify explicitly the absence of zero-divisors.

**B3.5.** The correct answer is (b) as 27 is not a prime number. For instance, class [3] is a zero-divisor in $\mathbb{Z}_{27}$.

**B3.6.** We have $725843 \equiv 3$ (mod 10), so

$$(725843)^{594} \equiv 3^{594} \quad \text{(mod 10)}.$$

Moreover,

$$3^{594} = 3^{4 \cdot 148+2} = (3^4)^{148} \cdot 3^2 \equiv 1^{148} \cdot 3^2 = 9 \quad \text{(mod 10)}.$$

So the last digit is 9.

**B3.7.** As $74 \equiv 2$ (mod 9), then $74^{6h} \equiv 2^{6h}$ (mod 9). Moreover, $2^{6h} = (2^6)^h$ and $2^6 \equiv 1$ (mod 9), so, for all $h \in \mathbb{N}$, we find that $74^{6h} \equiv 1^h \equiv 1$ (mod 9) and the required congruence class is 1.

**B3.8.** We have $43816 \equiv 6$ (mod 10), and further $6^2 \equiv 6$ (mod 10). Then, $6^k \equiv 6$ (mod 10) for all $k > 0$; it follows that

$$43816^{20321} \equiv 6 \quad \text{(mod 10)}.$$

**B3.9.** We have $29345 \equiv 5$ (mod 6). Moreover, $5^2 = 25 \equiv 1$ (mod 6), so

$$29345^{362971} \equiv 5^{362971} = (5^2)^{181485} \cdot 5^1 \equiv 1 \cdot 5 \equiv 5 \quad \text{(mod 6)}.$$

**B3.10.** We have $362971 \equiv 1$ (mod 6), so $362971^{29345} \equiv 1$ (mod 6).

**B3.11.** In class 1 modulo 9.

**B3.16.** The correct answer is (c).

**B3.17.** The correct answer is (a), as 4 is not relatively prime with 18.

**B3.18.** 39.

**B3.23.** As $9 \equiv 0$ (mod 3), we have that 3 divides an integer $n$ written in base 9 if and only if 3 divides the last digit of $n$.

**B3.24.** The correct answer is (a). Let us see why. First of all, 5 mod 4 = 1, so the congruence given is equivalent to $3x \equiv 1$ (mod 4). By Corollary 3.3.6, the congruence has exactly one solution modulo 4. Moreover, the solution is the inverse of 3 modulo 4, by the very definition of inverse. In order to find this inverse we may either proceed by trial and error, or with Bézout's identity, that is to say, computing the numbers $\alpha$ and $\beta$ such that $3\alpha + 4\beta = 1 = \text{GCD}(3, 4)$ using the Euclidean algorithm (see formula (1.14) on page 17 foll.). Indeed, from Bézout's identity it follows that $\alpha$ is the inverse of 3 modulo 4. In our case, we find $\alpha = -1 \equiv 3$ (mod 4) and $\beta = 1$. So we conclude that the only solution modulo 4 of the congruence is 3.

**B3.25.** The correct answer is (b). Indeed, by Proposition 3.1.8 the congruence is found to be equivalent (by dividing all coefficients by 3) to $x \equiv 3 \equiv 1$ (mod 2).

**B3.26.** The correct answer is (a). By Corollary 3.3.6 the congruence has exactly one solution modulo 9. We compute the inverse of 4 modulo 9, that is the solution of $4y \equiv 1$ (mod 9). A way of finding $y$ consists in computing Bézout's identity $4\alpha + 9\beta = 1$ using the Euclidean algorithm, finding $\alpha = 7$. Multiply both sides of the congruence by 7 we find the equivalent congruence $x \equiv 7 \cdot 7 \equiv 4$ (mod 9). On the other hand, $4 \equiv -5$ (mod 9).

**B3.27.** The correct answer is (c), by Proposition 3.3.4.

**B3.28.** The correct answer is (d).

**B3.36.** We have $190 \equiv 3 \pmod{17}$. So

$$190^{597} \equiv 3^{597} \pmod{17}.$$

As $\mathrm{GCD}(3, 17) = 1$, it follows that $3^{16} \equiv 1 \pmod{17}$ by Euler's Theorem, as $\varphi(17) = 16$ (verify this directly). So,

$$3^{597} = 3^{16 \cdot 37 + 5} = (3^{16})^{37} \cdot 3^5 \equiv 1^{37} \cdot 3^5 = 3^5 \equiv 5 \pmod{17}.$$

**B3.38.** As $\mathrm{GCD}(3, 7) = 1$ and $\varphi(7) = 6$, then Euler's Theorem says that $3^6 \equiv 1 \pmod{7}$, so $3^{13} = (3^6)^2 \cdot 3 \equiv 3 \pmod{7}$.

**B3.39.** The correct answer is (a).

**B3.40.** The correct answer is (c).

**B3.41.** The correct answer is (c).

**B3.42.** The correct answer is (c). Let us see why. By the Chinese remainder theorem 3.4.2, there exists exactly one solution modulo $5 \cdot 9 = 45$. The method for finding this solution is described in the proof of the theorem: with those notation we have $s = 2$, $r_1 = 5$, $r_2 = 9$, $c_1 = 3$ and $c_2 = 7$. So $R = 45$, $R_1 = 9$ and $R_2 = 5$. We have now to solve congruences $9x \equiv 3 \pmod{5}$ and $5x \equiv 7 \pmod{9}$. The only solution modulo 5 of the first one is $\bar{x}_1 = 2$, while the only solution modulo 9 of the second one is $\bar{x}_2 = 5$. So we may conclude that the solution of the congruence given is $\bar{x} = 9 \cdot 2 + 5 \cdot 5 = 43$.

**B3.43.** The correct answer is (d), because the system has solution $x \equiv 97 \pmod{120}$.

**B3.48.** It will happen on Saturday 31 March.

**B3.49.** There are 1786 books.

**B3.53.** The correct answer is (d), as may be verified by browsing any engagement diary, but we are sure the reader has applied instead the formula proved in the text, by substituting the values $g = 31$, $m = 10$ (as we are considering March as the first month of the year!), $s = 20$ and $y = 0$, because $2000 = 20 \cdot 100$, finding $x = 31 + 25 - 40 + 5 = 21 \equiv 0 \pmod{7}$.

**B3.54.** The correct answer is (c); indeed, in this case we have $g = 28$, $m = 12$, $2003 = 20 \cdot 100 + 3$, that is, $s = 20$ and $y = 3$, because we consider February 2004 as the last month of year 2003, so $x \equiv 28 + 31 - 40 + 3 + 5 = 27 \equiv 6 \pmod{7}$.

# Exercises of Chapter 4

**A4.4.** If $\sqrt{p} = a/b$, with $a$, $b$ integers and relatively prime, then $b^2 p = a^2$. Now the irreducible factor $p$ appears an odd number of times in the left-hand side and an even number of times in the right-hand side. This contradicts the Fundamental Theorem of Arithmetic.

**A4.5.** Notice that $\sum_{t=1}^{n} 1/t^s$ is the sum of the areas of $n$ rectangles, each of width 1 and heights $1, 1/2^s, \ldots, 1/n^s$. We may assume these rectangles to be located in the cartesian plane with the bases along the $x$-axis, on the line segments having as endpoints the points of abscissas $1, 2, \ldots, n$ and with the heights having as endpoints $(1, 1), (2, 1/2^s), \ldots, (n, 1/n^s)$. The graph of the function $y = 1/x^s$ is completely included in the union of these rectangles and the difference in the right-hand side of (4.2) is the area of the figure $\Sigma$ between the graph and the union of the rectangles. This figure is the union of the figures $\Sigma_1, \ldots, \Sigma_n$ such that $\Sigma_i$, $i = 1, \ldots, n$, is the figure between the $i$th rectangle and the segment of the graph of $y = 1/x^s$ which lies above the $i$th interval $[i, i+1]$. For all $i = 2, \ldots, n$, translate $\Sigma_i$ along the $x$-axis by a vector of length $i - 1$ with negative orientation. So we get a new figure $\Sigma_i'$ included in the first rectangle $R$, which is a square of area 1. Notice that the area of $\Sigma$ is equal to the area of the figure $\Sigma'$, the union of $\Sigma_1, \Sigma_2', \ldots, \Sigma_n'$, which is strictly included in $R$.

**A4.7.** The prime numbers less than or equal to $\sqrt{n}$ are approximately $2\sqrt{n}/\log n$. For each of these numbers it is necessary to delete all its multiples that are less than $n$, so it is necessary to carry out a number of operations that may be estimated by $\log^2 n$.

**A4.9.** Recall that $\binom{p}{k} = p!/k!(p-k)!$, and $p$ divides the numerator, but cannot divide the denominator, because it does not divide any of its factors.

**A4.10.** The binomial theorem yields

$$(x + y)^p = x^p + \sum_{k=1}^{p-1} \binom{p}{k} x^{p-k} y^k + y^p.$$

So we have to prove that the sum is divisible by $p$. But this is true, as in the sum we have both $k < p$ and $p - k < p$, so $\binom{p}{k}$ is an integer divisible by $p$, by Exercise A4.9.

**A4.13.** Let $n = pq$, where $p$ and $q$ are distinct primes. Then $\varphi(n) = \varphi(p)\varphi(q) = (p-1)(q-1) = n + 1 - (p+q)$. Vice versa, if we know $n$ and $\varphi(n)$, then $p$ and $q$ are the solutions of the second degree equation $x^2 - (n - \varphi(n) + 1)x + n$.

**A4.16.** Write out the factorisation of $n = p_1^{h_1} \cdots p_r^{h_r}$ di $n$. Then

$$\sum_{d|n} \mu(d) = \sum_{0 \le k_i \le 1, i=1,\ldots,r} \mu(p_1^{k_1} \cdots p_r^{k_r}) =$$

$$= 1 - r + \binom{r}{2} - \binom{r}{3} + \cdots + (-1)^r = (1-1)^r = 0.$$

**A4.17.** It suffices to verify that $((f * g) * h)(n)$ and $(f * (g * h))(n)$ both coincide with $\sum_{d_1 d_2 d_3 = n} f(d_1)g(d_2)h(d_3)$.

**A4.20.** Let $n, m$ relatively prime. We have

$$(f * g)(n) = \sum_{d|n} f(d)g\left(\frac{n}{d}\right), \quad (f * g)(m) = \sum_{d'|m} f(d')g\left(\frac{m}{d'}\right)$$

so, by the multiplicativity of $f$ and $g$,

$$((f * g)(n))((f * g)(m)) = \sum_{d|n,d'|m} f(d)f(d')g\left(\frac{n}{d}\right)g\left(\frac{m}{d'}\right) = \sum_{d|n,d'|m} f(dd')g\left(\frac{nm}{dd'}\right).$$

Hence the claim immediately follows.

**A4.22.** We have $\mu * E_f = \mu * (I * f) = (\mu * I) * f = II * f = f$.

**A4.24.** Möbius inversion theorem says that $\phi = \mu * \iota$ because $E_\varphi = \iota$. So if $n = p_1^{h_1} \cdots p_r^{h_r}$, then

$$\varphi(n) = \sum_{d|n} \mu(d)\frac{n}{d} = n - \sum_{i=1}^{r} \frac{n}{p_i} + \sum_{i,j=1}^{r} \frac{n}{p_i p_j} - \cdots$$

hence (4.4) may be imediately deduced.

**A4.26.** Apply Proposition 4.2.2. So it suffices to compute the functions $\nu$ and $\sigma$ on prime numbers, for which it is trivial to compute the functions.

**A4.30.** Keeping in mind the proof of Proposition 4.2.3, prove that the obvious mapping $[x]_{nm} \in \mathbb{Z}_{nm} \rightarrow ([x]_n, [x]_m) \in \mathbb{Z}_n \times \mathbb{Z}_m$ induces a group isomorphism $U(\mathbb{Z}_{nm}) \rightarrow U(\mathbb{Z}_n) \times U(\mathbb{Z}_m)$. The claim immediately follows.

**A4.31.** The claims follow from easy properties of cyclic groups. For instance, for the second identity, notice that $a$ and $b$ generate in $U(\mathbb{Z}_p)$ cyclic groups of order $\mathrm{Gss}(p,a)$ and $\mathrm{Gss}(p,b)$, respectively. As $\mathrm{GCD}(\mathrm{Gss}(p,a), \mathrm{Gss}(p,b)) = 1$, these cyclic groups only intersect in 1, so their direct product is in $U(\mathbb{Z}_p)$ and is a cyclic group of order $\mathrm{Gss}(p,a) \cdot \mathrm{Gss}(p,b)$, which is generated by $ab$.

**A4.36.** We have $m' = m + hn$. Then $x^m - x^{m'} = x^m(1 - x^{hn})$. Notice that $h(x)$ divides $x^{hn} - 1$.

**A4.37.** In base $a$ the number $a^{n-1} + a^{n-2} + \cdots + a^2 + a + 1$ is written as $(1 \ldots 1)_a$, where $n$ digits 1 appear. Multiplying by $a - 1$ we get the number $(a - 1 \ldots a - 1)_a$, where $n$ digits $a - 1$ appear, and this is exactly $a^n - 1$. This proves part (i). Part (ii) is proved analogously.

**A4.38.** Proceed by induction.

**A4.41.** Assume $M_p$ to be prime and $n = 2^{p-1} \cdot M_p$. Then, by Exercise A4.26, we have $\sigma(n) = 2^p \cdot M_p = 2n$. Vice versa, let $n = 2^s t$ be even and perfect, with $t$ odd. Then, again by Exercise A4.26 and by the multiplicativity of $\sigma$, we have $2^{s+1}t = 2n = \sigma(n) = (2^{s+1} - 1)\sigma(t)$; hence follows that $2^{s+1} \mid \sigma(t)$, so we may write $\sigma(t) = 2^{s+1}q$ and so $\sigma(n) = (2^{s+1} - 1)\sigma(t) = 2^{s+1}(2^{s+1} - 1)q$. As $\sigma(n) = 2^{s+1}t$, we have $t = (2^{s+1} - 1)q$. Moreover, $\sigma(t) = 2^{s+1}q = t + q$, so $q = 1$ as $1, q, t$ divide $t$. Thus, $t = 2^{s+1} - 1$ and $\sigma(t) = t + 1$, so $t$ is prime.

**A4.48.** If the claim were not true, there would be a decreasing sequence $\{h_n\}_{n\in\mathbb{N}}$ of positive integers such that $b = a^{h_n} b_n$ for all $n \in N$. Let $I$ be the ideal generated by the elements of the sequence $\{b_n\}_{n\in\mathbb{N}}$. As $A$ is Noetherian, $I$ is finitely generated. Assume $I = (b_1, \ldots, b_n)$, so $b_{n+1} = a_1 b_1 + \cdots + a_n b_m$ and $b = a^{h_{n+1}} b_{n+1} = a^{h_{n+1}}(a_1 b_1 + \cdots + a_n b_n)$. Hence deduce that $b = b(a_1 a^{h_{n+1}-h_1} + \cdots + b(a_n a^{h_{n+1}-h_n})$, so $1 = a^{h_{n+1}-h_n}(a_1 a^{h_n - h_1} + \cdots + a_n)$. Thus, $a$ would be invertible, yielding a contradiction.

**A4.49.** Verify for instance that it is closed under addition. If $a, b \in I$ there are positive integers $n, m$ such that $a \in I_n$ and $b \in I_m$. If we assume $n \leq m$, then $a, b \in I_m$, so $a + b \in I_m \subseteq I$.

**A4.50.** To construct the sequence, proceed inductively. Choose $x_1 \in I$, any element of $I$. Having chosen next $x_1, \ldots, x_n$, notice that $I_n = (x_1, \ldots, x_n) \neq I$ as $I$ is not finitely generated. So we may choose $x_{n+1} \in I - I_n$, and clearly $I_{n+1} \neq I_n$.

**A4.52.** Let $n$ be the degree of $f(x)$ and let $a$ be its leading coefficient. Notice that $\lim_{x \to +\infty} f(x)/x^n = a$. Hence deduce the claim.

**A4.61.** For all positive integers $m < n$ and for all pairs of polynomials of degree $m$ and $n - m$ with coefficients bounded by $N$, we have to take their product and check whether it is equal to $f(x)$. Multiplying two of these polynomials has complexity $\mathcal{O}((m + 1)(n - m + 1) \log^2 N)$. Notice however that the pairs of such polynomials are $\mathcal{O}(N^n)$.

**A4.62.** First of all, to compute $\bar{f}(x)$ it is necessary to divide the $n + 1$ coefficients by $p$ and to take the remainder. This has complexity $\mathcal{O}((n + 1)N \log p)$. Once $\bar{f}(x)$ has been found, for all positive integers $m < n$ and for all pairs of polynomials of degree $m$ and $n - m$ in $\mathbb{Z}_p[x]$ we must take their product and check whether it is equal to $\bar{f}(x)$. Multiplying two of these polynomials and reducing the result modulo $p$ has complexity $\mathcal{O}((m + 1)(n - m + 1) \log^3 p)$. Moreover, there are $p^n$ such pairs of polynomials.

**A4.63.** Consider the coefficients of $p(x)$ as indeterminates and interpret (4.20) as a system of $M + 1$ equations in $M + 1$ unknowns. So it certainly has some solution. If there were two distinct solutions, we would find two distinct polynomials $p(x)$, $q(x)$ of degree $n \leq M$ verifying (4.20). Then the non-zero polynomial $p(x) - q(x)$ of degree $n \leq M$ would have the $M + 1$ distinct roots $a_0, a_1, \ldots, a_M$, which is impossible (see Exercise A1.55).

**A4.64.** Notice that the determinant of the matrix of the system (4.20) is equal to $V(a_0, a_1, \ldots, a_M)$, so the uniqueness of the solution of (4.20) implies that it is not zero. Notice next that $V(x_1, \ldots, x_m)$, as a polynomial in $\mathbb{K}(x_1, \ldots, x_{m-1})[x_m]$, has the solutions $x_m = x_i$, $i = 1, \ldots, m - 1$, so $V(x_1, \ldots, x_m)$ is divisible by $x_m - x_i$, $i = 1, \ldots, m - 1$ in $\mathbb{K}(x_1, \ldots, x_{m-1})[x_m]$. The claim can be deduced using Gauss theorem.

**A4.65.** Proceed as in Exercise A4.63 considering the coefficients of $f(x)$ as indeterminates. The given conditions determine a system of $n + 1$ equations in $n + 1$ unknowns that always has a solution. It is unique, or else we would have a polynomial of degree $n$ having $h$ roots with multiplicities $m_1 + 1, \ldots, m_h + 1$, which is impossible.

**B4.2.** The correct answer is (a).

**B4.4.** The correct answer is (b).

**B4.5.** The correct answer is (d), because $1369 = 37^2$.

**B4.11.** The correct answer is (a).

**B4.12.** The correct answer is (b).

**B4.14.** The correct answer is (a), because the primes are 211, 223, 227, 229, 233, 239 and 241.

**B4.19.** $\mathrm{Gss}(8, a) = 2$.

**B4.23.** $\beta = 3, 6, 9$.

**B4.24.** $\beta = 2, 4, 5, 7, 8, 10$.

**B4.25.** $\beta = 23, 46, 92$.

**B4.29.** By Proposition 4.4.3, for every prime factor $p$ of $2^{11} - 1 = 2047$, we have $p \equiv 1 \pmod{22}$. As $45 < \sqrt{2047} < 46$, we see that $p = 23$, and $2047 = 23 \cdot 89$.

**B4.33.** The number 2 is not prime in $\mathbb{Z}[i]$. Indeed, $2 = (1+i)(1-i)$, so 2 divides the product $(1+i)(1-i)$, but 2 does not divide $1 + i$ nor $1 - i$: assume 2 divides $1+i$, that is, $1+i = 2(a+bi)$, $a, b \in \mathbb{Z}$. Denoting by $N(a+ib) = a^2 + b^2$ the complex norm of the number $a + ib$, we would have

$$N(1+i) = 2 = N(2(a+ib)) = N(2)N(a+ib) = 4(a^2 + b^2);$$

now, the relation $2 = 4(a^2 + b^2)$ is a clearly impossible to be satisfied in $\mathbb{N}$. Analogously for $1 - i$. On the other hand, $2 + 3i$ is prime.

**B4.35.** For instance, 2, 5 and $2 \pm \sqrt{-6}$ are irreducible. Which of them are prime?

**B4.36.** The correct answer is (c), because $10 = (2 + \sqrt{-6})(2 - \sqrt{-6})$.

**B4.37.** The correct answer is (d), because no one of the three polynomials is associate to the given polynomial; the primitive polynomials are $\pm(100x^3 + 36x^2 - 15x)$.

**B4.40.** The correct answer is (a). Let us see why. The polynomial $x^3 - 3x + 1$, as every third degree polynomial, is irreducible over $\mathbb{Q}$ if and only if it has no roots, because if it were reducible it would necessarily have a factorisation into a first degree polynomial and a second degree polynomial (or into three degree one polynomials); see page 190. But 1 and $-1$ are not roots of $x^3 - 3x + 1$, so by Proposition 4.5.37 this polynomial has no roots, so it is irreducible.

Let us see now what happens by considering the coefficients in $\mathbb{Z}_2$, $\mathbb{Z}_3$ or $\mathbb{Z}_{19}$. Over $\mathbb{Z}_2$ the polynomial becomes $x^3 + x + 1$, which is irreducible over $\mathbb{Z}_2$ because it has no roots, as may be verified by substituting $x = 0$ and $x = 1$ in the polynomial. As the polynomial is irreducible over $\mathbb{Z}_2$, it must be so a fortiori over $\mathbb{Q}$ (see page 190). It may be verified that the polynomial is reducible over $\mathbb{Z}_3$ and over $\mathbb{Z}_{19}$; indeed, over $\mathbb{Z}_3$ it becomes $x^3 + 1$ which admits $-1$ as a root, while 3 is a root over $\mathbb{Z}_{19}$.

**B4.42.** The correct answer is (a). Indeed, by substituting $x + 1$ for $x$, we find the polynomial $x^4 + 5x^3 + 10x^2 + 10x + 5$, which satisfies Eisenstein's criterion of irreducibility (Proposition 4.5.41 on page 188), so it is irreducible, and the given polynomial must be as well. We have argued exactly as in Example 4.5.42 on page 188 (with $p = 5$).

**B4.47.** The correct answer is (b), because $x^3 - 3x + 1 = x^3 + 1 = (x+1)^3$ over $\mathbb{Z}_3$.

**B4.52.** The correct answer is (d); indeed, the polynomial in (d) verifies the required conditions, has degree two and may be computed using the Lagrange interpolation polynomial for this degree. Notice that there are infinitely many degree three polynomials that verify those conditions.

# Exercises of Chapter 5

**A5.3.** Every field containing $A$ and $b_1, \ldots, b_n$ must also contain all rational expressions in $b_1, \ldots, b_n$ with coefficients in $A$. Conclude by verifying that these expressions form a field.

**A5.5.** A basis of $C$ as a vector space over $A$ is also a basis of $C$ as a vector space over $B$, so $[C : B]$ is finite. On the other hand, $B$ is a vector subspace of $C$ as a vector space over $A$, so $[B : A]$ is finite, because $[C : A]$ is finite.

**A5.6.** To prove that the given elements are linearly independent, suppose we have a relation of the form $\sum_{i,j} \alpha_{ij} a_i b_j = 0$, with $\alpha_{ij} \in A$. Then we have $\sum_{j=1}^{m} (\sum_{i=1}^{n} \alpha_{ij} a_i) b_j = 0$. By the linear independence of $\{b_1, \ldots, b_m\}$ over $B$ we have $\sum_{i=1}^{n} \alpha_{ij} a_i = 0$ for all $j = 1, \ldots, m$. By the linear independence of $\{a_1, \ldots, a_n\}$ over $A$ we have $\alpha_{ij} = 0$ for all $i = 1, \ldots, n, j = 1, \ldots, m$.

To prove that this is a system of generators, notice that for all $c \in C$ we have $c = \sum_{j=1}^{m} \beta_j b_j$, with $\beta_j \in B$ for all $j = 1, \ldots, m$. Use now the fact that $\{a_1, \ldots, a_n\}$ is a basis of $B$ as a vector space over $A$ to express every $\beta_j$ as a combination of $\{a_1, \ldots, a_n\}$ with coefficients in $A$ and conclude.

**A5.8.** Let $f_b(x) = f_1(x) \cdot f_2(x)$ with $f_1(x), f_2(x) \in A[x]$ be monic polynomials of positive degree. We have $f_1(b) \cdot f_2(b) = f_b(b) = 0$ so either $f_1(b) = 0$ or $f_2(b) = 0$; thus, either $f_1(x) \in I_b$ or $f_2(x) \in I_b$. Hence $f_b(x)$ divides either $f_1(x)$ or $f_2(x)$, leading to a contradiction. The fact that $I_b$ is prime may be proved analogously.

**A5.10.** If $A$ is a field and $I$ is a non-zero ideal of $A$, there is a non-zero element $a \in I$. Then $1 = a^{-1} \cdot a \in I$, so $I = A$ (see Exercise A5.9). Vice versa, let $a \neq 0$ be an element of $A$. Then $(a)$ is a non-zero ideal, so $(a) = A$. Hence, $1 \in (a)$, so there is a $b$ such that $ab = 1$. This proves that every non-zero element of $A$ has an inverse, so $A$ is a field.

**A5.11.** $B$ is an integral domain (see Exercise A5.8 and Exercise A3.12). Let $I$ be an ideal of $B$. Let $J$ be the preimage of $I$ under the natural mapping $A[x] \to B$. Verify that $J$ is an ideal and observe that $f(x) \in J$. Let $g(x)$ be the monic generator of $J$. Then $g(x)$ divides $f(x)$. Conclude that either $g(x) = f(x)$ or $g(x) = 1$, so $J = I$ or $J = A[x]$; hence either $I = (0)$ or $I = B$. Conclude by applying Exercise A5.10.

**A5.13.** Hint: use Lemma 5.1.8 and proceed by induction.

**A5.14.** Let $c \in C$. As $B \subset C$ is algebraic, there are $b_0, \ldots, b_n \in B$ not all zero such that $b_0 + b_1 c + \cdots + b_n c^n = 0$. Then $c$ is algebraic over $A(b_0, \ldots, b_n)$, so $[A(b_0, \ldots, b_n, c) : A(b_0, \ldots, b_n)]$ is finite. As $A \subset B$ is algebraic, $[A(b_0, \ldots, b_n) : A]$ is also finite, so by the multiplicativity of degrees $[A(b_0, \ldots, b_n, c) : A]$ is finite. Conclude by applying Exercise A5.13.

**A5.15.** Let $a, b$ be elements of $B$ that are algebraic over $A$. Then $[A(a, b) : A]$ is finite. As $a \pm b$, $ab$, and $a^{-1}$, if $a \neq 0$, are in $A(a, b)$, this implies that they are algebraic over $A$. Hence we deduce that $C$ is a field. If $c \in B$ is algebraic over $C$, it is algebraic over $A$ as well, so $c \in C$.

**A5.16.** Prove first that $A[x]$ is countable. Using this fact, prove that the set $X \subset B \times A[x]$ of pairs $(\alpha, f(x))$ such that $f(\alpha) = 0$ is countable. Conclude by observing that the projection of $X$ on $B$ has as its image the algebraic closure of $A$ in $B$.

**A5.19.** Observe that $f(x) = (x - \alpha_1) \cdots (x - \alpha_n)$. Then carry out the product.

**A5.22.** Apply the previous exercise.

**A5.23.** If $n = mh$ and if $\xi^m = 1$ then $\xi^n = (\xi^m)^h = 1$. Vice versa, let $R_m \subseteq R_n$. If either $\mathbb{K}$ has characteristic 0 or $p$ is relatively prime with $n$ and $m$, then $R_n$ has order $n$ and $R_m$ has order $m$ and we conclude by applying Lagrange's theorem (see Exercise A3.36).

**A5.24.** Suppose it has order $d < n$. Then $d = q_1^{\mu_1} \cdots q_h^{\mu_h}$ with $\mu_1, \ldots, \mu_h$ non-negative integers and $\mu_i \leq m_i$, $i = 1, \ldots, h$, where for at least one $i$ the inequality holds. Set $d_i = q_i^{\mu_i}$ and $e_i = d/d_i, i = 1, \ldots, h$. We have $\xi^d = \prod_{i=1}^h (\xi_i^{d_i})^{e_i}$. At least one of the elements $\xi_i^{d_i}$ is not 1: assume this happens exactly for the indices $i = 1, \ldots, k$. So, $\xi^d = \prod_{i=1}^k (\xi_i^{d_i})^{e_i}$. Notice that $\xi_i^{d_i}$ has order $\delta_i = n_i/d_i = q_i^{m_i - \mu_i}$ and $\delta_i \nmid e_i$, for $i = 1, \ldots, k$. So, for all $i = 1, \ldots, k$ we have $\xi_i^d = (\xi_i^{d_i})^{e_i} \neq 1$. In conclusion, notice that $\xi_1^d \xi_2^d \neq 1$ as $R_{n_1} \cap R_{n_2} = (1)$. Analogously, $\xi_1^d \xi_2^d \xi_3^d \neq 1$ as $R_{n_1 n_2} \cap R_{n_3} = (1)$, and so on.

**A5.29.** The roots of $x^{p^f} - x$ are 0, plus those of the polynomial $x^{p^f - 1} - 1$. As $\mathbb{F}^*$ has order $p^f - 1$, every non-zero element $a \in \mathbb{F}$ satisfies $a^{p^f - 1} = 1$, so is a root of $x^{p^f - 1} - 1$.

**A5.31.** Mimic the proof of Proposition 5.1.37.

**A5.32.** Mimic the proof of Corollary 5.1.38.

**A5.33.** Keeping in mind Exercise A5.32 and Möbius inversion theorem (see Exercise A4.22), one can find the formula $n_{d,q} = (1/d) \sum_{h|d} \mu(h) q^{d/h}$; hence $n_{d,q} > (1/d)(q^d - q^{d/2} - q^{d/3} - \cdots) > (1/d)(q^d - \sum_{i=0}^{[d/2]} q^i)$. From here, the claim immediately follows.

**A5.34.** Every automorphism of $\mathbb{F}$ fixes $\mathbb{F}'$. So we have an obvious restriction homomorphism $r : \mathrm{Aut}(\mathbb{F}) \to \mathrm{Aut}(\mathbb{F}')$. Recall that $\mathrm{Aut}(\mathbb{F})$ [$\mathrm{Aut}(\mathbb{F}')$, respectively] is cyclic of order $f$ [$f'$, resp.] generated by $\phi_{\mathbb{F}}$ [$\phi_{\mathbb{F}}$, resp.]. As obviousy $r(\phi_{\mathbb{F}}) = \phi_{\mathbb{F}'}$, the homomorphism $r$ is surjective. Its kernel, which is the group we are looking for, has order $f/f'$ and is generated by $(\phi_{\mathbb{F}})^{f'}$.

**A5.35.** We have $\mathbb{F}_{p^d} = \mathbb{Z}_p[x]/(f(x)) = \mathbb{Z}_p(\alpha) \subset \mathbb{F}$. Conclude keeping in mind Theorem 5.1.35.

**A5.36.** Every element $g(x) \in \mathbb{K}[x]$ is congruent modulo $(f(x))$, to the remainder of its division by $f(x)$.

**A5.38.** Keep in mind Exercise A2.20 and Proposition 5.1.44.

**A5.39.** Keep in mind Exercise A2.21 and Proposition 5.1.44.

**A5.40.** Keep in mind Exercise A2.22 and Proposition 5.1.44.

**A5.41.** In each division, $d^2$ multiplications are carried out, each having complexity $\mathcal{O}(\log^3 q)$. Moreover, at most $d$ division have to be performed. For more information, see Section 2.5.3.

**A5.42.** Use the method of completing the square discussed on page 234.

**A5.48.** Use Exercise A5.47 and Corollary 5.1.31.

**A5.49.** Notice that $\xi^4 = -1$, so $\xi^5 = -\xi, \xi^7 = -\xi^3$, then $G = 2(\xi - \xi^3)$. Hence, $G^2 = 4(\xi^2 - 2\xi^4 + \xi^6) = 8$. So we have $G^{p-1} = (G^2)^{(p-1)/2} = 8^{(p-1)/2}$. Using Proposition 5.2.22, we get $G^p = (8/p)G = (2/p)G$.

**A5.50.** We have $G^p = (\sum_{i=0}^{7} \epsilon(i)\xi^i)^p = \sum_{i=0}^{7} \epsilon(i)\xi^{pi} = \sum_{i=0}^{7} \epsilon(pi)\xi^i$. Keep in mind Exercise A5.58, we have $\epsilon(p)\epsilon(pi) = \epsilon(p^2 i) = \epsilon(p^2)\epsilon(i) = \epsilon(i)$. Hence, $\epsilon(p)G^p = \sum_{i=0}^{7} \epsilon(p)\epsilon(pi)\xi^i = \sum_{i=0}^{7} \epsilon(i)\xi^i = G$. Finally, notice that $G \neq 0$, as $G^2 = 8 \neq 0$.

**A5.52.** With the same idea as Exercise A5.50, we have $G^p = \sum_{i=0}^{q-1} (\frac{i}{q})^p \xi^{ip} = \sum_{i=0}^{q-1} (\frac{i}{q})\xi^{ip}$. Now notice that $(\frac{i}{q}) = (\frac{i}{q})(\frac{p^2}{q}) = (\frac{p^2 i}{q}) = (\frac{pi}{q})(\frac{p}{q})$. So $G^p = \sum_{i=0}^{q-1} (\frac{i}{q})\xi^{ip} = \sum_{i=0}^{q-1} (\frac{pi}{q})(\frac{p}{q})\xi^{ip} = (\frac{p}{q})(\sum_{i=0}^{q-1} (\frac{pi}{q})\xi^{ip}) = (\frac{p}{q})(\sum_{i=0}^{q-1} (\frac{i}{q})\xi^{i}) = (\frac{p}{q})G$.

**A5.53.** Using Proposition 5.2.22, we have

$$G^2 = G \cdot G = \left(\sum_{i=1}^{q-1} \left(\frac{i}{q}\right)\xi^i\right)\left(\sum_{j=1}^{q-1} \left(\frac{j}{q}\right)\xi^j\right) = \left(\sum_{i=1}^{q-1} \left(\frac{i}{q}\right)\xi^i\right)\left(\sum_{j=1}^{q-1} \left(\frac{-j}{q}\right)\xi^{-j}\right) =$$

$$= \left(\frac{-1}{q}\right)\sum_{i=1}^{q-1}\sum_{j=1}^{q-1} \left(\frac{ij}{q}\right)\xi^{i-j} = (-1)^{(q-1)/2}\sum_{i=1}^{q-1}\sum_{j=1}^{q-1} \left(\frac{ij}{q}\right)\xi^{i-j}.$$

Notice that we may consider the indices $i$, $j$ as non-zero elements of $\mathbb{Z}_q$. For every index $i$ in the external sum, perform in $\mathbb{Z}_q^*$ the variable change $j = ik$. This may be done, as when $k$ ranges in $\mathbb{Z}_q^*$, also $j$ ranges in $\mathbb{Z}_q^*$. So we have $G^2 = (-1)^{(q-1)/2}\sum_{i=1}^{q-1}\sum_{k=1}^{q-1}(i^2 k/q)\xi^{i(1-k)} = (-1)^{(q-1)/2}\sum_{i=1}^{q-1}\sum_{k=1}^{q-1}(k/q)\xi^{i(1-k)}$. Keeping in mind Exercise A5.46, we get

$$G^2 = (-1)^{(q-1)/2}\sum_{i=0}^{q-1}\sum_{k=0}^{q-1} \left(\frac{k}{q}\right)\xi^{i(1-k)} = (-1)^{(q-1)/2}\sum_{k=0}^{q-1} \left(\frac{k}{q}\right)\left(\sum_{i=0}^{q-1}\xi^{i(1-k)}\right).$$

For every $k \neq 1$ the internal sum equals zero: indeed, $\xi$ is a primitive $q$th root of unity, and being $q$ prime, every power $\xi^h$ with $q \nmid h$ is too, so for the values $i = 0, \ldots, q-1$, the powers $\xi^{i(1-k)}$ span the whole set $R_q$ (see Exercise A5.22). In conclusion, we have $G^2 = (-1)^{(q-1)/2}\sum_{i=0}^{q-1}\xi^0 = (-1)^{(q-1)/2}q$.

**A5.54.** We have $G^p = (G^2)^{(p-1)/2}G = ((-1)^{(q-1)/2}q)^{(p-1)/2}G$. Conclude keeping in mind Proposition 5.2.22 and Exercise A5.52.

**A5.57.** The claim is trivial if $\alpha = 1$. Proceed next by induction on $\alpha$.

**A5.58.** Hint: notice that

$$\epsilon(n) = \begin{cases} 1 & \text{if} \quad n \equiv \pm 1 \pmod 8, \\ -1 & \text{if} \quad n \equiv \pm 3 \pmod 8, \end{cases}$$

and examine separately the different cases for $n, m$ modulo 8.

**B5.2.** The correct answer is (a).

**B5.3.** The correct answer is (a), because $64 = 2^6$ and $\mathbb{F}_{64}$ may be constructed, for instance, as a quotient of $\mathbb{Z}_2[x]$ with respect to an irreducible polynomial of degree 6.

**B5.4.** The correct answer is (d), because $323 = 17 \cdot 19$ is not a prime number, but this is not sufficient to rule out the existence of a field of order 323. It is necessary to remark that 323 is neither a prime nor a prime power.

**B5.8.** The correct answer is (a).

**B5.13.** The correct answer is (c), as $16 = 2^4$.

**B5.20.** The correct answer is (c).

**B5.21.** The correct answer is (d), as $x^4 + x^2 + 1 = (x^2 + x + 1)^2$ over $\mathbb{Z}_2$. Let us see how to get this factorisation.

First of all, we check whether $x^4 + x^2 + 1$ has roots in $\mathbb{Z}_2$, but we do not find any. Verify next if the polynomial splits into two degree two polynomials. We may assume that there is a factorisation of the form: $x^4 + x^2 + 1 = (x^2 + ax + 1)(x^2 + bx + 1)$ where $a$ and $b$ are unknowns. In general, we should have written $x^4 + x^2 + 1 = (cx^2 + ax + d)(ex^2 + bx + f)$ where $c$, $d$, $e$ and $f$ are further unknowns. But a factor of a monic polynomial with coefficients in a field may always be chosen to be monic, so we may assume $c = 1$. Then also $e = 1$, as in the right-hand side the highest degree term is $cex^4$. Moreover, the constant term must be $1 = df$, which in $\mathbb{Z}_2$ is possible only if $d = f = 1$. Back to the factorisation of $x^4 + x^2 + 1$, by carrying out the product in the right-hand side we get $x^4 + (a+b)x^3 + (1+ab+1)x^2 + (a+b)x + 1$, so $0 = a + b$ and $1 = ab$. Both equations are satisfied only if $a = b = 1$. So we get the factorisation given at the beginning.

Another possible way of finding this factorisation consists in noticing that the degree two factors, if they exist, have to be irreducible, otherwise they would have a degree one factor which would also be a factor of the original polynomial. There are only four degree two polynomial over $\mathbb{Z}_2$: $x^2$, $x^2 + 1$, $x^2 + x$, $x^2 + x + 1$. The first three of them are reducible, because they have a root equal to 0, 1 and 0, respectively. So $x^2 + x + 1$ is the only degree two irreducible polynomial over $\mathbb{Z}_2$. If we divide our polynomial by $x^2 + x + 1$, we find a zero remainder and a quotient equal to $x^2 + x + 1$, so we get again the above factorisation.

**B5.22.** The correct answer is (a).

**B5.25.** The correct answer is (d). First of all, notice that $x^{27} - x$ has linear factors, because it has roots 0, 1 and $-1$ ($= 2$). Next, dividing by $x(x-1)(x+1)$, we find that the quotient is

$$x^{24} + x^{22} + x^{20} + x^{18} + x^{16} + x^{14} + x^{12} + x^{10} + x^8 + x^6 + x^4 + x^2 + 1,$$

and we must try to factor this. Verify that it has no linear factors, because 0, 1 and $-1$ are not roots of this polynomial. Then we have to look for its factors among the monic, irreducible polynomials of degree two or greater. We may verify that there are three monic, irreducible polynomials of degree two over $\mathbb{Z}_3$, but they do not divide our polynomial. So we look at the monic, irreducible polynomials of degree three. After some calculations, we find the following polynomials:

$$x^3 + 2x + 1, \qquad x^3 + 2x + 2, \qquad x^3 + x^2 + 2, \qquad x^3 + x^2 + x + 2,$$
$$x^3 + x^2 + 2x + 1, \qquad x^3 + 2x^2 + 1, \qquad x^3 + 2x^2 + x + 1, \qquad x^3 + 2x^2 + 2x + 2$$

and they are exactly all the factors of the degree 24 polynomial (so also of $x^{27} - x$), as may be verified by carrying out the divisions. So we have found 11 factors: three linear ones and eight of degree three.

**B5.27.** The correct answer is (a).

**B5.28.** The correct answer is (a), as only $\mathbb{Z}_5$ is a subfield of $\mathbb{F}_{125}$.

**B5.43.** The correct answer is (a).

**B5.44.** The correct answer is (d), as the solutions are $x = -1$, $x = -3$, $x = 4$ and $x = 6$ modulo 14.

**B5.49.** The correct answer is (a), as 13 divides 65.

**B5.50.** The correct answer is (d), as Legendre symbol is defined only if the denominator is a prime number.

**B5.51.** The correct answer is (c).

**B5.54.** The correct answer is (b). Let us see why. First of all, $973 = 7 \cdot 139$, so $\left(\frac{1003}{973}\right)$ is equal, by definition of Jacobi symbol, to the product of the Legendre symbols $\left(\frac{1003}{7}\right)$ e $\left(\frac{1003}{139}\right)$. Now, $1003 \bmod 7 = 2$, so $\left(\frac{1003}{7}\right) = \left(\frac{2}{7}\right) = 1$, where the last equality follows from Proposition 5.2.27. On the other hand, $1003 \bmod 139 = 30$, so $\left(\frac{1003}{139}\right) = \left(\frac{30}{139}\right)$. By part (4) of Proposition 5.2.22, we have $\left(\frac{30}{139}\right) = \left(\frac{2}{139}\right)\left(\frac{3}{139}\right)\left(\frac{5}{139}\right)$. Proposition 5.2.27 tells us also that $\left(\frac{2}{139}\right) = -1$, while the law of quadratic reciprocity (Theorem 5.2.28) implies that $\left(\frac{3}{139}\right) = -\left(\frac{139}{3}\right)$ and $\left(\frac{5}{139}\right) = \left(\frac{139}{5}\right)$. Finally, $139 \bmod 3 = 1$ and $139 \bmod 5 = -1$, so $\left(\frac{139}{3}\right) = \left(\frac{1}{3}\right) = 1$ and $\left(\frac{139}{5}\right) = \left(\frac{-1}{5}\right) = 1$. So we may conclude that $\left(\frac{1003}{973}\right) = (-1)(-1) = 1$.

## Exercises of Chapter 6

**A6.1.** As $n$ is not a prime, it is clear that neither is $m$ (see Section 4.4.1). As $n$ is a pseudoprime in base 2, we have $m - 1 = 2^n - 2 = kn$ for some integer $k$. So $2^{m-1} - 1 = 2^{kn} - 1 \equiv 0 \pmod{m}$.

**A6.3.** As $n$ is a pseudoprime in bases $a_1$ and $a_2$, then $a_1^{n-1} \equiv 1 \pmod{n}$ and $a_2^{n-1} \equiv 1 \pmod{n}$, so $(a_1 a_2)^{n-1} = a_1^{n-1} a_2^{n-1} \equiv 1 \pmod{n}$. Moreover, $(a_2^{-1})^{n-1} \equiv a_2^{1-n} \equiv 1^{-1} \equiv 1 \pmod{n}$.

**A6.8.** As for Exercise A6.1, we have that $m$ is not prime. We have $2^{n-1} - 1 = nk$ and $k$ is odd. So $m - 1 = 2^n - 2 = 2kn$ with $kn$ odd. Moreover, $2^{(m-1)/2} = 2^{nk} = (2^n)^k \equiv 1 \pmod{m}$. Indeed, $2^n = m + 1 \equiv 1 \pmod{m}$.

**A6.10.** Write $n = 2^s t + 1$ with odd $t$. From the hypotheses, it follows that $b^{2^{s-1}t} \equiv -1 \pmod{n}$.

**A6.12.** It suffices to observe that for all positive integers $k$ we have $5^k \not\equiv -1 \pmod{4}$.

**A6.15.** Use the properties of cyclic groups (see Exercises A3.23–A3.35).

**A6.16.** We have to compute the product of all elements of the group $U(\mathbb{Z}_n)$. As two reciprocal elements cancel out in the product, the result is $\prod_{x \in G} x$, where $G$ is the subgroup of the elements of order 2 of $U(\mathbb{Z}_n)$. Consider $G$ as an additive group, so the above product becomes the sum of the elements. The group $G$ is a $\mathbb{Z}_2$-vector space as well. If it has dimension 1, then $G$ has order 2; this corresponds exactly to the cases $n = 2, 4, p^h, 2p^h$, and the sum of its elements is of course 1. If the dimension of $G$ is greater than 1, we see that the sum of its elememts is 0. Indeed, we may interpret $G$ as $\mathbb{Z}_2^d$, $d > 1$. Put the $2^d$ elements of $\mathbb{Z}_2^d$ in a $2^d \times d$ matrix. Every column has $2^{d-1}$ entries equal to 0 and as many equal to 1. Summing the elements of $\mathbb{Z}_2^d$ columnwise, it is clear that the sum of the entries in each column is 0.

**A6.19.** We are working in $U(\mathbb{Z}_{2^l})$. So we may write $x = (-1)^s 5^t$, $m \equiv (-1)^\sigma 5^\tau$ (mod $2^l$) (see Remark 6.2.10). Then the equation to be solved is translated into the system consisting of the two equations

$$sh \equiv \sigma \pmod 2, \quad th \equiv \tau \pmod{2^{l-2}}$$

in $s$ and $t$. If $h$ is odd, then this system admits a unique solution. If $h$ is even, the first equation admits solutions, and exactly two of them, if and only if $\sigma$ is even, that is, $m \equiv 1 \pmod 4$. If $h$ is even, the second equation admits solutions, and exactly $d$ of them, if and only if $d \mid \tau$ that is, if and only if $\tau = dk$, that is $m = 5^{dk}$, so $m^{2^{l-2}/d} \pmod 5$)$^{2^{l-2}} \equiv 1 \pmod{2^l}$.

**A6.21.** Assume, for instance, $n$ to be odd. Let $n = p_1^{l_1} \cdots p_s^{l_s}$ be its factorisation. Then $U(\mathbb{Z}_n)$ is the direct product of $U(\mathbb{Z}_{p_i^{l_i}})$, $i = 1, \ldots, s$. Let $r_i$ be a generator of $U(\mathbb{Z}_{p_i^{l_i}})$, $i = 1, \ldots, s$. Then for every integer $m$ that is relatively prime with $n$ the class of $m$ in $U(\mathbb{Z}_n)$ may be written uniquely as $r_1^{d_1} \cdots r_s^{d_s}$ with $0 \le d_i \le p_i^{l_i-1}(p_i-1)$, $i = 1, \ldots, s$. The vector $(d_1, \ldots, d_s)$ is called *index system* of $m$ with respect to $(r_1, \ldots, r_s)$.

**A6.22.** Proceed by induction.

**A6.25.** If $n = m^k$, then $k \le \log_2 n$. For all $k$ find an estimate for the $k$th root of $n$ and then compute a $k$th power. The latter computation has complexity $\mathcal{O}(\log^3 n)$, while for the first estimate the complexity is $\mathcal{O}^\sim(\log^2 n)$.

**A6.26.** Proceed as in § 3.3.1.

**A6.28.** A pair $(x, f)$ is found by assigning arbitrarily $x_1 = f(x)$, in $k$ ways, $x_2 = f(x_1)$ arbitrarily in $X \setminus \{x_1\}$, in $k - 1$ ways, $\ldots$, $x_m = f(x_{m-1})$ arbitrarily in $X \setminus \{x_1, \ldots, x_{m-1}\}$, in $k - m$ ways, and the remaining values of $f$ arbitrarily without restrictions.

**B6.4.** The correct answer is (c).

**B6.20.** For instance, 5.

**B6.23.** The answer is $2, 5, 25, 121$.

**B6.24.** The group $U(\mathbb{Z}_{15})$ is the product of a cyclic group of order 2 and of a cyclic group of order 4.

**B6.25.** The group $U(\mathbb{Z}_{16})$ is the product of a cyclic group of order 2 and of a cyclic group of order 4.

**B6.26.** The group $U(\mathbb{Z}_{17})$ is cyclic of order 16.

**B6.27.** The group $U(\mathbb{Z}_{18})$ is cyclic of order 6.

**B6.28.** $|U(15)| = |U(16)| = 8$, $|U(17)| = 16$, $|U(18)| = 6$.

**B6.35.** The correct answer is (a).

**B6.36.** The factorisation is found to be $906113 = 13 \cdot 47 \cdot 1483$.

## Exercises of Chapter 7

**A7.10.** The solution is trivial if $n$ is even. Let $n$ be odd. Then $\varphi(n) = (p-1)(q-1) = n+1-(p+q)$. So we know the sum of $p$ and $q$, that is $p+q = n+1-\varphi(n) = 2b$, which is even, and their product $n = pq$. Thus, $p$ and $q$ are equal to $b \pm \sqrt{b^2 - n}$. Now there is a simple algorithm, having complexity $\mathcal{O}(\log^3 n)$, which computes $\lfloor \sqrt{n} \rfloor$. Indeed, if $n$ has $k+1$ binary digits, a first approximation $m_1$ of $\lfloor \sqrt{n} \rfloor$ is given by 1 followed by $\lfloor k/2 \rfloor$ zeros. If $m_1$ is not the correct value, change its second digit from left, a 0, into a 1, obtaining a value $m_2$. If it is too large, put the second digit back to 0 and repeat the above with the third digit, obtaining $m_3$. If on the other hand $m_2$ is too small, change its second digit into a 1 obtaining a different $m_3$, and so on.

**A7.11.** By the Chinese remainder theorem, it suffices to prove that $a^{de} \equiv a \pmod{p}$ for all $a$ and for all prime $p \mid n$. This is obvious if $p \mid a$, else it follows from Fermat's little theorem.

**A7.16.** If $g(x)y - f(x)$ were reducible, we would have $g(x)y - f(x) = a(x,y) \cdot b(x,y)$ where the two polynomials $a(x,y), b(x,y)$ have positive degree and at least one of them does not depend on $y$. Let $a(x,y) = a(x)$ be such a polynomial. Then $a(x)$ is a common factor of $f(x)$ and of $g(x)$, which is impossible.

**A7.17.** Hint: study the case in which the conic curve contains the point $O = (0,1)$ first; define the projection of the conic on the $x$-axis, associating with each point $P \neq O$ of the conic the intersection of the line through $P$ and $O$ with the $x$-axis.

**A7.27.** Here is the proof in the case $p \neq q$. We have already shown in the text that the line through $p$ and $q$ has equation $y - y_1 = (y_2 - y_1)(x - x_1)/(x_2 - x_1)$, so the ordinate $y_r$ of the point $r$ collinear with $p$ and $q$ is

$$y_r = y_1 + \frac{y_2 - y_1}{x_2 - x_1}(x_3 - x_1),$$

where $x_3$ is the abscissa of $r$; hence follows the formula for $y_3$, because $p + q$ is the symmetric point of $r$ with respect to the $x$-axis. On the other hand, the fact that $r$ lies on the elliptic curve says that $(x_3, y_r)$ is a solution of the following system:

$$\begin{cases} y = y_1 + \dfrac{y_2 - y_1}{x_2 - x_1}(x - x_1), \\ y^2 = x^3 + ax + b. \end{cases}$$

Squaring the first equation, we find that the right-hand side of the second equation is equal to the square of the right-hand side of the first equation, so we find a third degree equation in $x$, equivalent to

$$x^3 - \left(\frac{y_2 - y_1}{x_2 - x_1}\right)^2 x^2 + (\text{terms of degree 1 and 0 in } x) = 0,$$

whose solutions are $x_1$, $x_2$ and $x_3$. So the left-hand side of the last equation is a polynomial equal to

$$(x - x_1)(x - x_2)(x - x_3) = x^3 - (x_1 + x_2 + x_3)x^2 + (\cdots)x - x_1x_2x_3;$$

hence follows that

$$x_1 + x_2 + x_3 = \left(\frac{y_2 - y_1}{x_2 - x_1}\right)^2,$$

giving the formula for $x_3$.

**A7.34.** By Exercise A7.32, $x$ and $z$ are odd. Set $r = (z + x)/2$, $s = (z - x)/2$. Using Exercise A7.31, prove that $GCD(r, s) = 1$. As $y^2 = 4rs$, deduce that $rs$ is a square and, by Exercise A7.33, that there exist two integers $n$, $m$ such that $r = m^2, s = n^2$. Deduce (7.31).

**B7.1.** The correct answer is (d).

**B7.5.** The correct answer is (d).

**B7.7.** The correct answer is (d).

**B7.9.** The correct answer is (c).

**B7.11.** The correct answer is (a).

**B7.14.** The correct answer is (a).

**B7.15.** The correct answer is (c).

**B7.16.** The correct answer is (b).

**B7.20.** Consider the plaintext of the message sent to Edgar Allan Poe (see page 325). By applying Vigenère enciphering using as key word *UNITED STATES*, for instance using the program of Exercise C7.3, the following ciphertext is found:

GE IEIASGDXV, ZIJ QL MW LAAM XZY ZMLWHFZEK EJLVDXW KWKE TX LBR ATQH LBMX AANU BAI VSMUKHSS PWN VLWKAGH GNUMK WDLNRWEQ JNXXVV OAEG EUWBZWMQY MO MLW XNBX MW AL PNFDCFPXH WZKEX HSSF XKIYAHUL? MK NUM YEXDM WBXZ SBC HV WZX PHWLGNAMIUK?

It is straightforward to check that in the ciphertext given by Poe there are exactly 16 transcription errors: the third letter should be I rather than J, the fifth letter (a I) was omitted. The reader might want to check the remaining errors.

**B7.23.** The correct answer is (d).

**B7.24.** The correct answer is (d), as the plaintext is take a day out.

**B7.26.** The correct answer is (a).

**B7.27.** The correct answer is (d).

**B7.28.** The correct answer is (d), as the inverse matrix exists and is $\begin{pmatrix} 12 & 1 \\ 7 & -11 \end{pmatrix}$.

**B7.30.** The correct answer is (a).

**B7.31.** The correct answer is (d).

**B7.33.** The correct answer is (b).

**B7.35.** The correct answer is (d), because the image of $f$ is $\{1, 4\}$, so $f$ is not surjective. It follows that $f$ is not injective either, because the domain and the codomain have the same (finite) size: when this happens, the function is bijective if and only if it is surjective, and this happens if and only if it is injective. Or, more simply, $f$ is not injective because $f(1) = 4 = f(3)$.

**B7.38.** We have $2^{70} = 2^{50+20} = 2^{50}2^{20} = (-\bar{1})(-\bar{6}) = \bar{6} = \bar{2}\cdot\bar{3}$, so $2^{69} = \bar{3}$; therefore $\overline{69}$ is the required discrete logarithm.

**B7.39.** The correct answer is (d).

**B7.40.** The correct answer is (b).

**B7.41.** The correct answer is (b), because the only solution is $34 = 20 + 13 + 1$.

**B7.43.** The correct answer is (a).

**B7.45.** The correct answer is (a).

**B7.47.** The correct answer is (a).

**B7.49.** The correct answer is (d).

**B7.51.** The correct answer is (c), because 7927 is a prime number.

**B7.53.** The correct answer is (a).

**B7.55.** The correct answer is (a).

**B7.60.** The correct answer is (b).

**B7.61.** The correct answer is $p + q = (3, -2\sqrt{6})$, so (d).

**B7.64.** The correct answer is (a).

**B7.67.** The correct answer is (c). A way of solving this exercise is by trial and error. In the affine plane with coordinates in $\mathbb{F}_7$ there are $7 \cdot 7 = 49$ points. In particular, fixing an abscissa (which is a number modulo 7), there are exactly 7 points having that abscissa (and different ordinates). We shall see if any of these poins lies on the elliptic curve, and how many (at most two).

Consider first $x = 0$. Then, of the 7 points having abscissa 0, only $(0,0)$ belongs to the elliptic curve, because by substituting $x = 0$ in the curve equation we find $y^2 = 0$ that has 0 as its only solution. Consider now $x = 1$. Substituting it in the curve equation we find $y^2 = 0$, so among the points with abscissa 1 only $(1,0)$ belongs to the curve. For $x = 2$, we find $y^2 = 6$, which has no solutions, so there are no points having abscissa 2 on the elliptic curve. Analogously, by substituting $x = 3$ we find $y^2 = 3$, which admits no solutions either. For $x = 4$, on the other hand, we find $y^2 = 4$, which admits two solutions: $y = 2$ e $y = 5$, so $(4, 2)$ and $(4, 5)$ are points of the elliptic curve (the only ones with abscissa 4). For $x = 5$, we have $y^2 = 1$, which has two solutions: $y = 1$ and $y = 6$, that is $(5, 1)$ and $(5, 6)$ are points of the elliptic curve. Finally, for $x = 6$ we find $y^2 = 0$, that is, $(6, 0)$ is a further point of the elliptic curve. And this is all: in the affine plane we have found 7 points, so the curve has 8 points, including the point at infinity.

**B7.68.** The correct answer is (d). Proceed as in example 7.9.18.

# Exercises of Chapter 8

**A8.1.** This a variation of the classic *binomial probability distribution* (see [29], p. 63). Fix a value of $i$, $0 \le i \le n$. The probability of $n - i$ errors occurring in $n - i$ fixed positions of a binary string of length $n$ is $p^{n-i}(1 - p)^i$. So the probability of $n - i$ occurring in any given string of length $n$ is $\binom{n}{i} p^{n-i}(1-p)^i$ (see Exercise A1.15). If we have more 0's than 1's and decode as HEADS the incoming string, the probability of having made a mistake is that of at least $\lfloor n/2 \rfloor$ errors having occurred, and the formula follows from the above. Notice that $1 = (p+(1-p))^n = \sum_{i=0}^{n} \binom{n}{i} p^{n-i}(1-p)^i$. Hence follows that $0 < P_n < 1$. Finally, we have (see Exercise A1.23)

$$P_n < [p(1 - p)]^{n/2} \sum_{0 \le i < \lfloor n/2 \rfloor} \binom{n}{i} < 2^{n-1}[p(1 - p)]^{n/2}.$$

Notice that $(2p - 1)^2 > 0$, that is, $4p^2 - 4p + 1 > 0$, so $p(1 - p) < 1/4$ and from this it immediately follows that $P_n$ tends to zero when $n$ tends to infinity.

**A8.2.** Using the theorems about linear systems over a field, prove that there is exactly one solution of the system (8.3) when the values of $x_1, x_2, x_3, x_4$ are assigned arbitrarily.

**A8.4.** The triangle inequality is the only property whose truth it is interesting to verify. Notice that, if $x$ and $x'$ differ in the $i$th coordinate, then in that coordinate either $x$ differs from $x''$ or $x'$ differs from $x''$.

**A8.5.** A code detects $k$ errors if and only if, by modifying a codeword in $h \le k$ coordinates, we never get another codeword. This happens if and only if two codeword always have distance at least $k + 1$. This proves (i). A code corrects $k$ errors if and only if by modifying a codeword in $h \le k$ coordinates, the $n$-tuple we get has distance greater than $k$ from every other codeword. This happens if and only if two codewords always have distance at least $2k + 1$. This proves (ii).

**A8.6.** See the solution of Exercise A8.1.

**A8.8.** Let $i$ be a positive integer smaller than $n$. The elements of $\mathbb{F}_q^n$ having Hamming distance exactly $i$ from $x = (x_1, \ldots, x_n)$ are those differing from $x$ in exactly $i$ coordinates. They can be obtained as follows: choose in $\binom{n}{i}$ ways the coordinates $x_{h_1}, \ldots, x_{h_i}$, with $1 \le h_1 < \cdots < h_i$, of $x$ to be modified and, for each $j = 1, \ldots, i$, modify the coordinate of $x$ in position $h_j$ in $q - 1$ ways, as many as the elements of $\mathbb{F}_q$ different from $x_{h_j}$.

**A8.11.** By the properties of the binomials (see Exercise A1.14), we have $2^d = \sum_{i=1}^{d} \binom{n}{i} = 2 \cdot \sum_{i=1}^{k} \binom{n}{i}$.

**A8.13.** If Singleton bound is better than Hamming one, we have $\sum_{i=1}^{k} \binom{n}{i} < 2^{d-1}$, where $k = (d - 2)/2$. By proceeding as in the previous solution, verify that this relation is equivalent to $\sum_{i=0}^{k} \binom{d}{i} + 1/2\binom{d}{k+1} > \sum_{i=0}^{k} \binom{n}{i}$. Notice that $n \ge d$ and that the previous inequality holds for $n = d$. Prove instead that it does not hold for $n = d + 1$, and deduce that it does not hold for $n \ge d + 1$. Indeed, for $n = d + 1$, by using Equation (1.51), the inequality may be written as $1/2\binom{d}{k+1} > \sum_{i=0}^{k} [\binom{d+1}{i}) -$

$\binom{d}{i}] = \sum_{i=1}^{k} \binom{d}{i-1}$ and notice that for $d \geq 12$ we have $2\binom{d}{k-1} > \binom{d}{k+1}$. Directly verify cases $d = 8, 10$.

**A8.16.** Suppose **x** different from zero and notice that the degree two polynomial $f(t) = (t\mathbf{x} + \mathbf{y}) \times (t\mathbf{x} + \mathbf{y})$ only assumes positive or zero values. So its discriminant is non-positive.

**A8.17.** If the sequence $\{x_n\}_{n \in \mathbb{N}}$ is bounded from above, its least upper bound is also an upper bound for the sequence $\{b_n\}_{n \in \mathbb{N}}$, so $\xi = \limsup_{n \to \infty} x_n$ is finite.

**A8.20.** Set $n! = n^n e^{-n} a(n)$. It suffices to prove that $a(n)$ is $\mathcal{O}(n)$. Notice that $a(n+1)/a(n) = e(1 + 1/n)^{-n}$ and that we have the inequality $(1 + 1/n)^{n+1/2} > e$. The latter follows from the well-known formula

$$\log \frac{1+x}{1-x} = \log(1+x) - \log(1-x) = 2\left(x + \frac{x^3}{3} + \frac{x^5}{5} + \cdots\right) > 2x,$$

which holds for $0 < x < 1$, setting $x = 1/(2n+1)$. Then we have $a(n+1)/a(n) < (1 + 1/n)^{1/2}$, that is, the function $b(n) = a(n)/\sqrt{n}$ is positive and decreasing, so it tends to a finite, positive limit. Hence follows the claims.

**A8.21.** Set $r = [\delta n]$. The last term in the sum that appears in (8.6) is the largest one. So we have

$$\binom{n}{r}(q-1)^r \leq V_q(n,r) \leq (1+r)\binom{n}{r}(q-1)^r.$$

Take the logarithm in base $q$ and divide by $n$. Applying Stirling formula we get

$$\frac{1}{n} \log_q \binom{n}{r}(q-1)^r = \delta \log_q(q-1) + \log_q n - \delta \log_q r - (1-\delta)\log_q(n-r) + \mathcal{O}(1),$$

so

$$\frac{1}{n} \log_q \binom{n}{r}(q-1)^r = H_q(\delta) + \mathcal{O}(1).$$

Hence follows the claim.

**A8.22.** By the Gilbert–Varshamov bound, we have

$$\alpha(\delta) = \limsup_{n \to \infty} \frac{\log_q A_q(n, n\delta)}{n} \geq \lim_{n \to \infty}\left(1 - \frac{\log_q V_q(n, n\delta)}{n}\right).$$

The claim follows from Lemma 8.4.7.

**A8.23.** We have $(q-1)/q \leq \delta$ if and only if $dq - nq + n > 0$. In this case the Plotkin bound implies $A_q(n, d) \leq qd/(qd - n(q-1))$. Then

$$0 \leq \alpha(\delta) = \limsup_{n \to \infty} \frac{\log_q A_q(n, n\delta)}{n} \leq \limsup_{n \to \infty}\left(\frac{\log_q(n\delta q)}{n} - \frac{\log_q(n\delta q - nq + n)}{n}\right) = 0.$$

Suppose now $0 \leq \delta \leq (q-1)/q$, which implies that $dq - nq + n \leq 0$. Set $m = [q(d-1)/(q-1)]$ and notice that $m < n$. Let $C$ be a code of type $(n, M, d)_q$. Considering the mapping $p : C \to \mathbb{F}_q^{n-m}$ which, to each $x \in C$, associates the vector in $\mathbb{F}_q^{n-m}$ consisting of its last $n - m$ coordinates, notice that there is a subset $C'$ of

$C$ of size $M' \geq q^{m-n}M$ such that all elements of $C'$ have the same image in $p$. We may consider $C'$ as a code of type $(m, M', d)_q$. We may apply the Plotkin bound, which yields $q^{m-n}M \leq M' \leq qd/(qd - mq + m) \leq d$. So we have $q^{m-n}A_q(n, d) \leq d$, hence, by taking $d = n\delta$ and $n \gg 0$ we obtain the claim.

**A8.26.** The rows of a generating matrix $\mathbf{X}$ are linearly independent. So there is a non-zero minor of order $k$. Up to renaming the coordinates, we may assume that this minor is determined by the first $k$ columns. Let $\mathbf{A}$ be the submatrix of $\mathbf{X}$ determined by the first $k$ columns. The matrix $\mathbf{A}^{-1} \cdot \mathbf{X}$ is in standard form and is again a generating matrix for the same code, obtained with a basis change.

**A8.29.** It is sufficient to verify that $\mathbf{X} \cdot \mathbf{H}^t = \mathbf{0}$, where $\mathbf{0}$ is here the $k \times (n - k)$ zero matrix.

**A8.33.** The mapping $\mathbf{x} \in C \to \mathbf{e} + \mathbf{x} \in \mathbf{e} + C$ is a bijection between the words of $C$ and the elements of the coset $\mathbf{e} + C$. Show, further, that every element of $\mathbb{F}_n^q$ lies in exactly one coset.

**A8.40.** Consider the two binary codes

$$C = \{(1, 0, 1, 1, 1, 0, 1, 1, 1, 0), (1, 0, 1, 1, 0, 1, 1, 1, 0, 1), (0, 1, 1, 1, 0, 1, 0, 1, 0, 1),$$
$$(1, 1, 0, 1, 1, 0, 1, 1, 0, 1), (0, 0, 0, 0, 1, 1, 1, 1, 0, 0)\},$$
$$C' = \{(1, 1, 1, 1, 1, 0, 0, 1, 1, 0), (1, 0, 0, 0, 0, 0, 0, 0, 0, 0), (0, 1, 1, 0, 0, 0, 0, 0, 0, 0),$$
$$(0, 1, 1, 1, 1, 1, 1, 0, 0, 0), (1, 1, 1, 1, 0, 1, 0, 0, 0, 1)\}.$$

Verify that $\mathbf{D}(C) = \mathbf{D}(C')$. To prove that $C$ and $C'$ are not equivalent, show that there is a coordinate such that all the elements of $C$ have the same symbol on that coordinate, while for no coordinate the same happens in $C'$.

**A8.45.** Hint: prove that the ring of classes modulo $x^n - 1$ admits as a representative system the set of polynomials $\{a_0 + a_1x + \cdots + a_{n-1}x^{n-1}, \ a_i \in \mathbb{F}_q, \ 0 \leq i < n\}$.

**A8.46.** Let $x^n - 1 = a(x) \cdot b(x)$. A polynomial $c(x)$ in $\mathbb{F}_q[x]/(x^n - 1)$ is in $C$ if and only if it is divisible by $a(x)$ in $\mathbb{F}_q[x]$, so if and only if its product by $b(x)$ is zero modulo $x^n - 1$.

**B8.1.** The correct answer is (b). Let us see why. The vector $(0, 1, 1, 1, 1, 1, 1)$ is a word of the Hamming code if and only if it satisfies the system of three linear equations (8.3) on page 410, where the calculations are carried out in $\mathbb{Z}_2$. As the given vector does not satisfy any of the three equations of the system, this means that it is not a word of the Hamming code. Moreover, by the reasoning on page 410, it follows that the error is in the first position, so the correct word is $(1, 1, 1, 1, 1, 1, 1)$. To check that the calculations are correct, it is useful to verify that $(1, 1, 1, 1, 1, 1, 1)$ is actually a word of the Hamming code, that is, it satisfies the system (8.3). If this happens, as in our case, we may be sure that no mistake has been made.

**B8.2.** The correct answer is (a); indeed the vector satisfies all three equations of the system (8.3).

**B8.3.** The correct answer is (d), because the vector satisfies the first two equations of the system (8.3), but not the third one, so the error is in the seventh position, by what has been argued on page 410. To double-check this, notice that the correct vector $(0, 1, 0, 0, 0, 1, 1)$ verifies the system (8.3), that is to say, it is a word of the Hamming code.

**B8.4.** The correct answer is (c); indeed Hamming distance is by definition the number of different coordinates, which in this case are the second, third, fifth and sixth ones.

**B8.6.** The correct answer is (c). Indeed, a subset of a vector space is a vector subspace if: (1) it is closed under addition, (2) it is closed under the product by scalars. When we consider binary codes, the scalars are only 0 and 1, so condition (2) is automatically verified. We have only to check condition (1). Notice now that $(0, 0, 1) + (0, 1, 0) = (0, 1, 1)$; hence follows that all other possible sums, that is, $(0, 0, 1) + (0, 1, 1) = (0, 1, 0)$ and $(0, 1, 0) + (0, 1, 1) = (0, 0, 1)$, always give elements belonging to the subset, so it is a vector subspace. Finally, the relation we have found among the elements of the subspace says that the dimension is 2, because $(0, 1, 0)$ and $(0, 0, 1)$ clearly are linearly independent vectors.

**B8.11.** The correct answer is (b); indeed, $(1, 1, 2, 0) = (1, 0, 1, 1) + (0, 1, 1, 2)$, so there are no more than two linearly independent generators (recall that the dimension of a vector space is equal to the greatest number of linearly independent generators).

**B8.12.** The correct answer is (d). Let us see why. To compute the minimum distance of the code, write all codewords and compute their weight, that is, the number of non-zero coordinates. By Proposition 8.5.3, the minimum distance is equal to the minimum weight of the non-zero words. In our case, we find that the minimum distance is $d = 3$. On the other hand, we know that $n = 4$ and that $k = 2$ by Exercise B8.11, so $d = 3 = n - k + 1$, that is to say, the code is maximum-distance separable, according to Definition 8.5.2 on page 420. Finally, $d = 3$ implies that the code detects two errors and corrects one, by Theorem 8.3.4 on page 412.

**B8.18.** The correct answer is (d). Theorem 8.5.9 on page 424, indeed, says that the minimum distance of the linear code is equal to the minimum number of linearly independent columns of the parity check matrix. So, examine the columns of the matrix given in the exercise. Considering all possible combinations, we see that there are no three columns whose sum is zero (thus being linearly dependent), while there are four columns, for instance, the third, fourth, seventh and eighth one, whose sum is zero (so they are linearly dependent). So we may conclude that the minimum distance of the linear code is four.

# Exercises of Chapter 9

**A9.1.** Notice that every key is an ordered $r$-tuple of elements of the alphabet (see Exercise A1.21).

**A9.5.** Start as in the solution of Exercise A8.16.

**A9.6.** Again, start as in the solution of Exercise A8.16.

**A9.10.** The claim is trivial if the vector space has dimension 1. Moreover, one implication is trivial. Finally, assume that $AB = BA$. Deduce that $A$ and $B$ have a common eigenvector $v$. So the subspace $\langle v \rangle$ is invariant under $A$ and $B$. But then the orthogonal subspace $\langle v \rangle^\perp$ is also invariant under $A$ and $B$. Then proceed by induction.

**A9.11.** Every observable of the form

$$\frac{1}{2} \begin{pmatrix} \lambda_1 + \lambda_2 & \lambda_1 - \lambda_2 \\ \lambda_1 - \lambda_2 & \lambda_1 + \lambda_2 \end{pmatrix}$$

with $\lambda_1 \neq \lambda_2$.

**B9.1.** The solution is $HQRC$.

**B9.2.** The solution is $QMIA$.

**B9.6.** The answer is NO.

**B9.7.** We have

$$[B, A] = \begin{pmatrix} 0 & -2 \\ 2 & 0 \end{pmatrix}.$$

# References

1. Adleman, L.M., Rivest, R.L., Shamir, A.: *A method for obtaining digital signatures and public–key cryptosystems.* Communications of the ACM, **21**, 120–126 (1978)
2. Agrawal, M., Kayal, N., Saxena, N.: *PRIMES in P.* Ann. of Math., **160**, n. 2, 781–793 (2004)
3. Alford, W.R., Granville A., Pomerance, C.: *There are infinitely many Carmichael numbers.* Ann. of Math., **139**, n. 3, 703–722 (1994)
4. Artin, M.: *Algebra.* Prentice Hall, Englewood Cliffs, NJ, USA (1991)
5. Baldi, P.: *Introduzione alla probabilità con elementi di statistica.* McGraw-Hill, Milano (2003)
6. Baylis, J.: *Error–correcting codes.* Chapman and Hall Math., Londra (1998)
7. Bennet, C.H., Brassard, G.: *Quantum cryptography: public key distribution and coin tossing.* Proceedings of IEEE International Conference on Computers, Systems and Signal Processing, Bangalore, India, December 1984, 175–179 (1984)
8. Bennet, C.H., Bessette, F., Brassard, G., Salvail, L., Smolin, J.: *Experimental quantum cryptography.* J. Cryptology, **5**, 3–28 (1992)
9. Bennet, C.H., Brassard, G., Ekert, A.: *Quantum cryptography.* Scientific American October 1992, 50–57 (1992)
10. Boyer, C.B.: *A History of Mathematics.* Wiley, New York (1968).
11. Burton, D.M.: *Elementary number theory.* Allyn and Bacon, Inc., Boston, Mass.-Londra (1980)
12. Canuto, C., Tabacco, A.: *Analisi Matematica 1.* Unitext, Springer-Verlag, Milano (2003)
13. Ciliberto, C.: *Algebra lineare.* Bollati Boringhieri, Torino (1994)
14. Coppersmith, D.: *Fast evaluation of logarithms in fields of characteristic two.* IEEE Transactions on Information Theory IT, **30**, 587–594 (1984)
15. Curzio, M.: *Lezioni di algebra.* Liguori, Napoli (1970)
16. Davenport, H.: *The higher arithmetic. An introduction to the theory of numbers.* Cambridge University Press, Cambridge (1999)
17. Deutsch, D.: *The fabric of Reality.* Allen Lane, Londra (1977)
18. Deutsch, D., Ekert, A.: *Quantum Computation.* Physics World, **11**, n. 3, 33–56 (1998)
19. Diffie, W., Hellman, M.E.: *New directions in cryptography.* IEEE Transactions on Information Theory IT, **22**, 644–654 (1976)

508    References

20. Dirac, P.A.M.: *The principles of quantum mechanics.* Oxford University Press, New York (1958)
21. Doxiadis, A.: *Uncle Petros and Goldbach's Conjecture.* Bloombsbury Publ., New York and London (2000)
22. Ebbinghaus, H.D., et al.: *Numbers.* Springer-Verlag, Berlin Heidelberg New York (1991)
23. Garey, M.R., Johnson, D.S.: *Computers and intractability. A guide to the theory of $\mathcal{NP}$-completeness.* W. H. Freeman & C., San Francisco, Calif. (1979)
24. Grimaldi, R.: *Discrete and Combinatorial Mathematics.* Addison–Wesley, 5th ed., Reading, Mass. (1988)
25. Hardy, G. H.: *A Mathematician's Apology.* Cambridge University Press (1940)
26. Hardy, G.H., Wright, E.M.: *An introduction to the theory of numbers.* Oxford Science Publ., 5th ed., New York (1979)
27. Herstein, I.N.: *Topics in Algebra.* Wiley, New York (1975)
28. Hodges, A.: *A. Turing: the Enigma of Intelligence.* Unwin Paperbacks, Londra (1983)
29. Isaac, R.: *The pleasures of probability.* Springer-Verlag, Berlin Heidelberg New York (1995)
30. Koblitz, N.: *A course in number theory and cryptography.* Springer-Verlag, Berlin Heidelberg New York (1994)
31. Kraitchick, M.: *Recherches sur la théorie des nombres.* Gauthiers-Villars, Parigi (1929)
32. Lang, S.: *Algebra.* Addison Wesley, New York (1978)
33. Lenstra, A., Jr., Lenstra, H.W., Jr. (ed.): *The development of the number field sieve.* Springer-Verlag, Berlin Heidelberg New York (1993)
34. Lenstra, H.W., Jr.: *Primality testing.* In: Studiezweek Getaltheorie en Computers, 1–5 September 1980, Stichting Mathematisch Centrum, Amsterdam (1982)
35. Lenstra, H.W., Jr.: *Factoring integers with elliptic curves.* Ann. of Math., **126**, n. 2, 649–673 (1987)
36. van Lint, J.H.: *Introduction to coding theory.* II ed., Springer-Verlag, Berlin Heidelberg New York (1992)
37. van Lint, J.H., van der Geer, G.: *Introduction to coding theory and algebraic geometry.* DMV Seminar 12, Birkhäuser, Basel (1988)
38. Lomonaco, S.J.: *A talk on quantum cryptography or how Alice outwits Eve.* Proc. Sympos. Appl. Math. 58, American Math. Soc., Providence, R.I., 237–264 (2002)
39. McEliece, R.J.: *The theory of information and coding.* Encyclopedia of Math. and its Appl., vol. 3. Addison–Wesley, Reading, Mass. (1977)
40. McEliece, R.J., Ash, R.B., Ash, C.: *Introduction To Discrete Mathematics.* McGraw-Hill, New York, (1989)
41. MacWilliams, F.J., Sloane, N.J.A.: *The theory of error–correcting codes.* North Holland, Amsterdam (1977)
42. Menezes, A., Okamoto, T., Vanstone, S.A.: *Reducing elliptic curves logarithms to logarithms in a finite field.* IEEE Transactions on Information Theory IT, **39**, 1639–1646 (1993)
43. Monk, J. D.: *Introduction to set theory,* McGraw-Hill, New York (1969)
44. Odlyzko, A.M.: *Discrete logarithms in finite fields and their cryptographic significance.* Advances in Cryptology, Proc. Eurocrypt, **84**, 224–314 (1985)
45. Piacentini Cattaneo, G.M.: *Algebra, un approccio algoritmico.* Decibel Zanichelli, Bologna (1996)

46. Pomerance, C., Selfridge, J.L., Wagstaff, S.S.: *The pseudoprimes to* $25 \cdot 10^9$. Math. Comp., **35**, 1003–1026 (1980)
47. Quarteroni, A., Saleri, F.: *Introduzione al calcolo scientifico*. Unitext, Springer-Verlag, Milano (2004)
48. Ribenboim, P.: *The new book of prime numbers records*. Springer-Verlag, Berlin Heidelberg New York (1996)
49. Rosen, K.H.: *Elementary number theory*. Addison–Wesley, Reading, Mass. (1988)
50. Schoof, R.: *Elliptic curves over finite fields and the computation of square roots mod p*. Math. Comp., **44**, 483–494 (1985)
51. Sernesi, E.: *Geometria I*. Bollati Boringhieri, Torino (1989); published in English as Sernesi, E., Montaldi J.: *Linear Algebra: A Geometric Approach*. Kluwer Academic Publishers Group (1992)
52. Shamir, A.: *A polynomial time algorithm for breaking the basic Merkle–Hellman cryptosystem*. Proc. 23rd annual symposium on the foundation of computer science (Chicago, Ill., 1982), IEEE, New York, 145–152 (1982)
53. Shannon, C.E.: *Communication theory of secrecy systems*. Bell Systems Technical Journal, **28**, 656–715 (1949)
54. Shor, P.W.: *Polynomial–time algorithms for prime factorization and discrete logarithm on a quantum computer*. SIAM J. Computing, **26**, 14–84 (1997)
55. Siegel, C.L.: *Topics in complex function theory, Vol. I*. Wiley, New York (1969)
56. Silverman, J.H.: *The arithmetic of elliptic curves*. Springer-Verlag, Berlin Heidelberg New York (1985)
57. Singh, S.: *Fermat's Last Theorem*. Anchor Books, New York (1998)
58. Singh, S.: *The Code Book: The Science of Secrecy from Ancient Egypt to Quantum Cryptography*. Anchor Books, New York (2000)
59. Tenenbaum, G., Mendès France, M.: *The prime numbers and their distribution*. Student Math. Library, vol. 6, American Mathematical Society (2000)
60. Tsfaman, M.A., Vladut, S.G., Zink, T.: *On Goppa codes which are better than the Varshamov–Gilbert bound*. Math. Nachr., **109**, 21–28 (1982)
61. Weil, A.: *Number Theory. An approach through history from Hammurabi to Legendre*. Birkhäuser, Boston (1983)
62. Wiesner, S.: *Conjugate coding*. SIGACT News, **15**, n. 1, 78–88 (1983; original manuscript, around 1970)
63. Wiles, A.: *Modular elliptic curves and Fermat's Last Theorem*. Ann. of Math., **142**, 443–551 (1995)

# Index